FUZZY LOGIC AND ITS APPLICATIONS TO ENGINEERING, INFORMATION SCIENCES, AND INTELLIGENT SYSTEMS

THEORY AND DECISION LIBRARY

General Editors: W. Leinfellner (*Vienna*) and G. Eberlein (*Munich*)

Series A: Philosophy and Methodology of the Social Sciences

Series B: Mathematical and Statistical Methods

Series C: Game Theory, Mathematical Programming and Operations Research

Series D: System Theory, Knowledge Engineering and Problem Solving

SERIES D: SYSTEM THEORY, KNOWLEDGE ENGINEERING AND PROBLEM SOLVING

VOLUME 16

Editor: R. Lowen (Antwerp); *Editorial Board:* G. Feichtinger (Vienna), G. J. Klir (New York) O. Opitz (Augsburg), H. J. Skala (Paderborn), M. Sugeno (Yokohama), H. J. Zimmermann (Aachen).

Scope: Design, study and development of structures, organizations and systems aimed at formal applications mainly in the social and human sciences but also relevant to the information sciences. Within these bounds three types of study are of particular interest. First, formal definition and development of fundamental theory and/or methodology, second, computational and/or algorithmic implementations and third, comprehensive empirical studies, observation or case studies. Although submissions of edited collections will appear occasionally, primarily monographs will be considered for publication in the series. To emphasize the changing nature of the fields of interest we refrain from giving a clear delineation and exhaustive list of topics. However, certainly included are: artificial intelligence (including machine learning, expert and knowledge based systems approaches), information systems (particularly decision support systems), approximate reasoning (including fuzzy approaches and reasoning under uncertainty), knowledge acquisition and representation, modeling, diagnosis, and control.

The titles published in this series are listed at the end of this volume.

FUZZY LOGIC AND ITS APPLICATIONS TO ENGINEERING, INFORMATION SCIENCES, AND INTELLIGENT SYSTEMS

edited by

Z. BIEN

Korea Advanced Institute of Science and Technology (KAIST), Taejon, Korea

and

K. C. MIN

Yonsei University, Seoul, Korea

KLUWER ACADEMIC PUBLISHERS
DORDRECHT / BOSTON / LONDON

A C.I.P. Catalogue record for this book is available from the Library of Congress

Published by Kluwer Academic Publishers,
P.O. Box 17, 3300 AA Dordrecht, The Netherlands.

Kluwer Academic Publishers incorporates
the publishing programmes of
D. Reidel, Martinus Nijhoff, Dr W. Junk and MTP Press.

Sold and distributed in the U.S.A. and Canada
by Kluwer Academic Publishers,
101 Philip Drive, Norwell, MA 02061, U.S.A.

In all other countries, sold and distributed
by Kluwer Academic Publishers Group,
P.O. Box 322, 3300 AH Dordrecht, The Netherlands.

ISBN-13: 978-94-010-6543-6   e-ISBN-13: 978-94-009-0125-4
DOI: 10.1007/978-94-009-0125-4

Printed on acid-free paper

Softcover reprint of the hardcover 1st edition 1995

TABLE OF CONTENTS

Chapter 3. Mathematical Foundations

Chapter 4. Information Sciences

PREFACE

As we all have been witnessing lately, the fuzzy technology has emerged as one of the most exciting new technologies at hand. It is responsible for improving many industrial and consumer systems and profoundly affects our daily lives. To exchange ideas and research results of this important area of fuzzy logic, more than 500 researchers and system engineers from 36 countries had gathered at the Fifth IFSA World Congress in Seoul, Korea. Two objectives were set up for the meeting : one was to encourage communications between researchers of fuzzy logic and related systems and the second to explore industrial applications of fuzzy technology, implementing human intelligent in machines and systems. The present book is an outgrowth of contributions by the participants at the Congress. Selections were made so that this volume reflects important current directions of Fuzzy Logic and Technology. All the authors were requested to extend and/or update their original contributions in the proceedings.

The contents are divided into four chapters ; (i) Intelligent Systems, (ii) Engineering, (iii) Mathematical Foundations and (iv) Information Sciences. The first chapter on intelligent systems contains topics including approximate reasoning, knowledge representation, computer vision and pattern recognition. The second chapter on engineering deals with intelligent robotics, fuzzy logic control, industrial systems, manufacturing, image signal processing, hardware devices, neural network, genetic algorithm and nuclear science. The third chapter on mathematical foundations includes topics on non-classical logics, category theory, algebra, topology, probability and statistics, relational equations. The fourth chapter on information sciences handles information retrieval, fuzzy database system, management science, operation research and decision making.

We hope that this volume would serve the academia as well as the technical community in a way that fuzzy logic is appreciated as a methodology for enlarging concepts, design and manufacture of intelligent systems and, ultimately, attaining improved understanding of the basic science and the foundations of human reasoning.

We take this opportunity to express our sincere thanks to Prof. L. A. Zadeh, Prof. T. Terano and many other colleagues throughout the world for their thoughtful support to make the Seoul Congress a great success. With their fruitful contributions, many scientists, engineers and people in the government sector are convinced that the development of fuzzy logic is an essential step in preparing the 21st century.

We would like also to thank all the authors of this book for their contributions as well as Prof. R. Lowen at University of Antwerp and Mr. P. Leenards at Kluwer Academic Publishers for their kind encouragement in the preparation of this volume. Finally, we express our deep appreciation to Dr. J. W. Park for his kind help during the editorial process of this book.

Z. Bien and K. C. Min

KAIST, Taejon and Yonsei University, Seoul

Chapter 1.

INTELLIGENT SYSTEMS

EXPRESSING FUZZY MEASURE BY A MODEL OF MODAL LOGIC: A DISCRETE CASE[†]

ZHENYUAN WANG*, GEORGE J. KLIR*,
and GERMANO RESCONI**
*Department of Systems Science and Industrial Engineering,
Thomas J. Watson School of Engineering and Applied Science,
State University of New York at Binghamton,
Binghamton, New York 13902-6000, U. S. A.
**Department of Mathematics,
Catholic University, Brescia, Italy

1. Introduction

It is argued by Resconi et al. (1992, 1993) that propositional modal logic is a convenient and sufficiently broad framework within which various theories of uncertainty can be studied in a unified way. The argument is based upon a demonstration that fuzzy sets and various well-established types of fuzzy measures (or semicontinuous fuzzy measures), such as probability measures, belief measures, plausibility measures, and λ-fuzzy measure (Wang and Klir, 1992), can be constructed within the framework of modal logic.

While the initial papers by Resconi et al. (1992, 1993) are restricted to models of modal logic with finite sets of worlds, the aim of this paper is to extend the previous results to any discrete case, where the set of worlds is not necessarily finite, and a weighted model is introduced, such that the determined belief measures and plausibility measures can take any real numbers in [0, 1]. Furthermore, by solving the inverse problem, we show that any belief measure and plausibility measure can be expressed by a suitable model of modal logic.

To provide the basic knowledge on fuzzy measures and semicontinuous fuzzy measures, especially on belief measures and plausibility measures, including λ-fuzzy measures, we present a summary in Section 2. Further details can be found in books by Wang and Klir (1992) and Shafer (1976).

Two books by Hughes and Cresswell (1968, 1984) are recommended as a useful background on modal logic, but they are not essential. Relevant concepts of modal logic may be found in Resconi et al. (1992, 1993). Some recent results on modal logic interpretation of Dempster-Shafer theory are presented in Harmanec and Klir (1994) and

---

[†]This work is partially supported by the ONR Grant No. N00014-94-1-0263.

Z. Bien and K. C. Min (eds.),
Fuzzy Logic and its Applications, Information Sciences, and Intelligent Systems, 3–13.

Harmanec, Klir, and Wang (1995).

2. Fuzzy Measures

Let X be a nonempty set, $\mathcal{F}$ be a σ-algebra of subsets of X. A set function $\mu: \mathcal{F} \rightarrow [0, \infty]$ is called a lower (upper) semicontinuous fuzzy measure iff it is monotone, continuous from below (above), and vanishing at $\varnothing$. Both lower and upper semicontinuous fuzzy measures are referred to as semicontinuous fuzzy measures (SC-fuzzy measures, for short). μ is called a fuzzy measure iff it is both lower and upper semicontinuous fuzzy measure. A fuzzy measure (or semicontinuous fuzzy measure) μ is called regular iff $\mu(X) = 1$.

The power set $\mathcal{P}(X)$ is a σ-algebra. A set function $m: \mathcal{P}(X) \rightarrow [0, 1]$ is called a basic probability assignment iff

(BPA 1) $$m(\varnothing) = 0;$$

(BPA 2) $$\sum_{E \subset X} m(E) = 1.$$

It is known that there exists a countable subset $\mathcal{D}$ of $\mathcal{P}(X)$ such that $m(E) = 0$ whenever $E \notin \mathcal{D}$.

If m is a basic probability assignment, then the set function Bel: $\mathcal{P}(X) \rightarrow [0, 1]$ determined by

$$\mathrm{Bel}(E) = \sum_{F \subset E} m(F) \qquad \forall E \in \mathcal{P}(X)$$

is called a belief measure on $(X, \mathcal{P}(X))$, and the set function Pl: $\mathcal{P}(X) \rightarrow [0, 1]$ determined by

$$\mathrm{Pl}(E) = \sum_{F \cap E \neq \varnothing} m(F) \qquad \forall E \in \mathcal{P}(X)$$

is called a plausibility measure on $(X, \mathcal{P}(X))$.

A belief measure, Bel, is an upper semicontinuous fuzzy measure satisfying

(BM 1) $$\mathrm{Bel}\left(\bigcup_{i=1}^{n} E_i\right) \geq \sum_{I \subset \{1,\cdots,n\}, I \neq \varnothing} (-1)^{|I|+1} \mathrm{Bel}\left(\bigcap_{i \in I} E_i\right),$$

where $\{E_1, \dots, E_n\}$ is any finite subset of $\mathscr{P}(X)$, while a plausibility measure, Pl, is a lower semicontinuous fuzzy measure satisfying

(PM 1)
$$\mathrm{Pl}(\bigcap_{i=1}^{n} E_i) \le \sum_{I \subset \{1, \cdots, n\}, I \ne \varnothing} (-1)^{|I|+1} \mathrm{Pl}(\bigcup_{i \in I} E_i) \, ,$$

where $\{E_1, \dots, E_n\}$ is any finite subset of $\mathscr{P}(X)$.

The belief measure Bel and the plausibility measure Pl induced from a given basic probability assignment m are dual to one another in the sense that

$$\mathrm{Bel}(E) = 1 - \mathrm{Pl}(\bar{E}) \qquad \forall E \in \mathscr{P}(X).$$

If X is finite, then any nonnegative set function μ on $\mathscr{P}(X)$ satisfying $\mu(\varnothing) = 0$, $\mu(X) = 1$, and (BM 1) is a belief measure. Replacing (BM 1) with (PM 1), we can obtain a similar conclusion for plausibility measures.

A belief measure Bel coincides with its dual plausibility measure Pl if and only if it is a probability measure. In this case, the corresponding basic probability assignment m focuses on singletons, and inequalities (BM 1) and (PM 1) become equalities.

A λ-fuzzy measure, g_λ is a fuzzy measure on $(X, \mathscr{F})$ for which

$$g_\lambda(E \cup F) = g_\lambda(E) + g_\lambda(F) + \lambda g_\lambda(E) g_\lambda(F) \qquad \forall E, F \in \mathscr{F},$$

where $\lambda \in (-1/\mu(X), \infty) \cup \{0\}$ is a parameter. A regular λ-fuzzy measure is also called Sugeno's measure. On a finite space, any Sugeno measure possessing a nonnegative parameter value is a belief measure, while that possessing a nonpositive parameter value is a plausibility measure. Therefore, when $\lambda = 0$, any Sugeno measure is a probability measure. The dual of a Sugeno measure with parameter λ is still a Sugeno measure with parameter $\lambda' = -\lambda/(\lambda + 1)$.

Lemma 1. Let Bel_1 and Bel_2 be set functions on $\mathscr{P}(X)$. If Bel_1 satisfies (BM 1) and $\mathrm{Bel}_2(A) = \mathrm{Bel}_1(A) + c$ for any $A \in \mathscr{P}(X)$, where c is a constant, then Bel_2 satisfies (BM 1) too.

Proof. The conclusion can be obtained from the equality

$$\sum_{I \subset \{1, \cdots n\}, I \ne \varnothing} (-1)^{|I|+1} = 1 \qquad \forall n = 1, 2, \cdots. \qquad \blacksquare$$

Theorem 1. Let X be finite. If Bel is a belief measure on $\mathscr{P}(X)$, then there exists a family of probability measures $\{P_s \mid s \in S\}$ on $\mathscr{P}(X)$ such that $\mathrm{Bel} = \inf_{s \in S} P_s$, that is,

$$\mathrm{Bel}(A) = \inf_{s \in S} P_s(A)$$

for any $A \in \mathscr{P}(X)$. The cardinality of S is at most 2^{n-1}, where n is the cardinality of X.

Proof. Since

$$\mathrm{Bel}(X) = \inf_{s \in S} P_s(X)$$

is always true for any belief measure Bel and any family of probability measures $\{P_s \mid s \in S\}$, we just need to prove the following statement: if Bel is a nonnegative set function on $\mathscr{P}(X)$ satisfying $\mathrm{Bel}(\varnothing) = 0$ and (BM 1), then there exists a family of classical additive measures $\{P_s \mid s \in S\}$ such that

$$\mathrm{Bel}(A) = \inf_{s \in S} P_s(A) \qquad \forall E \in \mathscr{P}(X).$$

The restriction on the cardinality of S is realized in its proof.

The mathematical induction on the cardinality of X is used now. When $n = 1$, the above statement is obviously true, and the cardinality of S is 1, since Bel is a classical measure itself in this case. Now, assuming that the statement is true for n, we show that it is also true for $n + 1$. Let X consist of $n + 1$ points and Bel be a nonnegative set function on $\mathscr{P}(X)$ satisfying $\mathrm{Bel}(\varnothing) = 0$ and (BM 1). If we take $x \in X$ and denote $X' = X - \{x\}$, then the cardinality of X' is n. According to the assumption, there exists a family of classical measures $\{P_{1s} \mid s \in S_1\}$ on $\mathscr{P}(X')$ such that

$$\mathrm{Bel}(A) = \inf_{s \in S_1} P_{1s}(A) \qquad \forall A \in \mathscr{P}(X'),$$

and the cardinality of S_1 is at most 2^{n-1}. Defining Bel_1 on $\mathscr{P}(X')$ by

$$\mathrm{Bel}_1(A) = \mathrm{Bel}(A \cup \{x\}) - \mathrm{Bel}(\{x\}) \qquad \forall A \in \mathscr{P}(X'), \tag{I}$$

we know that Bel_1 is nonnegative and $\mathrm{Bel}_1(\varnothing) = 0$. By Lemma 1, we also know that Bel_1 satisfies condition (BM 1). Hence, from the assumption, there exists a family of classical measures $\{P_{2s} \mid s \in S_2\}$ on $\mathscr{P}(X')$ such that

$$\mathrm{Bel}_1(A) = \inf_{s \in S_2} P_{2s}(A) \qquad \forall A \in \mathscr{P}(X'),$$

and the cardinality of S_2 is also at most 2^{n-1}. For any $s \in S_2$, P_{2s} can be extended onto $\mathscr{P}(X)$ to be a classical measure in such a way:

$$P_{2s}(B) = P_{2s}(B - \{x\}) + \text{Bel}(\{x\}) \qquad \forall B \in \mathscr{P}(X) - \mathscr{P}(X') \qquad \text{(II)}$$

Thus,

$$\text{Bel}(B) = \inf_{s \in S_2} P_{2s}(B) \qquad \forall B \in \mathscr{P}(X) - \mathscr{P}(X')$$

For any $s \in S_1$, P_{1s} also can be extended onto $\mathscr{P}(X)$ to be a classical measure in a similar way:

$$P_{1s}(B) = P_{1s}(B - \{x\}) + \text{Bel}(\{X\}) - \text{Bel}(\{X'\}) \qquad \forall B \in \mathscr{P}(X) - \mathscr{P}(X'). \qquad \text{(III)}$$

By using condition (BM 1) for Bel, it is not difficult to verify that

$$P_{1s}(B) \geq \text{Bel}(B) \qquad \forall B \in \mathscr{P}(X) - \mathscr{P}(X') \text{ and } s \in S_1$$

and

$$P_{2s}(A) \geq \text{Bel}(A) \qquad \forall A \in \mathscr{P}(X') \text{ and } s \in S_2$$

Taking $S = S_1 \cup S_2$ and denoting

$$P_s = \begin{cases} P_{1s} & \text{if } s \in S_1 \\ P_{2s} & \text{if } s \in S_2 \end{cases},$$

we obtain a family of classical measures $\{P_s \mid s \in S\}$ on $\mathscr{P}(X)$ satisfying

$$\text{Bel}(A) = \inf_{s \in S} P_s(A) \qquad \forall A \in \mathscr{P}(X).$$

The cardinality of S is at most 2^n.
    The proof is now complete. ∎

By using the duality of belief measures and plausibility measures, we can obtain the following corollary immediately.

Corollary 1. If Pl is a plausibility measure on a finite measurable space, then there exists a family of probability measures such that $\text{Pl} = \sup_{s \in S} P_s$.

The proof of Theorem 1 provides us with a recursive method for constructing the desired family of probability measures.

Example 1. Let $X = \{a, b, c\}$. Belief measure Bel, its dual plausibility measure Pl, and the corresponding basic probability assignment m are listed in Table 1. By using the notation and the recursive method given in Theorem 1, we take $X' = \{a, b\}$ and assume that P_1 and P_2 have been well defined on $\mathscr{P}(X')$ whose values are listed in the first 4 rows under P_1 and P_2. Now, $x = c$, $\mathscr{P}(X) - \mathscr{P}(X') = \{\{c\}, \{a, c\}\ \{b, c\}, X\}$, and $n = 2$. According to equation (I) in Theorem 1, we obtain the values of Bel_1 on $\mathscr{P}(X')$ as listed in the last column of Table 1 and, therefore, by the assumption, we can find the values of additive measures P_3 and P_4 defined on $\mathscr{P}(X) - \mathscr{P}(X')$ as shown in the first 4 rows under P_3 and P_4 in the table. Since $\text{Bel}(\{x\}) = 1/5$, by using equation (II), we obtain the values of P_3 and P_4 on $\mathscr{P}(X) - \mathscr{P}(X')$ which are shown in the last 4 rows under P_3 and P_4. Finally, since $\text{Bel}(\{X\}) - \text{Bel}(\{X'\}) = 1/3$, by using equation (III), we obtain the values of P_1 and P_2 on $\mathscr{P}(X) - \mathscr{P}(X')$ and complete the table. P_1, P_2, P_3, and P_4 are probability measures and satisfy the requirement

$$\text{Bel}(A) = \min_{i=1, 2, 3, 4} P_i(A)$$

as well as

$$\text{Pl}(A) = \max_{i=1, 2, 3, 4} P_i(A)$$

for any $A \in \mathscr{P}(X)$.

Table 1. Belief measure and plausibility measure expressed by the infimum and the supremum of a family of probability measures

a	b	c	Bel	Pl	m	P_1	P_2	P_3	P_4	Bel_1
0	0	0	0	0	0	0	0	0	0	0
1	0	0	1/3	1/2	1/3	1/3	5/12	2/5	1/2	2/5
0	1	0	1/4	2/5	1/4	1/3	1/4	2/5	3/10	3/10
1	1	0	2/3	4/5	1/12	2/3	2/3	4/5	4/5	4/5
0	0	1	1/5	1/3	1/5	1/3	1/3	1/5	1/5	
1	0	1	3/5	3/4	1/15	2/3	3/4	3/5	7/10	
0	1	1	1/2	2/3	1/20	2/3	7/12	3/5	1/2	
1	1	1	1	1	1/60	1	1	1	1	

Essentially, in this example, Bel is a regular λ-fuzzy measure with $\lambda = 1$, while Pl is a regular λ-fuzzy measure with $\lambda = -1/2$.

3. Semicontinuous Fuzzy Measures Determined by Modal Logic

Let $M = [W, R, Q, V]$ be a model of modal logic, where $W = \{w_1, w_2, ..., w_n\}$ is a finite set of worlds, R is a reflexive relation on W (called accessibility relation), Q is the smallest set of relevant propositions including "a given point ϵ is classified in set A" (denoted by e_A) for any $A \in \mathscr{P}(X)$ and being closed under the necessity operator ($\square$) and the possibility operator ($\lozenge$) as well as the usual logic operators such as negation ($\neg$), disjunction ($\vee$), conjunction ($\wedge$), implication ($\rightarrow$), and equivalence ($=$), and V is a value assignment function

$$V: W \times Q \rightarrow \{T, F\}$$
$$(w, q) \; \longmapsto \; V_w(q)$$

satisfying, for any $w \in W$,

(VAF 1) $$V_w(e) = F;$$

(VAF 2) $$V_w(e_{\{x\}}) = \neg[\vee_{y \neq x} V_w(e_{\{y\}})], \; \forall x \in X;$$

(VAF 3) $$V_w(e_A) = \vee_{x \in A} V_w(e_{\{x\}}), \; \forall A \in \mathscr{P}(X).$$

The set function Bel defined on $\mathscr{P}(X)$ by

$$\mathrm{Bel}(A) = \sum_{i=1}^{n} \chi_T[V_{w_i}(\square \; e_A)]/n, \qquad \forall A \in \mathscr{P}(X)$$

where

$$\chi_T(t) = \begin{cases} 1, & \text{if } t = T \\ 0, & \text{if } t = F, \end{cases}$$

is a rational-valued belief measure on $\mathscr{P}(X)$. In fact, for each $w \in W$, there exists a unique $u_w \in X$ such that $V_w(e_{\{u_w\}}) = T$; let

$$B_w = \{u_\alpha | (w, \alpha) \in R, \alpha \in W\},$$

then the set function m defined by

$$m(E) = i/n \qquad \text{if } E = B_w \text{ for } i \text{ worlds } w \in W$$

is the basic probability assignment inducing the above-mentioned belief measure Bel. Similarly, the set function Pl defined on $\mathscr{P}(X)$ by

$$Pl(A) = \sum_{i=1}^{n} \chi_{\mathrm{T}}[V_{w_i}(\Diamond e_A)]/n, \qquad \forall A \in \mathscr{P}(X)$$

is the dual (rational-valued) plausibility measure of Bel on $\mathscr{P}(X)$.

4. A Weighted Model

To obtain a real-valued belief measure (or, a real-valued plausibility measure), we should introduce a weighted model of modal logic.

Let $[W, R, Q, V]$ be a model of modal logic, where W is not necessarily finite: $W = \{w_t |$ $t \in T\}$. The weight of each $w_t \in W$ is denoted by $^t\omega$ satisfying $\sum_{t \in T} {}^t\omega = 1$. We know that there exist at most countably many $^t\omega$ such that $^t\omega > 0$. So, we can use $\{^i\omega \mid i \in I\}$, where I is a countable index set, to denote them, and the corresponding worlds are denoted by $\{w_i |$ $i \in I\}$. The set function

$$\mathrm{Bel}(A) = \sum_{i \in I} {}^i\omega \, \chi_{\mathrm{T}}[V_{w_i}(\Box \, e_A)], \qquad \forall A \in \mathscr{P}(X)$$

$$(\text{or, } Pl(A) = \sum_{i \in I} {}^i\omega \, \chi_{\mathrm{T}}[V_{w_i}(\Diamond e_A)], \qquad \forall A \in \mathscr{P}(X))$$

is a belief measure (or, a plausibility measure, resp.) induced by the basic probability assignment that is defined as

$$m(E) = \sum_{t:B_{w_t} = E} {}^t\omega, \qquad \forall E \in \mathscr{P}(X)$$

where B_{w_t} has the same meaning as is given previously.

5. An Inverse Problem

Let X be a nonempty set (not necessary to be finite). Given a belief measure Bel (or a plausibility measure Pl) on $\mathscr{P}(X)$, can we construct a model of modal logic $[W, R, Q, V]$ by which Bel (or Pl) is determined in the way shown in the previous section? The following theorem gives an affirmative answer.

Theorem 2. For any given belief measure Bel (or plausibility measure Pl), there exists a weighted model of modal logic which determines Bel (or Pl).

Proof. Let X be a nonempty set and Bel be a belief measure on $\mathscr{P}(X)$ with basic probability assignment m, whose focal sets are $E_1, E_2, \ldots$. Taking $W = \mathscr{P}(X) - \{\varnothing\}$, we define an accessibility relation $R = (r_{EF})$ on W as follows: for any $E, F \in W$,

$$
r_{EF} = \begin{cases} 1 & \text{if } E = F, \text{ or } F \text{ is a singleton and } F \subset E \\ 0 & \text{otherwise} \end{cases}
$$

R is reflexive. For each $E \in W$, since E is nonempty, we can choose a point $x_E \in E$ as the representative of E. For any given $A \in \mathscr{P}(X)$, value assignment function V is defined by

$$
V_E(e_A) = \begin{cases} T & \text{if } x_E \in A \\ F & \text{otherwise} \end{cases}
$$

for any $E \in W$. V can be extended onto Q according to modal logic operators. The weight of each $E \in W$ is taken as $m(E)$. Denoting $m(E_i) = {}^i\omega$, $i = 1, 2, \ldots$, we have

$$
\sum_{i \in I} {}^i\omega \, \chi_T[V_{E_i}(\Box e_A)] = \sum_{E_i \subset A} m(E_i) = \text{Bel}(A) \qquad \forall A \in \mathscr{P}(X),
$$

where $I = \{1, 2, \ldots\}$. That is , $[W, R, Q, V]$ is the required model of modal logic.

Example 2 (Example 1 continued). Nonempty set X and belief measure Bel is given in Example 1. Take $W = \{\{a\}, \{b\}, \{c\}, \{a, b\}, \{a, c\}, \{b, c\}, X\}$ and define accessibility relation R on W by using Table 2.

Table 2. accessibility relation R

	$\{a\}$	$\{b\}$	$\{c\}$	$\{a, b\}$	$\{a, c\}$	$\{b, c\}$	X
$\{a\}$	1	0	0	1	1	0	1
$\{b\}$	0	1	0	1	0	1	1
$\{c\}$	0	0	1	0	1	1	1
$\{a, b\}$	0	0	0	1	0	0	0
$\{a, c\}$	0	0	0	0	1	0	0
$\{b, c\}$	0	0	0	0	0	1	0
X	0	0	0	0	0	0	1

We chose $a \in \{a\}$, $b \in \{b\}$, $c \in \{c\}$, $a \in \{a, b\}$, $a \in \{a, c\}$, $b \in \{b, c\}$, $c \in X$ as their represetatives respectively. Then value assignment function V is given in Table 3.

Table 3. Value assignment function V

		Set A							
		$\emptyset$	$\{a\}$	$\{b\}$	$\{c\}$	$\{a, b\}$	$\{a, c\}$	$\{b, c\}$	X
	$\{a\}$	F	T	F	F	T	T	F	T
	$\{b\}$	F	F	T	F	T	F	T	T
	$\{c\}$	F	F	F	T	F	T	T	T
World	$\{a, b\}$	F	T	F	F	T	T	F	T
E	$\{a, c\}$	F	T	F	F	T	T	F	T
	$\{b, c\}$	F	F	T	F	T	F	T	T
	X	F	F	F	T	F	T	T	T

Thus, model of modal logic $[W, R, Q, V]$ satisfies equation (IV).

References

Harmanec, D. and Klir, G. J. (1994) On modal logic interpretation of Dempster-Shafer theory of evidence, *International Journal of Intelligent Systems* 9, 941-951.

Harmanec D., Klir, G. J., and Wang, Z. (1995) Modal logic interpretation of Dempster-Shafer theory: an infinite case, (submitted to *International Journal of Approximate Reasoning*).

Hughes, G. E. and Cresswell, M. J. (1968) *An Introduction to Modal Logic*, Methuen, London and New York.

Hughes, G. E. and Cresswell, M. J. (1984) *A Companion to Modal Logic*, Methuen, London and New York.

Resconi, G., Klir, G. J., and St. Clair, U. (1992) Hierarchical uncertainty metatheory based upon modal logic, *International Journal of General Systems* 21(1), 23-50.

Resconi, G., Klir, G. J., and St. Clair, U. (1993) On the integration of uncertainty theories, *International Journal of Uncertainty, Fuzziness and Knowledge-Based Systems* 1(1), 1-18.

Shafer, G. (1976) *A Mathematical Theory of Evidence*, Princeton University Press, Princeton.

Wang, Z. and Klir, G. J. (1992) *Fuzzy Measure Theory*, Plenum Press, New York.

FUZZINESS REDUCTION METHOD
FOR A COMBINATION FUNCTION

S. TANO, T. ARNOULD, Y. KATO AND T. MIYOSHI
LIFE: Laboratory for International Fuzzy Engineering Research,
89-1 Yamashita-cho, Nakaku, Yokohama, 231 JAPAN

Abstract

Conventional fuzzy reasoning has a well known problem that the fuzziness of inferred results gradually increases according to the progress of the inference. Essential problems are (P1) the membership value of the combined result never approaches to grade 0, but always increases, and (P2) lack of reinforcement property. We proposed a new combination function which resolves the problems by introducing equilibrium E and dependency factors α and β represented by stochastic rules. Its behavior is consistent with that of a human. Moreover, it covers the range of conventional combination functions.

1. Introduction

At the Laboratory for International Fuzzy Engineering Research in Japan, we have developed an expert system called FOREX (FOReign exchange trade support EXpert system) as an application of fuzzy theory to a real problem[1]. The experience proved that fuzzy methodology is feasible and quite effective. However, at the same time, we found several serious problems in fuzzy inference. We classified the problems and made up six research themes to study at LIFE and we are now developing a fuzzy inference software environment FINEST. The six research themes include (1) generalization of and/or/aggregation[11], (2) categorization of implication functions and its parametrization, (3) fuzziness reduction method for a combination function, (4) formalization of fuzzy backward reasoning[12], (5) automatic tuning of inference method[10] including above (1) to (4) and, (6) software environment for fuzzy reasoning equipped with above (1) to (5).

Conventional fuzzy reasoning has a well known problem that the fuzziness of inferred results gradually increases according to the progress of the inference. It is sometimes called explosion of fuzziness. This paper describes (3) fuzziness reduction method for a combination function , as a solution to the problem.

2. Fuzziness of a Fuzzy Set

First of all, it is necessary to clarify the fuzziness of a fuzzy set before discussing about problems of a combination function. It is said that a fuzzy set has two types of

15

Z. Bien and K. C. Min (eds.),
Fuzzy Logic and its Applications, Information Sciences, and Intelligent Systems, 15–24.
© 1995 *Kluwer Academic Publishers.*

fuzziness which directly come from the membership function, i.e. height and width of a fuzzy set which correspond to (a) and (b) in Fig.1, respectively.

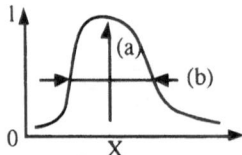

Fig.1 Two types of fuzziness

The fuzziness tends to be measured by the following two criteria.
    (a)if the height of a fuzzy set is tall, the fuzziness is small.
    (b)if the breadth of a fuzzy set is narrow, the fuzziness is small.
But a fuzzy set should be judged to have small fuzziness even when the membership value becomes close to zero. Since the criterion (a) is not precise enough to represent this fuzziness, the criterion (a) is rewritten as the following criteria (a1) and (a2), where (a1) is equivalent to (a) and (a2) is newly added.
    (a1)if the membership value is close to 1, the fuzziness is small.
    (a2)if the membership value is close to 0, the fuzziness is small.
    As a measure of such fuzziness, fuzzy entropy[2,3] and specificity[4] were proposed, which correspond to the criteria (a1,a2) and (b) respectively. Paper[5] analyzed the fuzziness of various operators by using these two measures.
    However, the breadth that the criterion (b) take account of is not suitable for the measure of fuzziness. Let's think over the extreme case of a fuzzy set whose membership grade is everywhere equal to 1. Although the fuzzy set is often denoted as total ignorance, generally speaking, it is impossible to affirm that the fuzzy set is vague. Therefore (b) should be removed from the criteria.
    Finally, we have criteria (a1) and (a2) to measure the fuzziness of a fuzzy set.

3. Problems of Combination Function

A combination operation is defined as a method concerning how to get one result A''' when two fuzzy sets A' and A'' are deduced by two inference processes. We call A''' the combined result, A' and A'' the premise results. Since it is obvious that $\mu_{A'''}(x)$ depends on $\mu_{A'}(x)$ and $\mu_{A''}(x)$ only, a combination function can be represented as follows:

$$\mu_{A'''}(x)=\text{combine}(\mu_{A'}(x), \mu_{A''}(x))$$

As the combination process can be considered to be a sort of logical disjunction, T-conorms are preferably used and in many cases, the Max operator is chosen as a combination function. In case of Max, the following formulas hold.

$$\mu_{A'''}(x)\geqq \mu_{A'}(x) \quad , \mu_{A'''}(x)\geqq \mu_{A''}(x)$$

Judging from the criterion (a1), the fuzziness of the result is reduced. It seems to be no problem at all. However, the combined result obtained with Max always becomes close to the total ignorance. In other words, the combined result obtained with a conventional combination function becomes close to one of the two non-fuzzy values, that is, grade 1, and never approaches to the other non-fuzzy value, that is, grade 0. Namely, conventional functions do not possess the property (a2). Because Max is a lower bound of T-conorms, this analysis is true for any T-conorm function.

The next problem is caused by the idempotent property of Max. The fuzziness of the combined result is not reduced, even if the same results are deduced by two inference processes which are independent from each other. This happens due to the idemponent nature, i.e. max(x,x)=x for any x and it is contrary to human intuition. When a person solves a problem by using fuzzy data and knowledge, he/she considers the problem from many points of view and synthesizes all the results in order to reduce the fuzziness of the result. The Max function is used in conventional fuzzy applications. In possibility theory, Min is used in case of reliable knowledge and Max for unreliable knowledge[6,7]. But, the use of Min and Max means nothing more than the simple selection of the grade of one result. They never use the result synergically. In other words, the results never reinforce each other. So the fuzziness is not reduced as a human expects. All idempotent combination functions have this problem.

Essential problems discussed here can be summarized as follows:

(P1) the membership value of the combined result never approaches to grade 0, but always increases.

(P2) lack of reinforcement property

4. Degree of Positive/Negative Belief

In this paper, "$\mu_S(x)=a$" means that the degree of belief of "an element x is S" is a, that is to say, it is an interpretation based on degree of belief. Grade 1 expresses complete belief and grade 0 expresses complete disbelief.

Now assume that "$\mu_S(x)=a$" and "$\mu_S(x)=b$" are derived from two independent evidences respectively. If both degrees a and b are close to 1, each result should be reinforced. So we get combine(a,b)>max(a,b).

On the contrary, if the degrees a and b are close to 0, what should happen ? Of course, each result should be reinforced, but the direction of reinforcement depends on the following interpretation.

(1) "an element x is S" is a little bit believed.

By this interpretation, combine(a,b) > max(a,b).

(2) "an element x is S" is strongly disbelieved.

By this interpretation, combine(a,b) < min(a,b).

Interpretation of (1) is a degree of positive belief, interpretation of (2) is a degree of negative belief. Although a grade value close to 1 can be interpreted as a degree of positive belief, a value close to 0 can not be uniformly interpreted. In some cases, it may represent a degree of positive belief. In other cases, it may represent a degree of negative belief.

Then, an equilibrium E is introduced to represent the tradeoff point which is interpreted as neither a positive belief nor a negative belief. As shown in Fig.2(a), the meaning of grade is divided at E. Values greater then E correspond to positive belief whereas values lower than E correspond to negative belief. When E is equal to 0, only positive meaning is represented(Fig.2(b)). When E is equal to 1, only negative meaning is represented(Fig.2(c)).

18

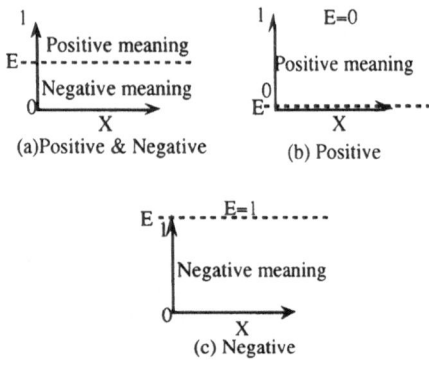

Fig.2 Equilibrium E

5. Dependency Represented by Stochastic Rule

As mentioned above, the property of reinforcement is indispensable for combination operation. It is also important to take account of the dependency of evidences. For example, premise results derived from independent evidences can be reinforced, but if the evidences are not independent, reinforcement should not be done. In this way, the dependency is an important factor for reinforcement.

Since it is possible to think of dependency as the underlying derivation relation, a pair of stochastic rules can represent the dependency. The following stochastic rules represent the dependency between the evidences A and B.

$$A \rightarrow B : \alpha$$
$$B \rightarrow A : \beta$$

The first rule says that evidence A induces evidence B with the probability α. The values α and β are interpreted as the ratio of common portion against A and B, i.e. $|A \cap B| / |A|$ and $|A \cap B| / |B|$ respectively. For example, "A -> B : 0.7" implies that 70% of evidence A is contained in evidence B.

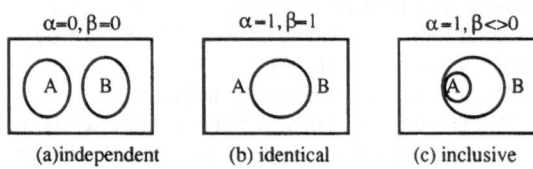

(a)independent   (b) identical   (c) inclusive

Fig.3 Examples of dependency

6. New Combination Operation

Assumption: "$\mu_S(x)=a$" was deduced from an evidence A, and the equilibrium is E1. "$\mu_S(x)=b$" was deduced from the other evidence B, and the equilibrium is E2. The dependency between A and B is represented by A -> B : α, B -> A : β. The equilibrium of the combined result is E3.

Fig.4 illustrates the main flow of the combination process.

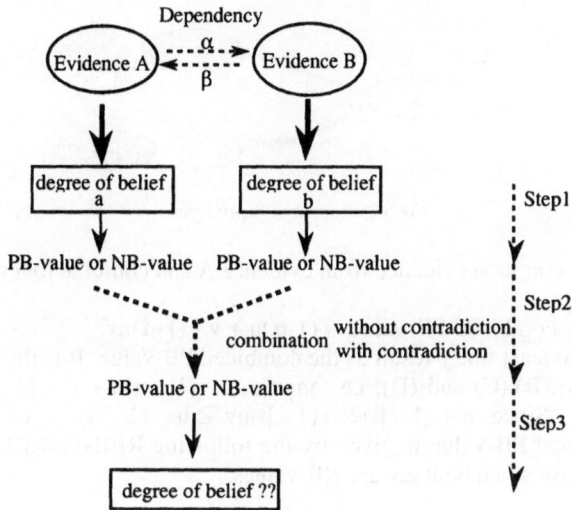

Fig.4 Combination of degrees of belief

Step1: The degrees of belief of the premises, i.e. a and b, are transformed into normalized degrees of positive or negative belief. A grade bigger than E, i.e. in [E,1], is transformed into a PB-value in [0,1] by the function PB. The function NB transforms a grade in [0, E] into a NB-value in [0,1]. These functions are defined as follows:

$$PB(x:E) = \frac{x - E}{1 - E}$$
$$NB(x:E) = \frac{E - x}{E}$$

In this step, a and b are transformed into u and v respectively.

Step2: In this step two cases have to be considered.
Case-I: Both u and v are PB-values or NB-values.

Assume that both u and v are PB-values. If the evidences A and B are independent, the result is u + v - uv, by using DS theory[8] or probability theory. For example, $\mu_S(x)=a$ can be interpreted as $m_A(\{x\}) = u$, $m_A(\Omega) = 1 - u$, where Ω is the universe of discourse. By application of Dempster's combination rule, $m(\{x\}) = u + v - uv$ is obtained.

Next, four possible approaches are shown to cope with the dependency in case the evidences are not independent.
(A) Use only the result derived by the evidence A. The combined PB-value becomes u.
(B) Use only the result derived by the evidence B. The combined PB-value becomes v.
(C) Abandon the common evidence from evidence B and combine with the modified B and A.

As shown in Fig.5, evidence space A∪B is divided into A and B', which is obtained

20

by throwing away the common area from B. Notice that A and B' become independent. Because the ratio of common area against B is b, the result derived from B' is modified as $(1-\beta)v$. The other result is still u. So the combined grade is $u + (1-\beta)v - (1-\beta)uv$.

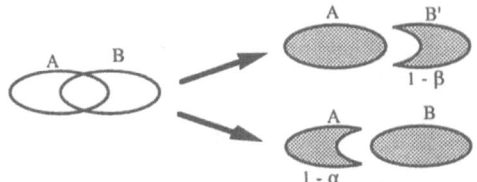

Fig.5 Treatment of dependency

(D) Abandon the common evidence from evidence A and combine together the modified A and B.

Similarly, the combined PB-value is $(1-\alpha)u + v - (1-\alpha)uv$.

We choose the least fuzzy result as the combined PB-value: It is the maximum value of the results (A),(B),(C) and (D), i.e. $\max(u, v, (1-\alpha)u + v - (1-\alpha)uv, u + (1-\beta)v - (1-\beta)uv)$. Since $u + (1-\beta)v - (1-\beta)uv \geq u$, $(1-\alpha)u + v - (1-\alpha)uv \geq v$, then the combined PB-value is given by the following REINFORCE function. This also covers the case when both u,v are NB-values.

$$\text{REINFORCE}(u{:}\alpha, v{:}\beta) = \max\left((1-\alpha)u+v-(1-\alpha)uv, u+(1-\beta)v-(1-\beta)uv\right)$$

Case-II: One of u and v is a PB-value and the other one is a NB-value.

Assume that u is a PB-value and v is a NB-value. This case is a sort of contradiction as shown in Fig.6. We use the intuitive combination method, similar to MYCIN's method[9], because there is no standard rational theory.

In case the evidences A and B are independent (Fig.6 (a)), it is natural to conclude that the combined result is a PB-value with the value u-v, if u is bigger than v.

What should be done in case that the evidences A,B are identical (Fig.6 (b))? It again seems to be reasonable to conclude that the combined result is a PB-value with the value u-v, if u is bigger than v.

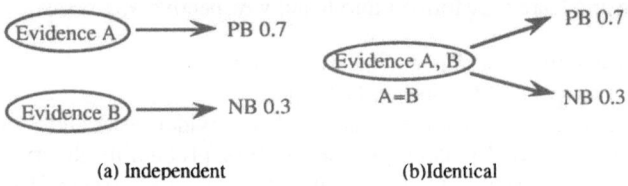

(a) Independent   (b)Identical

Fig. 6 Combination with contradiction

The dependency of the evidences has no influence on the combination with contradiction ! This interpretation can be justified by considering these opposite results u and v as two completely distinct results. Generally speaking, in case two different results are derived, whether the evidences are identical or dependent does not matter in the treatment of the results. For the same reason, the dependency has no influence on the combination operation.

Step3: PB^{-1} and NB^{-1} defined below transform PB-value and NB-value into a grade value.

$$PB^{-1}(x:E) = E + (1-E)x$$

$$NB^{-1}(x:E) = E - Ex$$

The combination operation explained here is summarized in Fig.7 as a function COMBINE($a:E1:\alpha$, $b:E2:\beta$, E3).

COMBINE($a:E_1:\alpha$, $b:E_2:\beta$, E_3) =

$$
\begin{cases}
a \geq E_1, \ b \geq E_2 \\
\quad PB^{-1}(\text{ REINFORCE(} PB(a:E_1):\alpha, \ PB(b:E_2):\beta) : E_3) \\[2ex]
a \geq E_1, \ b \leq E_2 \\
\quad PB^{-1}(PB(a:E_1) - NB(b:E_2) : E_3) \quad \text{if } PB(a:E_1) \geq NB(b:E_2) \\
\quad NB^{-1}(NB(b:E_2) - PB(a:E_1) : E_3) \quad \text{if } NB(b:E_2) \geq PB(a:E_1) \\[2ex]
a \leq E_1, \ b \geq E_2 \\
\quad NB^{-1}(NB(a:E_1) - PB(b:E_2) : E_3) \quad \text{if } NB(a:E_1) \geq PB(b:E_2) \\
\quad PB^{-1}(PB(b:E_2) - NB(a:E_1) : E_3) \quad \text{if } PB(b:E_2) \geq NB(a:E_1) \\[2ex]
a \leq E_1, \ b \leq E_2 \\
\quad NB^{-1}(\text{ REINFORCE(} NB(a:E_1):\alpha, \ NB(b:E_2):\beta) : E_3)
\end{cases}
$$

Fig.7 COMBINE($a:E1:\alpha$,$b:E2:\beta$, E3)

7. Characteristics of the Functions REINFORCE and COMBINE

(1) The REINFORCE function evolves from the logical sum(max) to the algebraic sum(u+v-uv) according to the dependency parameters α and β as shown in Fig.8.

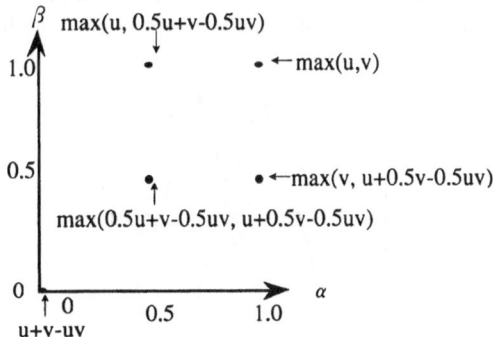

Fig.8 Range of the REINFORCE function

22

(2) Fig.9 shows the influence of equilibrium E by drawing COMBINE(μ_A(x):E:0, μ_A(x):E:0, E).

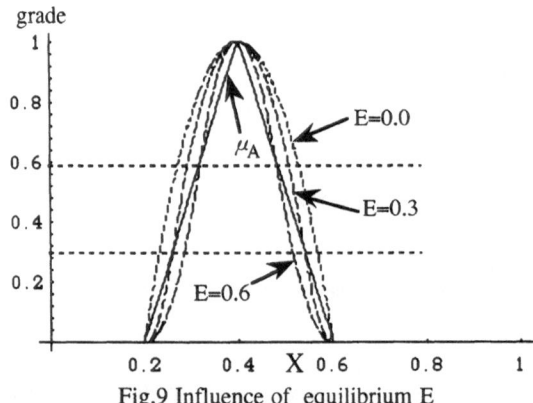

Fig.9 Influence of equilibrium E

(3) Fig.10 illustrates a graph of z=COMBINE(x:0.5:y, x:0.5:y, 0.5) - x. The weakening effect and strengthening effect are inverted at the equilibrium E. The level of the weakening and strengthening effect increases with the degree of the independence. For $\alpha=\beta=1$, the effect disappears.

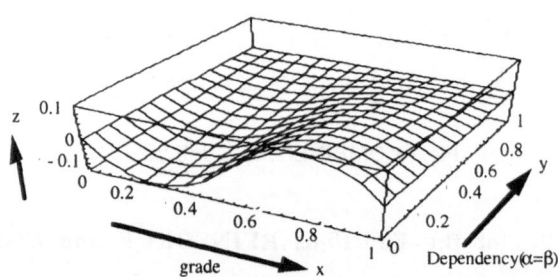

Fig.10 Z=COMBINE(x:0.5:y, x:0.5:y, 0.5) - x

(4) As shown in Fig.11, in case E=0 and $\alpha=\beta=1$, the combination function is identical to the popular max-based combination which is used in most applications.

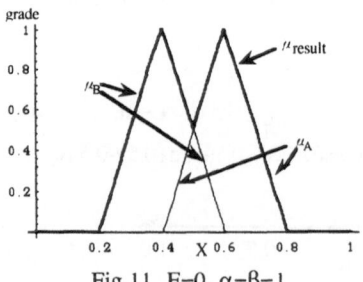

Fig.11 E=0, $\alpha=\beta=1$

(5) In case of E=1 and α=β=1, the combination function is identical to the conjunctive combination, which is used in possibility theory when the knowledge is reliable(min). This analysis suggests that a possibility distribution can be interpreted as a fuzzy set whose equilibrium E is equal to 1.

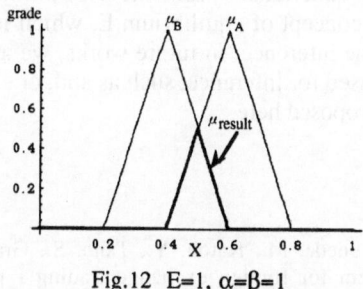

Fig.12 E=1, α=β=1

(6) Fig.13 shows three examples . Two fuzzy sets μ_A and μ_B, are combined into μ_result. E is equal to 0.4 for the three cases.

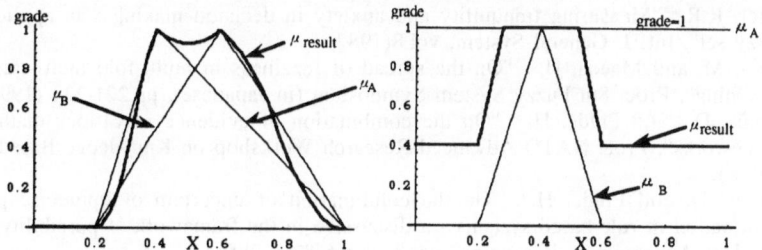

(a)Combination of two fuzzy sets μ_A and μ_B   (b)Combination of triangular μ_B and grade=1

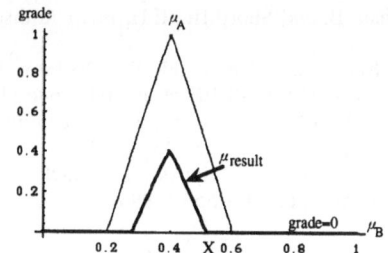

(c)Combination of triangular μ_A and grade=0

Fig.13 Examples

8. Summary

We proposed a new combination function which resolves the problems (P1) and (P2). Its behavior is consistent with that of a human. Moreover, it covers the range of conventional combination functions. From the viewpoint of human intuition, it is quite natural to adopt the concept of equilibrium E, which is also quite significant for the other components of the inference. In future works, we are going to make a similar change on the operations used for inference, such as and, or , implication, etc., to adjust the semantics to the idea proposed here.

References

1. Yuize, H., Yagyu, T., Yoneda, M., Katoh, Y., Tano, S., Grabisch, M., and Fukami, S.: "Decision support system for foreign exchange trading - practical implementation -", Proc. of IFES'91 , pp.971-982(1991).
2. Luca, A.D. and Termini, S.: "A definition of a nonprobabilistic entropy in the setting of fuzzy set theory", Information Control, vol.20, pp.301-312(1972).
3. Kosko, B. : "Fuzzy entropy and conditioning", Information Science, Vol.40, pp.165-174(1986).
4. Yager, R.R.: "Measuring tranquility and anxiety in decision making : an application of fuzzy set", Int. J. General System, vol.8(1982).
5. Imura, M. and Maeda, H. : "On the spread of fuzziness in multi-fold multi-stage fuzzy reasoning", Proc. 8th Fuzzy System Symposium (in Japanese), pp.221-224 (1992).
6. Dubois, D. and Prade, H. : "On the combination of evidence in various mathematical frameworks", Proc. NATO Advanced Research Workshop on Knowledge Based Control (1988).
7. Dubois, D. and Prade, H. : "On the combination of uncertain or imprecise pieces of information in rule-based systems - a discussion in the framework of possibility theory". Int. J of Approximate Reasoning, vol.2, pp.65-87 (1988).
8. Shafer, G. : "A mathematical theory of evidence", Princeton University Press (1976).
9. Buchanan, B. and Shortliffe, E.H.: "Uncertainty and evidential support", in Rule-Based Expert System (Buchanan, B. and Shortliffe, E.H. Eds,) Addison-Wesley, Reading, Mass.. pp.209-232(1984).
10. Miyoshi, T., Tano, S., Kato, Y., and Arnould, T. : Operator Tuning in Fuzzy Production Rules Using Neural Networks, FUZZ-IEEE'93, pp. 641- 646 (1993).
11. Kato, Y., Arnould, T., Miyoshi, T., and Tano, S. : Conjunction and disjunction with synergistic effect, FUZZ-IEEE'93, pp. 225- 230 (1993).
12. Arnould, T., Tano, S., Kato, Y., and Miyoshi, T. : Backward-chaining with fuzzy "if... then..." rules, FUZZ-IEEE'93, pp. 548- 553 (1993).

A COMPARISON BETWEEN TWO METHODS TO OPTIMIZE CROSS-DETECTING LINES FOR A FUZZY NEURON

MASUO FURUKAWA
Nagano National College of Technology
Nagano 381, Japan

AND

TAKESHI YAMAKAWA
Kyushu Institute of Technology
Iizuka, Fukuoka 820, Japan

Abstract.

We have described a design algorithm of membership functions for a fuzzy neuron using example-based learning with optimization of cross-detecting lines(Yamakawa *et al.*, 1992). The optimization discussed in (Yamakawa *et al.*, 1992) is called the inefficient cross-detecting line elimination method. We have also described the efficient cross-detecting line selection method as an advanced optimization method(Furukawa *et al.*, 1993). This paper shows a comparison between the inefficient cross-detecting line elimination method and the efficient cross-detecting line selection method, and the typical examples of membership function obtained by both methods. In comparison with the elimination method, the selection method can reduce the number of the common cross-detecting lines (i.e. the number of the sensor arrays) and the CPU time for design of the membership functions of a fuzzy neuron.

1. Introduction

The artificial neural network employing conventional neuron model has a massively parallel structure which is composed of many analog threshold elements connected each other through weights. A neural network has faster response and a higher performance then those of a sequential digital

25

Z. Bien and K. C. Min (eds.),
Fuzzy Logic and its Applications, Information Sciences, and Intelligent Systems, 25–34.
© 1995 *Kluwer Academic Publishers.*

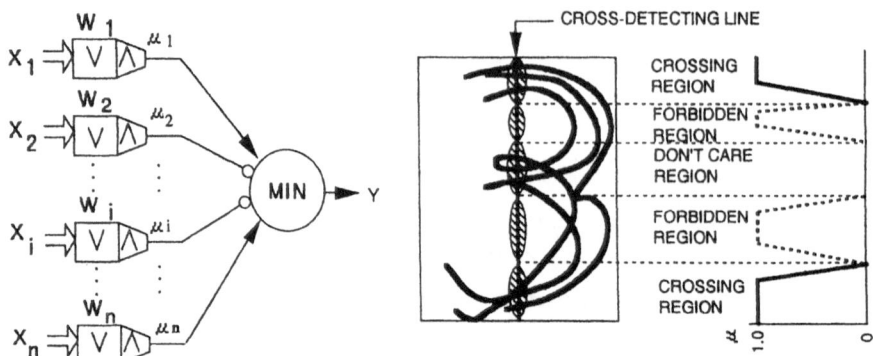

Figure 1. A schematic model of a fuzzy neuron.

Figure 2. Membership function on a cross-detecting line.

computer in emulating the capabilities of the human brain. But network structure and weighting factors are not designable. To cope with this problem, a *fuzzy neuron* has been proposed by T.Yamakawa(Yamakawa *et al.*, 1989) and implemented in an IC chip called *fuzzy neuron chip*.

The fuzzy neuron has been applied to a high speed recognition hardware system of hand-written characters(Yamakawa, 1990). The recognition by the fuzzy neuron seems to be similar to recognition by the zonde method which was developed by T.L.Dimond(Dimond, 1957)(Dimond, 1958). The advanced distinctions of a fuzzy neuron from the zonde method are no regulation of writing characters and more flexibility by membership functions.

Though a membership function makes a fuzzy neuron flexible and designable, know-hows of human experts are necessary for this design. So, a design algorithm of membership functions for a fuzzy neuron using example-based learning with optimization of cross-detecting lines has been described in (Yamakawa *et al.*, 1992). We call the optimization of cross-detecting lines, which have been discussed in (Yamakawa *et al.*, 1992), the inefficient cross-detecting line elimination method (the elimination method for short). We have also described an advanced method to optimize cross-detecting lines which is called the efficient cross-detecting line selection method (the selection method for short)(Furukawa *et al.*, 1993). This paper shows a comparison between the elimination method and the selection method, and the typical examples of membership function obtained by both methods.

2. A Fuzzy Neuron and It's Application for Pattern Recognition of Hand-Written Characters

Figure 1 shows a schematic model of a fuzzy neuron, in which X_1, X_2, ... , X_i, ... , X_n are vectors of crisp sets and/or fuzzy sets to inputs, w_1, w_2,

... ,w_i, ... , w_n are membership functions of the synaptic weight and Y is an output of a fuzzy neuron ranging from 0 through 1. Excitatory and inhibitory connections give the grade of soft matching μ_i between the input X_i and the synaptic weight w_i. The grade μ_i is fed to a MIN block of a fuzzy neuron model. Excitatory connections are represented by MIN, and inhibitory connections are represented by fuzzy logic complements followed by MIN. The output of a fuzzy neuron is a minimum grade of connections and gives the grade of soft matching between all of inputs and all of membership functions.

Pattern recognition of hand-written characters by a fuzzy neuron has been discussed(Yamakawa *et al.*, 1989), in which cross-detecting lines are employed to get features of characters as shown in Figure 2. There are three kinds of regions on the cross-detecting line, a *crossing region*, a *forbidden region* and a *don't care region*. The crossing region is the portion of the cross-detecting line where the character should cross. The forbidden region is the portion of the cross-detecting line where the character should not cross. The don't care region is the portion of the cross-detecting line where the character will uncertainly cross. One or more cross-detecting lines are assigned in the input frame. The cross-detecting lines of all characters, which are put into the same input frame, are referred to common cross-detecting lines. Sensor arrays should be located on the common cross-detecting lines.

The boundaries of these three kinds of regions are fuzzy, so that they should be defined by membership functions for each cross-detecting lines as shown in Figure 2. The membership functions of crossing regions are used for excitatory connections. The membership functions of forbidden regions are used for inhibitory connections. The don't care region is not used for recognition of characters, because it has no information.

Membership functions for crossing regions and forbidden regions can be assigned in the common cross-detecting lines. The appropriate number of cross-detecting lines are located horizontally and vertically in the frame. Cross-points detected by sensor arrays which are arranged on cross-detecting lines are applied to input busses of a fuzzy neuron. An output of each fuzzy neuron represents the grade of the possibility of the character.

3. Extraction of Membership Functions Using Example-Based Learning with Optimization of Cross-Detecting Lines

The algorithm of extraction of membership functions using example-based learning with optimization of cross-detecting lines has been discussed in (Yamakawa *et al.*, 1992). This algorithm is summarized as follows.

First, pattern samples of each categories are classified to adequate classes by their outlines.

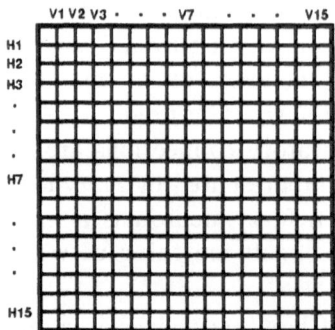

Figure 3. Assignment cross-detecting line.

Second, A fuzzy neuron is assigned to each classes and membership functions for each fuzzy neuron are extracted from the cross-points which are provided by classified pattern samples on cross-detecting lines.

A frame carries many cross-detecting lines. It means that the fuzzy neuron has the same amount of synaptic membership functions or more. However, all of the membership functions are not necessary for a reasonable pattern recognition depending upon the complexity of the pattern and the number of the categories to be distinguished. Thus, thirdly, the optimization of the number and the location of cross-detecting lines is applied to each membership functions. We have developed two optimization methods which are called the elimination method and the selection method. The procedures of the elimination method and the selection method for optimization of cross-detecting lines are described in this section.

In this paper, the number of horizontal cross-detecting lines and vertical cross-detecting lines, initially assigned in a frame uniformly as shown in Figure 3, are fifteen and fifteen for saving CPU time, respectively.

3.1. OPTIMIZATION BY THE ELIMINATION METHOD

To explain optimization by the elimination method, following criterion for eliminating of cross-detecting line is presented. This criterion is called C.E.L. for short. The following notation is assigned for definition of C.E.L.

- "$X(k)$"; a pattern or a character X of the k-th class which should be recognized.

– $x_i(k)$; the i-th pattern sample in the k-th class "$X(k)$", i.e. $x_i(k) \in X(k)$.
– $F.N$."$X(k)$"; a fuzzy neuron designed compatible to a class of character "$X(k)$".
– $\text{OUT}(X(k); x_i(k))$; an output of $F.N$."$X(k)$" when $x_i(k)$ is inputted.
– "$Y(l)$"; a pattern or a character Y of the l-th class which is out of the category including a class "$X(k)$".
– $y_j(l)$; the j-th pattern sample in a class "$Y(l)$", i.e. $y_j(l) \in Y(l)$.
– $\text{OUT}(X(k); y_j(l))$; an output of $F.N$."$X(k)$" when $y_j(l)$ is inputted.
– d; a minimum distance between $\text{OUT}(X(k); x_i(k))$ and $\text{OUT}(X(k); y_j(l))$. (typically:$d = 0.9$)

The criterion: If the cross-detecting line under the test is omitted and

$$\bigwedge_i \text{OUT}(X(k); x_i(k)) \; - \; \bigvee_Y \bigvee_l \bigvee_j \text{OUT}(X(k); y_j(l)) \; \geq \; d \qquad (1)$$

then the cross-detecting line should be omitted. Otherwise, the cross-detecting line should be held.

Checking of cross-detecting lines in case of Figure 3 is achieved in the following sequence using the C.E.L. If the number of the cross-detecting lines which are initially assigned in a frame is changed, the order of checking of cross-detecting lines should be determined in the similar manner.

STEP 1 The horizontal cross-detecting line of the center of the frame is examined by the C.E.L.

STEP 2 The vertical cross-detecting line of the center of the frame is examined by the C.E.L.

STEP 3 The horizontal cross-detecting line of the center of the upper half of the frame is examined by the C.E.L.

STEP 4 The vertical cross-detecting line of the center of the left half of the frame is examined by the C.E.L.

STEP 5 The horizontal cross-detecting line of the center of the lower half of the frame is examined by the C.E.L.

STEP 6 The vertical cross-detecting line of the center of the right half of the frame is examined by the C.E.L.

STEP 7 The horizontal cross-detecting line of the center of the top quarter of the frame is examined by the C.E.L.

STEP 8 The vertical cross-detecting line of the center of the most left quarter of the frame is examined by the C.E.L.

$\vdots$

STEP 29 The horizontal cross-detecting line of the center of the bottom eighth of the frame is examined by the C.E.L.

STEP 30 The vertical cross-detecting line of the center of the most right eighth of the frame is examined by the C.E.L.

3.2. OPTIMIZATION BY THE SELECTION METHOD

The selection method is achieved by two procedures. First, the procedure for optimization of common cross-detecting lines is carried out. In this procedure, common cross-detecting lines are selected out from initial cross-detecting lines as efficient cross-detecting lines for recognition. Second, the procedure for optimization of cross-detecting lines of each fuzzy neuron is carried out. In this procedure, efficient cross-detecting lines for each fuzzy neuron are selected out from common cross-detecting lines. To explain these procedures, the following notation is assigned.

- $CDL(m_{init})$; the m_{init}-th initial cross-detecting line.
- $CDL(m_{com})$; the m_{com}-th common cross-detecting line.
- $"X(k)"$; a pattern or a character X of the k-th class which should be recognized.
- $x_i(k)$; the i-th pattern sample in the k-th class $"X(k)"$, i.e. $x_i(k) \in X(k)$.
- $CDL("X(k)", m_{opt})$; the m_{opt}-th optimum cross-detecting line for the k-th class $"X"$.
- $F.N."X(k)"CDL(m_{init})$; a fuzzy neuron using only one cross-detecting line $CDL(m_{init})$ of a class of character $"X(k)"$.
- $OUT(X(k)CDL(m_{init}); x_i(k))$; an output of $F.N."X(k)"CDL(m_{init})$ when $x_i(k)$ is applied to the input.
- $"Y(l)"$; a pattern or a character Y of the l-th class which is out of the category including a class $"X(k)"$.
- $y_j(l)$; the j-th pattern sample in a class $"Y(l)"$, i.e. $y_j(l) \in Y(l)$.
- $OUT(X(k)CDL(m_{init}); y_j(l))$; an output of $F.N."X(k)"CDL(m_{init})$ when $y_j(l)$ is applied to the input.
- $F.N."X(k)"$; a fuzzy neuron designed compatible to a class of character $"X(k)"$.
- d'; a minimum difference between $OUT(X(k)CDL(m_{init}); x_i(k))$ and $OUT(X(k)CDL(m_{init}); y_j(l))$. (typically:$d' = 0.9$)
- $NUM(X(k)CDL(m_{init}); Y(l))$; the number of the classes $Y(l)$ including pattern samples which are not accepted as a character $"X(k)"$ by $F.N."X(k)"CDL(m_{init})$. i.e. the number of the classes $Y(l)$ which satisfy the equation(2).

$$\bigwedge_i OUT(X(k)CDL(m_{init}); x_i(k)) - \bigvee_j OUT(X(k)CDL(m_{init}); y_j(l)) \geq d'$$

$$(2)$$

- NUM$(X(k)CDL(m_{com}); Y(l))$; the number of the classes $Y(l)$ including pattern samples which are not accepted as a character $"X(k)"$ by F.N.$"X(k)"CDL(m_{com})$. i.e. the number of the classes $Y(l)$ which satisfy the equation(3).

$$\bigwedge_i \text{OUT}(X(k)CDL(m_{com}); x_i(k)) - \bigvee_j \text{OUT}(X(k)CDL(m_{com}); y_j(l)) \geq d' \tag{3}$$

The procedure for optimization of common cross-detecting lines is as follows:

STEP1 Calculate the sum of NUM$(X(k)CDL(m_{init}); Y(l))$ for all m_{init}'s by equation (4).

$$N_{m_{init}} = \sum_X \sum_k \text{NUM}(X(k)CDL(m_{init}); Y(l)) \tag{4}$$

STEP2 Pick up the cross-detecting line $CDL(m_{init})$ which has the maximum value of $N_{m_{init}}$ in initial cross-detecting lines as one of the $CDL(m_{com})$.

STEP3 Discard the class $Y(l)$, which is satisfy equation(3), from $"Y(l)"$.

STEP4 Repeat STEP1~STEP3 until $"Y(l)"$ becomes empty.

The procedure for optimization of cross-detecting lines of F.N.$"X(k)"$ is described by the following steps:

STEP1 Calculate the sum of NUM$(X(k)CDL(m_{com}); Y(l))$ for all m_{com}'s by equation (5).

$$N_{m_{com}} = \text{NUM}(X(k)CDL(m_{com}); Y(l)) \tag{5}$$

STEP2 Pick up the cross-detecting line $CDL(m_{com})$ which has the maximum value of $N_{m_{com}}$ in all m_{com}'s as one of the $CDL("X(k)", m_{opt})$.

STEP3 Discard the class $Y(l)$, which is satisfy equation(6), from $"Y(l)"$.

$$\bigwedge_i \text{OUT}(X(k)CDL("X(k)", m_{opt}); x_i(k))$$
$$- \bigvee_j \text{OUT}(X(k)CDL("X(k)", m_{opt}); y_j(l)) \geq d' \tag{6}$$

STEP4 Repeat STEP1~STEP3 until $"Y(l)"$ becomes empty.

4. Experimental Results

The elimination method and the selection method are implemented by an EWS. The algorithm of pattern recognition is implemented by a PC. Patterns for recognition are ten figures which are from "0" through "9".

The membership functions of ten figures which were extracted by example-based learning were optimized by both methods. Twenty pattern samples

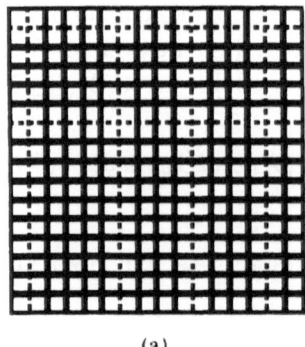
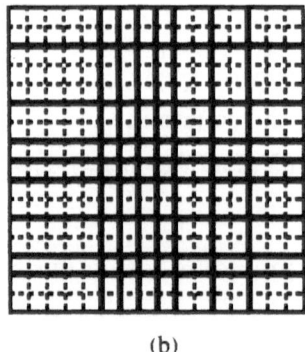

(a)   (b)

Figure 4. Common cross-detecting lines after optimization: (a) by the elimination method; (b) by the selection method.

for each figure were used and the pattern samples were classified into thirty-five classes by their outline. Then, the recognition of hand-written characters was carried out using the membership functions after optimization.

Figure 3 shows locations of cross-detecting lines which are assigned initially to optimize cross-detecting lines. There are fifteen cross-detecting lines for each of horizontal and vertical before optimization. Figure 4 shows locations of the common cross-detecting lines after both optimizations. Sensor arrays should be located as shown in Figure 4. The number of the common cross-detecting line was reduced from that of the initial cross-detecting line as a result of both optimizations, so both optimization methods can choose the common cross-detecting line which are necessary for effective recognition.

Figure 5 and Figure 6 show location of cross-detecting lines and membership functions for the typical class after optimization. In Figure 5, optimized cross-detecting lines by the selection method are more than optimized cross-detecting lines by the elimination method. On the other hand, in Figure 6 optimized cross-detecting lines by the selection method are less than optimized cross-detecting lines by the elimination method. It seems that the cross-detecting lines are located suitably and the membership functions which are assigned for each crossing region and forbidden region have suitable shape and allocation for the recognition.

Figure 7 shows the number of the cross-detecting lines of each fuzzy neuron after both optimizations. The selection method decreased the number of the cross-detecting lines of each fuzzy neuron from the elimination method in thirteen classes. But, in three classes, the selection method increased it. The number of the sensor arrays by the selection method was 38 percent less than the elimination method.

The processing time of the elimination method was 3 hours and 47 minutes and that of the selection method was 1 hour and 27 minutes. Thus, the processing time of the selection method was 62 percent shorter.

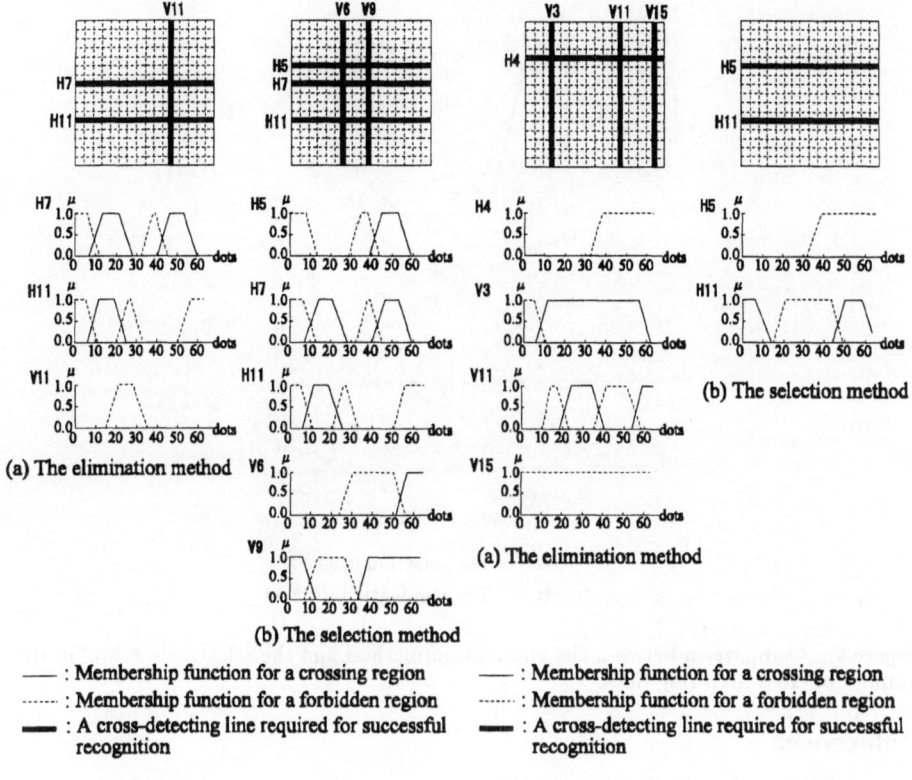

Figure 5. Example of the membership function of class1 for "0" after optimization: (a) by the elimination method; (b) by the selection method.

Figure 6. Example of the membership function of class4 for "6" after optimization: (a) by the elimination method; (b) by the selection method.

5. CONCLUSION

This paper shows a comparison between the inefficient cross-detecting line elimination method and the efficient cross-detecting line selection method, and the typical examples of membership function obtained by both optimization methods. It seems that the cross-detecting lines are located suitably and the membership functions obtained by both optimization methods have suitable shape and allocation for a reasonable recognition. So, fuzzy neuron can achieve a reasonable recognition of patterns employing the optimized membership functions.

As compared with the elimination method, the advantages of the selection method are that the amount of sensor arrays, the amount of the cross-detecting lines for each fuzzy neuron and the CPU time are reduced.

34

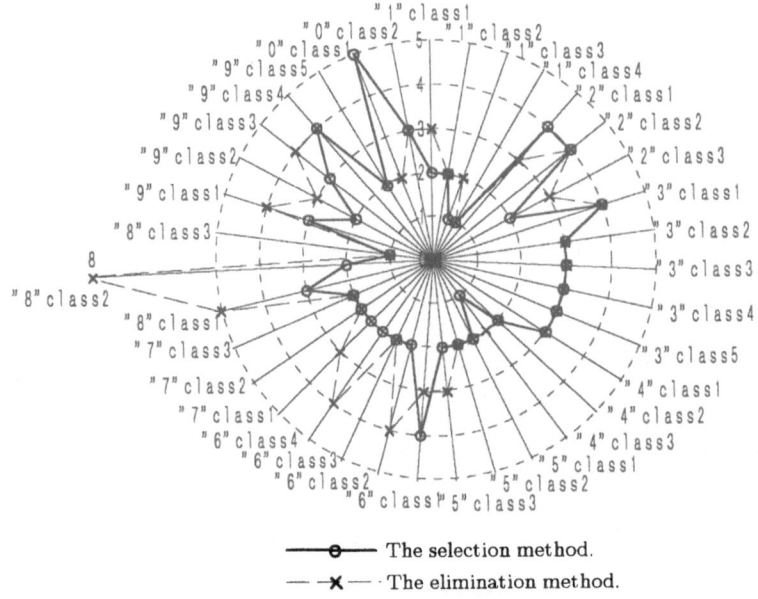

———⊖——— The selection method.

— —✕— · The elimination method.

Figure 7. Comparison between the elimination method and the selection method in the number of cross-detecting lines.

References

Dimond,T.L. (1957) Devices for Reading Handwritten Characters, *Proc. the eastern computer conference* pp. 232–237

Dimond,T.L. (1958) Reading Handwritten Characters, *Bell Lab. Record* **vol.36, no.1** pp. 34–35

Furukawa,M. and Yamakawa,T. (1993) The Advanced Method to Optimize Cross-Detecting Lines for A Fuzzy Neuron, *Proc. the Fifth IFSA World Congress* **vol.1** pp. 66–69

Yamakawa,T. and Tomoda,S. (1989) A Fuzzy Neuron and Its Application to Pattern Recognition, *Proc. the Third IFSA Congress* pp. 30–38

Yamakawa,T. (1990) Pattern Recognition Hardware System Employing a Fuzzy Neuron, *Proc. the International Conference on Fuzzy Logic & Neural Networks* pp. 943–948

Yamakawa,T. and Furukawa,M. (1992) A Design Algorithm of Membership Functions for A Fuzzy Neuron using Example-Based Learning, *Proc. IEEE International Conference on Fuzzy Systems 1992* pp. 75–82

MODELLING OF HYSTERETIC CURVES BASED ON FUZZY REASONING AND ITS APPLICATION

IKUTARO KUMAZAKI   EIICHI WATANABE   HITOSHI FURUTA
Research & Development Center   Kyoto Univ.   Kansai Univ.
Chubu Electric Power Co., Inc.   Yoshida Honmachi,   Ryozenji-cho,
Odaka-cho, Midori-ku,   Sakyo-ku,   Takatsuki City,
Nagoya 459 Japan   Kyoto 606-01 Japan   Osaka 569 Japan

1. Abstract

The first purpose of this study is to verify the validity of the identification method based on the successive learning of the rule parameters and to demonstrate that various shapes of curves can be expressed by the fuzzy inference rules even if the number of the rules is rather few as five.

The second purpose is to show that complex and nonlinear hysteretic curves of simple plate elements can be represented with Fuzzy Constitutive Relation defined by the fuzzy inference rules.

The third purpose is to present a modelling of the hysteretic curves based on the fuzzy reasoning and to verify the effectiveness of the Fuzzy Constitutive Relation of thin steel plate elements applying it to the prediction of the cyclic behavior of thin-walled hollow segments under constant axial compressive force and repetitive bending.

2. Introduction

In order to grasp and predict the ultimate state of a structure/member, its nonlinear hysteretic behavior is important, which is found out of the constitutive relation of a material and the general load-displacement relation considering the local instability.

Recently, design methods proceed to the Limit State Design Method so that a method of modeling which is simplified and capable of precision is expected. So far, on various nonlinear behaviors, many experimental data and case studies using finite element analysis of structural member and element are accumulated.

Consequently if an appropriate modelling method is developed with the aid of the accumulation and is effectively applied to the numerical analysis, it is convenient toward

Z. Bien and K. C. Min (eds.),
Fuzzy Logic and its Applications, Information Sciences, and Intelligent Systems, 35–43.
© 1995 *Kluwer Academic Publishers.*

solving the problems to improve the accuracy or the efficiency for the nonlinear analysis.

Of late, because of the progress of large computer, the finite element method etc. which are discretization methods are used in analyzing the strength of structures. Composite nonlinear analysis has been enabled taking account of the initial imperfections such as the initial deformation and the residual stress considering the geometrical and material nonlinearity. It is, however, quite time-consuming and expensive to carry out an over-all structural analysis in which local instability is considered under the repetitive load even if an advanced technique is utilized.

Therefore, the main purpose of this study is to present a modelling of the hysteretic curves based on the fuzzy reasoning and to show the effectiveness of Fuzzy Constitutive Relation of thin steel plate elements, namely, flange plates, applying it to the prediction of the cyclic behavior concerned with a section of thin-walled hollow segments under the combined action of constant axial compressive force and repetitive bending.

3. Definition of Fuzzy Constitutive Relation

Each membership function of the antecedent part is expressed by an isosceles triangle as shown in Fig. 1. The parameters determining the shape of the triangle membership function are the central point value a_i and the width b_i.

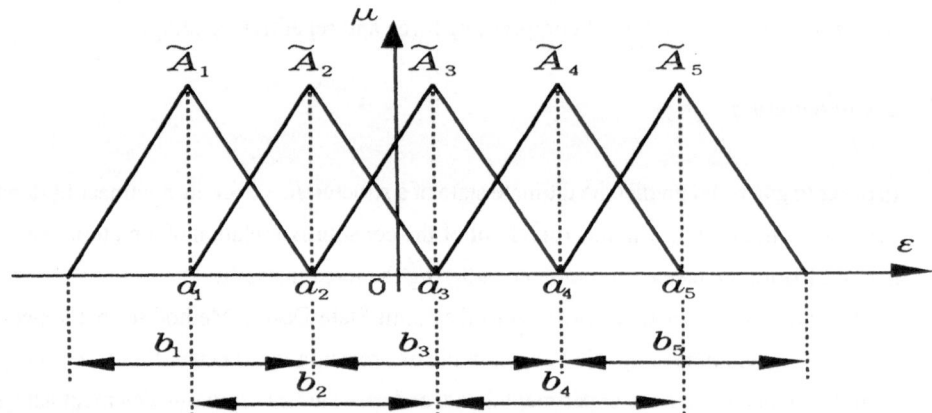

Figure 1. Membership functions.

It is assumed that the antecedent part of fuzzy inference rules used in this research has a variable, the number of membership functions is five and hence the number of the fuzzy inference rules is five. It may be said that a large number of membership functions result in being hard to deal with a model. On the other hand, if there are very few membership functions, the accuracy of approximation by the model will not be fine. That is why we assume that the number of membership functions is five.

If the input is expressed by ε, and the output is expressed by σ, the inference rules of the fuzzy reasoning can be expressed by the following expressions;

$$
\begin{aligned}
&\text{Rule 1: If } \varepsilon \text{ is } \widetilde{A}_1 \text{ then } \sigma = f_1 \\
&\text{Rule 2: If } \varepsilon \text{ is } \widetilde{A}_2 \text{ then } \sigma = f_2 \\
&\text{Rule 3: If } \varepsilon \text{ is } \widetilde{A}_3 \text{ then } \sigma = f_3 \\
&\text{Rule 4: If } \varepsilon \text{ is } \widetilde{A}_4 \text{ then } \sigma = f_4 \\
&\text{Rule 5: If } \varepsilon \text{ is } \widetilde{A}_5 \text{ then } \sigma = f_5 \\
&\text{Input: } \quad \varepsilon_0
\end{aligned}
\tag{1}
$$

$$\text{output:} \qquad\qquad\qquad\qquad z_0$$

where the symbol $\sim$ indicates fuzzy quantity and f_i is a real number. It is noted that the consequent part is defined by crisp numbers.

In this study, we call one set of these fuzzy inference rules "Fuzzy Constitutive Relation (FCR)", because the constitutive relation of a material is expressed by the input-output relation through the fuzzy inference rules as shown above.

4. Identification Method of Fuzzy Inference Rules[1]~[3]

It is assumed that the learning of a real number of the consequent part is expressed by the following formula [2];

$$
f_i^{new} = f_i^{old} + \alpha h_i (Z_c - z_0)
\tag{2}
$$

while the learning of the antecedent parts are derived as formulae [1];

$$a_i^{new} = a_i^{old} + 2\beta \frac{\text{sgn}(\varepsilon - a_i)(Z_e - z_0)(f_i - z_0)}{b_i \sum\limits_{i=1}^{n} h_i} \tag{3}$$

$$b_i^{new} = b_i^{old} + 2\gamma \frac{(Z_e - z_0)(f_i - z_0)|\varepsilon - a_i|}{b_i^2 \sum\limits_{i=1}^{n} h_i} \tag{4}$$

where α, β, γ are learning coefficients which concern the computation time and Z_e is a desirable output and z_0 is the output by fuzzy reasoning and h_i is the membership grade for the i-th membership function on the ε-axis for the value of ε.

5. Expression of Arbitrary Curves by Fuzzy Inference Rules

If k set of input values ε_j are given for the identified fuzzy inference rules, the fuzzy reasoning is conducted for each of inputs so as to derive k set of output values z_j. It is noted that k is an arbitrary natural number. The arbitrary curves can be expressed by depicting the data (ε_j, z_j) $(j = 1, 2, ..., k)$ which are obtained like this.

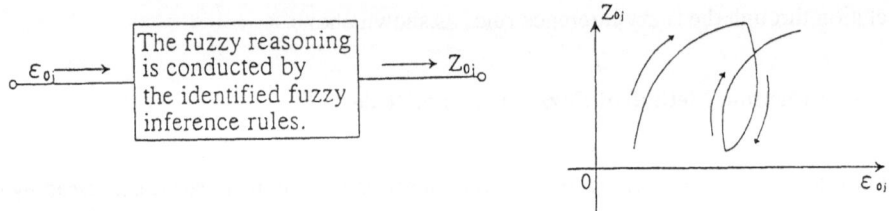

Figure 2. Expression of curve by rules

6. Modelling of Fuzzy Constitutive Relation

The learning of the parameters of the antecedent and consequent parts of the rules is carried out with repetitive compressive-tensile experimental data on a thin steel plate as the learning data. The learning of the parameters is done on each curve which corresponds to half loading cycle. It is noted that the curve corresponding to half cycle is defined as each curve between one edge point and next edge point of the repetitive hysteretic curves.

Stress-strain curves obtained by a compressive-tensile experiment on a specimen of the thin steel plate are shown in Fig. 3. Figs. 4 to 6 show variations of the parameters of identified fuzzy inference rules, according to the advance of loading cycle. The parameters whose variations are depicted in Figs. 4 to 6 are identified with the experimental data as the learning data. Next, variation of the identified parameters of the antecedent and consequent parts accompanied with the progress of loading cycle is observed. Then qualitative traits of the variation of the parameters are grasped by observation, and a modelling of the constitutive relation of a simple plate element is attempted based on the fuzzy inference rules. The inference rules are empirically derived by simple liner lines shown in Figs. 7 to 9 which represent the variations of the antecedent and consequent parameters of the fuzzy inference rules in accordance with progress of the loading cycle. The simple linear lines are established so that they can give expression to the simplified qualitative variation of the rule parameters.

Consequently, the Fuzzy Constitutive Relation on the simple plate is shown in Fig. 10.

7. Application of Fuzzy Constitutive Relation to Simplified Analytical Modelling [4]

The Fuzzy Constitutive Relation of the simple plate element is applied to three types of analytical models so as to evaluate the cyclic behavior of thin-walled hollow segments under the constant axial compressive thrust and the repetitive bending combined.

The analysis is done in such a way that the strains of flange and web plates must satisfy dynamic conditions, namely, the increment of axial force must become zero when the increment of curvature is given, and general restorative characteristics on composite members is evaluated. At this time, the proposed Fuzzy Constitutive Relation is utilized as the stress-strain relations on the thin plates with upper and lower flanges.

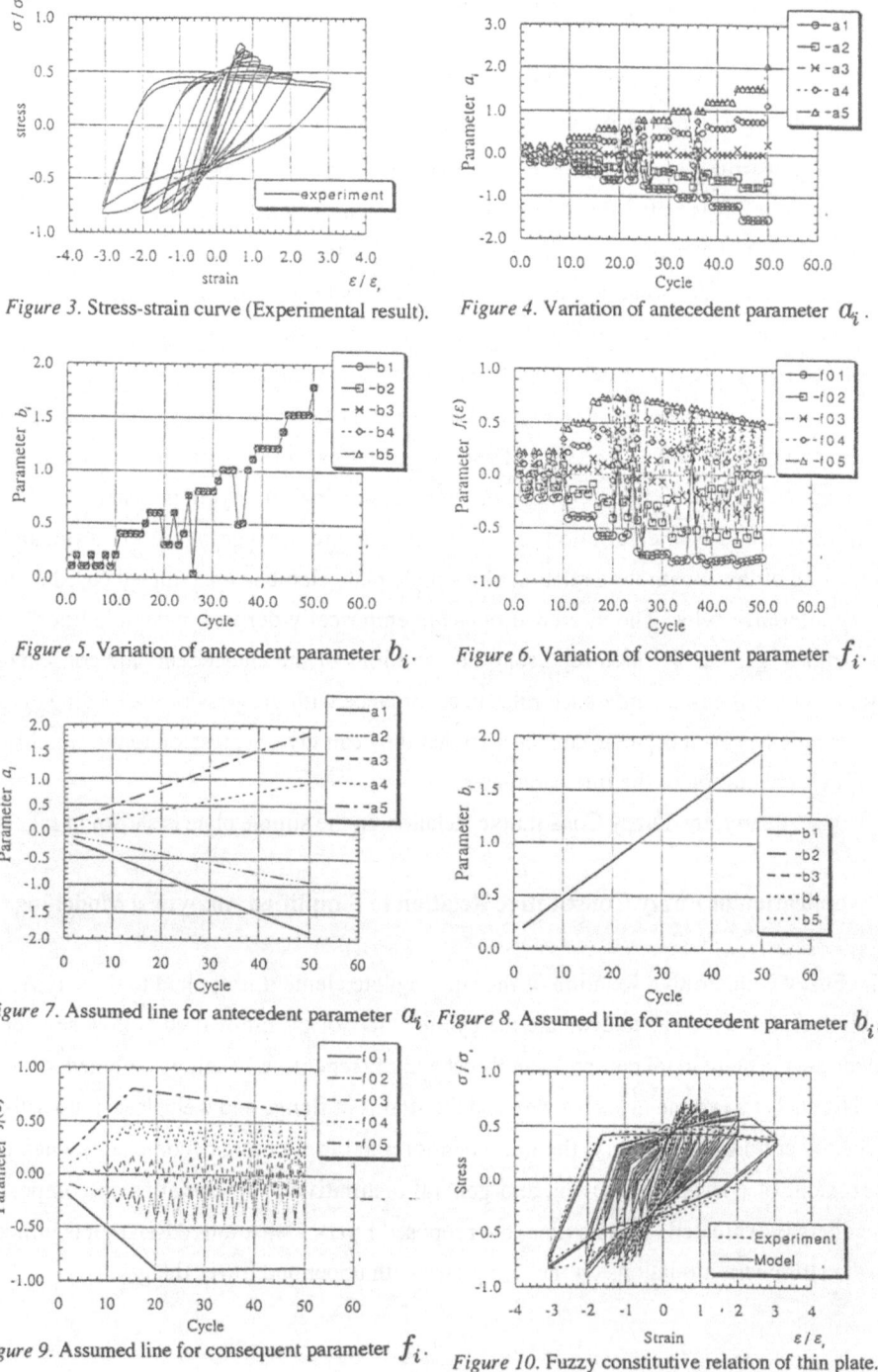

Figure 3. Stress-strain curve (Experimental result).

Figure 4. Variation of antecedent parameter a_i.

Figure 5. Variation of antecedent parameter b_i.

Figure 6. Variation of consequent parameter f_i.

Figure 7. Assumed line for antecedent parameter a_i.

Figure 8. Assumed line for antecedent parameter b_i.

Figure 9. Assumed line for consequent parameter f_i.

Figure 10. Fuzzy constitutive relation of thin plate.

The three types of models are Double Flange Model, Elastic Web Model and Elastoplastic Web Model. In the Double Flange Model, it is assumed that the web elements are fictitious elements which have no area and they are attached for only one purpose that they sustain the shapes of the beam. In the Elastic Web Model, resistance of web element is taken into account on a condition that the web element always behaves elastically and yields no local buckling, that is to say, the deterioration of the stress does not occur in the web plates, while the elastoplastic behavior is taken into consideration by use of a yield surface in the Elastoplastic Web Model. Accordingly, it may be said that the more realistic behavior of the web plates than Elastic Web Model can be expressed by introducing the formula of the yield surface.

Figs. 11 to 13 show analytical results by the three types of models.

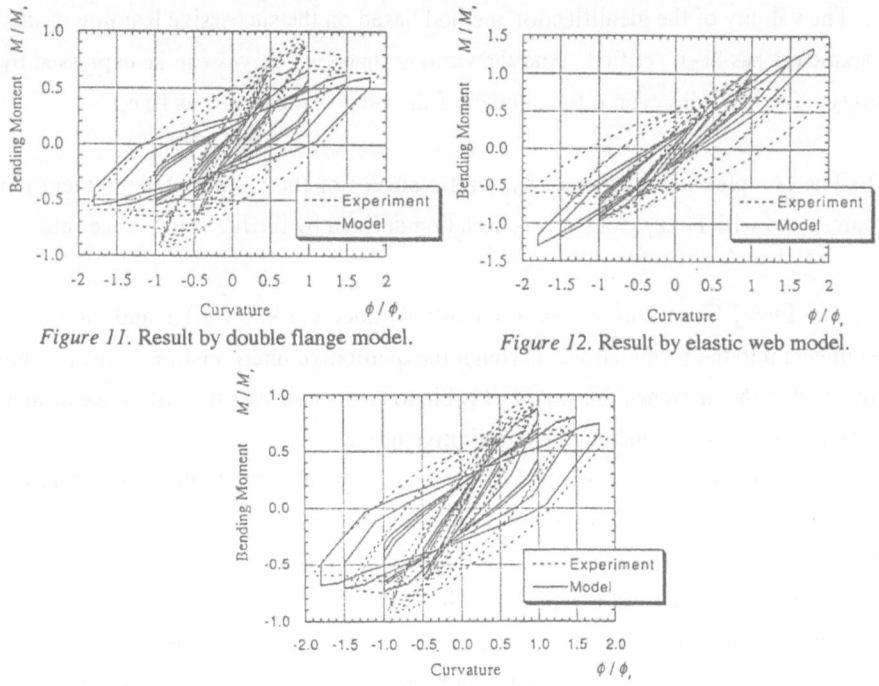

Figure 11. Result by double flange model.

Figure 12. Result by elastic web model.

Figure 13. Result by elastoplastic web model.

8. Concluding Remarks

In this research, the tuning method has been formulated to express the arbitrary curves with the fuzzy inference rules and the efficiency and the applicability of the method have been demonstrated through several numerical calculations.

The method to represent the arbitrary curves based on the fuzzy reasoning has been applied to the modelling of the hysteretic curves on the simple plate element under the repetitive compressive-tensile load. Using this method, the so-called "Fuzzy Constitutive Relation" has been derived. Then, the Fuzzy Constitutive Relation has been applied to the prediction of the cyclic behavior concerned with a section of the thin-walled hollow segments under the constant compressive thrust and repetitive bending.

Consequently, the following conclusions are derived:

(a) The validity of the identification method based on the successive learning of the rule parameters has been verified. And the various shapes of curves can be expressed by the fuzzy inference rules even if the number of the rules is rather few as five.

(b) The complex and nonlinear hysteretic curves of the simple plate element can be represented with Fuzzy Constitutive Relation defined by the fuzzy inference rules.

(c) The Fuzzy Constitutive Relation can introduce the knowledge and the insight of engineers into the formulation. Through the qualitative interpretation of the parameters involved in the inference rules, it is possible to easily consider the difference of material characteristics in establishing the constitutive model.

In the analysis of the thin-walled hollow segments subjected to the constant thrust and repetitive bending, the cyclic behavior concerned with a section of the segments can be predicted by Elastoplastic Web Model with a certain degree of success comparing with Double Flange Model and Elastic Web Model.

Fuzzy Constitutive Relation Model works in the above three analytical models. Furthermore, if the experimental data can be sufficiently collected, the constitutive model can be improved through the learning process.

It can be said that the proposed model, i. e. , the Fuzzy Constitutive Relation Model, lies between the ordinal analytical model based on physical characteristics and the model based on the Neural Network. Although the Neural Network Model is superior to interpolate experimental data, it is difficult to show a logical base to reach the final result. On the contrary, the proposed method can provide us of both the interpretation of the parameters and the satisfactory accuracy gained by the learning ability.

9. References

1) Nomura, H., Hayashi, I. and Wakami, N. (1990) A self tuning method of fuzzy reasoning by method of steepest descent and its application to moving obstacle avoidance, *Proc. of 6th Fuzzy System Symposium*, 423-426. (in Japanese)

2) Ichihashi, H. (1990) Back propagation error learning with hierarchical fuzzy models -online GMDH algorithms- , *Proc. of 6th Fuzzy System Symposium*, 539-542. (in Japanese)

3) Kosko, B. (1992) Fuzzy systems as universal approximators, *Proc. of IEEE International Conference on Fuzzy Systems*, 1153-1162.

4) Mori, T., Watanabe, E., Sugiura, K. and Oshima, K. (1991) A simplified analysis on thin-walled steel box beam-column under compression and cyclic bending, *Proc. of Symposium on Computational Methods in Structural Engineering And Related Fields, Volume* 15, 169-172. (in Japanese)

ESTIMATION OF MEMBERSHIP FUNCTIONS FOR PATTERN RECOGNITION AND COMPUTER VISION

SWARUP MEDASANI, JAESEOK KIM, AND
RAGHU KRISHNAPURAM
Department of Electrical and Computer Engineering
University of Missouri, Columbia, MO 65211, USA

Abstract

Membership functions are of vital importance to applications of fuzzy theory. In this paper, we consider several methods for generating membership functions for pattern recognition and computer vision applications. The methods we discuss include heuristic methods, methods based on probability-possibility transformations, methods using feed-forward neural networks, the Fuzzy C Means algorithm, and the Possibilistic C Means algorithm. A qualitative comparison is made.

1. Introduction

One of the fundamental issues associated with the application of fuzzy set theory is the estimation of membership functions. There has been much debate about the kind of data and methods that should be used for estimating membership functions. So far, there are no generally accepted membership generation methods which can handle all types of application problems. Moreover, all fuzzy researchers do not agree on what the "correct" interpretation of membership values should be. For example, Dubois and Prade [1] point out three different interpretations of the statement: "The membership value of George Bush in the class of tall men is 0.8." The three interpretations are: (a) 80% of the population declared that George Bush is tall (*likelihood view*), (b) 80% of the population described "tall" as an interval containing George Bush's height (*random set view*), and (c) George Bush's height is at a normalized distance equal to 0.2 from the closest ideal prototype (*typicality view*). Thus, one could think of a variety of methods to generate membership values depending on the interpretation. Nevertheless, only a few researchers have concentrated on the topic of membership functions. Much literature is oriented toward the determination of membership functions that reflect subjective perception about vague or imprecise concepts such as *tall persons, old men,* and *young men*. Unfortunately, these methods can not be directly applied to many practical problems. For example, in computer vision applications including image processing, pattern recognition, and scene interpretation, membership functions are not always subjective evaluations of vague concepts but rather a means to model the uncertainty contained in the input information such as images and/or features extracted from images. Therefore, appropriate methods for membership function generation in such practical applications must be suggested. It should be noted that no measures are available to evaluate the goodness or correctness of

45

Z. Bien and K. C. Min (eds.),
Fuzzy Logic and its Applications, Information Sciences, and Intelligent Systems, 45–54.
© 1995 *Kluwer Academic Publishers.*

the membership function generated using a particular method. This is a serious problem when membership functions are used to model concepts or features which have no physical meanings. Therefore, the evaluation of membership functions may rely on intuition, or results of an algorithm which uses membership functions as inputs (for example, classification rate or segmentation results).

This paper focuses on the methods of membership function generation for computer vision applications. This problem is of fundamental importance because the success of an algorithm depends on the membership functions used in its early stages. It might be impossible to come up with a single method which will work for most computer vision applications. Rather, several methods may have to be used and the choice of the method may depend on the type of the available data and the problem. Other methods to model uncertainty (such as probability) also use a variety of methods to estimate the underlying distributions depending on the situation. Therefore, we suggest several methods to estimate membership functions and make some qualitative comparisons.

2. Heuristic Methods

Heuristic methods use predefined shapes for the membership functions. Heuristic methods have been used very successfully in control approaches. In computer vision, heuristic membership functions may be used to describe certain spatial relations (such as "above" and "to the left of"[2]) and certain properties (such as lightness or darkness of a pixel value, position of a pixel, narrowness of a region [3]). Here we present three of the most frequently used shapes for heuristic membership functions in the literature.

2.1 PIECE-WISE LINEAR FUNCTIONS

The membership functions may be chosen to be linearly increasing, linearly decreasing or a combination of these.

Example 1:

$$\mu(x) = \begin{cases} 1 - \dfrac{|\alpha - x|}{\alpha} & \text{if } \alpha-a \leq x \leq \alpha+a \\ 0 & \text{otherwise} \end{cases}$$

The membership functions may also be chosen as piece-wise linear functions, i.e., they have linearly increasing, decreasing and flat regions.

Example 2:

$$\mu(x) = \begin{cases} 0 & \text{if } x < a_1 \\ ax + b & \text{if } a_1 \leq x \leq a_2 \\ 1 & \text{if } a_2 < x \end{cases}$$

where $a = 1/(a_2 - a_1)$ and $b = a_1/(a_1 - a_2)$

2.2 PIECE-WISE MONOTONIC FUNCTIONS

The membership functions have a (piece-wise) smooth transition between non-membership and full-membership regions. The smooth transition may be described by functions such as x^2, $\sin(x)$, $\arctan(x)$, and $\exp(x)$.

Example 3: The S-function

$$S(x;a,b,c) = \begin{cases} 0 & x \leq a \\ 2\left(\dfrac{x-a}{c-a}\right)^2 & a < x \leq b \\ 1 - 2\left(\dfrac{x-a}{c-a}\right)^2 & b < x \leq c \\ 1 & x > c \end{cases} \qquad b = (a+c)/2.$$

Example 4: The Π-function

$$\Pi(x;b,c) = \begin{cases} S(x;c-b,c-\dfrac{b}{2},c) & x \leq c \\ 1 - S(x;c,c+\dfrac{b}{2},c+b) & x > c \end{cases}$$

Most of membership functions in the heuristic category seem to work well for the specific problems for which they are intended. The linear and piece-wise linear membership functions have the following advantages: they provide a reasonably smooth transition, they are easily manipulated by fuzzy operators, and they are easily implement in hardware if speed is crucial. However, the shapes of the heuristic membership functions are not sufficiently flexible for many applications. Moreover, the parameters associated with the membership functions must be provided by experts. In some applications, the parameters need to be "tweaked" until the performance is acceptable, and this is a tedious process.

3. Transformation of Probability Distributions to Possibility Distributions

In image processing applications, normalized histograms have been traditionally treated as probability distributions. If we have a large number of samples, the normalized histogram of the samples can be assumed to approximate the pdf. If we treat membership functions as possibility distributions [4], then methods that transform probabilities to possibilities can be used to generate membership functions from histograms.

Let $X = \{x_i \mid i = 1,...,n\}$ be the universe of discourse. The x_i's are ordered such that $P1 \geq P2 \geq ... \geq Pn$, where p_i is the probability of occurrence of x_i, i. e., $p_i = P(\{x_i\})$. Let p_i denote the corresponding possibility value. A bijective transformation between probabilities and possibilities may be defined as [5]:

$$\pi_i = \sum_{j=1}^{n} \min(p_i,p_j) = ip_i + \sum_{j=i+1}^{n} p_j \qquad \text{and} \qquad p_i = \sum_{j=1}^{n} \frac{(\pi_j - \pi_{j+1})}{j} \qquad (1)$$

with the convention $\pi_{n+1} = 0$. This mapping was derived from the definition that the degree of necessity of event A in X is the extra amount of probability of elementary events in A over the amount of probability assigned to the most frequent elementary event outside A. From the above equation, it is seen that the overall shape of the possibility distribution is the same as that of probability distribution and vice versa, i.e.,

$$\pi_i = \pi_{i+1} \Leftrightarrow p_i = p_{i+1}, \ \pi_i > \pi_{i+1} \Leftrightarrow p_i > p_{i+1}.$$

Theorem: The possibility distribution represented by π_i is greater or equal to the corresponding normalized probability distribution. That is, $\pi_i \geq p_i / p_{max} \ \forall i \in \{1,2,...,n\}$, where $p_{max} = \max_i(p_i)$.

proof: Multiplying both sides of (1) by p_1/p_i,

$$\frac{p_1}{p_i}\pi_i = ip_1 + \frac{p_1}{p_i}\sum_{j=i+1}^{n} p_j \geq \sum_{j=1}^{i} p_j + \sum_{j=i+1}^{n} p_j = \sum_{j=1}^{n} p_j = 1$$

since $p_1 \geq p_i$ for all i. Hence, $\pi_i \geq p_i/p_1$.

4. Neural network based method

Feed-forward neural networks can be used to generate membership functions [6] from labeled training data. The number of input nodes in the neural network is chosen to be equal to the number of features, and the number of output nodes is chosen to be equal to the number of class labels. The desired value of the output for a input feature vector is 1 for the node representing the label associated with the feature vector and 0 for all the other output nodes. In order to generate class membership values, the multilayer network is trained using a suitable training algorithm such as the back-propagation algorithm. After the training procedure converges, the resulting network can be treated as a membership generation network, where the inputs are feature values and the outputs are membership values in the different classes. This method allows fairly complex membership functions to be generated because the network is highly nonlinear in general. Also, since the membership functions are generated from a classification point of view, this is highly desirable for pattern recognition applications, although the membership values may not be necessarily indicative of the degree of typicality of a feature with respect to a class.

5. Fuzzy c-means method

The Fuzzy C-Means (FCM) algorithm is one of the most popular fuzzy clustering algorithms [10]. Let $X = \{x_1, x_2, \ldots, x_n\}$ denote a data set. Let N be the number of data vectors and C be the number of classes. The FCM algorithm does the partitioning by minimizing the objective function [8]

$$J_m(U,V) = \sum_{j=1}^{N} \sum_{i=1}^{C} u_{ij}^m d_{ij}^2 \quad \text{subject to} \quad \sum_{i=1}^{C} u_{ij} = 1 \text{ for all } j,$$

where u_{ij} is the membership of x_j in cluster i, d_{ij} is the distance from x_j to prototype v_i, V is the collection of all prototypes v_i and $m \in [1,\infty)$ is a weighting exponent called the fuzzifier. The matrix $U = [u_{ij}]$ represents the fuzzy partition generated by the algorithm.

It has been shown that for $m>1$, under the assumption that x_j is not equal v_i for all i,j, that (U,V) may be a global minimum for J_m only if the following conditions hold.

$$u_{ij} = \frac{1}{\sum_{k=1}^{C}\left(\frac{d_{ij}^2}{d_{kj}^2}\right)^{\frac{1}{m-1}}} \quad \forall \in i,j, \tag{2}$$

$$v_i = \frac{\sum_{j=1}^{N} (u_{ij})^m x_j}{\sum_{j=1}^{N} (u_{ij})^m} \quad \forall i. \tag{3}$$

Starting with arbitrary mean vectors v_i, (2) and (3) are used in an alternating fashion to generate a fuzzy partition of the data. The advantages of the FCM algorithm are: (i) it can be used as an unsupervised algorithm, (ii) it can be used to generate multi-dimensional membership functions, and (iii) the shape of the membership functions can be controlled by using different types of distance measures. However, the number of classes must be provided to run the algorithm. Another problem is caused due to the probabilistic constraint: the memberships of a data point across classes must sum to one. Consider a two class problem with two compact clusters except that one data point is located exactly equidistant from the two compact clusters. The FCM assigns a membership of 0.5 in the two clusters for this data point no matter how far it is located from the two clusters, as long as the point is equidistant from the two clusters. Thus, the membership values are not representative of the degree of belonging (or typicality), and they cannot distinguish between a moderate outlier and an extreme outlier. This occurs because the memberships generated by this constraint are relative numbers and they represent "degrees of sharing". On the other hand, the membership functions generated by the FCM may be acceptable from a classification point of view, provided there are no noise points in the data set.

The FCM algorithm is an unsupervised algorithm in that it does not require that the feature vectors are labeled. However, it may also be used in supervised situations with certain modifications. For example, we could compute the class mean of each class using

$$v_i = \frac{1}{N_i} \sum_{x_j \in \text{class } i} x_j \,,$$

where N_i is the number of feature vectors with the class i label. After the class means are computed, we can compute the membership values using (2).

6. Possibilistic clustering method

One promising modification to the FCM algorithm is the possibilistic c-means suggested by Krishnapuram and Keller [9]. The objective function used is

$$J_m(U,V) = \sum_{j=1}^{N} \sum_{i=1}^{C} u_{ij}^m d_{ij}^2 + \sum_{i=1}^{C} \eta_i \sum_{j=1}^{N} (1-u_{ij})^m \tag{4}$$

where $\eta_i > 0$ are parameters. Note that there is no constraint on membership values except the constraint that $u_{ij} \in [0,1]$. The first term of J_m requires that the distance from the feature vector to the prototypes be low, whereas the second term forces the u_{ij} to be as large as possible. It has been shown that for (U,V) may be a global minimum of J_m only if

$$u_{ij} = \frac{1}{1+\left(\dfrac{d_{ij}^2}{\eta_i}\right)^{\frac{1}{m-1}}} \quad \forall \ i,j. \tag{5}$$

The update equation for the prototypes is the same as the one in (3). The value of m determines the fuzziness of the final partition and shape of the membership function. When $m \to 1$, the membership function is hard, and when $m \to \infty$, the membership function is maximally fuzzy. The value of η_i determines the distance at which the membership value of a point in a cluster becomes 0.5. Two possible ways to estimate the value of η_i are given below:

$$\eta_i = \frac{\sum\limits_{j=1}^{N} u_{ij}^m d_{ij}^2}{\sum\limits_{j=1}^{N} u_{ij}^m} \quad \text{or} \quad \eta_i = \frac{\sum\limits_{u_{ij}>\alpha} d_{ij}^2}{\sum\limits_{u_{ij}>\alpha} 1},$$

where α an appropriate α-cut. We first run the FCM algorithm and then use one of the above equations to estimate η_i. However, when the data is noisy, a better estimate of η_i may be given by $\eta_i = c \times MAD$, where c is a constant (usually $c=9$, see [11]), and

$$MAD = \text{median } \{ \mid d_{ij}^2 - \text{median } (d_{ij}^2) \mid \} \tag{6}$$

It is to be noted that the value of MAD is different for each cluster, since the medians in (6) are computed using only the distances corresponding to only those x_j that have their highest memberships (u_{ij}) in the cluster under consideration. In other words, the data set is crisply partitioned into C clusters, and the MAD value is computed for each cluster.

One advantage of this method is that the n.embership function u_{ij} depends only on the distance of the feature vector x_k from the prototype v_i and not on its distances to other clusters. This definition of membership values corresponds the intuitive notion of typicality [7]. This algorithm is relatively insensitive to noise and outliers [9]. Also, unlike in the case of the FCM algorithm, the PCM algorithm can be applied even when there is only one cluster in the data set, and it will not produce the trivial solution: (u_{ij}) = 1, $\forall j$. Thus, by considering feature vectors from one class at a time, it can be easily applied for labeled data in a supervised situation.

If we eliminate the memberships from (4) using (5), then the resulting objective function for the PCM can be written as

$$J(B;X) = \sum_{i=1}^{C} \sum_{j=1}^{N} w_{ij} (d_{ij}^2; \eta_i) d_{ij}^2, \quad \text{where} \quad w_{ij}(d_{ij}^2; \eta_i) = \left(\frac{1}{1 + \left[d_{ij}^2/\eta_i \right]^{1/(m-1)}} \right)^{m-1}.$$

It is to be noted that for $m>1$, w_{ij} is a monotinically decreasing function of d_{ij}^2. This suggests that in general we could choose any monotonically decreasing function for w_{ij} to obtain variations of the PCM algorithm. We will refer to this formulation the generalized PCM algorithm. In particular, if we choose

$$w_{ij} = \exp \{-d_{ij}^2\}$$

and

$$d_{ij}^2 = \frac{1}{2} \frac{(x_j - v_i)^T (x_j - v_i)}{K},$$

where K is the variance associated with the cluster, then update equations are

$$v_i = \frac{\sum\limits_{j=1}^{N} \mu_{ij} x_j}{\sum\limits_{j=1}^{N} \mu_{ij}} \quad \text{and} \quad \mu_{ij} = \exp \{-d_{ij}^2\} \left(1 - \frac{d_{ij}^2}{2K} \right)$$

It should be noted that the weights μ_{ij} can be negative if $d_{ij}^2 > 2K$. In order to avoid instabilities, we set them to zero whenever they are negative.

7. Experimental results

Due to space limitation, we present the results of only one data set, i.e., the IRIS data. We used features 3 and 4 of this data set. A scatter plot of the data set is shown in Fig. 1.

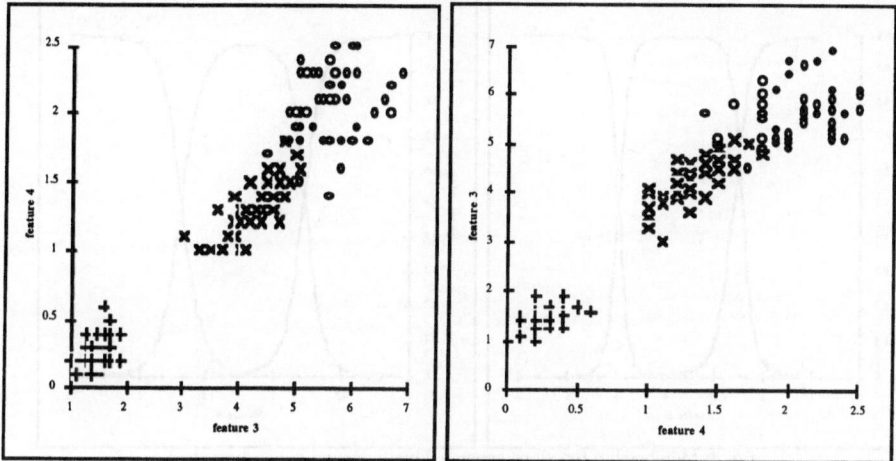

Fig. 1: Scatter plot of IRIS data

In the transformation method, the histograms of each class were constructed by counting the number of data points in nine equally spaced intervals, and then smoothed by a binomial window of size 5. The probability densities were obtained by normalizing the histograms. Fig. 2 shows the resulting membership functions. It can be seen that the values of the memberships in an area are related to the density of the feature vectors in the area, as expected.

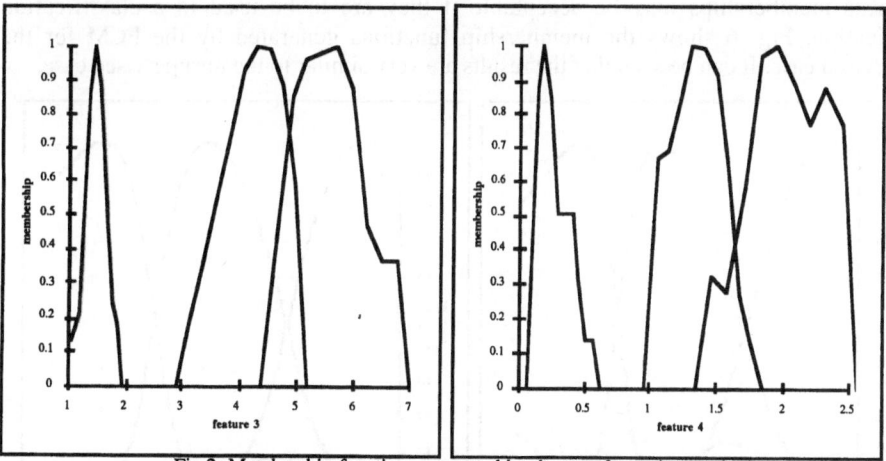

Fig.2: Membership functions generated by the transformation method

The neural network method was applied using the backpropagation network with one input unit (representing a feature), six hidden units and three output units (representing the three classes). As seen in Fig. 3, the membership functions show almost no gradation

52

within each class. This result implies that if a class is compact and well separated from the others, there is no difference between an ideal (typical) member and a less typical member. Another potential problem is that we cannot predict the shape of the membership function in regions where there are no training data points.

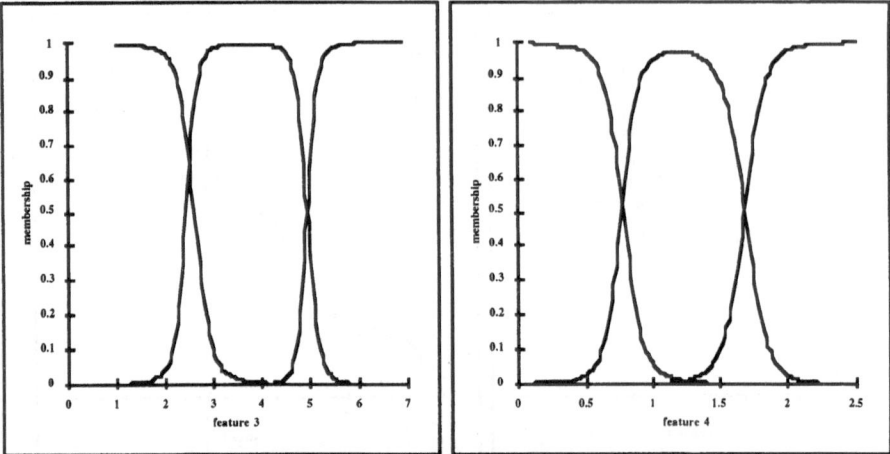

fig. 3: Membership functions generated by the neural network method

The FCM algorithm was run on one feature at a time, and the membership functions generated by this clustering algorithm are shown in Fig. 4 for the unsupervised case. The FCM seems to work well from a classification point of view. However, the membership assignments are counter-intuitive. Some of the membership functions have two peaks, which means that some points further away from the ideal member have larger membership values than some points closer to the ideal member. This happens because the memberships are forced to satisfy the probabilistic constraint. As mentioned in Section 6, these memberships may be acceptable if they are to be used in a classification application. Fig. 6 shows the membership functions generated by the FCM for the supervised case. It can be seen that the results are very similar to the unsupervised case.

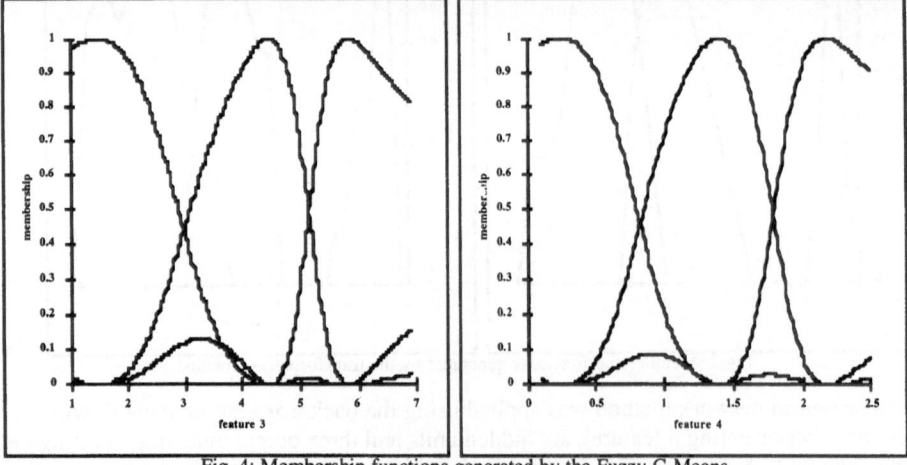

Fig. 4: Membership functions generated by the Fuzzy C-Means

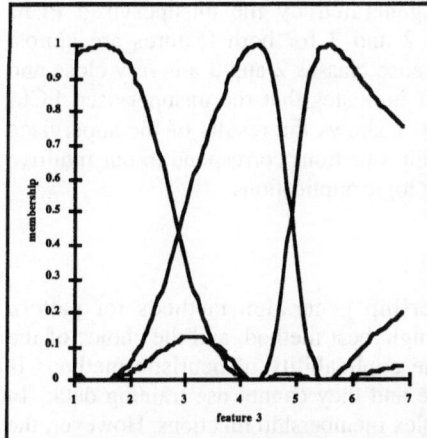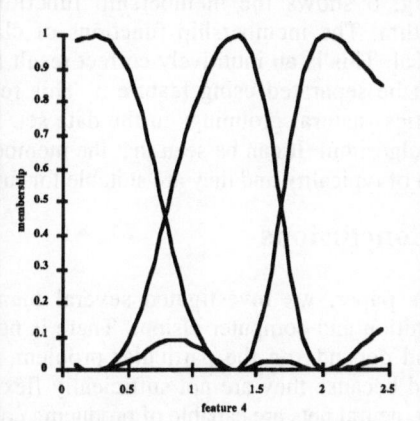

Fig. 6: Membership Functions generated by the Fuzzy C-Means via Supervised clustering.

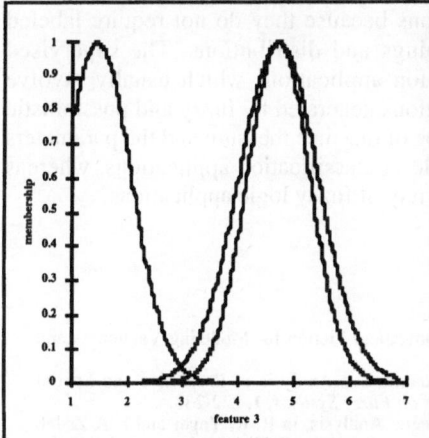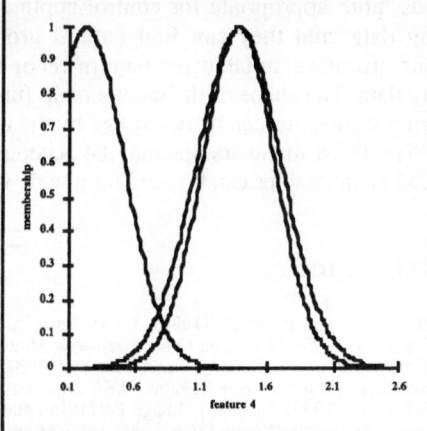

Fig. 5: Membership functions generated by the Possibilistic C-Means

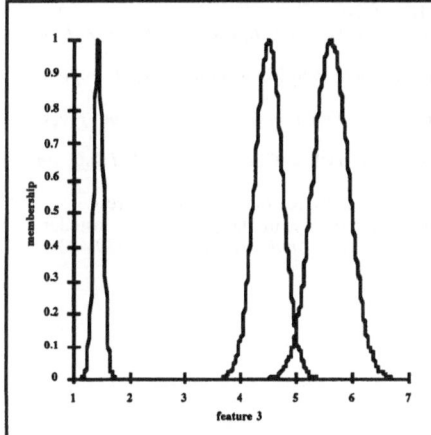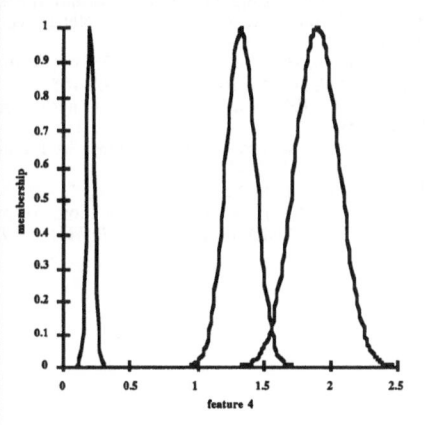

Fig. 7: Membership Functions generated by the Possibilistic C-Means via supervised clustering.

Fig. 6 shows the membership functions generated by the unsupervised PCM algorithm. The membership functions of class 2 and 3 for both features are almost identical. This is an intuitively correct result because classes 2 and 3 are very close and cannot be separated using feature 3. This result indicates that the unsupervised PCM identifies natural groupings in the data set. Fig. 7 shows the results of the supervised PCM algorithm. It can be seen that the membership functions correspond to our intuitive notion of typicality and they are suitable for fuzzy logic applications.

8. Conclusions

In this paper, we investigated several membership generation methods for pattern recognition and computer vision. There is no single best method, and the choice of the method depends on the particular problem. The applicability of heuristic methods is limited because they are not sufficiently flexible and they cannot use training data. In theory, neural nets are capable of producing complex membership functions. However, the shape of the membership function is unpredictable in regions where there is no training data. Also, they do not produce gradations within each class. Unsupervised clustering methods more appropriate for control applications because they do not require labeled training data, and they can find natural groupings and distributions. The supervised versions are more suitable for pattern recognition applications which usually involve training data. The shape of the membership functions generated by fuzzy and possibilistic clustering algorithms can be controlled by the type of distance measure and the parameters used. The FCM memberships may be acceptable in classification applications, whereas the PCM memberships can be used for a wide variety of fuzzy logic applications.

9. References

1. Dubois, D. and Prade, H. (1994) "Fuzzy Sets - A Convenient Fiction for Modeling Vagueness and Possibility," *IEEE Trans. on Fuzzy Systems*, **2**, 16-21.
2. Krishnapuram, R., Keller, J., and Ma, Y. (1993) Quantitative Analysis of Properties and Spatial Relations of Fuzzy Image Regions, *IEEE Transactions on Fuzzy Systems*, **1**, 222-233.
3. Pal, S. K. (1992) Fuzziness, Image Formation and Scene Analysis, in R. R. Yager and L A Zadeh (eds.), *An Introduction to Fuzzy Logic Application*, Kluwer Academic Publishers, Norwell, 185-200.
4. Zadeh, L. A. (1978) "Fuzzy sets as a basis for a theory of possibility," *Fuzzy Sets and Syst.*, **1**, 3-28.
5. Dubois, D. and Prade, H. (1983), "Unfair coins and necessity measures: Toward a possibilistic interpretation of histograms," *Fuzzy Sets and Systems*, **10**,, 15-20.
6. Takagi, H. and Hayashi, I. (1991) "NN-Driven Fuzzy Reasoning," *International Journal of Approximate Reasoning*, **5**, 191-212.
7. Zimmermann, H. J. and Zysno, P. (1985) "Quantifying Vagueness in Decision Models," *European Journal of Operations Research*, **22**, 148-158.
8. Bezdek, J.C.(1981) *Pattern Recognition with Fuzzy Objective Function Algorithms*, Plenum Press, New York.
9. Krishnapuram, R. and Keller, J. (1993) "A Possibilistic Approach to Clustering," *IEEE Trans. on Fuzzy Systems*, **1**, 98-110.
10. Bezdek, J. C and Pal, S. K. (1992) (Eds.), "Fuzzy Models for Pattern Recognition," IEEE Press.
11. Goodall, C. (1983) M-estimator of location: An outline of the theory, in D.C.Hoaglin, F.Mosteller, and J.W.Tukey, (eds), *Understanding Robust and Exploratory Data Analysis*, New York, 339-403.

RANGE IMAGE SEGMENTATION THROUGH FUZZY AND POSSIBILISTIC SHELL CLUSTERING

RAGHU KRISHNAPURAM, HICHEM FRIGUI,
AND OLFA NASRAOUI
Department of Electrical and Computer Engineering
University of Missouri, Columbia, MO 65211, USA

Abstract

Segmentation and surface approximation (i. e., description of surfaces in terms of parametrized surface patches) is an important step in range image analysis. In this paper, we show that the surface approximation problem can be formulated in terms of a C-means type fuzzy clustering algorithm with suitable (shell-like) prototypes and distance measures. We present an unsupervised fuzzy clustering algorithm to obtain an approximation of a range image in terms of planar and quadric surface patches.

1. Introduction

Many three-dimensional vision systems use surface descriptions in terms of linear or quadric surface patches to model objects. In order to recognize objects through model matching and to estimate their pose, one must first segment a range image into such surface patches. It has been shown recently that boundary description can be achieved through objective-function-based fuzzy clustering with suitable prototypes and distance measures [1,2]. In this paper, we extend this mathod to surface approximation and present an unsupervised fuzzy clustering algorithm to obtain an approximation of range data in terms of planar and quadric surface patches.

2. Fuzzy Objective Functions for Surface Fitting

Let $x_j = [x_j, y_j, z_j]$, $j = 1,..., N$, be N points in 3-dimensional feature space. Let $B = (\beta_1,...,\beta_C)$ represent a C-tuple of prototypes of C clusters. The prototypes β_i are represented by the parameter vectors p_i, which define the equations of the quadric surfaces as shown below

$$[p_{i1}, p_{i2}, \ldots, p_{i10}] \; [x^2, y^2, z^2, xy, yz, zx, x, y, z, 1]^T = p_i^T q = 0. \tag{1}$$

We now briefly describe three algorithms based on three different objective functions that can be used to segment a range image into surface patches, the FCQS [3] algorithm, the MFCQS [3] algorithm, and the FCPQS [4] algorithm.

Z. Bien and K. C. Min (eds.),
Fuzzy Logic and its Applications, Information Sciences, and Intelligent Systems, 55–64.
© 1995 *Kluwer Academic Publishers.*

2.1. THE FUZZY C-QUADRIC SHELLS (FCQS) ALGORITHM

The FCQS [3] algorithm minimizes the following objective function

$$J_Q(B,U,X) = \sum_{i=1}^{C} \sum_{j=1}^{N} (u_{ij})^m \, d_{Qij}^2 \qquad (2)$$

In (2), $B = (\beta_1,...\beta_C)$, $m \in [1,\infty)$ is a weighting exponent called the fuzzifier, u_{ij} is the membership of x_j in cluster β_i and $U = [u_{ij}]$ is the $C \times N$ fuzzy C-partition matrix [5] such that:

$$u_{ij} \in [0,1] \; \forall \, i, j; \; \sum_{i=1}^{C} u_{ij} = 1 \; \forall \, j; \; \text{and} \; 0 < \sum_{j=1}^{N} u_{ij} < N \; \forall \, i. \qquad (3)$$

The algebraic (or residual) distance from a point x_j to the prototype β_i represented by d_{Qij}^2 is defined as:

$$d_{Qij}^2 = (p_i^T q_j)^2 = p_i^T M_j \, p_i, \qquad (4)$$

where

$$M_j = q \, q_j^T \text{ with } q_j = [x_j^2, y_j^2, z_j^2, x_j y_j, y_j z_j, z_j x_j, x_j, y_j, z_j, 1]^T$$

to avoid the trivial solution for p_i the following constraint is imposed:

$$\| \, p_{i1}^2 + p_{i2}^2 + p_{i3}^2 + (1/2)p_{i4}^2 + (1/2)p_{i5}^2 + (1/2)p_{i6}^2 \, \| = 1 \qquad (5)$$

If we define

$$a_{ik} = \begin{cases} p_{ik} & 1 \le k \le 3 \\ \dfrac{p_{ik}}{\sqrt{2}} & 4 \le k \le 6 \end{cases}, \quad \text{and} \quad b_{ik} = p_{ik} \;\; 7 \le k \le 10$$

then it is easily verified [6] that minimization of (2) with respect to β subject to (5) yields the following update equations:

$$\left. \begin{aligned} a_i &= \text{eigenvector of } (F_i - G_i^T H_i^{-1} G_i) \text{ associated} \\ & \qquad \text{with the smallest eigenvalue} \\ b_i &= - H_i^{-1} G_i a_i \end{aligned} \right\} \qquad (6)$$

where

$$F_i = \sum_{j=1}^{N} (u_{ij})^m R_j, \qquad G_i = \sum_{j=1}^{N} (u_{ij})^m S_j, \qquad H_i = \sum_{j=1}^{N} (u_{ij})^m T_j,$$

$$R_j = r_j r_j^T, S_j = r_j t_j^T, T_j = t_j t_j^T,$$

$$r_j^T = [x_j^2, y_j^2, z_j^2, \sqrt{2}x_j y_j, \sqrt{2}y_j z_j, \sqrt{2}x_j z_j], \text{ and } t_j^T = [x_j, y_j, z_j, 1].$$

Minimization of (2) with respect to U subject to the constraint in (3) gives us[5]

$$\left. \begin{aligned} u_{ij} &= \dfrac{1}{\sum\limits_{k=1}^{C} \left(\dfrac{d_{Qij}}{d_{Qik}} \right)^{\frac{2}{m-1}}} \qquad && \text{if } I_j = \varnothing \\[2em] u_{ij} &= 0 \;\; i \notin I_j \\ \sum_{i \in I_j} u_{ij} &= 1 \;\; i \in I_j \end{aligned} \right\} \qquad \text{if } I_j \ne \varnothing \right\} \qquad (7)$$

where $I_j = \{i \mid d_{Qij}^2 = 0\}$.

The FCQS algorithm is summerized below:

```
                    FCQS   ALGORITHM
Fix the number of clusters C; fix m ∈ [1,∞) ;
Initialize the fuzzy C partition U^(0) ;
set counter l = 1;
REPEAT
      Update the parameter of each cluster prototype using (6);
      Update the partition matrix U^(l) using (7);
      l = l+1;
UNTIL  ‖ U^(l) - U^(l-1) ‖ < ε .
```

The FCQS algorithm is applied to the synthetic range image shown in Fig. 1(a) with $C = 6$. This image consists of two crossing cylinders with one sphere at each end. As seen in Fig. 1(b), the FCQS algorithm succeeds in fitting the 6 different surfaces.

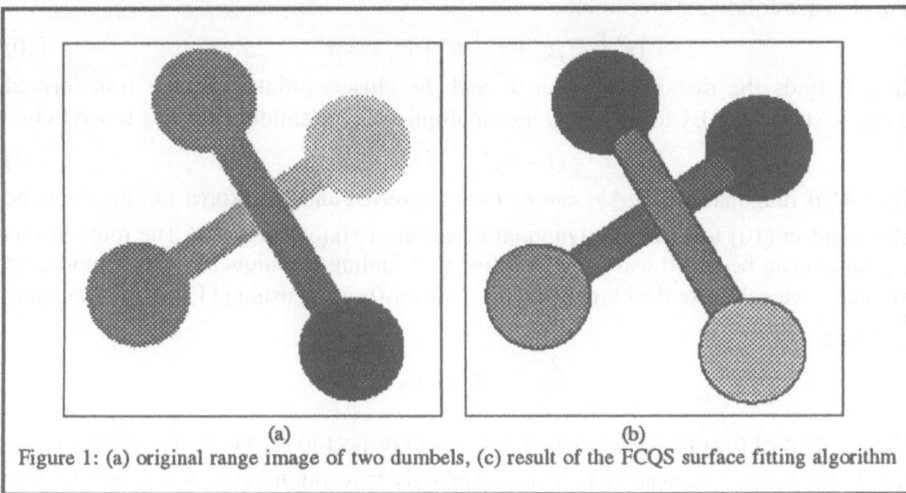

(a)   (b)

Figure 1: (a) original range image of two dumbels, (c) result of the FCQS surface fitting algorithm

Since the FCQS algorithm uses the algebraic distance given by (4) which is highly nonlinear in nature, the membership assignments are not very meaningful. Moreover, when there are surfaces of highly-varying sizes, the algebraic distance is biased towards smaller surfaces, and for a particular surface it is biased towards points inside the surface as opposed to points outside. Thus, the distance measure gives rather eccentric and highly curved fits if the data is scattered, because the prototypes try to enclose more points inside the surface. The distance is also sensitive to the placement of the feature point with respect to the surface.

2.2. THE MODIFIED FUZZY C-QUADRIC SHELLS (MFCQS) ALGORITHM

To alleviate the problem due to the nongeometric nature of d_{Qij} , Krishnapuram et al. [3] proposed a modified version of the FCQS algorithm called the Modified Fuzzy C Quadric Shells (MFCQS) algorithm. In this algorithm equation (1) is rewritten as

$$x^T A_i x + x^T b_i + c_i = 0, \tag{8}$$

where

$$A_i = \begin{bmatrix} p_{i1} & p_{i4}/2 & p_{i5}/2 \\ p_{i4}/2 & p_{i2} & p_{i6}/2 \\ p_{i5}/2 & p_{i6}/2 & p_{i3} \end{bmatrix}, \quad b_i = \begin{bmatrix} p_{i7} \\ p_{i8} \\ p_{i9} \end{bmatrix}, \text{ and } c_i = p_{i10}.$$

If we apply a linear transformation to the feature point x_j so that the matrix A_i becomes diagonal, then (8) is equivalent to

$$x'^T A'_i x' + x'^T b'_i + c'_i = 0$$

where A'_i (a diagonal matrix), b'_i, c'_i are the transformed parameters and x' is the location of x after transformation. The MFCQS algorithm uses the geometric (perpendicular) distance defined by

$$d^2_{Pij} = \min_{z'} \| x'_j - z' \|^2 \tag{9}$$

subject to the following constraint

$$(z'^T A'_i z' + z'^T b'_i + c'_i) = 0 \tag{10}$$

Thus, it finds the distance between x' and the closest point z' on the transformed quadratic surface β_i. By using a Lagrange multiplier λ, the solution to (9) is found to be

$$z' = \frac{1}{2}(I - \lambda A'_i)^{-1}(\lambda b'_i + 2x'_j). \tag{11}$$

Since A'_i is diagonal, $(I - \lambda A'_i)$ can be easily inverted and the expression for z' can be substituted in (10) to yield a polynomial equation of sixth order in λ. The roots of this polynomial can be found using any iterative root finding technique. For each root λ_k so computed, we calculate the corresponding z' vector (say z'_k) using (11) then we compute d^2_{Pij} using

$$d^2_{Pij} = \min_k \| x'_j - z'_k \|^2.$$

Minimization of the objective function in (2) with respect to p_i, when d^2_{Pij} is used as the underlying distance measure, can be achieved only by using iterative techniques such as the Levenberg-Marquardt algorithm [7]. In [8], Frigui and Krishnapuram argued that if p_i are updated using (6), the final solution is a good approximation to the exact solution for most computer vision applications. This leads to the MFCQS algorithm, in which the memberships are computed using d^2_{Pij}, but the parameters are updated using d^2_{Qij}.

2.3. THE FUZZY C PLANO-QUADRIC SHELLS (FCPQS) ALGORITHM

The MFCQS algorithm is computationally expensive since it involves solving the roots of a sixth-order polynomial equation in each iteration. It is practical only for the 2-D case [6], because in the 2-D case we need to solve only a fourth-order polynomial for which the roots have a closed form solution. To overcome this drawback, Krishnapuram et al. [4] proposed another modification to the FCQS algorithm called the FCPQS algorithm. This algorithm uses the approximate (first order) distance [9] of a point x_j from the shell which is given by

$$d^2_{Aij} = \frac{d^2_{Qij}}{|\nabla d_{Qij}|^2} = \frac{d^2_{Qij}}{p_i^T D(q_j)D(q_j)^T p_i}, \tag{12}$$

where d^2_{Qij} is the algebraic distance defined in (4) and the matrix $D(q_j)$ is the Jacobian of q (see (1)) evaluated at x_j.

The FCPQS algorithm minimizes the following objective function

$$J_A(B,U,X) = \sum_{i=1}^{C} \sum_{j=1}^{N} (u_{ij})^m d^2_{Aij} = \sum_{i=1}^{C} \sum_{j=1}^{N} (u_{ij})^m \frac{p_i^T M_j p_i}{p_i^T D(q_j)D(q_j)^T p_i}, \tag{13}$$

with the following constraint to avoid the all-zero trivial solution for p_i:

$$p_i^T \left[\sum_{j=1}^{N} (u_{ij})^m [D(q_j)D(q_j)^T] \right] p_i = \sum_{j=1}^{N} (u_{ij})^m.$$

In short notation, the above constraint is

$$p_i^T G_i p_i = N_i, \tag{14}$$

where

$$G_i = \sum_{j=1}^{N} (u_{ij})^m [D(q_j)D(q_j)^T] \quad \text{and } N_i = \sum_{j=1}^{N} (u_{ij})^m \tag{15}$$

If we assume that each data point is close to one of the surfaces and that the magnitude of the gradient is constant for all data points belonging to a surface, i.e., $p_i^T D(q_j)D(q_j)^T p_i \approx$ constant, then the denominator in (13) can be ignored, and it can be shown [12] that p_i can be obtained by solving the following generalized eigenvector problem

$$F_i p_i = \lambda_i G_i p_i, \tag{16}$$

where

$$F_i = \sum_{j=1}^{N} (u_{ij})^m M_j. \tag{17}$$

Minimization of the objective function in (13) with respect to U subject to (3) gives

$$\left. \begin{array}{ll} u_{ij} = \dfrac{1}{\sum\limits_{k=1}^{C} \left(\dfrac{d_{Aij}}{d_{Aik}} \right)^{\frac{2}{m-1}}} & \text{if } I_j = \varnothing \\[2em] \left. \begin{array}{l} u_{ij} = 0 \quad i \notin I_j \\ \sum\limits_{i \in I_j} u_{ij} = 1 \quad i \in I_j \end{array} \right\} & \text{if } I_j \neq \varnothing \end{array} \right\} \tag{18}$$

where $I_j = \{ i \mid d_{Aij}^2 = 0 \}$.

2.4. A WEIGHTING PROCEDURE TO IMPROVE FITS

The assumption that the magnitude of the gradient is constant for all data points assigned to a surface is always true for a linear (planar) prototype regardless of the location of the feature point x_j with respect to the prototype. Otherwise, this assumption is acceptable only if all the feature points lie close to one of the shells and the shells are either spheres, cylinders, or rectangular hyperbolas. Thus, it is not true for points lying on ellipsoids and many other quadric shapes. Therefore, the distance measure defined in (12) will be biased towards points for which gradient magnitude $|\nabla d_{Qij}|$ is high. In the case of ellipsoids, for example, the fit will be poorer near the major axis, particularly when the data is scattered. The fit will also be poor at the singularities (such as the line of intersection of a pair of planes) of the distance function, where the gradient magnitude goes to zero. Some researchers have suggested weighting the distance measure to improve the fit [10]. We use the following objective function in the reweight procedure.

$$J_{AW}(B, U, W) = \sum_{i=1}^{C} \sum_{j=1}^{N} (u_{ij})^m w_{ij} \, p_i^T \, M_j \, p_i, \tag{19}$$

The constraint in (14) is changed to

$$p_i^T \left[\sum_{j=1}^{N} (u_{ij})^m w_{ij} [D(q_j)D(q_j)^T] \right] p_i = \sum_{j=1}^{N} (u_{ij})^m,$$

or, in short notation

$$p_i^T \, G_{Wi} \, p_i = N_i, \tag{20}$$

where

$$G_{Wi} = \sum_{j=1}^{N} (u_{ij})^m w_{ij} [D(q_j)D(q_j)^T] \quad \text{and } N_i = \sum_{j=1}^{N} (u_{ij})^m. \tag{21}$$

The purpose of introducing these weights is to reduce the bias in the distance measure introduced due to the omission of the denominator in the expression of the approximate distance. Therefore, ideally these weights should be chosen as:

$$w_{ij} = \frac{1}{p_i^T [D(q_j) D(q_j)^T] p_i}.$$

With this choice of w_{ij}, the objective function in (19) becomes identical to the one in (13), and its minimization is not easy. To simplify the problem, we may treat the w_{ij} as constants which are computed from the parameters of the prototypes found in the previous iteration. Using this assumption, it can be easily shown that the minimization of J_{AW} reduces to the following generalized eigenvector problem:

$$F_{Wi} \, p_i = \lambda_i G_{Wi} \, p_i, \tag{22}$$

where

$$F_{Wi} = \sum_{j=1}^{N} (u_{ij})^m w_{ij} M_j \tag{23}$$

Since the reweight procedure is heuristic, the fit obtained after reweighting is not always guaranteed to be better than the original fit. Therefore, we accept the prototype parameters resulting from the reweight procedure only when the error of fit decreases. The sum of approximate distances d_{Aij}^2 for each individual cluster is a good measure of the error of fit.

When solving (16) or (22), care must be exercised because matrices F and G are highly unbalanced. Several methods for balancing matrices are available in the literarure.

The FCPQS algorithm with the reweight procedure is summerized below:

> **FCPQS ALGORITH WITH THE REWEIGHT PROCEDURE**
> Fix the number of clusters C; fix $m \in [1, \infty)$;
> Set iteration counter $l = 1$;
> Initialize the fuzzy C-partition $U^{(0)}$;
> **REPEAT**
>   Compute F_i and G_i for each cluster β_i using (15) & (17);
>   Compute $p_i^{(l)}$ for each cluster β_i by solving (16);
>   Compute error of fit ε_1;
>   Compute the weighted matrices F_{Wi} and G_{Wi} using (21) & (23)
>   Compute $(p_W)_i^{(l)}$ for each cluster β_i by solving (22);
>   Compute error of fit ε_2;
>   **IF** $(\varepsilon_1 < \varepsilon_2)$ **THEN** Replace the parameters $p_i^{(l)}$ by $(p_W)_i^{(l)}$;
>   Update $U^{(l)}$ using (18);
>   Increment l;
> **UNTIL** $(\| U^{(l-1)} - U^{(l)} \| < \varepsilon)$.

3. Possibilistic Objective Functions for Surface Fitting

The algorithms discussed in the previous section are sensitive to outlier points. The main source of this problem is the probabilistic constraint used in fuzzy clustering which states that the memberships of a data point across all clusters must sum to one. This problem has been discussed in detail in [11]. To overcome this drawback, possibilistic versions of these algorithms [12] can be devised. This is very easily achieved by changing the membership update equation to:

$$u_{ij} = \left[1 + \left(\frac{d^2_{ij}}{\eta_i} \right)^{\frac{1}{m-1}} \right]^{-1}. \tag{24}$$

In (24) d^2_{ij} can be d^2_{Qij}, d^2_{Pij}, or d^2_{Aij} depending on the algorithm used.

One attractive choice of η_i in practice is the average fuzzy intra-cluster distance given by:

$$\eta_i = \frac{1}{N_i} \sum_{j=1}^{N} (u_{ij})^m d^2_{ij}$$

A better choice may be the Median Absolute Deviation (*MAD*) given by

$$MAD = \underset{i}{\text{median}} \left\{ \, | \, d_i^2 - \underset{j}{\text{median}} \, (d_j^2) \, | \, \right\} \tag{25}$$

It is to be noted that the value of *MAD* is different for each cluster, since the medians in (25) are computed using only the distances corresponding to x_j that have their highest memberships in the cluster under consideration. In other words, the data set is crisply partitioned into C clusters in each iteration, and the *MAD* value is computed for each cluster.

In (24) it is clear that the updated value of u_{ij} depends only on the distance of x_j from β_i and not on the distance of x_j from all other prototypes, which is a desirable result. The prototypes are updated in the same manner as in the corresponding fuzzy algorithms.

4. The Quadric Compatible Cluster Merging Algorithm

When the number of surfaces to be fitted is not known, the FCPQS algorithm cannot be used. To overcome this problem, we developed the quadric compatible cluster merging (QCCM) algorithm. This algorithm assumes that a surface approximation of the data in terms of planar surface patches is available. Two planar patches are considered compatible if the quadric fit for all the points in the two clusters is good. The compatibility condition is checked for all neighboring planar patches, and they are merged if it is satisfied. While estimating the error of fit, we use the reweight procedure for better accuracy. This procedure is repeated until no more clusters can be merged.

The parameter vector of cluster β_i is given by the smallest generalized eigenvector solution of (16) or (22). However, we noticed that the solution may correspond to an "overfitted" prototype or to uncommon surface prototypes that almost never occur in real images such as: hyperboloids of two sheets, hyperbolic cylinders, and imaginary quadric surfaces. Therefore, we accept the smallest generalized eigenvector solution only if it represents the parameter vector of an "acceptable" surface type, otherwise we keep checking the next eigenvectors (assuming that they are in ascending order of the eigenvalues) until we find one that represents an "acceptable" surface prototype. The different types of the quadric surfaces and their identification conditions may be found in [12]. By an "acceptable" surface prototype, we mean the following: real ellipsoids, hyperboloids of one sheet, real quadric cones, elliptic paraboloids, real elliptic cylinders, and parabolic cylinders. We have not included planes in the above list since we assume that we have already obtained the best planar approximation before initiating the QCCM algorithm, and hence, no more two planes can be merged into a single plane.

5. The Unsupervised Surface Fitting Algorithm

The unsupervised 3-D quadric shell algorithm consists of running the Gustafson-Kessel (GK) algorithm [13] to fit an overspecified number C_{max} of planar clusters to the data, then merging all compatible planes using the Compatible Cluster Merging (CCM) algorithm for planes [14]. The result of this step is a planar approximation of each quadric surface present in the range image. The next step consists of applying the QCCM algorithm to merge all compatible planes into quadric surfaces.

6. Examples

The examples used in this section consist of some synthetic range images. The size of the images was 200×200. A sampling rate of 3 in the x and y direction was used to reduce computations. In all the examples shown in this section, the GK algorithm was applied with $m = 1.5$ and the initial number of clusters $C_{max} = 15$. Two clusters were considered compatible if they can be fitted by a single cluster with an error of fit less than 0.05. After the algorithm converged and the prototypes were identified, the FCPQS algorithm was performed one last time for two iterations with the original data set (without sampling), using the computed prototypes as the initialization. In this step, the distance measure used to update the memberships was defined as a linear combination of the Euclidean distance d_E and the approximate distance d_A, i.e.,

$$d_C^2 = \alpha d_A^2 + (1 - \alpha) d_E^2.$$

The Euclidean distance d_E^2 is measured from the statistical center of the cluster. The reason for using the distance d_C is as follows. In the 3-D case, most quadric surfaces such as cones, cylinders, and planes have an infinite extent, i.e., they are not bounded surfaces. Therefore, a cluster having the prototype of one of these surfaces will grab all points lying on its infinite extension even if the points are far from its center and much closer to another cluster. The value of α is heuristically set to a small real number of the order of 10^{-4}. Fig. 2(a) shows a synthetic 3-D image consisting of a cylinder and a sphere. Fig. 2(b) displays the initial planar approximation. As can be seen in this figure, the cylinder is approximated by 4 planar surface patches while the sphere is approximated by 3 surface patches. The final result of the unsupervised algorithm with the correct surfaces is shown in Fig. 2(c). Fig. 3(a) shows a synthetic range image consisting of 2 planes, a sphere and an ellipsoid. The planar approximation of this example is shown in Fig. 3(c). As can be seen in this figure, the planar fit is good, the sphere is approximated by 2 planar patches, and the ellipsoid is approximated by 4 planar patches. The final result of the QCCM algorithm is shown in Fig. 3(c). Fig. 4(a) shows another synthetic range image of a lamp which consists of a cone, a cylinder, and a plane. The initial planar approximation is shown in Fig. 4(b). As can be seen, the bottom plane is correctly identified. The cylinder is approximated by 3 planar patches, and the cone is approximated by 7 planar patches. The QCCM algorithm, using this planar approximation, succeeds in merging all compatible surfaces. The final result is displayed in Fig. 4(c).

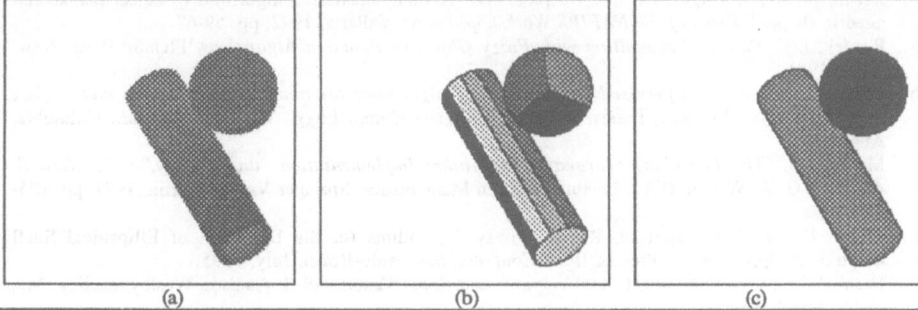

| (a) | (b) | (c) |

Figure 2: (a) original range image of a cylinder and a sphere, (b) planar approximation of image in (a), (c) result of the unsupervised surface fitting algorithm

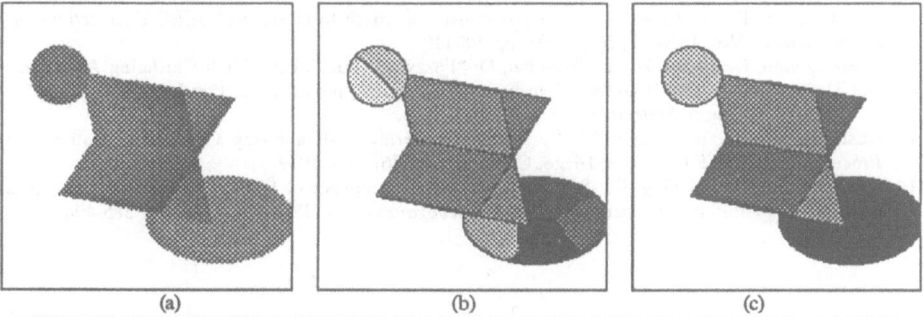

| (a) | (b) | (c) |

Figure 3: (a) original range image of two planes, a sphere and an ellipsoid, (b) planar approximation of image in (a), (c) result of the unsupervised surface fitting algorithm

64

 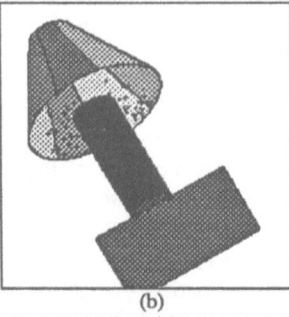 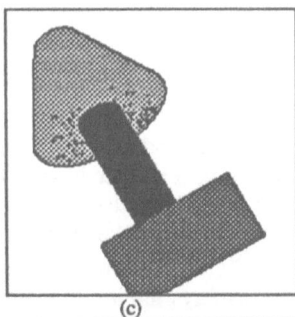

(a)   (b)   (c)

Figure 4: (a) original range image "lamp shade", (b) planar approximation of the image in (a), (c) result of the unsupervised surface fitting algorithm

7. References

1. Davé, R. N. and Bhaswan, K. "Adaptive Fuzzy C-Shells Clustering and Detection of Ellipses," *IEEE Trans. on Neural Networks*, vol. 3, no. 5, Sept. 1992, pp. 643-662.
2. Krishnapuram, R., Nasraoui, O., and Frigui, H. "The fuzzy C spherical shells algorithm: A new approach," *IEEE Trans. on Neural Networks*, vol. 3, no. 5, Sept. 1992, pp.663-671
3. Krishnapuram, R.,Frigui, H., and Nasraoui, O. "Quadratic Shell Clustering Algorithms and the detection of second degree curves," *Pattern Recognition Letters*, vol. 14, no. 7, July 1993, pp. 545-552.
4. Krishnapuram, R.,Frigui, H., and Nasraoui, O. "A fuzzy clustering algorithm to detect planar and quadric shapes," *Proc. of the NAFIPS Workshop*, Puerto Vallarta, 1992, pp. 59-69.
5. Bezdek, J. C. *Pattern Recognition with Fuzzy Objective Function Algorithms*, Plenum Press, New York, 1981.
6. Frigui, H. "*Robust unsupervised shell clustering algorithms for boundary description and surface approximation*", Master's Thesis, Dept. of Elec. and Comp. Engg., Univ. of Missouri, Columbia, Aug. 1992.
7. Moore, J. J. "*The Levenberg-Marquardt Algorithm: Implementation and Theory,*" in *Numerical Analysis*, G. A. Watson (Ed.), Lecture Notes in Mathematics, Springer Verlag, Berlin, 1977, pp. 105-116.
8. Frigui, H. and Krishnapuram, R. "On Fuzzy Algorithms for the Detection of Ellipsoidal Shell Clusters" to appear in the Proc of IFSA Congress, São Paulo-Brasil, July, 1995.
9. Haralick, R. M. and Shapiro, L. G. *Computer and Robot Vision*, vol. I, Addison Wesley, reading, MA, 1992, Appendices.
10. Taubin, G. "Estimation of planar curves, surfaces, and nonplanar space curves defined by implicit equations with application to edge and range image segmentation", *IEEE Trans. PAMI*, vol. 13, no. 11, 1991, pp. 1115-38.
11. Krishnapuram, R. and Keller, J. M. "A possibilistic approach to clustering," *IEEE Transactions on Fuzzy Systems*, Vol. 1, No. 2, May 1993, pp. 98-110.
12. Krishnapuram, R., Frigui, H. and Nasraoui, O. "Fuzzy and Possibilistic Shell Clustering Algorithms and Their Application to Boundary Detection and Surface Approximation: Part I and Part II," *IEEE Transactions on Fuzzy Systems*, vol. 3 No. 1 Feb. 1995.
13. Gustafson, E. E. and Kessel, W. C. "Fuzzy Clustering with a Fuzzy Covariance Matrix," *In Proceedings of IEEE CDC*, San Diego, California, pp. 761-766, 1979.
14. Krishnapuram, R. and Freg, C. -P. "Fitting an unknown number of lines and planes to image data through compatible cluster merging," *Pattern Recognition*, vol. 25, no. 4, 1992, pp. 385-400.

Chapter 2.

ENGINEERING

ON THE STRUCTURE AND LEARNING OF NEURAL-NETWORK-BASED FUZZY LOGIC CONTROL SYSTEMS

C. T. LIN
Department of Control Engineering
National Chiao-Tung University
Hsinchu, Taiwan, R.O.C.

AND

C. S. G. LEE
School of Electrical Engineering
Purdue University
West Lafayette, Indiana 47907, U.S.A.

Abstract. This paper addresses the structure and its associated learning algorithms of a feedforward multi-layered connectionist network, which has distributed learning abilities, for realizing the basic elements and functions of a traditional fuzzy logic controller. The proposed neural-network-based fuzzy logic control system (NN-FLCS) can be contrasted with the traditional fuzzy logic control system in their network structure and learning ability. An on-line supervised structure/parameter learning algorithm is proposed for constructing the NN-FLCS dynamically. The proposed dynamic learning algorithm can find proper fuzzy logic rules, membership functions, and the size of output fuzzy partitions simultaneously. Next, a Reinforcement Neural-Network-Based Fuzzy Logic Control System (RNN-FLCS) is proposed which consists of two closely integrated Neural-Network-Based Fuzzy Logic Controllers (NN-FLCs) for solving various reinforcement learning problems in fuzzy logic systems. One NN-FLC functions as a fuzzy predictor and the other as a fuzzy controller. Associated with the proposed RNN-FLCS is the reinforcement structure/parameter learning algorithm which dynamically determines the proper network size, connections, and parameters of the RNN-FLCS through an external reinforcement signal.

†This work was supported in part by the National Science Foundation under Grant CDR 8803017 to the Engineering Research Center for Intelligent Manufacturing Systems.

Z. Bien and K. C. Min (eds.),
Fuzzy Logic and its Applications, Information Sciences, and Intelligent Systems, 67–80.
© 1995 *Kluwer Academic Publishers.*

Furthermore, learning can proceed even in the period without any external reinforcement feedback.

1. Neural-Network-Based Fuzzy Logic Control System

During the past decade, fuzzy logic has found fruitful applications in various fields [15, 10, 5]. However, most control engineers are still frustrated with this technique due to a lack of systematic procedures for the design of fuzzy logic systems. The choice of membership functions and/or fuzzy logic rules remains heuristic and subjective, and a trial-and-error procedure is commonly used for the design of fuzzy logic systems. Recent direction of exploration is to design fuzzy logic systems that have the capability of learning from experience by itself [13, 11].

Distributed representation and learning capabilities are two major features of neural networks [3, 9]. In distributed representation, a value is represented by a pattern of activity distributed over many computing elements (CEs), and each CE is involved in representing many different values. So each CE has a receptive field, which is the set of all values that include all the patterns it represents. Therefore, each CE corresponds to a fuzzy set, and its receptive field corresponds to the membership function. Among the three classes of learning schemes, the unsupervised procedures [4] are suitable to find clusters of data indicating the presence of fuzzy rules. The supervised procedures and the reinforcement procedures are good to adapt the fuzzy rules or membership functions for the desired output in fuzzy logic systems. Hence, bringing the learning abilities of neural networks to fuzzy logic systems will provide a promising approach.

This paper presents a general Neural-Network-Based Fuzzy Logic Control System (NN-FLCS) for realizing the basic elements and functions of a traditional fuzzy logic control and decision system [15, 10, 5]. In this connectionist structure [6, 7, 8], the input and output nodes represent the input states and output control/decision signals, respectively, and in the hidden layers, there are nodes functioning as membership functions and rules. An on-line supervised structure/parameter learning algorithm is proposed to construct NN-FLCS dynamically. This algorithm blends *fuzzy similarity measure* with supervised gradient-descent learning to perform structure and parameter learning simultaneously. The fuzzy similarity measure is a tool to determine the degree to which two fuzzy sets are equal. Using this measure, a new output membership function may be added, and the rule-node connections (the consequence links of rule nodes) can be changed properly. In some learning environments, obtaining exact training data may be expensive. This motivates the desire of integrating two Neural-Network-Based Fuzzy Logic Controllers (NN-FLCs) into a Reinforcement Neural-Network-Based Fuzzy Logic Control System (RNN-FLCS) for solving various reinforcement learning problems. One NN-FLC functions as a fuzzy predictor and the other as a fuzzy controller. Structurally, these two NN-FLCs share the first two layers of the proposed NN-FLCS; that is, they use the same distributed representation of input patterns. This representation is the overlapping type and is dynamically adjustable through the learning process. Associated with the proposed RNN-FLCS is the reinforcement structure/parameter learning algorithm which dynamically determines the proper network size, connections, and parameters of the RNN-FLCS through an external reinforcement signal. Furthermore, learning can proceed even in the period without any external reinforcement feedback. The proposed RNN-FLCS makes the design of fuzzy logic controllers more practical for real-world applications since it greatly lessens the quality and quantity requirements of the feedback training signals.

Figure 1 shows the structure of our NN-FLCS which has five layers. Nodes at layer one are input nodes which represent input linguistic variables. Layer five is the output layer. Nodes at layers two and four are *term nodes* and act as membership functions to represent the terms of the respective linguistic variable. Each node at layer three is a rule node which represents one fuzzy logic rule. Thus, all layer-three nodes form a fuzzy rule base. Layer-three links define the preconditions of the rule nodes, and layer-four links define the consequences of the rule nodes. The links at layers two and five are fully connected between linguistic nodes and their corresponding term nodes. We shall next describe the functions of the nodes in each of the five layers of the proposed connectionist model. In the following, f is an integration function of a node, which combines activation from other nodes to provide net input for this node. a is an activation function of a node, which outputs an activation value as a function of net input. In the following equations, superscript is used to indicate the layer number.

- **Layer 1:** The nodes in this layer transmit input values directly to the next layer. That is,

$$f = u_i^1 \qquad \text{and} \qquad a = f. \qquad (1)$$

From Eq. (1), the link weight at layer one (w_i^1) is unity.

- **Layer 2:** If we use a single node to perform a simple membership function, then the output function of this node should be this membership function. For example, for a bell-shaped function,

$$f = M_{x_i}^j(m_{ij}, \sigma_{ij}) = -\frac{(u_i^2 - m_{ij})^2}{\sigma_{ij}^2} \qquad \text{and} \qquad a = e^f, \qquad (2)$$

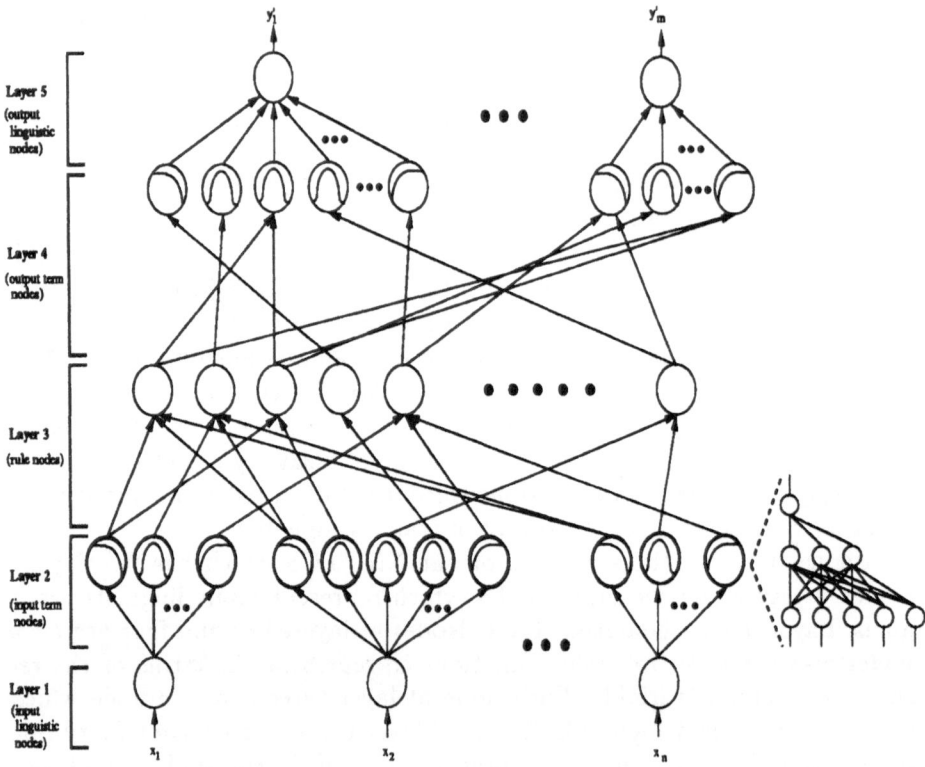

Figure 1. Proposed Neural-Network-Based Fuzzy Logic Control System (NN-FLCS).

where m_{ij} and σ_{ij} are, respectively, the center (or mean) and the width (or variance) of the bell-shaped function of the jth term of the ith input linguistic variable x_i. Hence, the link weight at layer two (w_{ij}^2) can be interpreted as m_{ij}.

■ **Layer 3:** The links in this layer are used to perform precondition matching of fuzzy logic rules. Hence, the rule nodes should perform the fuzzy AND operation,

$$f = \min(u_1^3, u_2^3, \ldots, u_p^3) \qquad \text{and} \qquad a = f. \qquad (3)$$

The link weight in layer three (w_i^3) is then unity.

■ **Layer 4:** The links at layer four perform the fuzzy OR operation to integrate the fired rules which have the same consequence,

$$f = \sum_{i=1}^{p} u_i^4 \qquad \text{and} \qquad a = \min(1, f). \qquad (4)$$

Hence the link weight $w_i^4 = 1$.

■ **Layer 5:** The nodes in this layer transmit the decision signal out of the network. These nodes and the layer-five links attached to them act as the defuzzifier. If m_{ij}^5's and σ_{ij}^5's are the centers and the widths of the membership functions, respectively, then the following functions can be used to simulate the *center of area* defuzzification method [5]:

$$f = \sum w_{ij}^5 u_i^5 = \sum (m_{ij}\sigma_{ij})u_i^5 \qquad \text{and} \qquad a = \frac{f}{\sum \sigma_{ij}u_i^5}. \qquad (5)$$

Here the link weight at layer five (w_{ij}^5) is $m_{ij}\sigma_{ij}$.

2. On-line Structure/Parameter Learning Algorithm

We first propose an on-line learning algorithm that can dynamically learn the network structure and parameters simultaneously. The proposed structure/parameter learning algorithm uses the fuzzy similarity measure [7] to perform the structure learning and the back-propagation algorithm to perform the parameter learning. Given the supervised training data, the proposed learning algorithm first decides whether or not to perform the structure learning based on the fuzzy similarity measure of the output membership functions. If the structure learning is necessary, then it will further decide whether or not to add a new output term node (a new membership function), and it will also change the consequences of some fuzzy logic rules properly. After the structure learning process, the parameter learning will be performed to adjust the parameters of current membership functions. This structure/parameter learning will be repeated for each on-line incoming training input/output data pair. After the structure/parameter training loop has been performed, rule combination is then initiated to find the minimum node representation of fuzzy logic rules as in [6].

An initial form of the network is first constructed before this network is trained. Then, during the learning process, new output term nodes may be added and some connections may be changed. Finally, after the learning process, some nodes and links of the network will be deleted or combined to form the final structure of the network. The initial form of the network is same as that described in [6], except that there is only one link between a rule node and an output linguistic variable. This link is connected to some term node of the output linguistic variable. The initial candidate (term node) of the consequence of a rule node can be assigned by an expert (if possible) or be chosen randomly. A suitable term in each output linguistic variable's term set will be chosen for each rule node after the learning process.

After the initialization process, the learning algorithm enters the training loop in which each loop corresponds to a set of training input data

$x_i(t), i = 1, \ldots, n$, and the desired output value $y_i(t), i = 1, \ldots, m$, at a specific time t. Basically, the idea of back-propagation [9] is used for this supervised learning to find the errors of node outputs in each layer. Then, these errors are analyzed by the fuzzy similarity measure to perform structure adjustments or parameter adjustments. The goal is to minimize the error function

$$E = \frac{1}{2}[y(t) - \hat{y}(t)]^2 \tag{6}$$

where $y(t)$ is the desired output, and $\hat{y}(t)$ is the current output. For each training data set, starting at the input nodes, a forward pass is used to compute the activity levels of all the nodes in the network. Then starting at the output nodes, a backward pass is used to compute $\partial E/\partial y$ for all the hidden nodes. Assuming that w is the adjustable parameter in a node (e.g., center of a membership function), the general learning rule used is

$$w(t+1) = w(t) + \eta(-\frac{\partial E}{\partial w}) \quad \text{and} \quad \frac{\partial E}{\partial w} = \frac{\partial E}{\partial f}\frac{\partial f}{\partial w} = \frac{\partial E}{\partial a}\frac{\partial a}{\partial f}\frac{\partial f}{\partial w} \tag{7}$$

where η is the learning rate. To show the learning rules, we derive the rules layer by layer using the bell-shaped membership functions with centers m_i's and widths σ_i's as the adjustable parameters for these computations.

■ **Layer 5:** Using Eqs. (5) and (6), the expected updated amount of the center parameter m_i and the width parameter σ_i are, respectively,

$$\Delta m_i(t) \overset{\triangle}{=} m_i(t+1) - m_i(t) = \eta[y(t) - \hat{y}(t)]\frac{\sigma_i u_i}{\sum \sigma_i u_i}. \tag{8}$$

$$\Delta \sigma_i(t) \overset{\triangle}{=} \sigma_i(t+1) - \sigma_i(t) = \eta[y(t) - \hat{y}(t)]\frac{m_i u_i(\sum \sigma_i u_i) - (\sum m_i \sigma_i u_i)u_i}{(\sum \sigma_i u_i)^2} \tag{9}$$

The error to be propagated to the previous layer is

$$\delta^5 = \frac{-\partial E}{\partial f} = \frac{-\partial E}{\partial a}\frac{\partial a}{\partial f} = y(t) - \hat{y}(t). \tag{10}$$

• **Fuzzy Similarity Measure:** In this step, the system will decide if the current structure should be changed or not according to the expected updated amount of the center and width parameters. To do this, the expected center and width are, respectively, computed as

$$m_{i-new} = m_i(t) + \Delta m_i(t) \quad \text{and} \quad \sigma_{i-new} = \sigma_i(t) + \Delta \sigma_i(t). \tag{11}$$

From the current membership functions of output linguistic variables, we want to find the one which is the most similar to the expected membership function by measuring their fuzzy similarity. Let $M(m_i, \sigma_i)$

represent the bell-shaped membership function with center m_i and width σ_i. Let

$$
\begin{aligned}
degree(i,t) &= E[M(m_{i-new}, \sigma_{i-new}), M(m_{i-closest}, \sigma_{i-closest})] \\
&= \max_{1 \le j \le k} E[M(m_{i-new}, \sigma_{i-new}), M(m_j, \sigma_j)], \quad (12)
\end{aligned}
$$

where $k = |T(y)|$, $E(\cdot, \cdot)$ is the fuzzy similarity. If A and B are two fuzzy sets with bell-shaped membership functions, The approximate fuzzy similarity measure of A and B, $E(A, B)$, can be computed as follow: Assuming $m_1 \ge m_2$,

$$
E(A, B) = \frac{M(A \cap B)}{M(A \cup B)} = \frac{M(A \cap B)}{\sigma_1 \sqrt{\pi} + \sigma_2 \sqrt{\pi} - M(A \cap B)}. \quad (13)
$$

Here

$$
M(A \cap B) = \frac{1}{2} \frac{h^2(m_2 - m_1 + \sqrt{\pi}(\sigma_1 + \sigma_2))}{\sqrt{\pi}(\sigma_1 + \sigma_2)} + \quad (14)
$$

$$
\frac{1}{2} \frac{h^2(m_2 - m_1 + \sqrt{\pi}(\sigma_1 - \sigma_2))}{\sqrt{\pi}(\sigma_2 - \sigma_1)} + \frac{1}{2} \frac{h^2(m_2 - m_1 - \sqrt{\pi}(\sigma_1 - \sigma_2))}{\sqrt{\pi}(\sigma_1 - \sigma_2)},
$$

where $h(x) = \max\{0, x\}$. After the most similar membership function $M(m_{i-closest}, \sigma_{i-closest})$ to the expected membership function $M(m_{i-new}, \sigma_{i-new})$ has been found, the following adjustment is made:

IF $degree(i, t) < \alpha(t)$,
    THEN
        create a new node $M(m_{i-new}, \sigma_{i-new})$ in layer 4
           and denote this node as the $i - closest$ node,
        do the structure learning process,
    ELSE IF $M(m_{i-closest}, \sigma_{i-closest}) \ne M(m_i, \sigma_i)$
        THEN
        do the structure learning process,
        ELSE
        do the following parameter adjustments:
           $m_i(t+1) = m_{i-new}$ and $\sigma_i(t+1) = \sigma_{i-new}$
        skip the structure learning process.

$\alpha(t)$ is a monotonically increasing scalar similarity criterion.

- **Structure Learning:** To find the rules whose consequences should be changed, we set a *firing strength threshold*, β. Only the rules whose firing strengths are higher than this threshold are treated as really firing rules. Only the really firing rules are considered to be changing their consequences, since only these rules are fired strongly enough to contribute to the above results of judgement. Assuming that the term node $M(m_i, \sigma_i)$ in layer 4 has inputs from rule nodes $1, \ldots, l$ in layer 3, whose corresponding firing strength are a_i^3's, $i = 1, \ldots, l$, then

IF $a_i^3(t) \geq \beta$, THEN change the consequence of the ith rule
node from $M(m_i, \sigma_i)$ to $M(m_{i-new}, \sigma_{i-new})$.

- **Layer 4:** There is no parameter to be adjusted in this layer. Only the error signals (δ_i^4's) need to be computed and propagated. From Eqs. (4) and (5), the error signal δ_i^4 is derived as:

$$\delta_i^4(t) = [y(t) - \hat{y}(t)] \frac{m_i \sigma_i (\sum \sigma_i u_i) - (\sum m_i \sigma_i u_i) \sigma_i}{(\sum \sigma_i u_i)^2}. \qquad (15)$$

- **Layer 3:** As in layer four, only the error signals need to be computed. According to Eq. (3), this error signal can be derived as: $\delta_i^3 = \delta_i^4$. If there are multiple outputs, the error signal becomes $\delta_i^3 = \sum_k \delta_k^4$. Here the summation is performed over the consequences of a rule node.
- **Layer 2:** Using Eq. (7) and Eqs. (2) and (3), the adaptive rule of m_{ij} and σ_{ij} are, respectively,

$$m_{ij}(t+1) = m_{ij}(t) - \eta \frac{\partial E}{\partial a_i} e^{f_i} \frac{2(u_i - m_{ij})}{\sigma_{ij}^2}, \qquad (16)$$

$$\sigma_{ij}(t+1) = \sigma_{ij}(t) - \eta \frac{\partial E}{\partial a_i} e^{f_i} \frac{2(u_i - m_{ij})^2}{\sigma_{ij}^3}, \qquad (17)$$

where $\frac{\partial E}{\partial a_i} = \sum_k q_k$. The summation here is performed over the rule nodes that a_i feeds into, and

$$q_k = \begin{cases} \delta_k^3, & \text{if } a_i \text{ is minimum in } k\text{th rule node's inputs} \\ 0, & \text{otherwise.} \end{cases} \qquad (18)$$

3. Structure/Parameter Learning Algorithm for RNN-FLCS

Unlike the supervised learning problem, the reinforcement learning problem has only very simple "evaluative" information called reinforcement signal available for learning. In this paper, the reinforcement signal $r(t)$ is defined as a value between -1 and 1 corresponding to various degrees of failure or success. We also assume that $r(t)$ is the reinforcement signal available at time step t and is caused by the input and actions chosen at time step $t-1$ or even affected by earlier inputs and actions. The objective of learning is to maximize the reinforcement signal. The proposed RNN-FLCS, as shown in

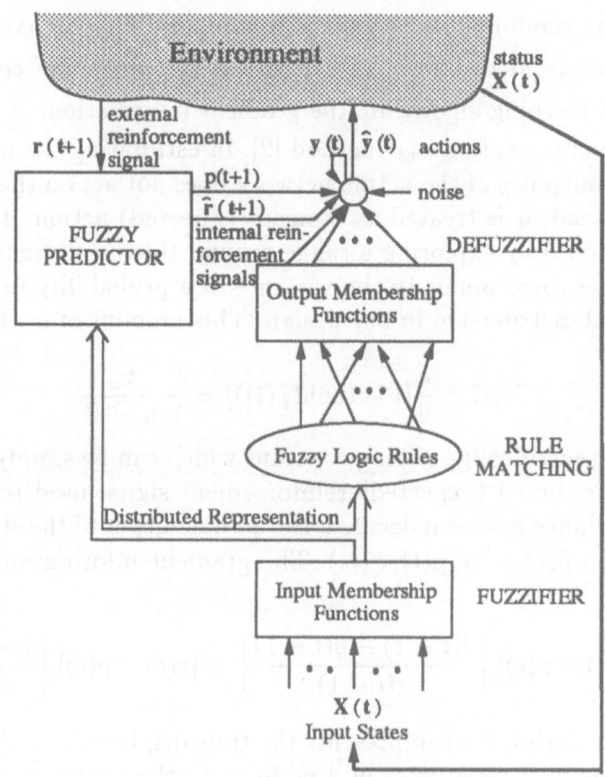

Figure 2. Proposed Reinforcement Neural-Network-Based Fuzzy Logic Control System (RNN-FLCS).

Fig. 2, integrates two NN-FLCs into a learning system: one NN-FLC for the fuzzy controller and the other for the fuzzy predictor. These two NN-FLCs share the same layers 1 and 2 and have individual layer 3 to layer 5. In this section, a reinforcement learning algorithm is proposed for the RNN-FLCS with a single-step fuzzy predictor to solve simpler reinforcement learning problems in which a reinforcement signal is only one time step behind its corresponding action. For the case that there is a long time delay between an action and the resulting reinforcement signal, a more powerful multi-step fuzzy predictor is necessary for the RNN-FLCS.

3.1. STOCHASTIC EXPLORATION

In this subsection, we first develop the learning algorithm for the action network. The goal of the reinforcement structure/parameter learning algorithm is to adjust the parameters (e.g., m_i's) of the action network or to change the connectionist structure or even to add new nodes, if necessary, such that the reinforcement signal is maximum. That is, $\Delta m_i \propto \frac{\partial r}{\partial m_i}$. To know $\frac{\partial r}{\partial m_i}$, we need to know $\frac{\partial r}{\partial y}$, where y is the output of the action network. In our learning algorithm, the gradient information, $\frac{\partial r}{\partial y}$, is estimated by the stochastic exploratory method [2]. In estimating the gradient information, the output y of the action network does not act on the environment directly. Instead, it is treated as a mean (expected) action. The actual action, $\hat{y}$, is chosen by exploring a range around this mean point. This range of exploration corresponds to the variance of a probability function, which is the normal distribution in our design. This amount of exploration, $\sigma(t)$, is chosen as

$$\sigma(t) = \frac{k}{2}[1 - \tanh(p(t))] = \frac{k}{1 + e^{2p(t)}}, \qquad (19)$$

where k is a search-range scaling constant which can be simply set to 1, and $p(t)$ is the predicted (expected) reinforcement signal used to predict $r(t)$. Once the variance has been decided, the actual output of the stochastic node can be set as $\hat{y}(t) = N(y(t), \sigma(t))$. The gradient information is estimated as

$$\frac{\partial r}{\partial y} \approx [r(t) - p(t)] \left[\frac{\hat{y}(t-1) - y(t-1)}{\sigma(t-1)} \right] \equiv [r(t) - p(t)] \left[\frac{\hat{y} - y}{\sigma} \right]_{t-1} \qquad (20)$$

where the subscript, $t-1$, represents the time displacement. Assuming that w is an adjustable parameter in a node (e.g., the center of a membership function), the general parameter learning rule used is (as in Eq. (7))

$$w(t+1) = w(t) + \eta(\frac{\partial r}{\partial w}) \quad \text{and} \quad \frac{\partial r}{\partial w} = \frac{\partial r}{\partial a} \frac{\partial a}{\partial f} \frac{\partial f}{\partial w}. \qquad (21)$$

- **Layer 5:** Using Eqs. (5), (20), and (21), the expected updated amount of the center parameter and the width parameter are respectively

$$\Delta m_i(t) = \eta[r(t) - p(t)] \left[\frac{\hat{y} - y}{\sigma} \right]_{t-1} \left[\frac{\sigma_i u_i}{\sum \sigma_i u_i} \right]_{t-1} \qquad (22)$$

$$\Delta \sigma_i(t) = \eta[r(t) - p(t)] \left[\frac{\hat{y} - y}{\sigma} \right]_{t-1} \left[\frac{m_i u_i (\sum \sigma_i u_i) - (\sum m_i \sigma_i u_i) u_i}{(\sum \sigma_i u_i)^2} \right]_{t-1} \qquad (23)$$

The error to be propagated to the preceding layer is

$$\delta^5(t) = \frac{\partial r}{\partial f^5} = \frac{\partial r}{\partial a} \frac{\partial a}{\partial f^5} = [r(t) - p(t)] \left[\frac{\hat{y} - y}{\sigma} \right]_{t-1} \qquad (24)$$

Fuzzy Similarity Measure: In this step, the system will decide whether the current structure should be changed or not according to the expected updated amount of the center and width parameters (in Eqs. (22) and (23)). This procedure of using the fuzzy similarity measure is the same as for the on-line learning algorithm.

■ **Layer 4:** There is no parameter to be adjusted in this layer. Only the error signals (δ_i^4's) need to be computed and propagated. From Eqs. (5) and (21), the error signal δ_i^4 is derived as

$$\delta_i^4(t) = [r(t) - p(t)] \left[\frac{\hat{y} - y}{\sigma}\right]_{t-1} \left[\frac{m_i \sigma_i (\sum \sigma_i u_i) - (\sum m_i \sigma_i u_i)\sigma_i}{(\sum \sigma_i u_i)^2}\right]_{t-1}$$
(25)

In the multi-output case, the computations in layers five and four are exactly the same as the above using the same internal reinforcement signals and proceeding independently for each output linguistic variable.

■ **Layer 3:** As in layer four, only the error signals need to be computed. According to Eqs. (4) and (21), this error signal can be derived as $\delta_i^3(t) = \delta_i^4(t)$. If there are multiple outputs, then the error signal becomes $\delta_i^3(t) = \sum_k \delta_k^4(t)$, where the summation is performed over the consequences of a rule node; that is, the error of a rule node is the summation of the errors of its consequences.

■ **Layer 2:** Using Eqs. (2) and (21), the adaptive rule of m_{ij} and σ_{ij} are respectively

$$m_{ij}(t+1) = m_{ij}(t) - \eta \left[\frac{\partial r}{\partial a_i}\right]_t \left[e^{f_i}\frac{2(u_i - m_{ij})}{\sigma_{ij}^2}\right]_{t-1}$$
(26)

$$\sigma_{ij}(t+1) = \sigma_{ij}(t) - \eta \left[\frac{\partial r}{\partial a_i}\right]_t \left[e^{f_i}\frac{2(u_i - m_{ij})^2}{\sigma_{ij}^3}\right]_{t-1},$$
(27)

where $\frac{\partial r}{\partial a_i} = \sum_k q_k(t)$ as in Eq. (18).

3.2. SINGLE-STEP FUZZY PREDICTOR

We shall use an NN-FLC to develop a single-step fuzzy predictor (evaluation network) as shown in Fig. 2. The function of the single-step fuzzy predictor is to predict the external reinforcement signal, $r(t)$, one time step ahead, that is, at time $t - 1$. Here, $r(t)$ is the real reinforcement signal resulting from the inputs and actions chosen at time step $t - 1$, but it can only be known at time step t. If the fuzzy predictor can produce a signal, $p(t)$, which is the prediction of $r(t)$ but is available at time step $t - 1$, then the time delay problem can be solved. With a correct predicted signal, $p(t)$, a better action can be chosen by the action network at time step $t - 1$, and the corresponding learning can be performed on the action network at time step t upon receiving the external reinforcement signal $r(t)$. As indicated in the last subsection, $p(t)$ is necessary for the stochastic exploration with a multi-parameter probability distribution (in Eq. (6)). The other internal reinforcement signal, $\hat{r}(t)$, in Fig. 2 is set as $\hat{r}(t) = r(t) - p(t)$, which is the prediction error for computing Eq. (7) by the action network. The single-step prediction is the extreme case of the multi-step prediction which will be presented in the next section. The goal to train the single-step fuzzy predictor is to minimize the squared error prediction:

$$E = \frac{1}{2}[r(t) - p(t)]^2, \tag{28}$$

where $r(t)$ represents the desired output (real external reinforcement signal), and $p(t)$ is the current output (predicted reinforcement signal). Then the gradient information can be easily derived as $\frac{\partial E}{\partial p} = p(t) - r(t)$. Similar to the learning rule developed in the last section, we can derive the structure/parameter learning algorithm for the single-step fuzzy predictor using the general parameter learning rule: $w(t+1) = w(t) + \eta(-\frac{\partial E}{\partial w})$, where w is the adjustable parameters in the fuzzy predictor. The learning equations are the same as Eqs. (21)-(26) if $\frac{\partial r}{\partial y}$ is replaced by $(-\frac{\partial E}{\partial p})$ and the effects caused by this replacement are properly updated, that is, all the terms $[r(t) - p(t)]\left[\frac{\hat{y}-y}{\sigma}\right]_{t-1}$ in Eqs. (21)-(26) are replaced with the term $[r(t) - p(t)]$.

3.3. MULTI-STEP FUZZY PREDICTOR

When both the reinforcement signal and input patterns from the environment may depend arbitrarily on the past history of the network output and the network may only receive a reinforcement signal after a long sequence of outputs, the credit assignment problem becomes severe. This *temporal credit assignment* problem results because we need to assign credit or blame to each step individually in such a long sequence for an eventual success or failure. To solve the temporal credit assignment problem, the technique based on the temporal-difference methods [1, 12], which are often closely related to the dynamic programming techniques [14], is used. Unlike the single-step prediction or the supervised learning method which assigns credit according to the difference between the predicted and actual output, the temporal-difference methods assign credit according to the difference between temporally successive predictions. See [8] for more details on this multi-step fuzzy predictor.

4. Conclusion

A general connectionist model of a fuzzy logic control system called NN-FLCS was proposed. To incorporate the NN-FLCS with on-line learning ability, an on-line structure/parameter learning algorithm was proposed. This on-line learning algorithm utilized the fuzzy similarity measure and the back propagation to provide a novel scheme to combine the structure learning and the parameter learning such that the whole network structure with correct parameters can be set up on-line. Then two NN-FLCs are closely integrated into a Reinforcement Neural-Network-Based Fuzzy Logic Control System (RNN-FLCS) for solving various reinforcement learning problems. Furthermore, by combining the techniques of temporal difference, stochastic exploration, and the proposed on-line supervised structure/parameter learning algorithm, a reinforcement structure/parameter learning algorithm was derived for the RNN-FLCS. Using the proposed reinforcement learning algorithm, a fuzzy logic controller to control a plant and a fuzzy predictor to model the plant can be set up dynamically through simultaneous structure/parameter learning for various classes of reinforcement learning problems. The proposed RNN-FLCS makes the design of fuzzy logic controllers more practical for real-world applications since it greatly lessens the quality and quantity requirements of the feedback training signals.

References

1. Barto, A. G., Sutton, R. S. and Anderson, C. W. (1983) "Neuronlike adaptive elements that can solve difficult learning control problems," *IEEE Trans. Syst. Man Cybern.*, Vol. 13, No. 5, 834-847.
2. Franklin, J. A. (1989) "Input space representation for reinforcement learning control," *Proc. of 1989 IEEE Int'l Conf. Intelligent Machine*, 115-122.
3. Hinton, G. E., McClelland, J. L. and Rumelhart, D. E. (1986) "Distributed representations," in *Parallel Distributed Processing*, Vol. 1, MIT Press, Cambridge, 77-109.
4. Kosko, B. (1990) "Unsupervised learning in noise," *IEEE Trans. on Neural Networks*, Vol. 1, No. 1, 44-57.
5. Lee, C. C. (1990) "Fuzzy logic in control systems: fuzzy logic controller – part I & II," *IEEE Trans. Syst. Man Cybern.*, Vol. SMC-20, No. 2, 404-435.
6. Lin, C. T. and Lee, C. S. G. Lee (1991) "Neural-network-based fuzzy logic control and decision system," *IEEE Trans. on Computers*, Vol. C-40, No. 12, 1320-1336.
7. Lin, C. T. and Lee, C. S. G. Lee (1992) "Real-Time Supervised Structure/parameter Learning for Fuzzy Neural Network," *Proc. of 1992 IEEE Int'l Conf. on Fuzzy Systems*, San Diego, CA, 1283-1290.
8. Lin, C. T. and Lee, C. S. G. Lee (1994) "Reinforcement Structure/Parameter Learning for an Integrated Fuzzy Neural Network," *IEEE Transactions on Fuzzy Systems*, Vol. 2, No. 1, 46-63.
9. Rumelhart, D. E., Hinton, G. E. and Williams, R. J. (1986) "Learning internal representations by error propagation," in *Parallel Distributed Processing*, Vol. 1, MIT Press, Cambridge, 318-362.
10. Sugeno, M. Ed. (1985) *Industrial Applications of Fuzzy Control*, Amsterdam: North-

Holland.

11. Sugeno, M. and Nishida, M. (1985) "Fuzzy control of model car," *Fuzzy sets Syst.*, Vol. 16, 103-113.

12. Sutton, R. S. (1988) "Learning to predict by the methods of temporal difference," *Machine Learning*, Vol. 3, 9-44.

13. Tanscheit, R. and Scharf, E.M. (1988) "Experiments with the use of a rule-based self-organizing controller for robotics applications," *Fuzzy Sets Syst.*, Vol. 26, 195-214.

14. Werbos, P. J. (1990) "A menu of design for reinforcement learning over time," in *Neural Networks for Control*, W. T. Miller, III, R. S. Sutton, and P. J. Werbos, eds, Cambridge: MIT Press.

15. Zadeh, L. A. (1988) "Fuzzy logic," *IEEE Computer*, 83-93.

OPTIMIZATION OF A FUZZY ADAPTIVE NETWORK FOR CONTROL APPLICATIONS

A.O. ESOGBUE AND J.A. MURRELL
School of Industrial and Systems Engineering
Georgia Institute of Technology
Atlanta, Georgia 30332-0205

Abstract. In this paper, we describe the use of certain optimization techniques, principally dynamic programming and high level computational methods, to enhance the capabilities of a fuzzy adaptive neural network controller which we had developed for on-line control and adaption of complex nonlinear processes. Potential applications to an array of processes from diverse fields are discussed.

Keywords. Self-learning controller, optimization, fuzzy-neuro control

1. Introduction

The appeal of fuzzy control as a tool for practical and cost-effective implementation of control strategies for complex, nonlinear, imprecisely-defined processes for which standard models and controls are impractical or cannot be derived has begun to attract the attention of various researchers and practitioners including those outside the fuzzy field and profession. However, deriving fuzzy control rules is often difficult and time-consuming. Furthermore, problems of high-dimensionality are incurred in the implementation of controls for systems with multiple inputs and outputs.

More efficient and systematic methods for knowledge acquisition and fuzzy controller synthesis are needed. Of particular interest are adaptive fuzzy controllers capable of learning from process data to automatically generate a set of fuzzy control rules and improve on them over time. A systematic review of these issues and problems as well as a discussion of the future of fuzzy adaptive control is given by Esogbue and Murrell [11].

Z. Bien and K. C. Min (eds.),
Fuzzy Logic and its Applications, Information Sciences, and Intelligent Systems, 81–89.
© 1995 *Kluwer Academic Publishers.*

It is quite apparent that a considerable amount of work has been done and is still being done in this area. Each of these efforts is targeted at the eradication, the minimization, or the mitigation of the problems attendant on fuzzy logic control methods and algorithms. Some of these include: i) the fuzzification process-how to partition the universe of discourse and assign membership functions, ii) the rule combination process- how to combine rules and/or ensure their completeness, mutual exhaustiveness, consistencies, etc. as well as interpolated control decisions especially when the method of means is inapplicable, iii) analysis- how to derive a meaningful analysis of any resulting control or the performance of the control system; for example, do the usual performance criteria of classical control theory apply? How can we make some sense of our fuzzy control system when applied to novel situations or when small perturbations of the original conditions are applied? In this area, questions of stability and robustness are being raised and addressed. When the situation becomes considerably complex and thus necessitating complex sets of rules, how can we ensure efficiencies both in modeling and computational aspects of our controller?

Solutions to aspects of the foregoing catalog of problems are being proferred with varying degrees of utility. A classic example of an appealing approach which is of topical interest is the use of various hybrid algorithms. An optimal wedding of these sundry algorithms is key to their successful use in solving these problems. One way to wrestle with these issues is to consider optimization a priori in our controller designs.

2. Resume of a Fuzzy Adaptive Controller

As a leitmotif for our discussion, we consider a fuzzy adaptive neural controller network which we have proposed as an approach to mitigating some of the problems of fuzzy logic controllers. The details are given in Esogbue and Murrell [12]. Briefly,the proposed controller has a unique combination of features and capabilities. It is adaptive and has the capability of learning from process data on-line. It performs a fuzzy discretization of the process state and control variable spaces and implements fuzzy logic control rules as a fuzzy relation. The membership functions of the fuzzy discretization are adjusted online and the fuzzy relation is learned using a performance measure as feedback reinforcement, requiring little prior knowledge about the process; no training data sets nor any error signal derived from knowledge of the desired plant trajectory are needed.

The fuzzy discretization procedure employs a statistical data compression technique permitting multivariable state vector inputs. Additional plant variables can be added without a geometric increase in the complexity of the controller structure. This procedure extracts the essential information from each variable needed to form fuzzy subsets of the process state space. While it adapts both the membership functions and the control rule state-control association, the controller primarily learns the control rule associations, unlike many other methods which fix the rule relationships and adjust the membership functions. The controller is implemented with neural networks, featuring a self-organizing neural network, a reinforcement learning neural network, and an associative memory network.

3. The Controller Operation

The operation of the controller is summarized here. At each interval of a discrete time sequence, the current process state vector is input to the controller. Its membership in each of several reference fuzzy subsets of the input space is calculated in terms of its similarity to the ideal, prototype member of each fuzzy set. Initially the locations of the prototype vectors in the state space are uniformly distributed. Throughout the time sequence, an adaptive algorithm adjusts these locations to reflect the actual clustering of the state vectors into fuzzy sets. The dispersion of the corresponding membership functions are also adapted to the state vector inputs by a similar algorithm.

Once an input state has been given its fuzzy characterization in terms of the reference fuzzy sets, the appropriate control fuzzy set is selected. Initially, the selection is arbitrary, but a learning algorithm based on the reinforcement of a performance measure is used to increase the frequency with which good controls are selected. In the process, the controller learns a fuzzy relation between the input state vector and output control vector which embodies the fuzzy control rules. After the learning phase, the fuzzy relation is used to calculate fuzzy control in terms of the reference fuzzy sets of the control space. From this, a crisp control vector is computed.

The controller has five subsystems: the Statistical Fuzzy Discretization Network (SFDN), Fuzzy Correlation Network (FCN), Stochastic Learning Correlation Network (SLCN), Control Activation Network (CAN), and the Performance Evaluation System (PES) . A block diagram of these and the plant is shown in Fig. 1.

4. Optimization of Networks

To illustrate how fuzzy logic networks can be optimized at the design phase, we outline the injection of optimization seeking methods in some of the various sub-systems. An example of its injection at the operation phase may be gleaned from the work of Smith and Takagi [21] dealing with optimal dynamic switching of reasoning methods. Such a perspective may be bene-

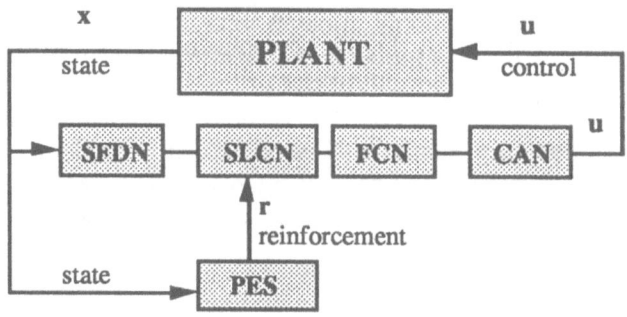

Figure 1. Controller Subsystems and Plant.

ficial to the problem of efficient switching from the learning to the control phases of our controller.

One of the ways to optimize the performance of the controller is to employ other methods of performing fuzzy partitioning of the state space in the SFDN . For example, one could use some techniques from the field of qualitative simulation as employed by Co and Narasimha [6]. Specifically, depending on the required model tolerance, the method based on the concept of landmark region will give a minimal partitioning of the universe of discourse.

The SFDN provides a means of aggregating similar plant states, thus permitting implementation of the control as a discrete relation. The adaptation update equations are of the simple delta-rule type. This is of the form

$$\vec{m}_i(k+1) = \begin{cases} \vec{m}_i(k) + \alpha(k)\,(\vec{x}(k) - \vec{m}_i(k)) & \text{for } i \in \eta(i^\star, k) \\ \vec{m}_i(k) & \text{otherwise} \end{cases} \quad (1)$$

where $\vec{m}_i(k)$ is a vector representing the location at time step k of the ith node in the SFDN, $\alpha(k) \in (0,1)$ is the location adaption rate which is a decreasing function of k, $\vec{x}(k)$ is the input state vector at time k, and $\eta(i^\star, k)$ is the set of SFDN nodes within a small (decreasing function of k) neighborhood of the maximally activated SFDN node $i^\star$. The spread parameters for each of the SFDN nodes are updated in a similar manner. There are problems with this model which tends to slow the learning process considerably for on-line application. These could be ameliorated by optimal clustering schema that respond dynamically. This is particularly advantageous when considering parallel distributed computation of neural networks. Even here, it is instructive to apply the techniques of distributed dynamic programming as well as optimal routing and flow control methods used in complex network processing as discussed, for example, by Bertsekas [3].

Pursuing the foregoing direction, we explore the invocation of optimal clustering algorithms such as those based on fuzzy dynamic programming as proposed by Esogbue [10]. The algorithm has two components. First, the optimal number of clusters is determined; next, the optimal clustering problem is addressed. The computational advantage of dynamic programming such as stability and rapidity of convergence could be beneficial to this phase of the controller.

Another section of the controller where optimization seeking methods could prove profitable relates to the Controller Activation Network (CAN). Here, the the fuzzy control output of the FCN is defuzzified. The controller algorithm currently supports the center of area (COA) and simplified max defuzzification (MAX) methods. Preliminary experience with the controller on sample classic benchmark problems indicate the superiority of the COA in certain problems. This is represented as

$$\vec{u} = \frac{\sum_{j=1}^{r} b_j \cdot \vec{u}_j}{\sum_{j=1}^{r} b_j} \tag{2}$$

where u_j is the prototype location vector and b_j is the inferred membership function value for the jth control fuzzy set.

Our experience suggests some role for optimality with respect to the choice of a defuzzification method in this phase of the controller operation. This issue has now become quite focal as researchers wrestle with the problem. An instance is the work of Smith and Takagi [21] with regards to optimal dynamic switching of reasoning methods, and furthered in Smith [22] where it is shown that fuzzy optimal control is dependent on the dynamic interaction between, as well as the sequencing of, the operators, defuzzification methods, and the reasoning methods.

Other illustrative concerns about the role of optimality in adaptive learning defuzzification techniques can be found in Song and Leland [23]. We summarize their basic result in the sequel. Defuzzification is treated as a mapping and the method that optimizes the performance index function is called the defuzzification mapping. Consider the following problem:

$$\text{optimize}_{DF \in D} \left\{ C \left(\Xi(DF(\tilde{A})) \right) \right\} \tag{3}$$

where D is the space formed by all possible defuzzification mappings, C is the performance index function, $\tilde{A}$ is the fuzzy set, DF is a defuzzification mapping, and Ξ is a transformation. The following theorem was proved: Given a fuzzy set $\tilde{A}$, if $C(\cdot)$ is a continuously differentiable convex function of Ξ and Ξ is a linear function of $DF(\tilde{A})$, then $DF^\star$ is an optimal defuzzification mapping for $\tilde{A}$ if and only if

$$\left.\frac{\partial C\left(\Xi(DF(\tilde{A}))\right)}{\partial DF(\tilde{A})}\right|_{DF=DF^\star} = 0. \tag{4}$$

We are currently investigating the application of this concept to the controller redesign.

The second area of improvement involves the FCN. The FCN implements the fuzzy control rules as a fuzzy relation G (learned by the SLCN) which associates the collection of fuzzy sets $X_1, \ldots, X_r$ for input vectors $x \in X$ with the fuzzy sets $U_1, \ldots, U_s$ for the controls $u \in U$. This is accomplished with a fuzzy associative memory (FAM) or correlation network. Again, see Esogbue and Murrell for the details. Of particular interest is optimal mapping when multi-input–multi-output systems are involved. Here, we propose the use of parallel architecturing schemes to implement the multi-fuzzy state variables. One form of this technique involves multiple level fuzzy bank arrays operating in multiple stage pipeline. The other uses fully parallel multiple level fuzzy bank arrays to enhance the performance of mapping processing. Details of the use of these techniques are given by Hwang and Tai [14].

5. Some Applications

The enhancements provided by the injection of optimization driven considerations to the type of control that this controller can provide are potentially applicable in many settings. It can be used in control situations in which there are multiple sensor measurements which may be noisy or imprecise or which require sensor fusion to generate a coherent picture of the process state. It can be used for high-level decision-making or control of data processing, intelligent system reconfiguration in response to changing conditions, or to direct flow in networks. Control of highly nonlinear dynamical systems (e.g., robot arms, etc.) for which it has been difficult to apply standard control theory methods is another application area where adaptive fuzzy controllers have proven effective. It can be used for failure detection and diagnosis or in a statistical process monitoring and control mode. Wherever intelligent decision-making in real-time is required for coping with an uncertain, noisy and/or changing environment, this type of automatic controller may be useful.

In particular, we feel that this optimized network and others with optimization in place will provide considerable enhancements to an array of interesting scenarios whose model configurations are depicted in the sequel. In the field of anesthesia, for example, the fuzzy controller, with feedback, used by Meier et al. [18] to control the depth of anesthesia during surgery with isoflurane; the adaptive closed circuit controller of Vishnoi [24]; the hierarchical multistage fuzzy model for the control of the depth of anesthesia of Linkens, et al. [17]; and our fuzzy dynamic programming model for Intra-operative anesthesia administration [8] are of special appeal. Our controller could be used in any of the five subsystems in [17] where a role is prescribed for the rule-based fuzzy logic controller. In particular, it should offer immediate advantages in the first and second phases where the depth of anesthesia is inferred and the expert–rule-based controller is introduced, respectively.

Other applications include the model of fuzzy medical diagnosis and patient treatment proposed by Esogbue and Elder [7] and, in the area of flexible manufacturing, the robust adaptive scheduler for an intelligent workstation controller applied to a shop floor control system consisting of three hierarchical control levels-shop, workstation, and equipment proposed by Cho and Wysk [5]. The problem of freeway traffic control studied by Ngo and Li [19] may also benefit from this model.

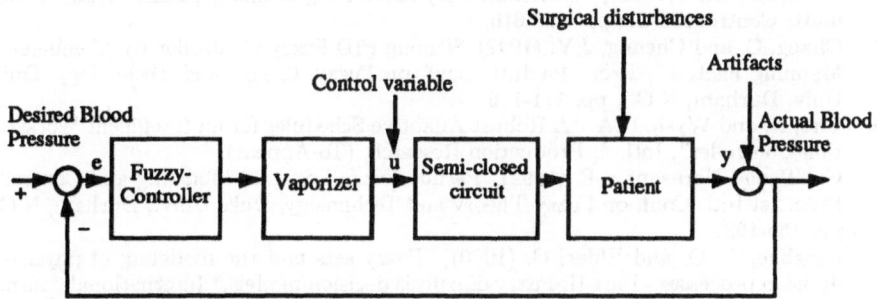

Figure 2. Block Diagram of the Control Loop for the Control of the Depth of Anesthesia (Meier, et al., 1992).

The above concerns and issues are now underway in our Laboratory. This is particularly the case as we essay to apply the controller to more complex scenarios than the inverted pendulum problem. A typical example is the power system stabilization problem [13].

6. Acknowledgment

This research is sponsored in part by the National Science Foundation under Grant ECS-9216004.

References

1. Barto, A.G., Sutton, R.S., and Anderson, C.W. (1983) "Neuron-like adaptive ele-

88

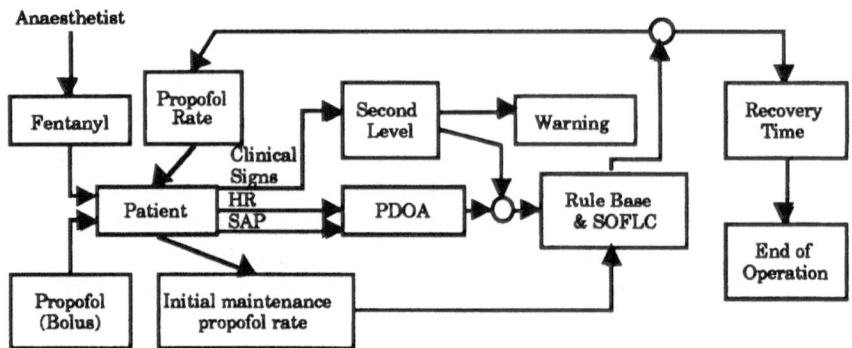

Figure 3. Block Diagram for Anesthesia Control (Linkens, et al., 1994).

ments that can solve difficult learning problems", IEEE Trans. on Systems, Man and Cybern., vol. 13, pp. 834-846.

2. Berenji, H.R. (1992) "A reinforcement learning-based architecture for fuzzy logic control," Int. J. of Approximate Reasoning, vol. 6, pp. 267-292.

3. Bertsekas, D.P. (1982), "Distributed Dynamic Programming", IEEE Trans. Automatic Control AC-27, pp. 610-616.

4. Chang, C. and Cheung, J.Y. (1992) "Tuning PID Fuzzy Controller By Membership Mapping Factors", Proc. 1st Intl. Conf. on Fuzzy Theory and Technology, Duke Univ. Durham, N.C. , pp. 171-175.

5. Cho, H. and Wysk, R.A. "A Robust Adaptive Scheduler for an Intelligent Workstation Controller", Intl. J. Production Research, (To Appear).

6. Co, T. and Narasimha, P. (1992) "Pseudo Fuzzy Logic for Modeling and Control", Proc. 1st Intl. Conf. on Fuzzy Theory and Technology, Duke Univ., Durham, N.C., pp. 156-162.

7. Esogbue, A. O. and Elder, C. (1980), "Fuzzy sets and the modeling of physician decision processes - Part II: Fuzzy diagnosis decision models," International Journal of Fuzzy Sets and Systems, vol. 3, no. 1, pp. 1-9.

8. Esogbue, A.O. (1983) "A Fuzzy Dynamic Programming Model of Intra-Operative Anesthesia Administration", in Management Decision Support Systems Using Fuzzy Sets and Possibility Theory, Verlag TUV Rheinland Gmbh, Koln, pp. 155-161.

9. Esogbue, A.O. and Bellman, R.E. (1984) "Fuzzy dynamic programming and its extensions," in Fuzzy Sets and Decision Analysis, TIMS Studies in Management Science, vol. 20, pp. 147-167.

10. Esogbue, A.O. (1986) "Optimal Clustering of Fuzzy Data Via Fuzzy Dynamic Programming." J. Fuzzy Sets and Systems, Vol. 18, pp. 283-298.

11. Esogbue, A.O. and Murrell, J.A. (1993) "Advances in Fuzzy Adaptive Control", Computers Math. Applic., vol. 27, no. 9/10, pp. 29-35.

12. Esogbue, A.O. and Murrell, J.A. (1993) "A Fuzzy Adaptive Controller Using Reinforcement Learning Neural Networks", Proc. 2nd IEEE Conf. on Fuzzy Systems, San Francisco, CA.

13. Esogbue, A.O., Song, Q. and Hearnes, W.E. (1995), "Application of a Self-Learning Fuzzy-Neuro Controller to the Power System Stabilization Problem," Proc. 1995 World Congress on Neural Networks, Washington, D.C.

14. Hwang, C.J. and Tai, Y.P. (1992) "Parallel Architectures of BIOFAM System", Proc. 1st Intl. Conf. on Fuzzy Theory and Technology, Duke Univ., Durham, N.C., pp. 119-125.

15. Jang, J.S.R. (1992) "Fuzzy controller design without domain experts," Proc. of

IEEE Int. Conf. on Fuzzy Systems 1992, pp. 289-296.

16. Lee, C.C. (1990) "Intelligent Control Based on Fuzzy Logic and Neural Net Theory," Proc. Int. Conf. Fuzzy Logic and Neural Networks, Iizuka, Japan , pp. 759-764.

17. Linkens, D.A., Shieh, J.S., and Peacock, J.E., (1994) "SADAP: A simulator for hierarchical fuzzy control of depth of anaesthesia", Proc. 1994 First Intl. Joint Conf. of NAFIPS, IFIS, and NASA, San Antonio, TX., pp. 386-390.

18. Meier, R., et al., (1992) "Fuzzy Control of Blood Pressure During Anesthesia With Isoflurane", Proc. IEEE Intl. Conf. on Fuzzy Systems, San Diego, Calif., pp. 981-987.

19. Ngo, C.Y. and Li, V.O. (1992) "Freeway Traffic Control Using Logic Controller," Proc. 1st Intl. Conf. on Fuzzy Theory and Technology, Duke Univ., Durham, N.C., 1992, pp. 214-218.

20. Patrikar, A. and Provence, J. (1990) "A self-organizing controller for dynamic processes using neural networks," Int. Joint Conf. on Neural Networks IJCNN , vol. 3, pp. 359-364.

21. Smith, M.H. and Takagi, H. (1993) "DSFS: Towards Optimal Switching in Reasoning Methods," Proc. 5th Intl. Fuzzy Systems Association World Congress, Seoul,, Korea.

22. Smith, M.H. (1994) "Optimization of Fuzzy Systems by Dynamic Switching of Reasoning Methods," Ph.D. thesis, Engineering–Electrical Engineering and Computer Science, University of California–Berkeley.

23. Song, Q. and Leland, R.P. (1994) "Adaptive Learning Defuzzification Techniques and Applications," submitted to Fuzzy Sets and Systems.

24. Vishnoi, R. and Roy, R.J. (1991) "Adaptive control of closed-circuit anesthesia," IEEE Transactions on Biomedical Engineering, vol. 38, pp. 39-46.

25. Zheng, J., et al. (1992) "Dynamic Adaptive Fuzzy Controller", Proc. 1st Intl. Conf. Fuzzy Theory and Technology, Duke Univ., Durham, N.C., pp. 164-167.

LEXICOGRAPHIC TUNING OF A FUZZY CONTROLLER

T. H. WHALEN AND B. SCHOTT
Decision Sciences Department
Georgia State University
Atlanta, Georgia
USA 30303-3083

Abstract

A fuzzy control system typically requires "tuning," or adjustment of the parameters defining its linguistic variables. Automating this process amounts to applying a second "metacontrol" layer to drive the controller and plant to desired performance levels. Current methods of automated tuning rely on a single crisp numeric functional to evaluate control system performance. A generalization of Box's complex algorithm allows more realistic tuning based on lexicographic aggregation of multiple ordinal scales of performance, such as effectiveness and efficiency. The method is presented and illustrated using a simple inverted pendulum control system.

1. Control and Metacontrol

Figure 1 presents the basic idea of a control system. The controller compares the output of some physical process (the "plant") against a desired value (the "control objective" or "set point") and applies a control signal to the plant. The goal is to drive its output toward the control objective. In general the plant's behavior is also subject to an uncontrolled disturbance. The disturbance may be an initial difference between the plant output and the control objective, it may consist of later fluctuations that tend to push the plant's output away from the control objective, or both. (More detailed treatments of control

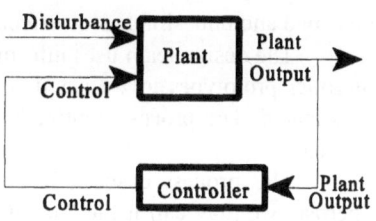

Figure 1:Basic Control System

dynamics make a distinction between the internal state of a plant and its external output. The present discussion suppresses this distinction without undue loss of generality since observable states can be mapped into the output vector while unobservable states can be approximated as disturbances.)

91

Z. Bien and K. C. Min (eds.),
Fuzzy Logic and its Applications, Information Sciences, and Intelligent Systems, 91–99.
© 1995 *Kluwer Academic Publishers*.

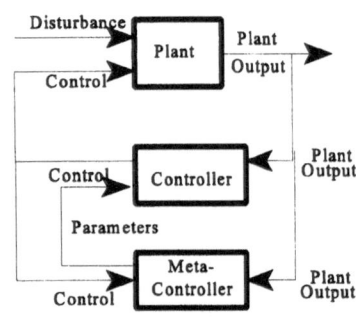

Figure 2: MetaControl System

Figure 2 adds another element to the picture. The "metacontroller" observes the inputs and outputs of the controller. These inputs and outputs are accumulated and compared against metacontrol objectives; on the basis of this comparison the metacontroller adjusts the parameters of the controller to bring the behavior of the system as a whole closer to the metacontrol objective.

Metacontrol is almost always an important part of the design and implementation of a control system. The process begins with choosing the basic design of the controller; examples include traditional Proportional Integral Differential (PID) controllers, neural nets, and fuzzy logic control systems. In any case, the controller design will typically include several parameters whose values collectively specify a particular member of a general family of controllers. These parameters define a mathematical space that must be searched to find a satisfactory or optimal controller for the system in question.

The "metacontroller" in the early stages of implementation is generally the members of the design team themselves, aided by general purpose hardware and software. The design team specifies various prototype versions of the controller, each of which defines a point in parameter space. They then test each prototype against a real or simulated plant. The evaluation criteria for judging prototype controllers can be conceptualized in terms of two broad categories. The first category, effectiveness, concerns how well the controlled system approximated the control objective. The second category, efficiency, concerns how well the controller itself performed in terms of resources consumed and undesirable side effects produced.

The design team uses information about the effectiveness and efficiency of the controller prototypes investigated so far to pick new points in parameter space to be investigated. The process continues iteratively until the system is good enough to be released.

The process described above relies very heavily on human judgment and expertise; we may call it the "manual metacontrol" paradigm. There are several good reasons to try to automate parts of the metacontrol process. One strong impetus comes from the area of adaptive control, in which the metacontroller is integrated with the controller. An adaptive system continually evaluates the effectiveness and efficiency of the control process and continually or intermittently modifies the values of the control parameters to improve them. It has become customary to refer to this updating of control parameters as "learning" by analogy with the way a human operator improves his or her control of a system with increasing experience.

Another reason to automate the metacontrol process is to be able to deliver more control systems to the market in less time. Advances in hardware and software have opened up many opportunities to market "smart" devices of all sorts, as soon as control systems exist to implement them. In this context, automated metacontrol systems become an important component of computer aided design, greatly increasing the productivity of control engineers.

A final reason for automated metacontrol is documentation. A controller that arises from a standardized search algorithm with a well defined stopping criterion may be easier to recognize as "good enough" than one which is simply the best one seen so far in an undocumented process of trial and error search.

2. Fuzzy Logic Control

Fuzzy controllers have received considerable attention, both practical and theoretical, because domain experts without special training in control engineering can qualitatively frame fuzzy rules for narrowly defined systems. [Sugeno and Yasukawa, 1991] Although the general structure of such rules can be acquired rather directly because of their linguistic flavor, tuning or calibration of the fuzzy variables can nevertheless be very challenging.

We consider standard fuzzy controllers which encode their knowledge as rules comprised of combinations of subrules. A typical subrule i has the form "If the value of x_i is X_i and the value of y_i is Y_i, then the value of z_i should be Z_i." Lower case letters x and y signify the names of antecedent variables such as position and velocity while X_i and Y_i are fuzzy linguistic values describing these variables. Similarly, z and Z_i are a consequent variable and its fuzzy value.

The rule contains subrules $i=1,...,i$ which are fused into the overall rule by the fuzzy operator minimum or maximum, depending on the multivalued logic employed in the system. The term set for the fuzzy values X, Y, and Z commonly includes: LARGE NEGATIVE; NEGATIVE; SMALL NEGATIVE; ZERO; SMALL POSITIVE; POSITIVE; and LARGE POSITIVE. A typical subrule is "If the error angle is SMALL NEGATIVE and the angular velocity is SMALL NEGATIVE, then the force of the push should be SMALL POSITIVE."

In operation, the fuzzy controller observes the actual data values for the antecedent variables x and y, denoted x and y. In a practical fuzzy control system, the actual values are observed in the form of crisp numeric singletons. Also the operational controller defuzzifies the rule's detached consequent value z into a crisp numeric singleton that is the actual control signal to the plant.

Common performance variables for mobile systems are safety, fuel economy, smoothness of ride, and speed of recovery. Performance factors of the controller itself include speed, robustness, memory needs, physical dimensions, and cost. We are concerned in this study with the effect of tuning decisions upon two performance factors: the length of time the pole remains balanced (effectiveness), and the smoothness of the control (efficiency). We attempt to optimize system performance in relation to these criteria, seeking effectiveness first, and then efficiency. The methodology employed does not assume that the controllable factors and the performance variable have continuous numeric values.

3. Metacontrol for Fuzzy Logic Controllers

Researchers have proposed and tried many approaches to searching parameter spaces of controllers in general and fuzzy logic controllers in particular. Nearly all the automated approaches involve defining a single numeric objective functional[1] or "figure of merit" to express both the effectiveness and the efficiency of the control process. Given such an objective functional, various workers have optimized it analytically [Kirk, 1970], using Response Surface Methodology [Schott & Whalen, 1992], using neural networks [Hayashi et al, 1992; Kosko, 1992; Berenji, 1992; Keller & Tahani, 1992], as well as other approaches. However, the use of a single numeric objective functional seems to work directly contrary to the principal advantages of fuzzy control systems over their non-fuzzy counterparts.

In a fuzzy control system, the control objectives themselves may be ordinal; they need not be restricted to an interval or ratio scale. Fuzzy control objectives can reflect and exploit the fact that many real situations are more or less tolerant of imprecision. For example, a fuzzy controller may strive to maximize a global assessment of "comfort" in a transportation system while a nonfuzzy system can only optimize some mathematical function combining acceleration, vibration, and noise. As a result, it seems questionable to tune a fuzzy controller using optimization procedures that attempt to estimate first and second derivatives of the degree to which the system meets its fuzzy control objectives.

Ordinal scales also facilitate lexicographic and other nonlinear approaches for dealing with multiple objectives. The use of a single numeric objective functional to capture all aspects of effectiveness and efficiency trade-offs can be problematic even in nonfuzzy control environments. And it is doubly questionable to represent the trade-offs between fuzzy effectiveness and fuzzy efficiency with a single crisp functional.

A classic algorithm, coincidently published in the same year as Zadeh's original article on fuzzy sets, provides a solution to both these problems. The algorithm is due to M.J. Box [1965; Himmelblau 1972 p.177-178]; it is called the "complex" algorithm not because it is especially intricate but because it involves a set of points in parameter space consisting of more than the minimum number of points necessary to span the space. Box called such a set of points a "complex" to distinguish it from a simplex which contains only the minimum number of points, which is one more than the number of dimensions.

The great advantage of Box's complex algorithm for tuning a fuzzy controller is that it uses only ordinal evaluations of the quality of points in parameter space. In fact, the complex algorithm can seek an optimum even when the quality of the points is incompletely ordered. (The algorithm also has some faults; when the objective is single and differentiable, derivative based approaches require fewer evaluations of the objective

---

[1]

The term "functional" is used rather than "function" to emphasize the fact that the argument of the objective functional is itself a function of time, which tracks system performance throughout the test period.

function [Himmelblau, 1972]. Also, the algorithm can fail when the minimum lies along a sharply curving valley.)

4. Tuning a Fuzzy Controller By Box's Complex Algorithm

The following discussion presents a generalization of Box's original algorithm as applied to tuning a fuzzy control system with some free parameters. The approach generalizes that of Box by explicitly considering the possibility that the quality of points might not be completely ordered. It is possible that two control systems may perform equally well within the limits of our ability to judge them. It is also possible that two control systems may be clearly different in their performance, but we are still unwilling or unable to say one is better and the other worse. This can happen when one is clearly more effective, but the other satisfies minimal effectiveness requirements and is much more efficient.

STEP 1:

To begin tuning a fuzzy controller using the complex algorithm, select an initial complex of points in the mathematical space defined by the parameters of the control system to be implemented. (Box suggests that the number of points be three times the number of parameters.) Each point defines a control system; run each control system with a real or simulated plant for a standard trial period. It is important that the points span the space of parameters; this can be accomplished either by using a design matrix as in [Schott & Whalen, 1992] or by random perturbation from an initial value as suggested by Box.

STEP 2:

Rank the points from best to worst with respect to the performance of the corresponding controllers. (Ties and incomparables are allowed.) For example, an engineer might rank the performance of control systems for a vehicle in terms of safety while a human factors expert ranked them in terms of comfort and an accountant ranked them in terms of cost. The final ranking might depend on safety, with ties on safety broken by trading off comfort and cost.

STEP 3:

Select the worst point in the current set. If there is no unique worst point, randomly select a point from among those that are not ranked better than any other point. Construct a line in parameter space from the worst point to the centroid of the other points. Multiply the distance from the worst point to the centroid by an "overexpansion factor" (Box suggests 1.3), and extend the line beyond the centroid by a distance equal to the result. This defines the new candidate point in parameter space.

STEP 4:

Run the corresponding control system with the real or simulated plant for a standard trial period, and compare its performance with the worst point in the complex. If the new point is better than the old worst point, replace the latter with the new point and return to step 2. If the new point is worse than the old worst point, if the two points are tied, or if the two points are not comparable, then create a new candidate point half way between the old candidate point and the centroid. Make this the new candidate point and repeat Step 4.

Continue these steps until the points in the complex are all within a predetermined radius of one another or until some other criterion is met. If the algorithm seems to be stuck in Step 4, check the performance of the control system defined by the centroid of the complex. If the centroid control system performs worse than any of the points in the complex, Box's algorithm cannot proceed. To get further improvement in this case, re-start the algorithm with a new complex in the vicinity of the best points seen so far.

5. Example: Tuning an Inverted Pendulum Controller

Control of an inverted pendulum has become a common benchmark problem among fuzzy researchers such as Geva and Sitte (1993). A cart on a straight track is pushed with varying degrees of force according to the controller's instructions. A sensor detects the angle Θ in radians that the pole makes with the vertical. The angular velocity of the pole angle, Θ', is computed approximately based on the change in Θ. Another sensor measures the cart's position δ relative to its starting position. A pushing force Γ is applied to the cart. Θ, Θ', δ, and Γ can take on positive and negative values depending on leftward or rightward orientation.

The fuzzy controller uses eleven sub-rules containing Θ and Θ' as antecedent variables, and with Γ as the consequent variable. Five terms were defined for each variable: NEGATVE; SMALL NEGATIVE; ZERO; SMALL POSITIVE; and POSITIVE. All fuzzy (linguistic) variables were represented as symmetrical trapezoids. The scales of all the trapezoids on each universe of discourse were uniform relative to one another, but the scales on different universes were independent.

The controller was tuned (metacontrolled) by calibrating the scale[2] of the axes of the three universes: Θ, Θ', and Γ. Two criteria were used for optimization. The most important goal of the system, its "effectiveness," was simply to balance the pole. The system was run for a simulated period of 5 seconds, divided into 250 "ticks" of the simulation clock. If the pole angle Θ passed out of the controllable range during this time,

---

[2] Each continuous variable's axis was discretized at 17 equidistant values. The "scale" value is the distance between adjacent points, with the median (eighth) point always anchored at zero.

the number of ticks remaining in the test period was reported. If the pole was still standing at the end of the simulation, this figure was equal to zero. Effectiveness scores are presented in Figures 1 and 2 in the column headed "time left;" the smaller this number, the more effective the control.

A second goal, "efficiency," was to achieve a smooth, steady balancing of the pole rather than a jittery or runaway one. (Cart-pole systems are subject to a "runaway" condition, in which the pole remains balanced at an angle while the cart accelerates continuously until it runs into the end of the track.) The second goal was represented by the product of two quantities: the integral of the absolute angle $|\Theta|$ and the integral of the absolute cart position $|\delta|$. This number is very large under runaway conditions and moderately large for an inefficient, jittery control system. Efficiency scores are presented in Tables 1 and 2 in the column headed "instability;" the smaller this number, the more efficient the control.

The two goals were combined lexicographically. In other words, any control system that balanced the pole for a longer period ranked better than any control system that balanced the pole for a shorter period, regardless of their relative efficiency ratings. If the control systems both balanced the pole for the entire experimental period, or if they both balanced the pole for equal periods of time before losing control, then the more efficient control system ranked higher. (The actual scores used were time remaining when control was lost and a measure of inefficiency, so optimization was by minimization.)

Following Box's suggestion, we used nine original sample points to tune the three parameters of the system. The triads (Θ, Θ', and Γ scales) for each of the 9 original sample points in Table 1 were set judgmentally to include a broad range of reasonable designs. Each point specifies the scale values of the 3 variables: pole angle scale in radians, pole angular velocity scale in radians per second, and pushing force scale in newtons.

The initial nine points consisted of the controller for which all three scales equalled 2.0 plus the eight controllers formed by all combinations of angle (Θ) scale = 0.03 or 1.0 radians; angular velocity (Θ') scale = 0.02 or 2.0 radians per second; and push (Γ) scale = 1 or 10 newtons.

The smaller the scale for Θ and Θ', the more sensitive the controller is with respect to changes in angular position and velocity. The larger the scale for Γ, the stronger the output of the controller. At each stage of tuning, the 9 points are presented in sorted order, so the ninth row is always the worst of the nine points currently under consideration.

Every experiment was run with a starting angle $\Theta = 0.05$, and all other transient variables set to 0. Time was incremented every 0.02 seconds, cart mass was 1.0 Kg, pole mass was 0.1 Kg, pole length was 0.5 m, and acceleration due to gravity was 9.8 m/s². The simulation was based on differential equations provided by Hamid Berenji [Berenji, 1992]. The simulation assumed a frictionless plant and certain other simplifications, and is not intended to precisely represent a real inverted pendulum. Box's complex algorithm was implemented as a Lotus 123 spreadsheet.

Table 1 shows the original set of nine points in parameter space along with the two criterion variables for each one. At the bottom of the table the coordinates of the

98

centroid point and the vector from the worst point to the centroid appear. In the box at the top of the table appear the coordinates of the next candidate point. The box also contains the locations where the user will enter the values of the criterion variables from the simulation using the controller defined by the candidate point.

Table 2 shows the spreadsheet after approximately 65 iterations. Note that all nine points are much closer together than in the original spreadsheet. The first of the nine points would define the system to be implemented and marketed if this were an actual design project.

6. Summary

Box's complex (tuning) method has been shown to produce an effective and efficient fuzzy controller when applied to the measurement scale of the controller's inputs and outputs. Controller tuning is sensitive to starting conditions and other system states. The ability to build robustness across states into such a controller is being actively studied.

References

Berenji, H. R.: A reinforcement learning-based architecture for fuzzy logic control, *International Journal of Approximate Reasoning* 6 2 (1992), 267-292.

Box, M.J.: A New Method of Constrained Optimization and a Comparison with Other Methods, *Computer Journal*. 8 (1965), 42-52.

Geva, S. and Sitte, J.: A cartpole experiment benchmark for trainable controllers, *IEEE Control Systems*, 13 5 (1993), 40-51.

Hayashi, I., Nomura, H., Yamasaki, H. & Wakami, N.: Construction of fuzzy inferences rules by neural network driven fuzzy reasoning and neural network driven fuzzy reasoning with learning functions, *International Journal of Approximate Reasoning* 6 2 (1992), 241-266.

Himmelblau, D.: *Applied Nonlinear Programming*, McGraw-Hill, New York, 1972.

Keller, J. M. & Tahani, H.: Implementation of conjunctive and disjunctive fuzzy logic rules with neural networks, *International Journal of Approximate Reasoning* 6 2 (1992), 221-240.

Kirk, D.: *Optimal Control Theory: An Introduction*, Prentice-Hall, Englewood Cliffs, 1970.

Kosko, B:. *Neural Networks and Fuzzy Systems*, Prentice-Hall, Englewood Cliffs, 1992.

Schott, B. and Whalen, T.: Tuning a fuzzy controller using quadratic response surfaces, *Proceedings of the North American Fuzzy Information Processing Society Conference* 1, (1992) 330-339.

Sugeno, M. & Yasukawa, T.: Linguistic modeling based on numerical data, *Proceedings of the 4th International Fuzzy Systems Association Congress*, (1991) 264-267.

Table 1: Initial Spreadsheet Tableau

	INPUT TO SIMULATOR			OUTPUT FROM SIMULATOR		
	0.582	0.954	3.606			Better
	newΘ	newΘ'	newΓ	newT	newI	
	Θ	Θ'	Γ	time left	instability	
1	0.03	0.2	1	0	83.2758	
2	0.03	0.2	10	0	100.012	
3	0.03	2	1	30	569	
4	1	0.2	10	154	123000	
5	1	0.2	1	213	2570000	
6	0.03	2	10	219	723000	
7	1	2	1	223	792000	
8	2	2	2	223	4600000	
9	1	2	10	223	5090000	
	badΘ	badΘ'	badΓ	badT	badI	
	0.64	1.1	4.5	centroid		
	-0.36	-0.9	-5.5	vector		0.1625 coefficient

Table 2: Final Spreadsheet Tableau

	INPUT TO SIMULATOR			OUTPUT FROM SIMULATOR		
	0.028	0.485	6.601			Better
	newΘ	newΘ'	newΓ	newT	newI	
	Θ	Θ'	Γ	time left	instability	
1	0.023	0.347	7.243	0	12.2303	
2	0.027	0.435	6.034	0	22.6592	
3	0.031	0.62	6.522	0	31.7899	
4	0.031	0.515	4.781	0	35.0634	
5	0.027	0.444	6.598	0	39.3420	
6	0.032	0.59	7.159	0	39.9405	
7	0.028	0.578	9.925	0	46.3209	
8	0.022	0.333	4.546	0	46.7025	
9	0.021	0.116	6.561	0	49.9000	
	badΘ	badΘ'	badΓ	badT	badI	
	0.027	0.485	6.601	centroid		
	0.007	0.369	0.040	vector		0.0013 coefficient

ON DESIGN OF THE SELF-ORGANIZING FUZZY LOGIC SYSTEM BASED ON GENETIC ALGORITHM

*Yong Ho Kim, **Hyun Chan Cho, *Young Keel Choi and *Hong Tae Jeon
*Department of Electronic Engineering Chung-Ang University
Seoul 156-756, KOREA Telephone:82-2-820-5297
**Department of Electronic Engineering KITE, KOREA

Abstract

This paper proposes the Self-organizing Fuzzy logic system based on genetic algorithm(GA-FLS). It is able to generate their optimal rules on owing to genetic algorithm. GA-FLS has many desirable advantages ; it is not to need precise information on the structure of the system and has adaptivity on changes in dynamic characteristics. In this paper, the method which fuzzy logic control rules are self-organized using genetic algorithm will be discussed and the effectiveness of the proposed method will be verified by computer simulation of the 2 d.o.f. planar robot manipulator.

1. Introduction

Recently, the scope of application of fuzzy logic has been widely extended to many fields because of its effectiveness in the uncertain information processing. Especially, in designing the control system, fuzzy logic rule-based controller draws lots of attention from the following desirable aspects; (1) it needs not the mathematical modelling of controlled plant, and (2) the fuzzy logic can make the linguistic control and parallel process computing possible.

However, there exist some difficulties in constucting the FLC. One of them is to construct the appropriate fuzzy rule base, which is usually acquried from the knowledge of experts or the experimental results. Futhermore if the controlled plant has the highly nonlinear dynamic characteristics, its implementation becomes more difficult and, even though rule base had been made, it could not be judged whether the rule base is proper or not. Also, it is tedious to verify the effectiveness of the rule base. Thus, recently many researches are going on to leap over these difficulties. Most of the researchs are classified into; 1) fusion with neural networking and 2) self-tuning using

101

Z. Bien and K. C. Min (eds.),
Fuzzy Logic and its Applications, Information Sciences, and Intelligent Systems, 101–110.
© 1995 *Kluwer Academic Publishers.*

look-up table.

This paper presents a novel method that can self-organize the effective and fine rules for a controlled plant. The proposed approach is based on the genetic algorithm of "natural selection and evolution". The genetic algorithm can find the optimal values of scaling factors, which transform the input variables into that of in universe of discourse of fuzzy variables. Also the genetic algorithm can determine the effective membership function of the fuzzy variable.

The organization of this paper is as follows. In section 2, the conventional genetic algorithm will be explained and the method about the self-organization of fuzzy control rules is proposed using the genetic algorithm in section 3. Then computer simulation results are shown using a 2 d.o.f. planar robot in section 4. Finally the conclustions are deferred to section 5.

2. Genetic Algorithm

The genetic algorithm proposed by J.H. Holland[4][5][6] comes from "Survival of the fittest" in the nature. The algorithm is a sort of simulated evolution search algorithm to find the optimal solution of the undefined function, $Y=G(X)$. And has three major operators; reporduction, crossover and mutation. The flow of genetic algorithm can be simply explained as follows;

[STEP 1] Determine the string length to represent the variable X in the form of binary code 1 and 0, and establish an initial string.

[STEP 2] Construct an initial population using the initial string in STEP 1.

[STEP 3] Transform each string of population into decimal code and then find the fitness value of function, $G(x)$.

[STEP 4] Select the proper strings according to the fitness value to form a gene pool.

[STEP 5] Obtain a new population through the evolution process of cossover and mutation between the genes of STEP 4.

[STEP 6] Repeat STEP 3 ~ STEP 5 until some measure is satisfied.

Fig.1-(a) shows the string representation of decimal factor as in STEP 1 and fig.1-(b) shows the crossover of two 8-bit strings, X(parent 1) and Y(parent 2) in STEP 5. At the crossover, crossing site is usually selected by random numbers. Meanwhile mutation (fig.1-(c)) is simultaneously accomplished with crossover and is done by a simple bit-transition with a selected random bits.

parameter	string
0.750000 ⟷	01100000

Fig. 1 - (a) Representation of parameter into string

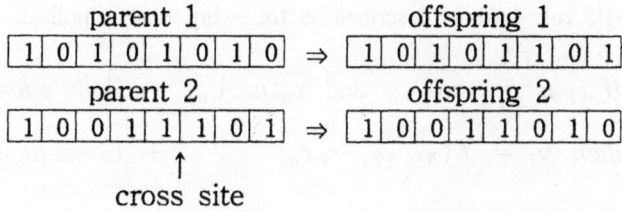

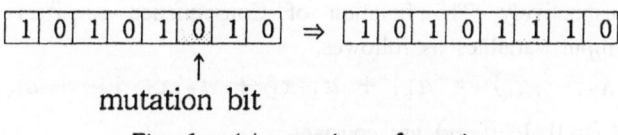

↑

cross site

Fig. 1 - (b) crossover between two parent strings

↑

mutation bit

Fig. 1 - (c) mutation of a string

Fig 1. Representation of parameter into string and its crossover

The above genetic algorithm has some distinct characteristics as follows;
1. In the genetic algorithm,the population of solution space is utilized.
2. Genetic algorithm is blind. That is, it needs not the mathematical information of optimizing functions, such as the possibility of derivative and continuity, etc.
3. Genetic algorithm is able to find the global optimum solution.

3. Self-Organization of fuzzy logic control rule.

The proposed GA-FLC is shown in Fig 2. It has the fuzzy rule of equation (1), and use the hybrid inference[Sugeno]. In Fig 2, GC and GE are the scaling factors of error and change of error, respectively.

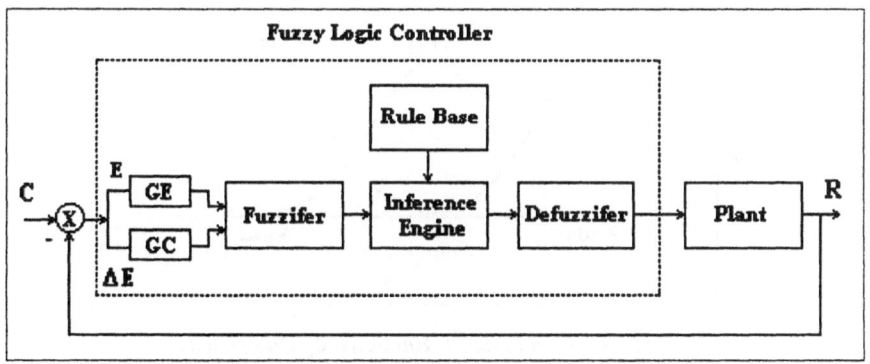

Fig 2. The proposed Fuzzy Logic Controller

Fuzzy logic rule base is described as following fuzzy implications,

$$R_j : \quad If\ x_1\ is\ A_{ij}\ and\ \cdots\ and\ x_n\ is\ A_{nj} \quad \rightarrow Antecedence$$

$$then\ y_j = f_j(x_1, x_2, \cdots, x_n) \qquad \rightarrow Consequence$$

(1)

where R_j is j-th rule, x's are input variables, A_{ij} is fuzzy variables with membership functions shown in the Fig. 3, and y_j is the output through the j-th rule, respectively. The function of Consequence consistes of the linear function of input variables as followes,

$$f_j(x_1, x_2, \cdots, x_n) = a_{0j} + a_{1j}x_1 + a_{2j}x_2 + \cdots + a_{nj}x_n \quad (2)$$

where $a_{ij}(i=0, 1, \cdots, n)$ are constant.

But it is, as above mentioned, not easy to implement the adquate rules to the controlled plant. Especially, if the controlled plant is a time-varying and nonlinear system, it becomes more difficult.

To overcome the above diffuculties and self-tune the rule-base according to the controlled plant, we have been to use the method that modify membership functions of fuzzy variables of consequence and adjust the scaling factors automatically[IFSA 93', Y.H.Kim, et. al]. In [Y.H.Kim, et. al], fine tuning of fuzzy rule base is not considered because membership functions of antecedence are fixed and singleton inference method is used. Thus, in this paper, membership functions of antecedence, constants in linear function of consequence and scaling factors are selected as main object to adjust(or optimization factors) to construct the optimal rule base fitted in the control environment. The procedure to tune each values is explained in the followings.

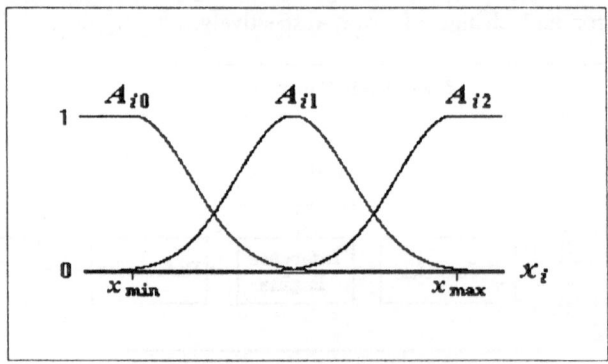

Fig. 3   Membership functions of Antecedence

3-1. Determination of string and fitness

As above mentioned, since the scaling factors (GE and GC) and membership functions affect massively on the performance of FLC, these parameters are chosen as the optimizatrion objects.

In order to apply the genetic algorithm, the scaling factors must be represented into the binary code. This can be easily accomplished by coding them as the unsigned binary.

But, it is not simple to represent the membership functions of the consequents as binary code. Generally, to characterize the bell type membership function on the universe of discourse, three points of membership function must be specified. In this paper, we use three values to describing the membership function ; right width, point of center, left width (fig.4).

Finally one string can be formulated by combining two sets of binary code. Fuzzy rule base and scaling factors can be adjusted by changing the contents of the string.

Next the fitness value must be determined to measure whether the the present fuzzy rule base and scaling factors (or the string) is proper or not. The fitness can be defined as follows.

$$Fitness = \frac{K_1}{Error} + \frac{K_2}{Change\ of\ error} + \frac{K_3}{Energy} \quad (3)$$

Here k1,k2 and k3 are constant values and energy is input to the plant.

In Eq. 3. the string of high fitness means that the corresponding rule base and scaling factors becomes more adequatre to the control purpose and yields the small error and energy.

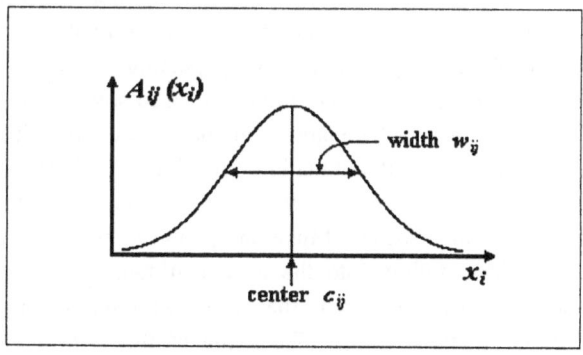

Fig. 4 Parameters of membership function.

3-2. Generation of new rule base by evolution

After determining the form of string and fitness, the next procedure to find the string of the highest fitness value is explained as follows.

At first, an initial string which implies the initial rule base and scaling factors is formulated by generating binary numbers at random. With the initial string, an initial population can be constructed from M strings which are randomly generated. Fig.5 showes an example of the initially generated string. This string represents a membership function of antecedence and a set of scaling factors, and constants of consequence.

A_{11}		~	A_{nq}			rule 1				~	rule m			
center	width	~	center	width	a_{01}	a_{11}	...	a_{n1}	~	a_{0m}	a_{1m}	...	a_{nm}
10001001	11001010		10101111	01011001	11001100 ~ 10011110					10101111 ~ 01011001			

Fig. 5 The representation of one string

After the initial population being consisted of initial strings such as the string in fig.5 , the fitness value of each rows of population are obtained by the following procedure.

[STEP 1] Find the membership function of consequent of each rule and the scaling factors by transforming each string into decimal value.

[STEP 2] Construct fuzzy logic controller using the rule base and scaling factors obtained from step 1.

[STEP 3] Find error, change of error and energy and then the fitness value

Finding the fitness values of all the strings, next population (or generation) can be constructed by using the good strings, (or more fitted strings) of the initial population. This procedure is done by giving more selection probability to the string of higher fitness value[4] and forming gene pool.

With a gene pool, next generation of population is constructed through crossover between each strings and mutation. Crossover can be done by exchanging arbitrary strings. And mutation can be occured by flipping some bits in the string. For example, if the mutation probability is 0.033 and the length of the string is 184 bits, then only 6 bits of all 184 bits are flipped(0→1 or 1 →0) for mutation of the string. Mutation plays a role of preventing the evoluation procedure from falling into the local extrimal.

As each generation are pregressed, the string with the highest fitness value increases and sets to a steady value. The string of the highest value is selected to be the rule base of the FLC, which yields to minial error, and change of error and energy to the plant.

4. Computer Simulation

4-1. Command trajectory and input

To show its effectiveness of the proposed method, two link planar robot manipulator (cf. Fig.6) is chosen for computer simulation.

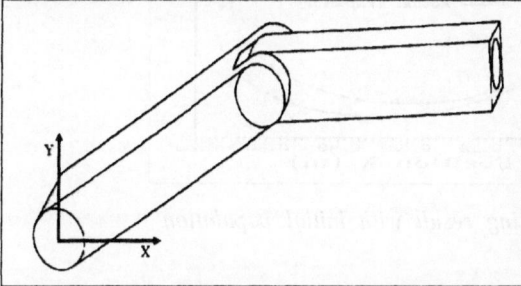

Mass of link 1 : 2.5Kg
Mass of link 2 : 2.0Kg
Length of link 1: 0.53M
Length of Link 2: 0.47M

Fig 6. two link planar manipulator

The trajectory of the robot end-effector is given as a circle in the Cartesian x-z plane as followes.

$$X_d(t) = \begin{pmatrix} P_x(t) \\ P_y(t) \\ P_z(t) \end{pmatrix} = \begin{pmatrix} 0.3 + 0.25\cos\theta(t) \\ 0.3 + 0.15\sin\theta(t) \\ 0.0 \end{pmatrix} \tag{4}$$

Also the string has a 450 bit length which consist of the 18 bit of for the scaling factors, 212 bits for the membership functions of the antecedence and 216 bits for the consequence. Mutation probability is set to 0.033 and the total number of strings in the population is 250.

4-2. Results

Computer simulation results are showed from Fig.7 to Fig.11. Fig.7 is a result of the highest fitness value in the initial population and fig.8 shows the joint error. Fig. 9 showes the trajectory which the robot travelled by the FLC obtained after 20 generations and fig 10 is its joint error. Fig.11 shows the variation of the fitness value at every generation. From this results, it can be easily concluded that FLC after 20 generations can effectively execute any kinds of trajectory without further generation.

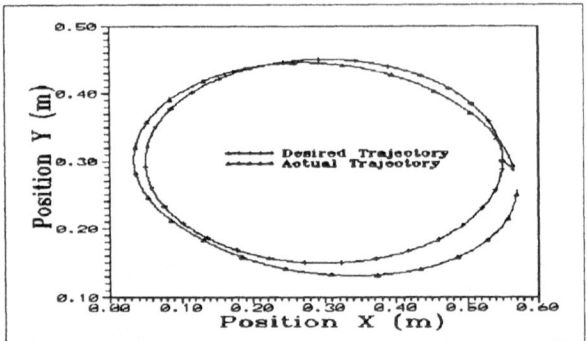

Fig. 7 *The tracking result with initial population*

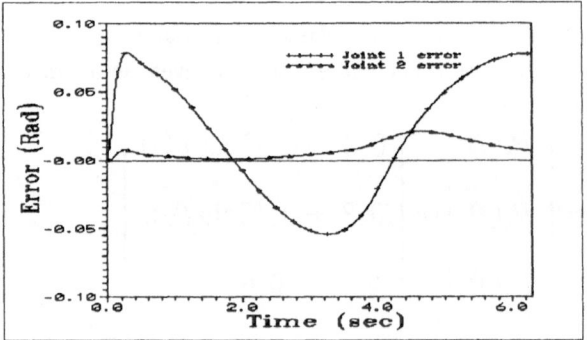

Fig. 8 *The joint error with initial population*

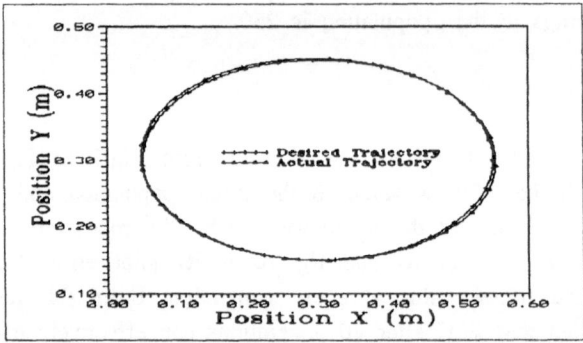

Fig. 9. *The tracking result with best string a fer evolution*

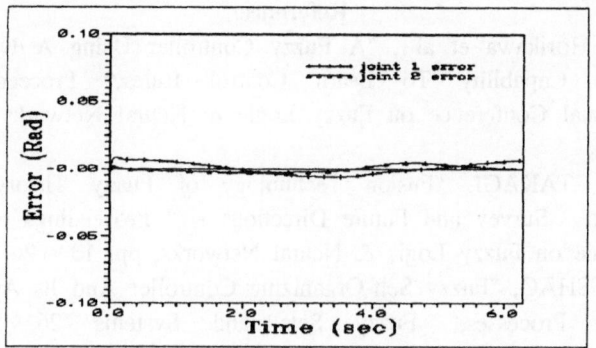

Fig. 10. The joint error with best string after evolution

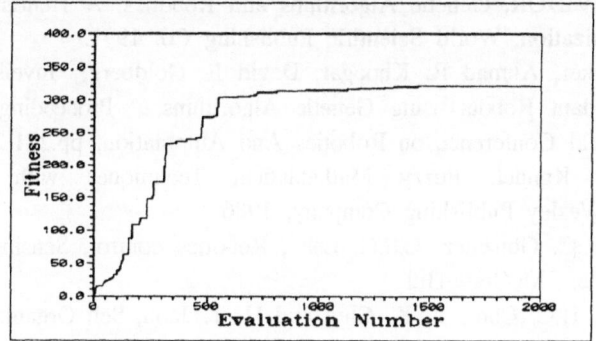

Fig. 8. Fitness variation in evolutionary process

VI. Conclusion

It is not easy to construct the efficient rule base of the FLC. Futhermore, if the controlled plant is highly nonlinear and time varying, it becomes much harder and tedious job to obtain the proper FLC.

In this paper an approach for the self-organization of fuzzy rule base has been proposed using the genetic algorithm, which is a powerful optimization search algorithm and recently attracts many attention. To demonstrate the effectiveness of the proposed method, the FLC of two-link planar is utilized. From the computer simulations, it can be concluded that the FLC using genetic algorithms can self-organize the optimal fuzzy rules and scaling factors.

Reference

[1] Shin-ichi Horikawa et al , "A Fuzzy Controller Using A Neural Network And Its Capability To Learn Control Rules," Proceedings of the International Conference on Fuzzy Logic & Neural Networks , pp.103-106, 1990

[2] Hideyuki TAKAGI, "Fusion Technology of Fuaay Theory and Neural NetWorks - Survey and Future Directions - ," Proceedings of International Conference on Fuzzy Logic & Neural Networks, pp. 13 - 26, 1990

[3] Shihuang SHAO, "Fuzzy Self-Organizing Controller And Its Application for Dynamic Processes," Fuzzy Sets and Systems 26, pp. 151-164, North-Holland , 1988

[4] David E. Goldberg , Genetic Algorithms in Search, Optimization & Machine Learning, Addison-Wesley Publishing Company, 1989.

[5] Yuval DAVIDOR, Genetic Algorithms and Robotics, A Heuristic Strateghy for Optimization, World Scientific Publishing Co. 1991.

[6] Joey K.Paker, Ahmad R. Khoogar, David E. Goldberg, "Inverse Kinematics of Redundant Robots using Genetic Algorithms, " Proceedings 1989 IEEE International Conference on Robotics And Automation, pp.271-276, 1989

[7] Abraham Kandel, Fuzzy Mathematical Techniques with Applications, Addison-Wesley Publishing Company, 1986

[8] K.S. Fu, R.C. Gonzalez, C.S.G. Lee , Robotics control, Sensing, Vision and Intelligence, McGraw-Hill

[9] Y.H. Kim, H.C. Cho , Y. K. Choi and H. T. Jeon,"Self-Organization of Fuzzy Rule Base Using Genetic Algorithm," Fifth IFSA Worls Congress,pp 881-886,1993.

FUZZY CONTROL AS SELF-ORGANIZING ADAPTIVE CONSTRAINT-ORIENTED PROBLEM SOLVING

OSAMU KATAI, MASAAKI IDA AND TETSUO SAWARAGI
Faculty of Engineering, Kyoto University

KIMINORI SHIMAMOTO
Electronics Research Labolatories, Nippon Steel Corp.

SOSUKE IWAI
Faculty of Science and Engineering, Meijo University

AND

MASAHIRO TERABE AND TADASHI HORIUCHI
Faculty of Engineering, Kyoto University

Abstract. By introducing the notion of constraint-oriented fuzzy inference, we will show that it provides us ways of fuzzy control methods that has abilities of adaptation, learning and self-organization. The basic supporting techniques behind these abilities are "hard" processing by Artificial Intelligence or traditional computational framework and "soft" processing by Neural Network, Genetic Algorithm and Reinforcement Learning techniques. In the former processing, Qualitative Reasoning and Instance Generalization by Symbolic Reasoning play important role, while by the latter processing, fuzzy control becomes capable of learning, adaptation and evolutional self-organization.

1. Introduction

We have already introduced a "constraint-oriented way of treating fuzziness" which can be applied to various kinds of problem solving including control and planning problems, etc [1], [2]. In this paper, we will first show the whole scope of our constraint-oriented approach to fuzzy information processing, particularly for the case of fuzzy control, and then show that this framework will provides us ways of fuzzy control that has abilities of

111

Z. Bien and K. C. Min (eds.),
Fuzzy Logic and its Applications, Information Sciences, and Intelligent Systems, 111–122.
© 1995 *Kluwer Academic Publishers.*

adaptation, learning and self-organization. The basic supporting techniques behind these abilities are "hard" processing by Artificial Intelligence and traditional computational framework and "soft" processing by Neural Network, Genetic Algorithm and Reinforcement Learning techniques. The reason that these techniques can be incorporated to fuzzy control systems is that the notion of constraints itself has two fundamental properties, that is, the "modularity" property due to its declarativeness and the "logicality" property due to its two-valuedness. From the former property, the modularity property, decomposing and integrating constraints can be carried out quite easily and efficiently, which enables us to do the above "soft" processing. From the latter property, the logicality property, Qualitative Reasoning and Instance Generalization by Symbolic Reasoning can be carried out, thus enabling the "hard" processing.

2. The Whole Scope of Constraint-Oriented Way of Fuzzy Control

Fig. 1 shows the way how the above mentioned abilities can be substantiated for fuzzy control. Suppose that we are given with experience of control, that is, a set of instances of recognition-action pairs with their results and evaluation. Then, instance generalization by the methods of "Qualitative Reasoning (QR)" or "Instance Generalization (IG)" can be applied to yield generalized constraint regions on recognition-action pairs. Their decomposition and further selection and refinement by Neural Network (NN) based techniques yield "betweenness rules" which are given as a general kind of constraint-oriented rules and are the basis of constraint-oriented fuzzy inference [3]. More precisely, the decomposition of the betweenness rules yield "constraint-interval fuzzy inference rules" that are also introduced by the authors. Moreover, we have another route to derive the betweenness rules, that is, by the use of pairing of the instances that is certified to satisfy certain constraint conditions through appropriate methods of Qualitative Reasoning and Instance Generalization, and then by the use of Neural Network-based refinement, we have the betweenness rules [4], [5].

These constraint-interval fuzzy inference rules together with the values of recognition (observation) variables enable us to carry out constraint-interval fuzzy inference to derive "interval constraints" on the values of the actions, i.e., the manipulating variables, whose conjunctive and/or disjunctive integration together with defuzzification yield the values of the manipulating variables through which generation of next action is done yielding another instance of recognition-action-result triplet. Moreover, for complex control problems, we can decompose the goal (objective) of control yielding constraint regions on recognition-action pairs on which we "decouple" the

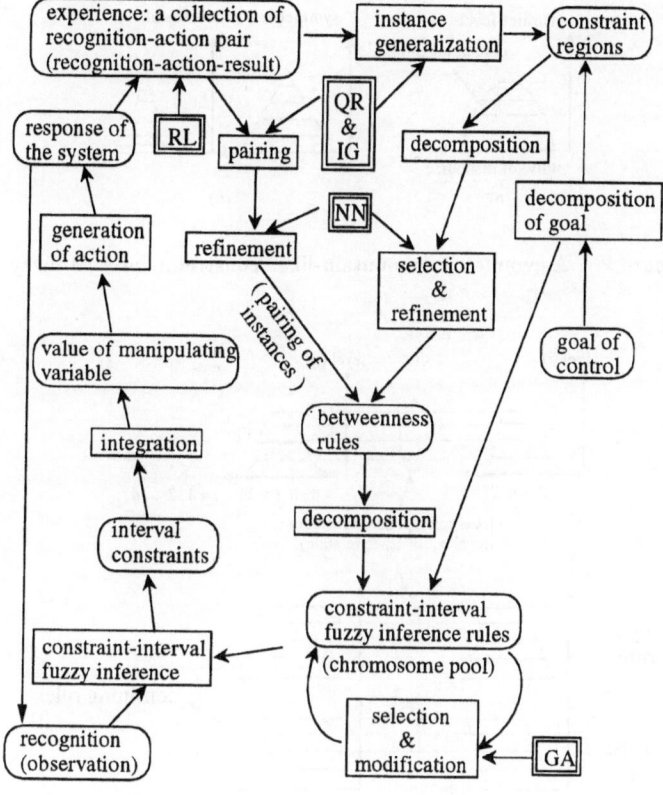

Figure 1. Whole scope of constraint-oriented fuzzy control

control problem into more simple subproblems.

Also, Genetic Algorithm (GA) techniques can be applied directly to the collection of constraint-interval rules, that is regarded to be constituting a pool of "chromosomes," i.e., a genetic pool, which are selected, modified and reproduced to refine the pool and the activated values of control variables by the recognition are used to derive the value of manipulating variables whose response from the environment is evaluated by the use of Reinforcement Learning (RL) techniques so as to be used in the GA-based selection, modification and refinement of the chromosomes [6].

3. Modularity-Logicality of Constraint-Oriented Fuzziness and Composition-Decomposition of Fuzzy Inferencce Rules

In this section, we will explain several key concepts in our framework. The first one is the notion of "constraint-interval fuzzy set." As shown in **Fig.**

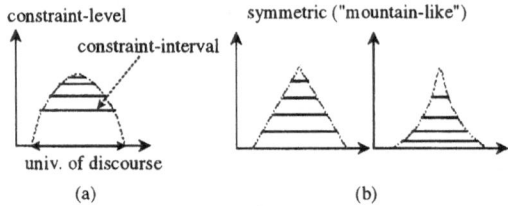

Figure 2. A symmetric "mountain-like" constraint-interval fuzzy set

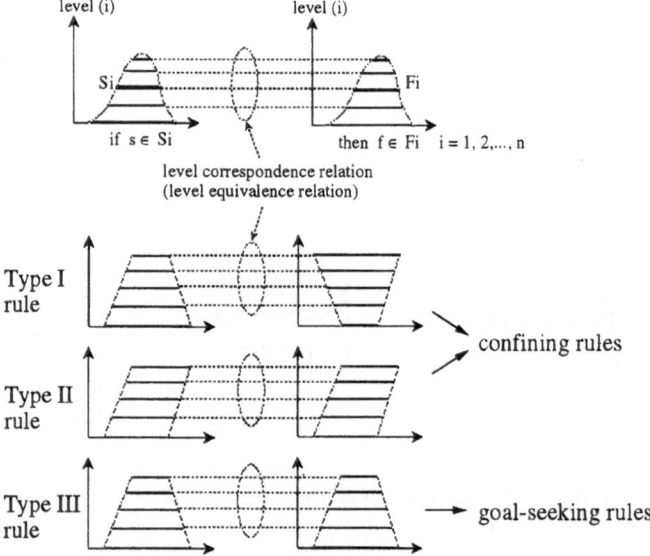

Figure 3. Constraint-interval fuzzy inference rules ans their types

2(a), it is given as an ordered collection of crisp intervals on the universe of discourse each of which represents a constraint called "constraint-interval". Namely, the grade axis (in the traditional Fuzzy Set Theory) is now regarded to be an ordinal scale axis, hence the sets in **Fig. 2(a)** are regarded to be the same as the ones in **Fig. 2(b)**.

Fuzzy inference rules whose "if (antecedent)" and "then (consequent)" parts are regarded respectively to be constraint-interval fuzzy sets are called "constraint-interval fuzzy inference rule" which says that if a variable s is in constraint-interval S_i, then a variable f should be in the constraint-interval F_i for $i = 1, 2, \cdots, n$, as shown in **Fig. 3**. In this kind of rule, we have three typical rules, that is, type I, type II and type III rules. The former two types of rules are used to confine the permissible area of

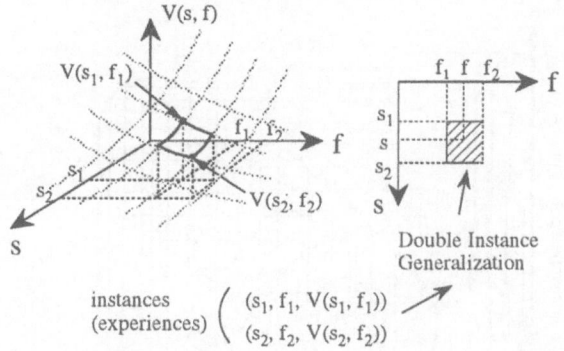

instances
(experiences)
$\left(\begin{array}{l} (s_1, f_1, V(s_1, f_1)) \\ (s_2, f_2, V(s_2, f_2)) \end{array} \right.$

Figure 4. Betweenness rule and its acquisition by Double Instance Generalization (DIG)

the values of control variables and hence are called "confining" rules. The remaining rule type, i.e. type III rule, is used to suggest the desirable area of control(manipulating variable) and hence is called "goal-seeking" rule.

More general types constraint-oriented rules are given as "betweenness rules" such as if s is between s_1 and s_2, then f should also be between f_1 and f_2, which is written as

$$bet(s : s_1, s_2) \rightarrow bet(f : f_1, f_2)$$

This kind of rule can be derived by generalizing instance information, particularly by the method of "Double Instance Generalization (DIG)" introduced by the authors [3]. For instance, if we are informed of the qualitative shape or tendency of the values $V(s, f)$ of the pair of the values of s and f as shown in **Fig. 4**, we can derive the following kinds of general rules from the triplets $(s_1, f_1, V(s_1, f_1))$ and $(s_2, f_2, V(s_2, f_2))$ such as

$$bet(s : s_1, s_2) \ \& \ bet(f : f_1, f_2) \ \& \ (s_1 < s_2 \ \& \ f_1 < f_2)$$
$$\rightarrow bet(V(s, f) : V(s_1, f_1), V(s_2, f_2))$$

Thus, for assuring the least value $V(s_1, f_1)$ of the recognition-action pair (s_1, f_1), we will arrive at the following rule of control:

$$if \ bet(s : s_1, s_2) \ and \ s_1 < s_2 \ \& \ f_1 < f_2 \ hold,$$
$$then \ set \ f \ such \ that \ bet(f : f_1, f_2) \ holds,$$

where $<$ represent the partial order on s, which is sometimes given as a product of the orders on componential variables such as

$$s < s' \equiv s_j < s'_j \ for \ j = 1, 2, \cdots, n$$

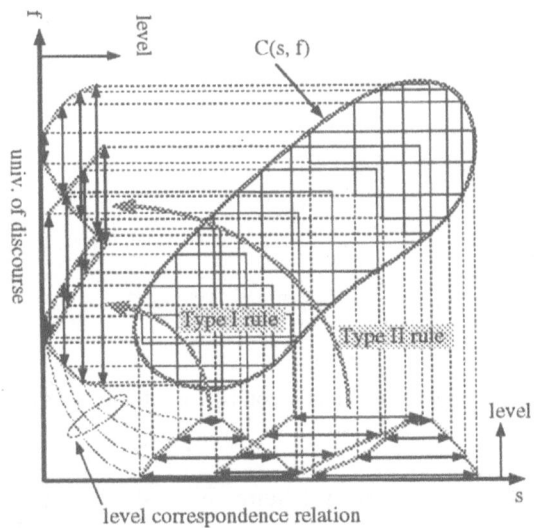

Figure 5. Approximation of a crisp constraint by a collection of cofining (Type I & Type II) rules

Also, the betweenness relation is sometimes given as

$$bet(s : s_1, s_2) \equiv s_1 < s < s_2 \ \ or \ \ s_2 < s < s_1$$

Thus the betweenness rule given above is reduced to the following constraint-interval fuzzy inference rule:

$$if \ \ s_{1_j} < s_j < s_{2_j}, \ \ j = 1, 2, \cdots, n \ \ and \ \ f_1 < f_2 \ \ hold,$$
$$then \ \ set \ f \ \ such \ that \ f_1 < f < f_2 \ \ hold,$$

Suppose that we are given with a crisp constraint region on the pair of s and f as shown in **Fig. 5**. Then, decomposition of this region into rectangular areas yields type I and type II rules as shown in the figure.

Along with the ways mentioned above, we will obtain the following procedure for deriving fuzzy control which assures the least evaluation of control result.

1. Select the instances of experience which have sufficiently good results.
2. Delete the instances which are subsumed by other instances.
3. Search for the pairing of the instances to compose constraint areas.
4. If the constraint is of multiple levels, that is "graded", we also have to organize the obtained area among different levels.

This selection can be carried out by Hopfield Networks or Boltzmann Machines with an energy function evaluating the area of coverage with preferably less number of rectangles, that is products of intervals, having better continuity among different levels of constraints [5].

4. Decomposed Fuzzy Control Scheme for Cart-Pole Systems

The modularity of constraints enables the "decomposition" and "integration" of constraints. For instance, the control of a cart-pole system can be decomposed into that of the pole and that of the cart. Each control system merely relies on respective sides of observation, i.e., the antecedent part of a fuzzy inference rule for the control of cart (pole) consistsof a constraint-interval fuzzy set on the position (angle) and that on the velocity (angle velocity) of the cart (pole). These results (consequent part) fuzzy inference are then integrated by using "AND" composition to derive the constraint interval fuzzy set on the manipulating variable, the external force to the cart, which is then defuzzified to yield the exact value of the variable.

We set two kinds of goals on the cart and the pole, i.e., confining rule and goal-seeking rule.

(Cg): goal-seeking for the cart,
(Cc): confining the cart,
(Pg): goal-seeking for the pole,
(Pc): confining the pole,

The confining rule limit the move of the cart or the pole in a certain prespecified region, hence it permits their swinging motions, while the goal-seeking rule insists on their convergent behavior to the origin of the rail or to the vertical position of the pole. Thus we will have the following four cases of the goal of control:

case (1): (Cg) & (Pg)
case (2): (Cg) & (Pc)
case (3): (Cc) & (Pg)
case (4): (Cc) & (Pc)

The result of control is shown in **Fig. 6**. In case (1), the pole and the cart respectively insist on their own goal-seeking activities, hence the compromise between them results in their swinging motions from left to right as shown here. In this case, the behavior of the system is rather stable; it is seldom that the cart runs out of the rail or the pole falls down unless the initial condition is too severely set. In case (2), the pole insists on its own goal-seeking activity and the cart accommodates or adapts itself to the pole's behavior, hence the pole is held vertically and the cart swings smoothly right and left by using the full range of the rail. In case (3), on

118

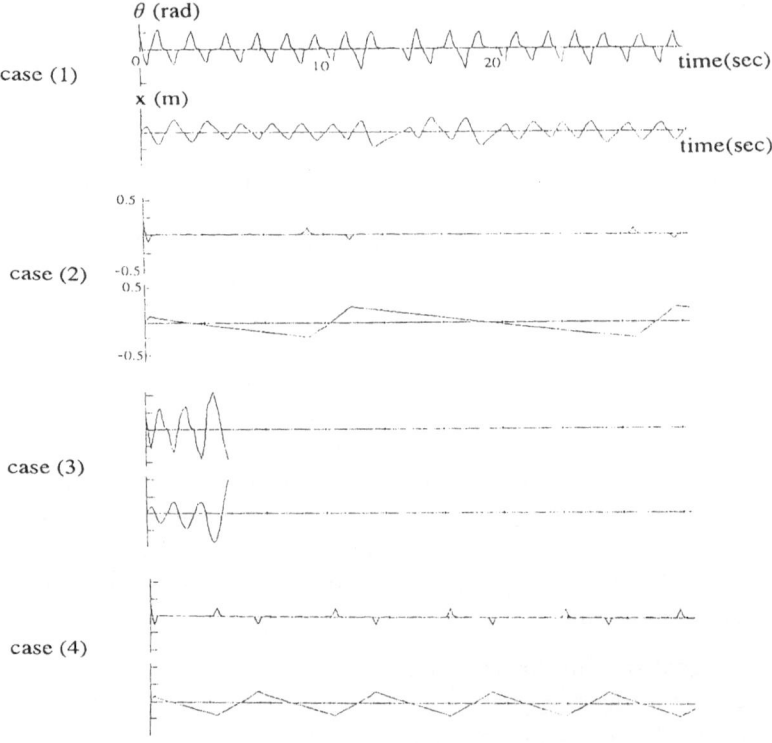

θ (rad)

case (1)

x (m)

time(sec)

time(sec)

case (2)

case (3)

case (4)

Figure 6. The behaviors of the cart-pole system by decomposed fuzzy control

the contrary, the pole accommodates itself to the activity of the cart which results in so tight a condition that it is impossible to prevent the pole from falling down. Hence both of them move rather rapidly at first but are stopped suddenly by the falling down of the pole in the early stage. In case (4), both of them accommodate themselves to each other's behavior, hence they move very smoothly. The range of the cart movement is a bit small compared to that of case (3).

5. Application of Genetic Algorithm and Reinforcement Learning Techniques for Constructing Fuzzy Control Systems

As mentioned in Section 2, the collection of constraint-interval fuzzy inference rules can be refined by the use of Genetic Algorithm due to the modularized structures of fuzzy rules. In Genetic Algorithm, trial and error experiences are used to refine the population of solution candidates which are coded into sequences of symbols called "chromosomes". The refinement of the population is done through a process that is derived from analogy to the evolutional genetic processes in creatures. These chromosomes are evaluated to calculate their degree of fitness to the environment (that is, the given problem). These chromosomes are then selected and reproduced by referring to their fitness values. A "crossover" between selected chromosomes subsequently takes place yielding chromosomes, some of which are expected to be better than the original ones. Finally, "mutation" on the chromosomes is done to yield novel chromosomes. The whole process of this evolution can be regarded as a multi-point search for the optimal solution. The crossover operation shifts the points of search in a global fashion, while the mutation operation does so in a local fashion [6].

We regard each componental crisp constraint-interval rule as a chromosome, that is, we will adopt the Michigan approach instead of the Pittsburgh approach . In Pittsburgh approach, the whole inference system itself is coded as a chromosome, thus the size of the chromosomes becomes huge compared to that in the former approach. The evolutional operations, on the contrary, becomes complex in the Michigan approach compared to the Pittsburgh approach. The main reason for adopting the former approach, the Michigan approach, is that we are searching for a method that will yield self-organizing mechanisms in the constraint-interval fuzzy inference systems. The self-organization is carried out by linking componental constraint-interval rules among different levels to yield a constraint-interval fuzzy rule.

In the Michigan approach, we have to evaluate the contribution of each fragmental componental rule making up the evolutional operations in the Michigan approach more difficult to be carried out than that of the Pittsburgh approach. Particularly, for the case of production systems for control, the evaluation on the whole control actions is usually done after a sequence of actions is applied, hence very complicated evaluation algorithms such as Bucket Brigade Algorithm are used. However, in our case, we use instantaneous evaluation of control.

We have applied this method to the cart-pole control problem where the cart position is disregarded. The crossover probability and mutation probability are set as 0.6 and 0.01, respectively, and the number of chromosomes is set to be 90. The initial condition is set randomly in the region: $-0.4 < \theta < 0.4(rad)$, $-0.4 < \dot\theta < 0.4(rad/sec)$.

We obtained various constraint-interval fuzzy inference rules ranging from type I to type III, that is, we have a large amount of diversity of chromosomes. This means that trial and error experience is still insufficient for converging to the optimal control rules. However, this diversity is also the origin of the adaptability of this system. **Fig. 7** shows the obtained value of manipulating variable (external force to the cart) versus various values θ and $\dot\theta$. It is observed that there still remains the area which has to be learned. This happens due to that the advance of learning incidentally

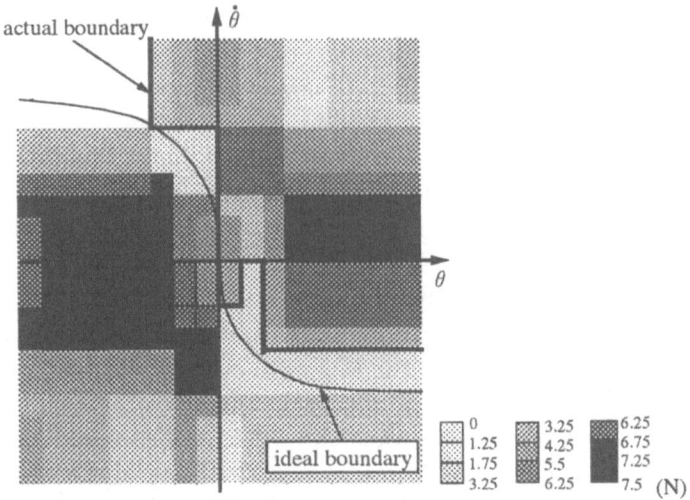

Figure 7. The obtained value of manipulating variable (external force to the cart)

hides this dangerous area from the learning system.

To confirm the adaptability of this control scheme, we also applied this scheme to the cart-pole control problem where an environment change such as an inclination of the cart track took place. In this case, the average of swing angle θ of the pole became smaller as the generation goes on after the change of environment, as shown in **Fig. 8**. This is because the system gradually altered the distribution of its divergent chromosome pool thus shifting its control action in accordance with the new environment. More precisely, as shown in **Fig. 9**, the chromosome fittest to the new environment plays the role of "pilot" thus leading the other chromosomes to move to be adapted to the new environment.

In Genetic Algorithm process, the fitness value of each chromosomes is calculated according to estimated evaluation value $V(s, f)$ of the pair of state s and action f which is estimated by incremental revision by the following way:

$$
\begin{aligned}
V(s, f) &= V(s) - C(f) - V(s(s, f)) \\
V(s) &= Min\{V(s(s, f)) + C(f)\}
\end{aligned}
$$

where $V(s)$, $C(f)$, and $s(s, f)$ stand for the value of cart-pole system being at s, the cost of using f, and the next state of s after applying f, respectively. This process of evaluation of state-action pairs is executed simultaneously with the learning process of fuzzy control rule based on the

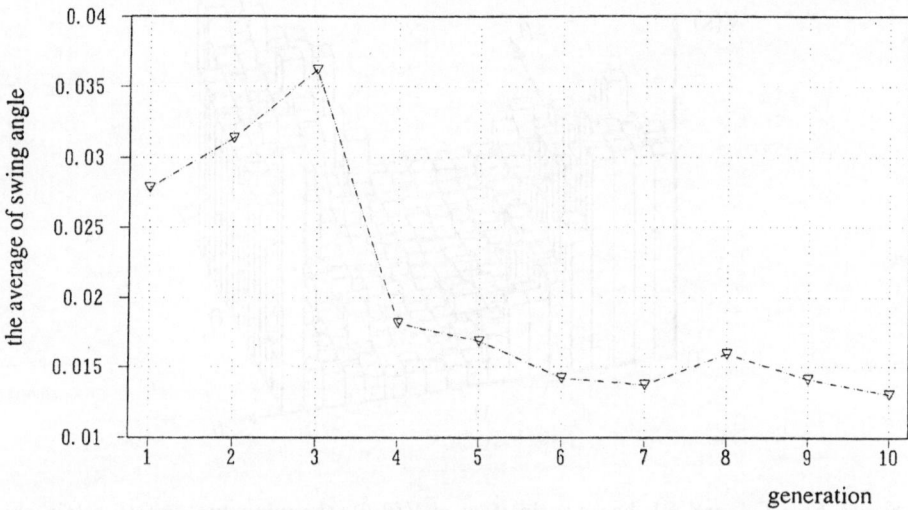

Figure 8. The adaptation of the evolutional control scheme to an environment change

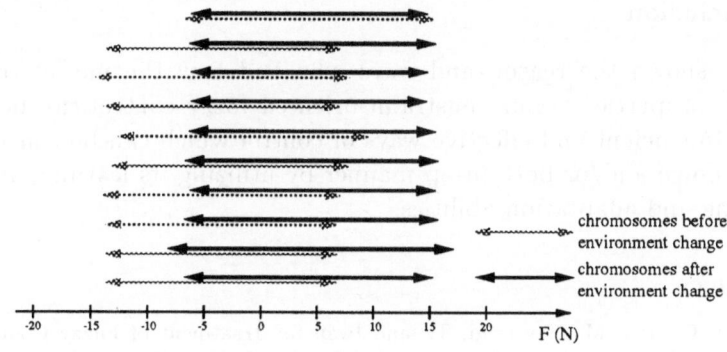

Figure 9. The adaptation of chromosome pool

Genetic Algorithm. **Fig. 10** shows the estimated value of $V(s)$, i.e., $V(\theta, \dot{\theta})$, for the case where the position and the velocity of the cart are disregarded.

The learning process of evaluation of state-action pairs can be viewed to have certain similarity to Q-learning, a well-known method of model free reinforcement learning [7]. Both methods calculate estimated evaluation values of state-action pairs based on the idea of dynamic programming (DP), but the learning parameters such as learning rate and discount factor, and the way of selecting action stochastically are not included in our method.

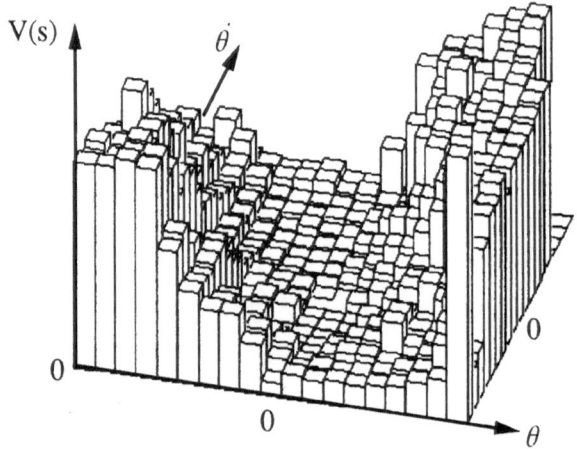

Figure 10. GA and RL based estimation of $V(\theta, \dot\theta)$, the minimum cost to attain the origin $(0,0)$

6. Conclusion

We have shown the reason and ways why and how the modularity and logicality properties of our constraint-oriented fuzzy control can be incorporated to efficient and effective ways of control which can be constructed in a top-down and/or bottom-up manner by utilizing its learning and self-organizing and adaptation abilities.

References

1. Katai, O., Ida, M., Sawaragi, T., and Iwai, S.: Treatment of Fuzzy Concepts by Order Relations and Constraint-oriented Fuzzy Inference, Proc. of NAFIPS '90, A Quarter Century of Fuzziness, pp.300-303 (1990)
2. Katai, O., Ida, M., Sawaragi, T., and Iwai, S.: Dynamic and Context-Dependent Treatment of Fuzziness from Constraint-Oriented Perspectives, Proc. of IFSA '91, Vol. on Artificial Intelligence, pp.101-104 (1991)
3. Katai, O., Sawaragi, T., and Iwai, S.: A Framework for learning and Reasoning via Order Relations, in Z. W. Ras (ed.), Methodologies for Intelligent Systems, Vol.4, Elsevier, pp.297-304 (1989)
4. Katai, O., Ida, M., Sawaragi, T., and Iwai, S.: Fuzzy Inference Rules and Their Acquisition from Constraint-oriented Perspectives, Proc. of IIZUKA '90 pp.211-216 (1990)
5. Katai, O., Ida, M., Sawaragi, T., Iwai, S.: Extracting Fuzzy Knowledge from Rank Ordered Sample Data via Instance Generalization and Connectionistic Computation, Proc. of IIZUKA '92, Vol. 2 pp.751-754 (1992)
6. Goldberg, D. E.: Genetic Algorithms, in Search, Optimization & Machine Learning, Addison Wesley (1989)
7. Watkins, C. J. and Dayan, P: Technical Note: Q-Learning, Machine Learnig, vol.8, pp.279-292 (1992)

LIPREADING USING THE FUZZY DEGREE OF SIMILARITY

K. KUROSU*, T. FURUYA**, M. SOEDA**, and S. TAKEUCHI***
*Kinki University
11-6, Kayanomori, Iizuka-shi, Fukuoka-ken, 820, JAPAN
** Kitakyushu College of Technology
5-20-1, Shii, Kokura-minami-ku, Kitakyushu, 803, JAPAN
*** NTT
3, Sakuramachi, Kumamoto-shi, 860, JAPAN

Abstract

Lipreading through visual processing techniques help provide some useful systems for the hearing impaired to learn communication assistance. This paper proposes a method to understand spoken words by using visual images taken by a camera with a video-digitizer. The image is processed to obtain contours of lips, which are converted into approximated hexagons. The pattern lists, consisting of lengths and angles of hexagons, are compared to get the fuzzy similarity between two hexagons. By similarity matching, the mouth shape is recognized as the one which has the pronounced voice. Some experiments, exemplified by recognition of the Japanese vowels, are given to show feasibility of this method.

1. Introduction

Lipreading is one of communication forms to understand a speaker without hearing; rather, with watching movements of lips, jaws and facial expressions. But it takes a long time to acquire the ability of lipreading. One of reasons for difficulties of learning lipreading is that the Japanese language has more than one hundred syllables with merely about fifteen mouth shapes which can be distinguished easily[1]. These special features in Japanese lead to the reality, where the same mouth shape corresponds to many pronounced syllables, that requires for the listeners to have high-level techniques of understanding. Many papers are reported for lipreading, some of which aim to realize systems to understand spoken words by recognizing visual images of lip movements[2],[3]. Some of researchers developed the systems to help people learn lipreading or teach it[4]. Also some papers are reported to construct some supplement systems for acoustic speech recognition[5]. Besides their purposes, there are many methods employed for recognition or visual data processing to realize lipreading, all of that stimulate engineers and researchers into studying varieties of implementation

123

Z. Bien and K. C. Min (eds.),
Fuzzy Logic and its Applications, Information Sciences, and Intelligent Systems, 123–131.
© 1995 Kluwer Academic Publishers.

methods toward more feasible lipreading[6]. We proposed lipreading methods by visual image processing with pattern classification[7], by Fuzzy logic[8], by using an X-Y tracker[9], and by using a neural network[10]. Though many aspects and problems were discussed, none of their works are easy to be implemented because of speed, memory capacities, unrealistic settings, or poor recognition rates. Therefore, the lipreading must be investigated to produce more feasible and simpler systems. We have proposed a method[11] related to pattern recognition, seeking better performance in the frame and shape recognition by a fuzzy similarity matching. This paper reports a method, combined with pentagon approximation and the fuzzy similarity, which intends to be technically realizable and effective. Many examples of vowel recognition through mouth shape are presented to show better performance.

2. Image processing

2.1. IMAGES OF MOUTH SHAPES

Some pictures taken by a full color video-digitizer are processed to obtain the mouth shapes of a subject in the system developed as shown in Fig.1. The subject, fixed in front of a camera, is illuminated by a white lamp, where his lips are colored by a lip stick. The camera gets the front views of his face, and they are digitized into a personal computer. Memorized video data 640×400 pixels are, first, compressed into 160×100 pixels, all of which are filtered to remove noise and excessive information by averaging the 3 by 3 surrounding points. After obtaining 160×100 pixels filtered data with 256 brightness levels, nonlinear transformation is applied to get binary black and white data. A nonlinear adaptive threshold method, whose algorithm determines the optimal threshold and binary pixels automatically, is used to obtain binary data to distinguish the lip shapes against the other backgrounds. After binary transformation, the contours of lips or mouth are estimated by the following procedures.
(1) From the contours of the image, some closed contours are kept after erasing the trace of the contours which do not become enclosed lines.
(2) If the area of the closed contour is too small, the closed contour is erased.
(3) In the remained contours, the contour placed at the bottom is selected as the one for mouth shapes.
By this algorithm illustrated in Fig.2, the mouth shape can be obtained.

2.2. HEXAGON APPROXIMATION

The characteristic values of the mouth shape is needed to distinguish mouth shapes depending on the pronounced voices. From the contours of the mouth shape, six characteristic points are determined by the following algorithm as shown in Fig.3. Imagine X and Y axis on the mouth shapes.
(1) Searching the leftmost point of the contour of the mouth shape, let the coordinate of this point by the X axis be x1. Finding the uppermost and the lowermost points of the contour whose coordinates are (x1+1) by the X axis, let the coordinate of the midpoint

of these two points by the Y axis be y1. Then the point A(x1, y1) is taken as the representative point of the mouth shape.

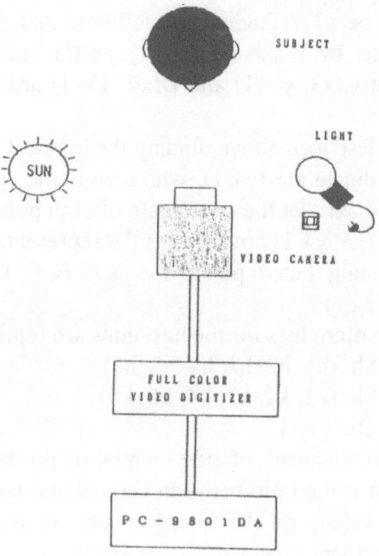

Fig.1 Experimenta apparatus

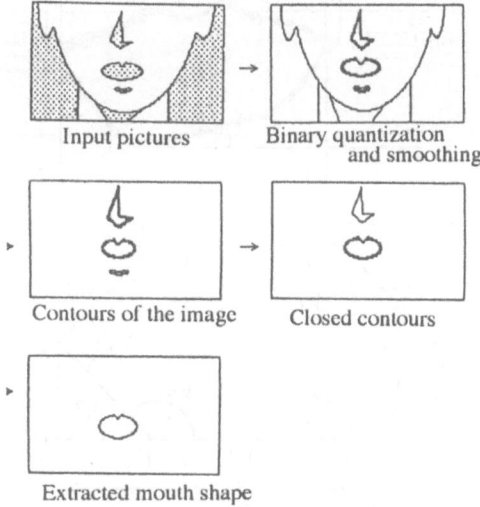

Input pictures

Binary quantization and smoothing

Contours of the image

Closed contours

Extracted mouth shape

Fig.2 Extraction of a mouth shape

(2) In the same manner as described above, finding two points of the contour whose coordinates are (x2-1), where x2 is the coordinate of the rightmost point of the contour by X axis, let the coordinate of the midpoint of these two points read by Y axis be y2. Then the point B(x2, y2) is taken as the representative point of the mouth shape.

(3) Searching the uppermost point of the contour of the mouth shape, let the coordinate of this point by Y axis be y3. Finding the leftmost and the rightmost points of the contour which coordinates by Y axis are (y3+1), let the coordinate of each point be x3 and x4. Then the point C(x3, y3+1) and D(x4, y3+1) are taken as the representative points.

(4) In the same manner described above, finding the leftmost and the rightmost points of the contour whose coordinate are (y4-1), where y4 is the coordinate of the lowermost point of the contour by Y axis, let the coordinate of each point be x5 and x6. Then two points E(x5, y4-1) and F(x6, y4-1) are taken as the representative points.

As shown in Fig.3, by joining these 6 points, that is A, B, C, D, E and F, the mouth shape is approximated as a hexagon.

The characteristic pattern lists for the hexagons are represented as

$$\text{Sh: (h1, h2, h3, h4, h5, h6)}$$
$$\text{Sk: (k1, k2, k3, k4, k5, k6)} \tag{1}$$
$$\text{Sty: (ty)}$$

where the list Sh has six elements of side lengths of the hexagon, the list Sk has six angles, and Sty's element is the ratio between vertical and horizontal lengths. The lists of these characteristic values of the hexagon are compared and computed to get similarity values between them.

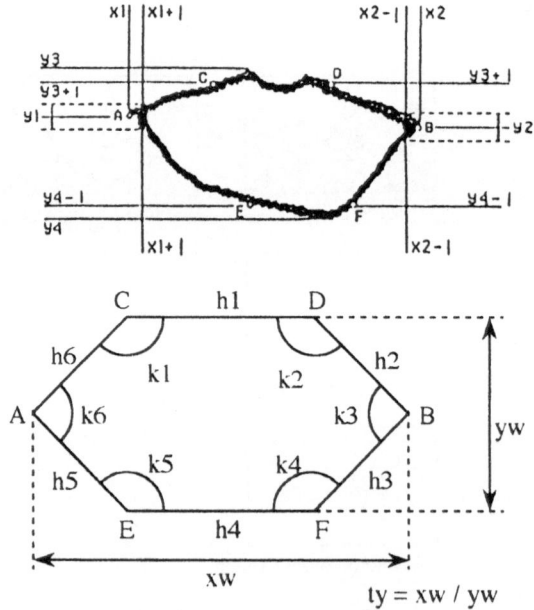

$$ty = xw / yw$$

Fig.3 Hexagonal approximation

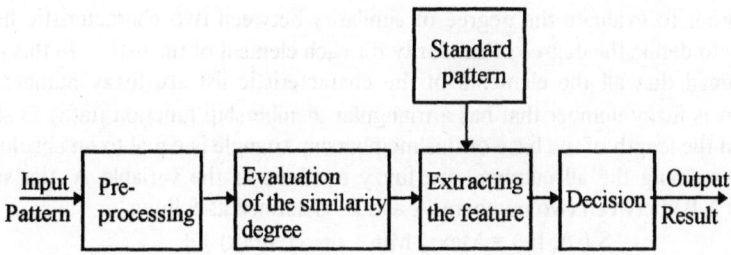

Fig.4 The process of the pattern recognition

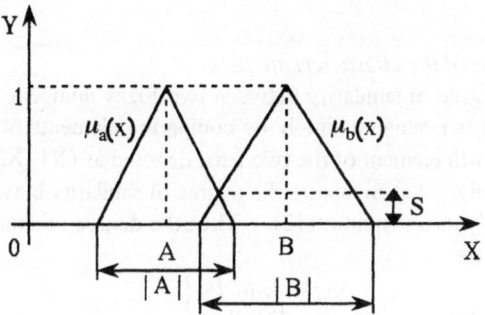

Fig.5 The similarity degree of two fuzzy numbers

3. The method for vowel recognition

3.1. DISTINCTION OF THE MOUTH SHAPE

The following pattern recognition method is applied to distinguish vowels by using symbolic characteristic lists of mouth shape approximated with hexagons. The basic procedure of the pattern recognition consists of five processes as shown in Fig.4. In this figure, the most essential and important part is the process to get the characteristic lists and to evaluate the similarity degree between the standard pattern and extracted characteristic pattern.

Using the list of these lengths and angles of the hexagons defined in 2.2, the similarity between two kinds of patterns, the standard patterns and the patterns of mouth shapes detected by a video camera, is evaluated. In the practical cases, it is necessary to consider the errors in observed mouth shapes and approximated hexagons. So we use a fuzzy degree of similarity based on the fuzzy theory and its idea to distinguish vowels more exactly.

3.2. THE FUZZY DEGREE OF SIMILARITY

3.2.1. *The degree of similarity between two numbers*

In order to evaluate the degree of similarity between two characteristic lists, it is necessary to define the degree of similarity for each element of the lists. In this research, it is assumed that all the elements of the characteristic list are fuzzy numbers. The variable A is fuzzy number that has a triangular membership function $\mu a(x)$ as shown in Fig.5, and the length of the base of this membership triangle is equal to an absolute value of A. Assuming the all numbers are fuzzy numbers as the variable A, the similarity degree $S(A,B)$ between two numbers, A and B, is defined as follows.

$$S (A, B) = Max \{ Min \{ \mu a(x), \mu b(x) \} \} \tag{2}$$

Here μa and μb are the membership function of A and B respectively. Fig.5 shows S(A, B). In this figure, the distance S is the degree of similarity between two numbers, where $0 \leq S \leq 1$.

3.2.2. Comparison of the characteristic lists

Using the degree of similarity between two fuzzy numbers, similarity between two characteristic lists is evaluated simply by comparing elements of the lists. Let Xn and Yn represent the n-th element of the two lists, denoted as (X1, X2, ... , Xn) and (Y1,Y2, ... ,Yn) respectively; Let Si denote the degree of similarity between Xi and Yi. Si can be calculated easily using equation (2). Then the degree of similarity between two lists is defined as follows.

$$SL = \min_{1 \leq i \leq n} \{Si\} \tag{3}$$

3.2.3. Similarity degree of the characteristic lists

The mouth shape is approximated as a hexagon, so each of the shape patterns is symbolized as the line length and turning angular degree that is represented as Xi and Ai, where Xi is the length of i-th line and Ai is the turning angle from the i-th line to (i+1)-th line. If the polygon has n edges, the shape patterns can be determined by using
(Xi, Ai)   i=1,2, ... ,n
Any polygon, which consists of n lines and n angles, can be expressed as 2n symbolic sequence such as

$$(X1, A1, X2, A2, ... , Xn, An) \tag{4}$$

These sequences are the characteristic lists of the mouth shape. Let L1=(X1, A1, X2, A2, ... , Xn, An) denote a characteristic list; Let L2=(Y1, B1,Y2, B2, ... ,Yn, Bn) denote the other characteristic list. Because the sequences (Xi, Ai, ... , Xn, An, X1, A1, ... , Xi-1, Ai-1) represent the same shape as (4), whose position is changed.
And so, when we compare a characteristic list L1 with the other list L2, we must consider n lists concerning L1 as follows;

L1:   L11=(X1, A1, X2, A2, ... , Xn, An)
        L12=(X2, A2, X3, A3, ... , Xn, An, X1, A1)

        L1n=(Xn, An, X1, A1, ... , Xn-1, An-1)

The degree of similarity between the list L1k and L2 is denoted as SLk (k=1,2, ... ,n), which is calculated using equation (3). Then the similarity degree of two characteristic patterns, L1 and L2, of mouth shape is defined as following equation,

$$SF = \underset{1 \le k \le n}{Max} \{SLK\} \qquad (5)$$

3.3. THE METHOD FOR RECOGNIZING THE VOWELS

As stated in the previous sections, the shape is symbolized as the side lengths and inner angles of the hexagon to make the characteristic lists.   Then two kinds of the characteristic lists - one is the list whose components are the lengths and the angles and the other is the list whose components are the ratios of the width to the height of the mouth - are used to calculate the fuzzy degree of similarity between the mouth patterns and the standard pattern to recognize the spoken vowels.

The procedures for discrimination of vowels are as follows.   The extracted pattern is compared to the standard pattern to recognize the spoken vowels by using the fuzzy similarity degree.   The degree of similarity of the ratio of length and width is calculated using the equation (2). Because the mouth shape is approximated by a hexagon, there are 6 kinds of the characteristic lists concerning the side lengths and the angles as in equation (4).   So the degree of the similarity of the lengths and angles is obtained by using equation (5).   In each case, the pattern, whose value of the similarity degree is the largest, scores two points and the pattern whose similarity degree is the second largest, scores one point.   At the end, the pattern, whose total score is largest, is regarded as the pattern of the spoken vowel.

Thus the spoken vowels are determined from the mouth shape.

4. Experimental results

The images of a subject, who pronounced the Japanese 5 vowels, A, I, U, E and O, were taken into a personal computer.   The four sets of samples taken 4 times from the same subject were processed into the characteristic pattern lists.   The same experiment was carried out for another subject to get one set of pattern.   An example of the pattern computation is shown in Fig.6.   The standard pattern for 5 vowels were computed by averaging 3 sets of the values chosen from the subject. recognition by matching was experimented with these 5 sets of vowel patterns, as follows.

Experiment 1:   Recognition between the standard pattern and the 3 patterns, which are used for constructing the standard pattern.

Experiment 2:   Recognition between the standard pattern and the fourth pattern, exclusive of the sets used to form the standard pattern.

Experiment 3:   Recognition between the standard pattern and the patterns of another subject.

The results are shown in Table1, where O corresponds to the correct recognition and × corresponds to wrong recognition.

5. Conclusion

It is shown that the spoken Japanese vowels can be recognized by the proposed lipreading method, where the fuzzy similarity method is applied to distinguish the mouth shapes which are approximated as hexagons. The results reveal that the perfect recognition is possible when the standard pattern are composed of the same person's mouth shapes. In order to improve recognition rates, construction of standard patterns must be examined more theoretically. In the future, some improvements are predicted for faster lipreading by carrying out efficient processing in extraction of mouth shapes and pattern matching.

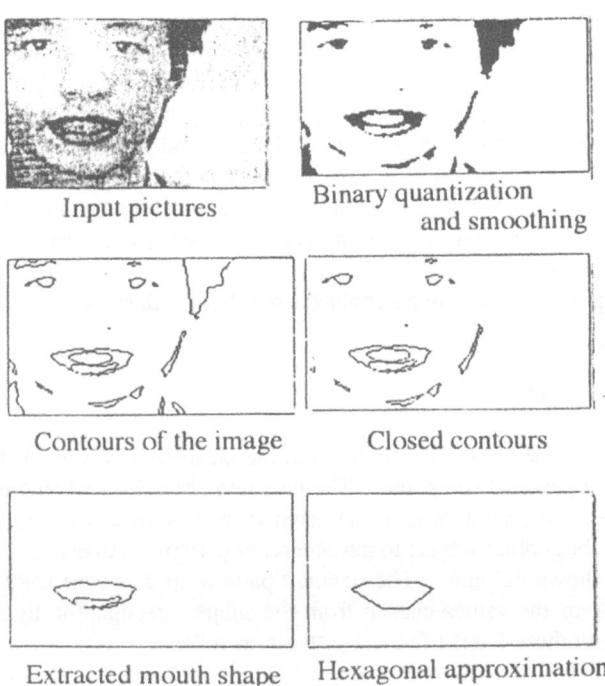

Input pictures   Binary quantization and smoothing

Contours of the image   Closed contours

Extracted mouth shape   Hexagonal approximation

Fig.6   Example of the pattern computation

TABLE 1. The recognition results

○: Sucess   ×: Failure

Input	Experiment 1	Experiment 2	Experiment 3
A	○	○	× O
I	○	○	○
U	○	○	○
E	○	○	○
O	○	○	× U

References

1. Fukuda, Y., Hiki, S.: Characteristics of the mouth shape in the production of Japanese stroboscopic observation, *J.Acoust.Soc.Japan (E)* **3-2** (1982), 75-91.

2. Mase, K., Pentland, A.: Automatic lipreading by optical flow analysis, *Trans. Inst. Electron. Inf. Commun. Eng.* **D-II-6** (1990), 796-803.

3. Okada, K., Ohira,C., Nakamura, H.: A method for lipreading, *Trans. Inst. Electron. Inf. Commun. Eng.* **D-II-9** (1989), 1532-1583.

4. Gulliiams, I., Segui, A.: Interactive videodisc for teaching and evaluating lipreading, *Proceedings of An. Intern. Conf. of IEEE Eng. in Medicine and Biology Soc.*, **Cat.88Ch2566-8** (1989), 1551-1552.

5. Pentajan, E.D.: Automatic lipreading to enhance speech recognition, *Proceedings of CVPR'85* **Cat.85Ch2145-I** (1985), 40-47.

6. Watanabe, T.: Machine lip reading of two concatenated vowels, *Trans. JPN. Soc. Mech. Eng. (C)* **55-509** (1989), 134-138.

7. Matsuoka, K., Furuya,T., Kurosu, K.: Speech Recognition by image processing of lip movements, *Trans. of SICE* **22-2** (1986), 67-74.

8. Kurosu, K., Furuya, T., Takeuchi, S.: Lip reading by fuzzy logic, *Proc. of 2nd Congress of IFSA* (1987), 23-4C.

9. Furuya, T., Soeda, M., Kurosu, K., Tamura, S.: Speech recognition with lip movement data using an X-Y tracker, *Trans. of SICE* **27-8** (1991), 958-965.

10. Kurosu, K., Furuya, T., etal.: Lipreading by a neural net, *Proc. of 8th European Annual Conf. on Human Decision Making and Manual Control* (1989), 329-339.

11. Weijing, Z., Furuya, T., Kurosu, K.: A list processing method on   Composition and decomposition of figure patterns, *Tran. of SICE* **26-1** (1990), 81-86.

References

INTELLIGENT RESTRUCTURING OF AUTOMATED PRODUCTION SYSTEMS

CLARENCE W. DE SILVA
NSERC Professor of Industrial Automation
Department of Mechanical Engineering
The University of British Columbia
Vancouver, Canada V6T 1Z4

Abstract

A dynamic-structure system is one that has the flexibility to change the system configuration automatically so as to operate in an optimal manner. A conceptual model for a dynamic-structure system is presented in this chapter [1]. In the present model, the interchangeable or shareable components of the overall system are grouped together. Their activity levels are evaluated by an intelligent preprocessor that is associated with the group. A knowledge-based task distribution system evaluates the activity levels and makes decisions as to how the components operating below capacity should be shared with those that are overloaded. Associated decision making can be effected through fuzzy logic. A simulation example is given to illustrate the application of dynamic restructuring.

1. Introduction

Considerable attention has been given to the use of decentralized or distributed control in complex, large-scale systems [2]. When various control functions in a large system can be ranked into different levels, a hierarchical control approach may be useful [3,4]. Also, in some situations, the components (both hardware and software) that perform the ranked functions could be arranged in a hierarchical manner. Published work on hierarchical systems has been limited primarily to those which are designed to possess fixed structures, and these structures do not change during operation of the system. This is the case, for example, in the systems described in [5] and [6].

Z. Bien and K. C. Min (eds.),
Fuzzy Logic and its Applications, Information Sciences, and Intelligent Systems, 133–145.
© *1995 Kluwer Academic Publishers.*

A dynamic-structure system is one which has the flexibility to change its communication and control structure automatically so that the architecture in which the system is integrated could be altered without having to redesign or retool the system. Even though the design, analysis, implementation, control, and operation of a fixed-structure system is usually simpler than those of a dynamic-structure system, it is generally less efficient and nonoptimal. For example, if two workcells have components that could be easily shared, it makes sense to share a component that operates well below its capacity with a workcell that has an overloaded component of the same type. Such sharing would be possible only if the system structure has the necessary flexibility to communicate with and control the shared component regardless of which workcell the component is associated with at a particular instant. Underlying here are the advantages of a dynamic-structure system [1,7].

Different levels of knowledge, expertise, and intelligence may be associated with different functions of a hierarchical control system, and dynamic restructuring will have to rely on such knowledge. Specifically, intelligent preprocessors may be necessary to evaluate, interpret, and transform various types of information available within and without the system. Furthermore, the decisions of system restructuring have to be made "intelligently" through a suitable knowledge system, by taking into consideration the available, preprocessed information. This is the basis of knowledge-based dynamic structuring. This chapter will describe a model for a knowledge-based dynamic-structure system [1]. An illustrative example will be drawn from the fish processing industry to indicate the utility of the model.

2. Knowledge-Based Hierarchical Systems

A system that has the flexibility to automatically change its structure should posses proper control means to provide that flexibility. Since a system of this type has to be able to provide a variety of different services during operation, it tends to be complex in general. Consequently, the task of controlling such a system would be intractable unless the associated architecture of control and communication is properly organized.

A layered architecture can facilitate the operation of a complex, flexible system [8,9]. Since the higher layers generally deal with low-resolution, imprecise, and incomplete information, more intelligence would be needed in the associated decision making process [10]. In contrast, in the lower layers, as in servo loops, information (e.g., signals from feedback sensors [11]) is used directly, without subjecting to intelligent preprocessing, in taking control actions. A hierarchy of intelligence may be identified in this connection [12, 13].

Specifically, "Knowledge" may be interpreted as *structured information*, within the context of computer-automated process control. Various means such as logic, semantic networks, frames, and production systems may be employed to represent and process knowledge [13]. Next, "Expertise" may be treated as *specialized knowledge*, and relates to in-depth knowledge that is needed to handle specialized situations. At the top level in this hierarchy rests "Intelligence" which cannot be defined precisely, but may be interpreted as the capacity to acquire and apply knowledge, thereby displaying some intelligent behavior. This is a somewhat circular interpretation. Intelligent characteristics include the ability to perceive, reason, learn, and make inferences, particularly in imprecise, vague, or fuzzy situations, and making use of incomplete information. It follows that, some intelligence is needed to gain knowledge from information, and to gain expertise by specializing that knowledge. This hierarchy is schematically represented in Figure 1. This is not a strict hierarchy because the separation of the knowledge layer and the expertise layer is somewhat fuzzy. Intelligence is needed for "preprocessing" that is involved in converting information into knowledge and expertise. The intelligence itself is fortified through the act of preprocessing of information and knowledge, with varying degrees of incompleteness, imprecision, uncertainty, vagueness, and fuzziness [13]. Preprocessing may include perception, reasoning, learning, and inference. It is the "outward appearance" of the hierarchical system in Figure 1, that is considered intelligent. Associated root-level computations would be hardly classified as intelligent. As an example in the process automation domain, one could model the execution of a routine task as knowledge-based and responding to a critical situation as expertise-based.

3. A Model for Dynamic Restructuring

Proper integration is crucial for efficient and cost-effective operation of a process control system. Since redesign of the system architecture and re-integration of the system are costly and time consuming, it is desirable to consider a flexible system whose structure could be automatically reconfigured according to various process requirements.

A model for a dynamic-structure system was conceptually proposed in [7] and further enhanced in [1] and [4]. This model is schematically represented in Figure 2. In this architecture, the workcell components of the processing system are grouped in such a

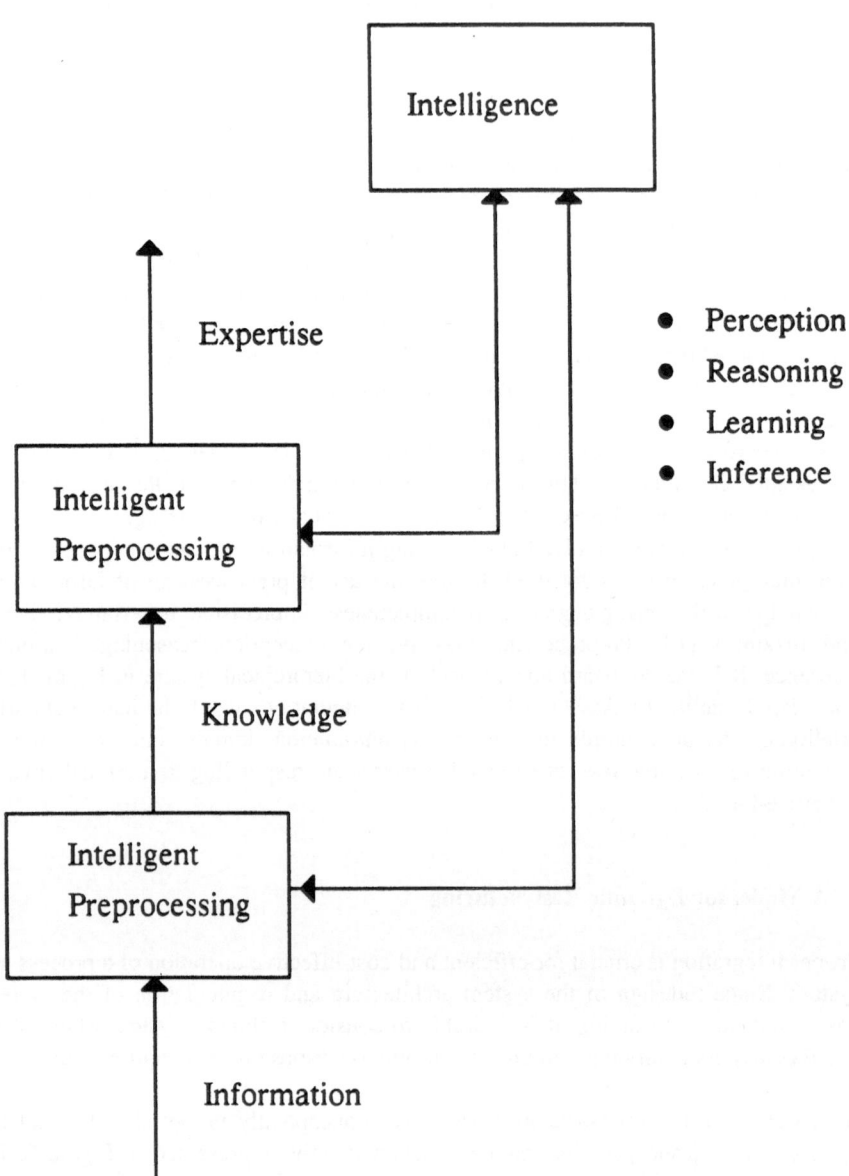

Figure 1. A Hierarchy of Intelligence

manner that the functions of the components within each group are similar and they are shareable or interchangeable without affecting the functionality of a workcell. In a general sense, both hardware devices and software modules may be considered as workcell components. Each workcell component has a controller that manages its operation. Also, each component has one or more sensors that will provide the necessary information to an "Intelligent preprocessor" which will determine the activity level of the particular component. In a fixed-structure system, the task distribution amounts to allocating tasks to various workcells in the system, and the constituent components of the workcells themselves are permanent. For a dynamic-structure system, however, a more intelligent task distribution system (TDS) would be needed. Here the TDS has to routinely monitor the activity levels of the workcell components, as provided by the corresponding intelligent preprocessors. It will then redistribute the constitution of the workcells by sharing some components that operate below their design capacity with workcells having components in the same group that are overloaded. These restructuring decisions are transmitted to the system restructuring controller (SRC) which activates the necessary communication and control links and provides the control strategies to effect the component sharing.

The activity levels of the components A_c in a workcell with a given structural configuration, depends on the process load of the workcell. If the workcell load (or demand) changes due to reasons such as supply-demand variations (e.g., new orders, new raw material), the activity levels of the workcell components will have to be modified. It follows that the load levels of the workcells have to be provided as inputs to the TDS, and these inputs will trigger the decision-making process for workcell restructuring, on

the basis of the required levels of component activity (L_c). Similarly, restructuring maybe needed if some components become disabled or degraded.

Reasoning associated with the restructuring decisions could be quite complex. For example, when more than one overloaded component and more than one component operating below capacity are present within a group, there arises a so-called conflict resolution problem [13]. Here, the decision of which components should be shared with which workcells and for what functions should be made by taking into consideration various factors such as the degree of overload and under-capacity, component characteristics, workspace geometry (e.g., proximity of a component to the workcell with which it is expected to share), ease of sharing, and the speed of restructuring. An intelligent (or knowledge-based) TDS is needed for this purpose. The knowledge base

138

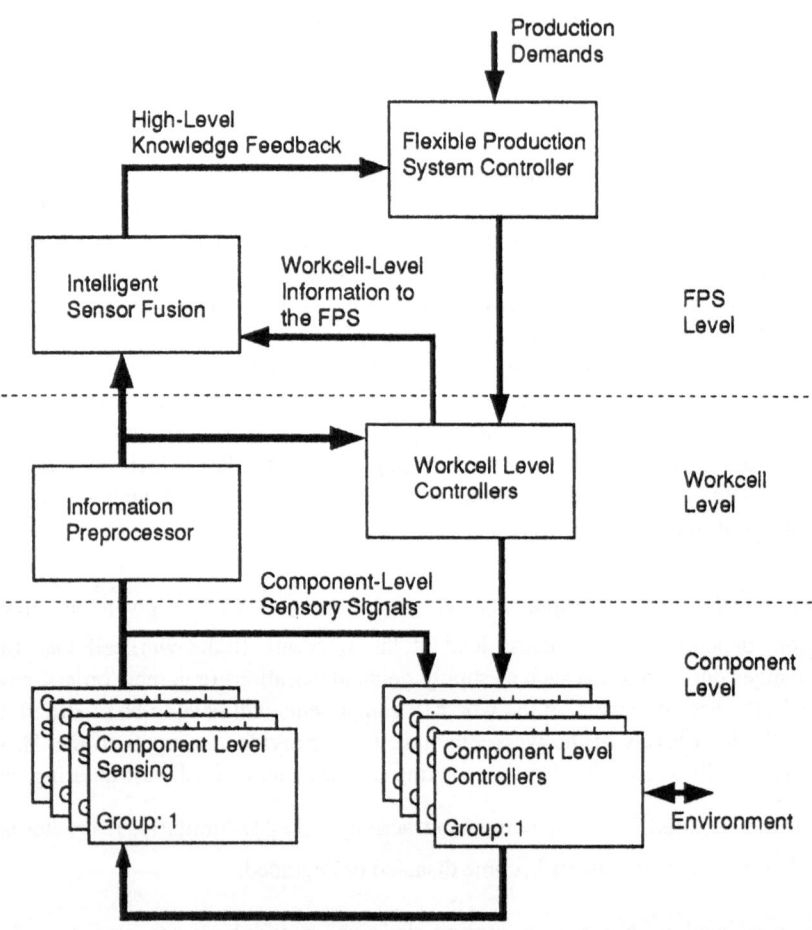

Figure 2. The Architecture of a Flexible Processing System with Dynamic Restructuring

and the context in the decision making process could be considered fuzzy, and the associated decisions can be determined by applying the compositional rule of inference [15].

Analytical representation of the dynamic restructuring model may be facilitated by the formulation given now. The activity levels A of the workcell components may be given by

$$A(\underline{w}, g, c) = \underline{F}(g) \otimes \underline{S}(\underline{w}, D_{\underline{w}}, g, c) \tag{1}$$

where c = identifier for a component in a workcell; g = identifier for the component group to which c belongs; $\underline{w}$ = set of workcells among which c is shared; $D_{\underline{w}}$ = task demand from each workcell; $\underline{S}$ = sensory signals from component c for determining the activity level; $\underline{F}$ = intelligent preprocessor for a group of similar components that infers the activity level of a component; and $\otimes$ = a suitable inference operator.

The intelligent preprocessor ($\underline{F}$) is typically a knowledge-based reasoning system. The associated variables of knowledge representation could be fuzzy and the knowledge base itself might be expressed as a set of linguistic statements. In this case $\underline{F}$ may be interpreted as a multidimensional membership function [10]. The sensory signals $\underline{S}$ from the workcell components are crisp and of high resolution. The context that is needed by $\underline{F}$ to infer the activity level would be higher level information of lower resolution. For example, peak values, averages, standard deviations, correlations, trends, quality measures, and times of certain critical values could be involved. The operator $\otimes$ can be quite subjective [10] and should be interpreted depending on the particular component and the specific need. For example, in many situations $\otimes$ could be a knowledge processing operation such as the application of the compositional rule of inference to a fuzzy rule base, combined with conventional procedures of task planning.

Once the activity levels of the components are available, an input trigger such as a change in a workcell load or component character should initiate the process of activity evaluation for workcell reconfiguration. This process may be formulated as

$$c(\underline{w}^*) = \underline{R} \otimes \left[\underset{c \in g}{\otimes} A(\underline{w}, g, c) \right] \tag{2}$$

where $\underline{w}$ = the workcell association of component c prior to reconfiguration; $\underline{w}^*$ = the workcell association of c after the inferred reconfiguration; $\underline{R}$ = knowledge system for task redistribution, and $\oplus$ = combinational operation. Note that the operations implied by equation (2) have to be performed for all components that are overloaded and all other components that fall into the component groups to which the overload components belong. Again, the knowledge base associated with the decision making process of equation (2) could be a set of fuzzy linguistic statements, and accordingly $\underline{R}$ could be interpreted as a multidimensional membership function. Then the inference operator $\otimes$ would correspond to the application of the compositional rule of inference. The combinational operator $\oplus$ too is subjective and situation-specific, and has to be performed separately for each component group. For example, it may constitute a simple comparison of activity levels of the components within a group and ranking them accordingly so as to pair, say, an overloaded component with one that operates below capacity. Analytically, this will incorporate the optimization problem [13, 14]

$$\text{minimize } J = J\left(\underline{\alpha}_c, \underline{C}_c, \underline{A}_c, \underline{L}_c\right)$$

$$\text{subject to } P_w\left(\underline{A}_c\right) = D_w\left(\underline{L}_c\right) \tag{3}$$

Here,

C_c = work capacity of component c

L_c = load assigned by the taskplanning procedure, to component c

A_c = actual activity level of component c

D_w = work demand from workcell w

P_w = production level of workcell w

α_c = vector of weighting parameters

Note that

If $L_c \leq C_c$ for $\forall$ $c \varepsilon$ w

then $A_c = L_c$ (4)

If $\exists$ some $c \varepsilon$ w such that $L_c > C_c$

then $A_c \leq C_c$

and $P_w < D_w$ (5)

Suppose that a fuzzy-logic rule base which suggests priorities (*PR*) for sharing various levels of component under-capacity (*UC*) with various levels of component overloads (*OC*) is available. Then, using fuzzy-logic decision making, the priority value (*PV*) for sharing a specific overloaded component with a specific under-capacity component may be computed by using

$$\mu_{PV}(y) = \max_{\substack{i,j \\ k(i,j)}} \left\{ min\left[\mu_{oc}^i(L_{oc} - A_{oc}), \mu_{uc}^j(C_{uc} - A_{uc})\right] \cdot \mu_{PR}^k(y) \right\} f(OC, UC) \quad (6)$$

where

$\mu_A^b(\cdot): \mathfrak{R} \to [0, 1]$ = membership function of the *b* th fuzzy-resolution of the fuzzy descriptor *A*, which is a function mapped from the real line $\mathfrak{R}$ on to the closed interval [0,1]

$f(c,d)$ = feasibility index of sharing an undercapacity component *d* with an overloaded component *c*.

The concept of feasibility index is discussed in [13].

Finally, the new workcell configuration $\underline{w}^*$ will provide the information for the SRC to activate the necessary communication and control links and to operate the workcells according to the new configuration.

4. Example of Application

An application in fish processing is considered now. Specifically, a system consisting of subsystems such as a fish cutting system, a grading system, and a packaging system will be reconfigured as workcells within a process plant, and will be implemented as a dynamic-structure control system. There exist many similarities in terms of hardware, software and processes in the subsystems. For example, system components such as sensors (including CCD cameras), image processing systems, actuators, grippers, and conveyors are similar. As a result, the nature of component interfacing, signal processing, and low-level control will be similar both in hardware and software. This means that there exists the prospect of sharing similar components among the workcells so as to reduce overloading and to achieve somewhat optimal operation.

Figure 3 presents a simulation example of dynamic restructuring. The overall processing system consists of three workcells -- for cutting, packaging and grading, as shown. Each workcell has common components such as robots for fish handling and processing, automated guided vehicles (AGVs) for the transfer of raw fish, processed fish and waste, and vision stations for detection, gauging and quality evaluation of objects. The activity levels of the components are shown as percentages of their designed capacities, and are presented as solid bars, with the dotted region indicating the available excess capacity. Also, each workcell has a demand level in a given phase, which represents the load on the workcell to achieve the necessary productivity.

It is seen that in Phase 1, the components in all three workcells operate below their capacities. Then due to a drop in the supply of fish, the cutting demand drops by 40% and the grading demand increases by 50% due to a glut of processed fish. Also, orders for processed and packed fish have remained unchanged and hence the 0% change in packaging load. These conditions result in associated changes in the activity levels of the workcell components, as shown by the Transition Phase. Now the workcells need to be reconfigured for improved operation. Possibilities of component sharing are indicated by broken lines in the Transition Phase. Once the reconfiguration is effected, in Phase 2, the activity levels of the components have changed, to achieve a somewhat balanced operation under the changed loading condition of the workcells. It is seen that, in Phase 2, none of the components are overloaded unlike in the Transition Phase. Also, an AGV has been completely released from the cutting workcell and has been allocated to the grading workcell. Similarly, a vision station has been completely released from the cutting workcell and has been kept aside.

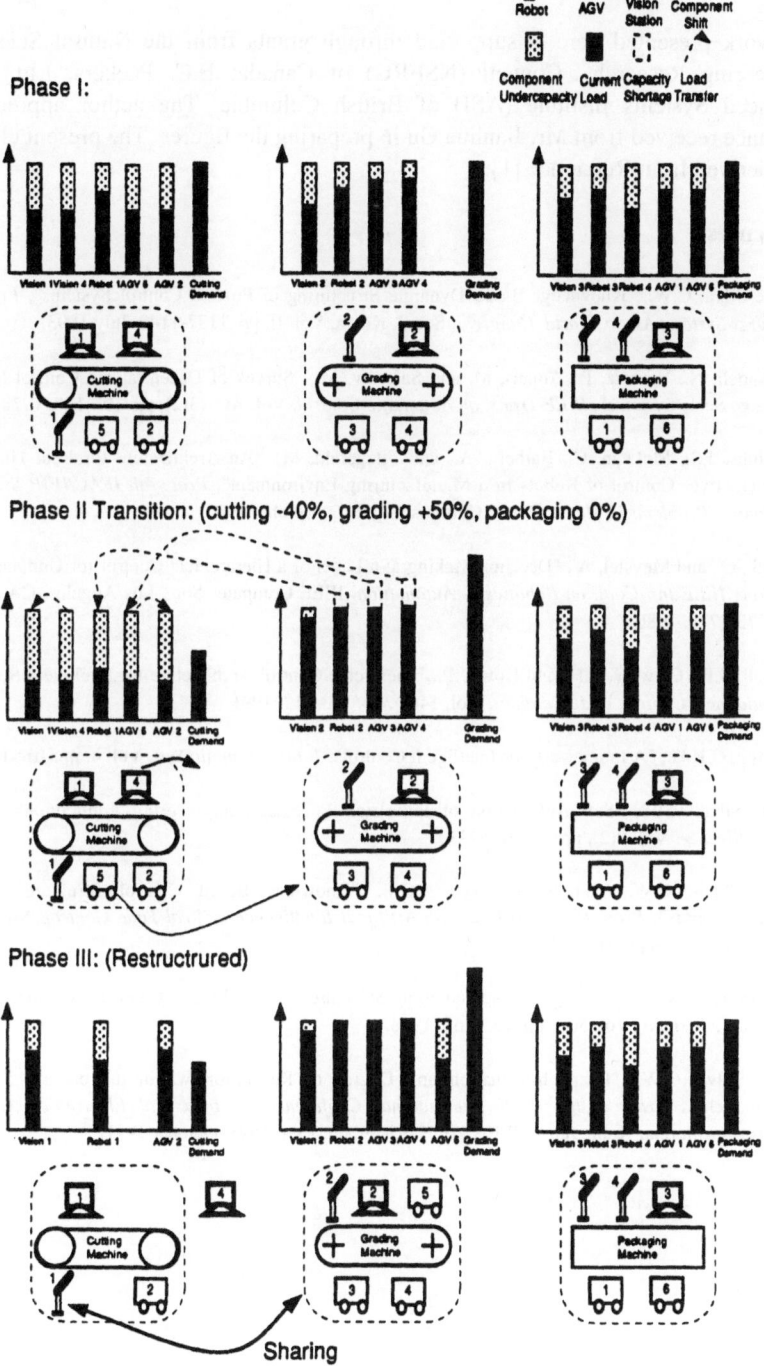

Figure 3. A Simulation Example of Dynamic Restructuring

144

Acknowledgments

The work presented here is supported through grants from the Natural Sciences and Engineering Research Council (NSERC) of Canada; B.C. Packers, Ltd.; and the Advanced Systems Institute (ASI) of British Columbia. The author appreciates the assistance received from Mr. Jianhua Gu in preparing the figures. The present chapter has been derived from Reference [1].

References

1. De Silva, C.W., "Knowledge-Based Dynamic Structuring of Process Control Systems", *Proc. 5th Int. Fuzzy Systems Assoc. World Congress*, Seoul, Korea, Vol. II, pp. 1137-1140, July 1993.

2. Sandell, N., Varaiya, P., Athans, M, and Safonov, M., "Survey of Decentralized Control Methods for Large Scale Systems", *IEEE Trans. on Automatic Control,* Vol. AC-23(2), pp. 108-128, 1978.

3. Albus, J.S., McLean, C., Barbera, A., and Fitzgerald, M., "An Architecture for Real-Time Sensory-Interactive Control of Robots in a Manufacturing Environment", *Proc. 4th IFAC/IFIP Symp. Inform. Control Problems in Manuf. Tech.*, Gaithersburg, MD, October 1982.

4. Isik, C., and Meystel, A., "Decision Making at a Level of a Hierarchical Control for Unmanned Robot", *Proc. IEEE Int. Conf. on Robotics & Automation,* IEEE Computer Soc., Los Angeles, CA, Vol. 3, pp. 1772-1778, 1986.

5. Hall, S.R., Crawley, E.F., and How, J.P., "Hierarchic Control Architecture for Intelligent Structures", *J. Guidance, Control, and Dynamics,* Vol. 14(3), pp. 503-512, 1991.

6. Pang, G.K.H., "A Framework for Intelligent Control", *J. Intel. Robotic Sys,.* Vol. 4, pp. 109-127, 1991.

7. De Silva, C.W., "Soft Automation of Industrial Processes", *Engineering Applications of Artificial Intelligence,* Vol. 6(2), pp. 87-90, 1993.

8. De Silva, C.W., and MacFarlane, A.G.J., "Knowledge-Based Control Structure for Robotic Manipulators", *Proc. IFAC Workshop on Artificial Intelligence in Real-Time Control,* Swansea, U.K., pp. 143-148, September 1988.

9. Saridis, G.N., "Knowledge Implementation Structures of Intelligent Control Systems", *Journal of Robotic Systems,* Vol. 5(4), pp. 255-268, 1988.

10. De Silva, C.W., "Fuzzy Information and Degree of Resolution within the context of a Control Hierarchy", *Proc. of the IEEE International Conference on Industrial Electronics, Control, and Instrumentation,* Kobe, Japan, IEEE 91CH2976-9, pp. 1590-1595, October 1991.

11. De Silva, C.W., *Control Sensors and Actuators*, Prentice-Hall, Englewood Cliffs, NJ, 1989.

12. De Silva, C.W., "Research Laboratory for Fish Processing Automation", *Int.l J. of Robotics and Computer-Integrated Manufacturing*, Vol. 9(1), pp. 49-60, 1992.

13. De Silva, C.W., *Intelligent Control : Fuzzy Logic Applications,* CRC Press, Boca Raton, FL, 1995.

14. De Silva, C.W., and Gu, J., "An Intelligent System for Dynamic Sharing of Workcell Components in Process Automation", *Int. J. Engineering Applications of Artificial Intelligence*, Vol. 7(5), pp. 571-586, 1994.

15. Zadeh, L.A., "Theory of Approximate Reasoning", *Machine Intelligence*, Hayes, J., et al. (eds) Vol. 9, pp. 149-194, 1979.

FUZZY SYSTEM MODELING AND ITS APPLICATION TO MOBILE ROBOT CONTROL

Y.H. JOO, H.S. HWANG, K.B. WOO
Dept. of Electrical Eng.,
Yonsei University, Seoul 120-749, Korea

K.B. KIM
Div. of Elec. & Inf. Technology,
KIST, Seoul 136-791, Korea

Abstract

A systematic identification method that realizes fuzzy modeling using the input and output data pairs of system is presented. Such a model is composed of fuzzy implications. The implications are automatically generated by the structure and parameter identification. In the structure identification the optimal or near optimal number of fuzzy implications is determined in view of valid partition of data set. The parameters defining the fuzzy implications re identified by a GA (Genetic Algorithm) hybrid scheme to minimize mean square errors globally. Numerical examples are provided to evaluate the feasibility of the proposed approach. Comparison shows that the suggested approach can produce a fuzzy model with higher accuracy and a smaller number of fuzzy implications than the ones achieved previously in other methods. The proposed approach has also been applied to construct a fuzzy model for the navigation control of a mobile robot. The validity of the resultant model is demonstrated by experimentation.

Keywords : Fuzzy modeling, Identification, Genetic Algorithm, Mobile robot

1. Introduction

Fuzzy modeling has been studied to deal with complex, ill-defined and uncertain systems, in which conventional mathematical models may fail to give satisfactory results. The studies on the fuzzy system modeling have largely been devoted to two approaches. One is based on composite fuzzy relational equations [5]. The approach is theoretically clear, but may suffer difficulties since the solution of a fuzzy relational equation is usually not unique, and sometimes it even does not exist at all. For this reason, the approach is not practical. The other is termed linguistic model [4, 6, 7, 9], in which a fuzzy model is composed of a set of fuzzy implications, and they are identified by optimization techniques from sample data. The model has been popular in industrial applications. Linguistic fuzzy modeling for control was first dealt with by

Z. Bien and K. C. Min (eds.),
Fuzzy Logic and its Applications, Information Sciences, and Intelligent Systems, 147–156.
© 1995 *Kluwer Academic Publishers.*

Tong [4, 9]. He proposed a logical examination method to construct linguistic models. In spite of its satisfactory result, it should be noted that the method is difficult to extend to high-dimensional systems. Pedrycz [10] proposed a new composition rule and identification algorithms of fuzzy systems. His fuzzy model is based on the concept of a referential fuzzy set and Zadeh's conditional possibility distribution. Utilizing the same type of fuzzy model, Xu [6] presented a general fuzzy system identification approach, which includes structure identification, parameter estimation and an associated self-learning algorithm. However, it requires that the number of referential fuzzy sets in each universe(r) be chosen empirically. A larger r will demand more computational efforts in building and using the model.

This paper is aimed at presenting a systematic approach in the identification procedure of a fuzzy system, and its flexible adaptability to complex and high-dimensional systems. The identification is classified into the structure identification and the parameter identification. The structure identification determines the number of fuzzy implications constituting a fuzzy model in view of valid partitions of data set using fuzzy c-means clustering(FCM). The shapes of membership functions in the fuzzy implications are also determined in this process. The parameters are identified by a GA hybrid scheme to minimize the mean square errors of the data set. GA, which is a global optimization technique, proves to be flexible, so it is able to identify simultaneously the parameters in the premise and the consequence of the fuzzy model. However, it does not guarantee convergence to a global optimum. In order to solve this problem, GA is combined with a conventional complex method [11] which exploits the convergence of problem-specific technique. It does guarantee global optimization and local convergence.

A numerical example is provided to evaluate the feasibility of the proposed approach. The proposed approach has also been applied to construct the fuzzy model for the navigation control of a mobile robot, which models human driver's control actions. The resultant model is validated by experimentation to show that it is suitable enough for industrial applications.

2. Fuzzy Model and Reasoning

We consider two implications consisting of fuzzy model as follows:

R^1: If x_1 is Small and x_2 is Big, then $y_1 = w_1 \cdot a_1 + b_1$.
R^2: If x_1 is Big and x_2 is Medium, then $y_2 = w_2 \cdot a_2 + b_2$.

where Small, Medium and Big are fuzzy labels, w_i the degree of fulfillment of the premise, and a_i and b_i consequent parameters.

Given input data x_1^0 and x_2^0, the output y^* inferred from the above two implications is obtained in terms of the average of y_1 and y_2 with the weights w_1 and w_2 as follows:

$$w_1 = \mu_{Small}(x_1^o) \cdot \mu_{Big}(x_2^o), \quad w_2 = \mu_{Big}(x_1^o) \cdot \mu_{Medium}(x_2^o) \tag{1}$$

$$y^* = \frac{w_1 \cdot y_1 + w_2 \cdot y_2}{w_1 + w_2} = \frac{w_1 \cdot (w_1 \cdot a_1 + b_1) + w_2 \cdot (w_2 \cdot a_2 + b_2)}{w_1 + w_2} \tag{2}$$

3. Identification of Fuzzy Model

3.1. STRUCTURE IDENTIFICATION

In order to carry out the structure identification which determines the number of fuzzy implications and the prototypes of the membership functions, fuzzy c-means clustering [1] is introduced and adapted. It is a classification oriented approach based on quantitative information of a system, and produces a fuzzy c-partition of data set. Our purpose is to find the number of clusters to effectively describe the relation between each input-output value of a system, via iterative optimization of the squared distances weighted by the square of the memberships. To illustrate this, we consider a system composed of two inputs x_1 and x_2 and one output y. If the optimal or sub-optimal cluster numbers of x_1-y and x_2-y are c_1 and c_2, respectively, the number of fuzzy implications is determined by $c_1 \cdot c_2$. It means that the total input space to output is acquired by all the combinations of the input subspaces partitioned for each input variable, under the assumption that input variables (x_1 and x_2) are mutually independent. We consider a q-input and one-output system. The procedure to identify the structure is summarized as follows:

Step 1: Construct a sample data set $X_k = \{x^j_{1l}, x^j_{2l}, ..., x^j_{kl}, ..., x^j_{nl}\}$ for fuzzy c-partition of input space in the fuzzy model. Here, x^j_{1l} is composed of x^j_{kl} and x^j_{k2} which are the kth data of input(x_j) and output (y). n is the total number of data items, and j is 1 to q. Set j=1.

Step 2: Set c=2 and determine mc, the practically permissible maximum number of fuzzy implications to be identified ($2 \leq c \leq$ mc). Set p=1 and initialize membership value matrix, $U^{(p-1)}$ for μ_{ik} as 1/c.

Step 3: Calculate the c cluster centers $v_i^{(p)}$ with $U^{(p-1)}$ and (3) for the ith cluster center.

$$v_{il}^{(p)} = \sum_{k=1}^{n} (\mu_{ik})^2 x_{ik} \left/ \sum_{k=1}^{n} (\mu_{ik})^2 \right. \qquad l = 1,2 \tag{3}$$

where $\mu_{ik} = \mu_{ik}(x^j_{kl})$ is the membership grade of x^j_{kl} in fuzzy set μ_i.

Step 4: Update $U^{(p)}$ for k=1 to n.

   ① Calculate I_k and I_k' by using the distance D_{ik} from the kth data x^j_k to the center of the ith cluster v_i.

$$I_k \equiv \{i \,|\, 1 \leq i \leq c, D_{ik} = \| X_k - v_i \| = 0\}$$
$$I_k' \equiv \{1, 2, ...,c\} - I_k$$

② For data item k, compute new membership values.

i) if $I_k = \emptyset$, $\mu_{ik} = D_{ik}^2$, $\mu_{ik} = \mu_{ik} \Big/ \sum\limits_{i=1}^{c} \mu_{ik}$

ii) if $I_k' \neq \emptyset$, $\mu_{ik} = 0$ for all $i \in I_k'$, and $\sum\limits_{i \in \Gamma_k} \mu_{ik} = 1$

Step 5: Compare $J_m^{(p)}$ and $J_m^{(p-1)}$. If $| J_m^{(p)} - J_m^{(p-1)} | \leq \varepsilon$ and $c \leq mc$, $c = c+1$ and go to step 2, else if $| J_m^{(p)} - J_m^{(p-1)} | > \varepsilon$, $p = p+1$ and go to step 3. If $c > mc$, go to step 6.

$$J_m^{(p)} = \sum_{k=1}^{n} \sum_{i=1}^{c} (\mu_{ik})^2 \cdot D_{ik}^2 \tag{4}$$

Step 6: For any c ($2 \leq c \leq mc$), a validity measure for the valid fuzzy partitioning of input space are calculated by using (3) [2].

$$S = J_m / (n \cdot d_{min}) \tag{5}$$

where $d_{min} = \min\limits_{i,k} \| v_i - v_k \|^2$ is the minimum distance between cluster centroids.

In (5), minimizing S corresponds to minimizing J_2 which is the goal of FCM, and the additional factor d_{min} means the separation measure between clusters. The more separate the clusters, the larger d_{min} and the larger S. From x_k, an appropriate partition number of the relation between the jth input variable and the output variable can be determined at lower value of S.

When the optimal number of clusters is founded, the prototypes of the membership functions, which are mapped into the partitioned area of input (x_{k1}^j), are determined from the values of $\mu_{ik} (x_{k1}^j)$. In order to eliminate the ripples in the membership functions calculated and make them convex types, we transform them into triangular or trapezoidal membership functions.

Increase j by one and go to step 1 until all the q input variables are considered.
Step 7: The final number of fuzzy implications in the fuzzy model is determined by the multiplication of the partition numbers selected in step 6 for all input variables.

3.2. PARAMETER IDENTIFICATION

The identification of parameters, which define the membership functions of the premise and coefficients of the consequence, is carried out by a GA hybrid scheme, which combines effectively the advantages of GA and the complex method [11]. Using this hybrid scheme, we are able to identify simultaneously parameters in the premise and the consequence of fuzzy implications in view of global optimization. The fact that GA exploits only the coding and the objective function value to determine plausible trials in the next generation, gives the flexibility for its application to optimization problems. However, GA is a blind search. Hence GA has disadvantages when compared to other methods that do make use of problem-specific information. When problem-specific information exists, it may be advantageous to combine problem-specific information with GAs to improve ultimate genetic search performance and guarantee convergence to a global optimum.

A GA hybrid scheme is proposed, in which GA runs to substantial convergence and then a local optimization procedure takes over searching from the points which display good off-line performance in the evolution of GA. In the hybrid scheme, GA finds the hills and the hill-climber(local optimization technique) goes and climbs them. As a hill-climber, the complex method is utilized. The procedure for the parameter identification by the GA hybrid-scheme is summarized as follows:

Step 1: Set the maximum generation number(max_gen), and population size. Fix crossover rate and mutation rate. Set the coding length of each parameter to be identified.

Step 2: Generate random populations P(0) composed of randomly generated binary-codes. They are decoded into real values which we want to identify. Evaluate the fitness of individuals in P(0) by (7).

$$E = \frac{1}{n}\sum_{i=1}^{n}(y_i^o - y_i^*)^2 \qquad (6)$$

Fitness function, $f = 1.0/E$ (7)

where n is the total number of data, y_i^o target output, and y_i^* output inferred from fuzzy implications.

Step 3: Set the generation number(gen) to 1. Generate P(t+1) from P(t) by selecting the fittest individuals from P(t), and recombining them by crossover and mutation with the crossover and mutation rates, respectively. Evaluate the fitness of individuals in P(t).

Step 4: Increase gen by 1 until gen is not greater than max_gen. Decode the fittest population into real values, and retain them and its fitness value.

Step 5: To guarantee convergence to a global optimum, the complex method to solve constrained minimization problems is introduced and adapted as follows:

   ① An initial feasible solution set $\{X_1, X_2, ..., X_k\}$ (k≥n+1, n is the number of parameters to be identified), which satisfies all the constraints imposed by membership functions and shows better fitness, is selected from the retained values in step 4. Here X_i is composed of n parameters to be identified.

   ② The objective function E defined in (6) is evaluated at the k feasible solution set. If X_h corresponds to the largest function value, the process of reflection is used to find a new solution, X_r as

$$X_r = (1+\alpha)\cdot X_o - \alpha\cdot X_h \qquad (8)$$

$$\text{where } X_o = \frac{1}{(k-1)}\cdot\sum_{l=1,l\neq h}^{k}X_l$$

③ If the solution, Xr is feasible (satisfies constraints) and $E(X_r) < E(X_h)$, the solution X_h is replaced by Xr, and go to step ②. If $E(X_r) \geq E(X_h)$, a new trial solution, X_r is recalculated by (8) with $\alpha = \alpha/2$. This procedure is repeated until $E(X_r) < E(X_h)$ and α becomes smaller than a prescribed small quantity, ε. If such a X_r can not be found, X_r is discarded and go to step ② with a new X_p which has the second highest function value instead of X_h.

④ If X_r found in step ③ violates the constraints, a new X_r is obtained by (9).

$$(X_r)_{new} = \frac{1}{2} \cdot (X_o + X_r) \tag{9}$$

⑤ Each time the worst solution, X_h of the current complex is replaced by X_r, the convergence of the process is tested by (10).

$$\{ \frac{1}{k} \cdot \sum_{j=1}^{k} [E(X') - E(X_j)]^2 \}^{1/2} \leq \delta \tag{10}$$

where X' is the centroid of all the k solution set of the current complex.

X_r that satisfies (10) is the final optimal feasible solution that we want to get. To get initial feasible solution set that is closer to the top of the hill, the reduction technique of search areas of the parameters is proposed. In GA, we code binarily parameters which we want to identify, and concatenate them. Each coding has its own sub-length L, and minimum and maximum values, P_{min} and P_{max}, and the precision of the decoded value is controlled by $(P_{max} - P_{min})/(2^L - 1)$. Therefore, the adjustment of P_{min}'s and P_{max}'s, which define search area for all the parameters, has an effect on the accuracy of solutions found by GA without increasing the coding length of each parameter. To do this, the following step 4' is added to step 4.

Step 4': Set the generation number, re_gen (1<re_gen<max_gen). When gen becomes re_gen, the reduction is carried out through the increment of P_{min} and the decrement of P_{max} of each parameter. New P_{max}'s and P_{min}'s define the ranges of the values decoded from the populations whose fitness values are greater than a value β. Until gen becomes max_gen, GA evolves searching the reduced area.

4. Application to the Identification of Gas Furnace

In this section, the feasibility of the proposed approach is illustrated through the identification of a fuzzy model of gas furnace. Our purpose is to obtain a fuzzy model which describes the relation between a gas flow u(t) and the combusted CO_2 concentration y(t) of the gas furnace using the 296 pairs of data presented by Box and Jenkins [3]. We consider u(t-4) and y(t-1) as input variables, and y(t) as the output variable. A small S in (5) indicates a partition in which all the clusters are overall compact, and separate to each other. The c_1 for u(t-4) and c_2 for y(t-1) with the smallest S are both 5. Those with the second smallest S are 3 and 2. To reduce the number of fuzzy implications, appropriate numbers of clusters which describe each input-output relation effectively are determined as 3 for c_1 and 2 for c_2, since the smallest S and the second smallest S for c_1 and c_2 do not considerably differ in their magnitudes. The membership function values of each input variable calculated in step 4 of section 3.1 are calculated, in which the membership functions of the input variables are labeled as Positive, Zero, Negative for u(t-4), and Small and Big for y(t-1), respectively. In order to eliminate the ripples of the membership functions , we transform them into triangular and trapezoidal membership functions. After the structure identification, the fuzzy implications are shown in Figure 2.

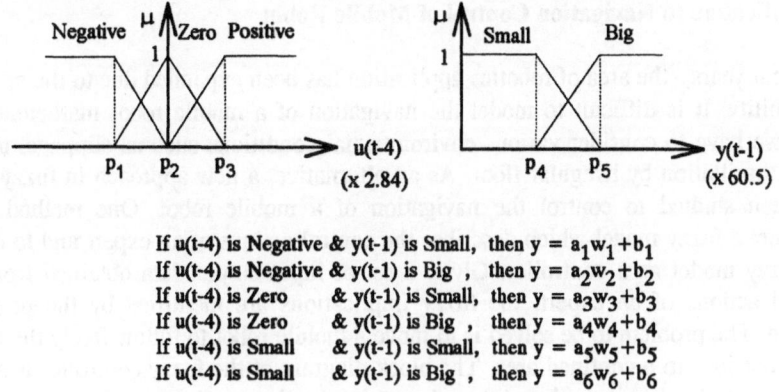

If u(t-4) is Negative & y(t-1) is Small, then y = $a_1w_1+b_1$
If u(t-4) is Negative & y(t-1) is Big , then y = $a_2w_2+b_2$
If u(t-4) is Zero & y(t-1) is Small, then y = $a_3w_3+b_3$
If u(t-4) is Zero & y(t-1) is Big , then y = $a_4w_4+b_4$
If u(t-4) is Small & y(t-1) is Small, then y = $a_5w_5+b_5$
If u(t-4) is Small & y(t-1) is Big , then y = $a_6w_6+b_6$

Figure 2. Fuzzy implications after structure identification

In the fuzzy implications, $p_1, p_2, ..., p_5$, and $a_1, b_1, ..., a_6, b_6$ are the parameters to be found by the parameter identification. Initial parameters for running the GA hybrid scheme are as follows: The population size is 50, length of individuals 10, crossover rate 0.6, and mutation rate 0.033. The reduction was carried out by the generation, 101. The reflection coefficient α is 1.3, its criterion ε, 10^{-6}, and the convergence criterion δ, 10^{-3}. 22 points which result in better off-line performances in GA by the reduction of search area are selected as initial feasible points of the complex method. The parameters identified by the GA hybrid scheme are shown in TABLE 1. In TABLE 2, the mean square errors calculated by (6) are compared with the results achieved previously in other approaches.

TABLE 1. Identified parameters for fuzzy implications

Premise		Consequent			
p_1	-1.635	a_1	8.062	b_1	58.555
p_2	0.622	a_2	4.848	b_2	53.240
p_3	0.94	a_3	0.407	b_3	47.192
p_4	0.79	a_4	9.765	b_4	51.564
p_5	0.968	a_5	2.056	b_5	44.148
		a_6	35.878	b_6	43.912

154

TABLE 2. Comparison of identification error with other fuzzy models

Literature	Mean square error	Rules	Note
[3]	0.71		ARMA model
[4]	0.469	19	Fuzzy model
[5]	0.776	20	Fuzzy model
[6]	0.328	25	Fuzzy model
[7]	0.190	6	Fuzzy model
Our result	0.166	6	Fuzzy model

5. Application to Navigation Control of Mobile Robot

In recent years, the area of robotics application has been expanded due to the addition of mobility. It is difficult to model the navigation of a mobile robot mathematically since we have to consider various environmental conditions such as slippage, tear of wheel, oscillation by irregular floor. As an alternative, a new approach in fuzzy logic has been studied to control the navigation of a mobile robot. One method is to construct a fuzzy model which describes the control actions of an expert and to utilize the fuzzy model as a controller. Given a set of input-output data obtained from the control actions of an expert, the fuzzy implications are identified by the proposed method. The problem to be solved is to move a mobile robot to follow freely the center of the corridor in a confined area. The block diagram of the fuzzy controller is shown in Figure 3. At every sampling interval, state evaluation block reads the measurement of sonar sensors to know the distances to the left and the right walls, and calculates the orientation of the mobile robot from the measured distances. The fuzzy controller inferences a new steering angle from the orientation and the difference between the distances to the left and the right wall, and the mobile robot is moved with the new steering angle.

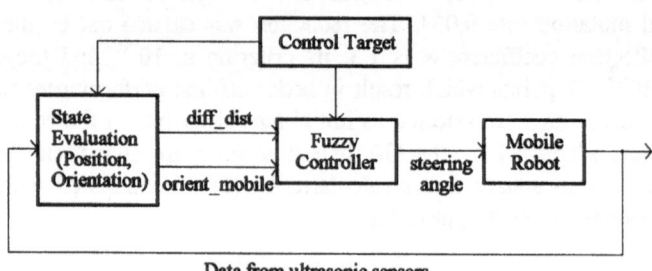

Figure 3. Block diagram of the fuzzy controller

The experimental mobile robot is equipped with two powered wheels that provide motion and steering, and four free wheels at the four corners of the vehicle. The main characteristics of the used mobile robot are as follows: The mobile robot is 45(cm)

height, 75(cm) length, 70(cm) width. The reference point is located in the center of 4 free wheels. Maximum velocity is 1(m/sec) with maximum acceleration of 1(m/sec^2). The experiment is carried out under the condition that the constant velocity is 0.4(m/sec), the turning speed 0.2(m/sec), and sampling time 0.12(sec). For the navigation control, the fuzzy model identified from an expert's control actions is shown in Figure 4. The navigation trajectory of the mobile robot controlled by the identified fuzzy model is displayed in Figure 5.

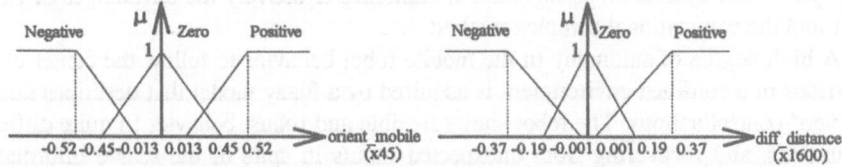

If diff_dist is Negative & orient_mobile is Negative, then y = -1.27·w1+52.0
If diff_dist is Negative & orient_mobile is Zero   , then y = -6.05·w2+26.4
If diff_dist is Negative & orient_mobile is Positive , then y = 9.50·w3-9.07
If diff_dist is Zero   & orient_mobile is Negative, then y = 0.11·w4+14.8
If diff_dist is Zero   & orient_mobile is Zero   ' , then y =-0.57·w5+0.57
If diff_dist is Zero   & orient_mobile is Positive , then y =-2.34·w6-13.8
If diff_dist is Positive & orient_mobile is Negative, then y =-5.99·w1+8.06
If diff_dist is Positive & orient_mobile is Zero   , then y = 5.45·w2-26.0
If diff_dist is Positive & orient_mobile is Positive , then y = 3.77·w3-53.9

Figure 4. The identified fuzzy model for the navigation control

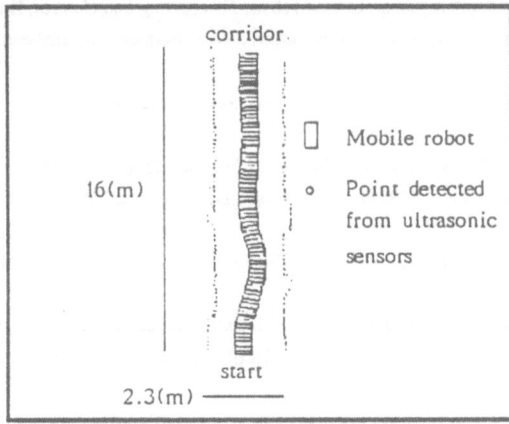

Figure 5. The navigation trajectory of the mobile robot

6. Conclusion

The reason why the accuracy of the identified fuzzy model of a gas furnace is superior to that of other fuzzy models is summarized as follows: The proposed GA hybrid scheme makes it possible to identify the optimal parameters globally in the premise and the consequent of the fuzzy implications at the same time. Besides the accuracy of the identification, another advantage of the proposed approach is that its procedure is comparatively simple and clear, so it can flexibly and robustly be applied to even more complex fuzzy system modeling since it combines effectively the advantages of FCM, GA and the conventional complex method.

A high degree of autonomy in the mobile robot behavior to follow the center of the corridor in a confined environment is acquired by a fuzzy model that describes human drivers' control actions. The robot shows flexible and robust behavior in quite different situations, and recovering from unexpected inputs in spite of the scarce information initially available. The resultant model is validated by experimentation, and proven to be suitable enough for industrial application.

Acknowledgment : This work was supported in part by the Samsung Electronics Co. We are most grateful to Dr. S.K. Kim for his inspiration and his encouragement in this work.

References

[1] Bezdek, J. C. (1981) *Pattern Recognition with Fuzzy Objective Function Algorithms*, New York Plenum.

[2] Xuanli, X. and erardo, B.(1991) A validity measure for fuzzy clustering, *IEEE Trans. Pattern Anal. Machine Intel.*, PAMI-13(8), 841-847.

[3] Box, G. E. P. and et al. (1970) *Time Series Analysis, Forecasting and Control*, Holden Day, San Francisco.

[4] Tong, R. M. (1980) The evaluation of fuzzy models derived from experimental data, *Fuzzy Sets and Systems.*, 13, 1-12.

[5] Pedrycz, W. (1984) An identification algorithm in fuzzy relational systems, *Fuzzy Sets and Systems*, 13, 153-167.

[6] Chen-Wei Xu (1989) Fuzzy systems identification, *IEE Proc.*, 136(4), 146-150.

[7] Sugeno, M. and Yasukawa, T. (1993) A Fuzzy-Logic-Based Approach to Qualitative Modeling, *IEEE Trans. on Fuzzy Systems*, 1(1), 7-31.

[8] Goldberg, D. E. (1989) *Genetic Algorithms in Search, Optimization, and Machine Learning*, Addison Wesley.

[9] Tong, R. M. (1978) Synthesis of fuzzy models for industrial processes, *Int. J. General Systems.* 4, 143-162.

[10] Pedrycz, W. (1983) Numerical and applicational aspects of fuzzy relational equations, *Fuzzy Sets and Systems.* 11, 1-18.

[11] Rao, S. S. (1984) *Optimization Theory and Application*, JOHN WILEY & SONS.

[12] Hwang, H. S., Joo, Y. H., and Woo, K. B. (1993) Generalized fuzzy modeling, *Fifth IFSA World Congress*, 1145-1150.

DEFUZZIFICATION WITH CONSTRAINTS

RONALD R. YAGER AND DIMITAR P. FILEV
Machine Intelligence Institute
Iona College, New Rochelle, NY 10801

Abstract

We look at the problem of defuzzification in situations where in addition to the usual fuzzy output of the controller there exists some ancillary restriction on the allowable defuzzified values. We provide two basic approaches to address this problem. In the first approach we enforce the restriction by selecting the defuzzified value through a random experiment in which the only values which have nonzero probabilities are in the allowable region, this method makes use of a nonmonotonic conjunction operator. In the second approach we convert the problem to one of constraint optimization.

1. Introduction

In a fuzzy logic controller the output of the knowledge base is a fuzzy subset F of the output space. The defuzzification process is then used to select a crisp value y* as the controller output. In previous papers on defuzzification [1, 2] we have shown that the COA and MOM methods of defuzzification are special cases of a more general method called **BADD defuzzification**. Implicitly it is assumed that the defuzzification step is carried out in an environment in which all elements of the output space are allowable values for the crisp output y*. Generally weighted averaging techniques, such as the COA method, work well. Two situations can be identified in which these types of techniques brake down. The first are those in which the output fuzzy set exhibits a kind of multi-modality. In these cases it becomes questionable whether it is meaningful to combine values which are far away from each other. In [3] we have suggested an approach to these kinds of situations using the idea of a combinability relationship. The second situation, though not completely distinct from the first, occurs when there exists some kind of restrictions, ancillary constraints or forbidden zones exist in the output space. [4] Again here pure weighted averaging techniques can brake down. For example, if we have a robot with an obstacle in its path two solutions exist. The robot can go to the left or right of the obstacle, this is the fuzzy set F. If one uses the COA defuzzification technique, which as we have noted is essentially a weighted averaging of the solutions suggested by F, one comes

157

Z. Bien and K. C. Min (eds.),
Fuzzy Logic and its Applications, Information Sciences, and Intelligent Systems, 157–166.
© 1995 *Kluwer Academic Publishers.*

up with a crisp foolish answer of going straight into the obstacle. In this work we investigate this problem of defuzzification in the presence of restrictions on the allowed defuzzified values.

2. A Generalized View of Defuzzification

In [2] Yager and Filev describe the general defuzzification process as consisting of three steps. In the first step the output fuzzy set F is transformed into a new fuzzy subset E. This new fuzzy subset is then normalized to give us a probability distribution P, where.$P(y_i) = \dfrac{E(y_i)}{\Sigma_i (y_i)}$. The selection of the defuzzified value y* is then made based upon P. The most commonly used process of selection is to take y* as the expected value with regard to the distribution P, thus $y^* = \Sigma_i p_i y_i$. This can be seen as essentially a weighted averaging of the elements in the output space based upon the membership grades in the transformed fuzzy set E.

In [1] Filev and Yager suggested a general formulation for the step of transforming F into E, called the BADD transformation. Specifically it is suggested that

$$E(y) = (F(y))^{\alpha} \qquad \alpha \geq 0.$$

They show that for $\alpha = \infty$, one gets the MOM method, $P(y) = 0$, for those y's not equal to the maximal membership grade and $P(y) = 1/n$ for those y's equal to the maximal membership grade in F (n is the number of elements attaining the maximal membership grade). For $\alpha = 1$, they show that the above method leads to the COA method. They also show that for $\alpha = 0$, $P(y_i) = P(y_j)$ for all i and j, and hence no information is used from the set F. In [1, 2] they associate the selection of α with the confidence in the fuzzy knowledge base, the larger α, the more confidence.

In [2] Yager & Filev suggest an alternative to the expected value approach of selecting y* called the RAGE defuzzification process. In this method we calculate the probability distribution P in the same manner as described above and then the RAGE defuzzification proceeds is as follows:

(1) We divide the unit interval into n intervals one for each output value y_i. We denote these intervals

$$R_i = [a_i, b_i] \qquad i = 1, \ldots, n$$

where we define these R_i as

$$a_1 = 0 \qquad b_1 = p_1$$
$$a_i = b_{i-1} \qquad b_i = a_i + p_i \qquad \text{for } i > 1.$$

(2) We perform a random experiment generating a random number $r \in [0, 1]$
(3) If $r \in R_i$ then we use $y^* = y_i$.

As shown in [2] this method always results in y* being some element for which $p_i \neq 0$, ie. $F(y_i) \neq 0$. It should also be clear that this random approach doesn't result in a averaged value for y*.

3. Defuzzification Under Constraints and Forbidden Zones

In the usual approach to defuzzification it is assumed that the allowable values for the defuzzified output was any point in the output space. The problem of defuzzification becomes more complicated if we consider the possibility that restrictions exist in the sense that from the whole universe of discourse only some values are allowed for the defuzzified value

Suppose we have a fuzzy logic control (FLC) knowledge base. Assume that the knowledge-base gives, for a particular input, an output consisting of a fuzzy subset F of the output universe of discourse X. The meaning of this fuzzy set F is that for each value $x \in X$ F(x) indicates the degree to which the rule base recommends x as the control value. Here we shall consider that our problem has some restrictions on the allowable defuzzified values. The allowed defuzzified values can be expressed as a another fuzzy subset H of X. In this case H(x) indicates the degree to which x is an allowed value.from the defuzzification process.

In this paper we don't discuss the nature of the allowable solutions. They can be defined from a technological point of view - e. g. certain type of control actions cannot be performed in a given setting. Allowable restrictions can be also set on the basis of control strategy - e. g. to preserve the control system from unallowable control actions. In some cases the fuzzy subset of allowable restrictions H can be a crisp set, defining a new discretization of the universe of discourse X.

The problem of defuzzification under restriction the becomes one of selecting a element y^* recommended by F and also allowed by H.

In the following we shall suggest some approaches to the solution of this problem. We note that in [4] Pflugar, Yen & Langari suggest a solution which is different then the approaches suggested here.

We recall that two basic methodologies have been described for the selection of a value once having obtained the probability distribution. In the first case we perform a random experiment to select the element. In the second approach the probabilities are used to generate an expected value. In the following we shall suggest defuzzification procedures under restrictions based on both these approaches

4. RAGE Defuzzification Under Restrictions

In the earlier section we described the RAGE defuzzification process. In this section we extend this idea to the environment where there is some restriction on the allowable values for the defuzzified value. We indicate the set of allowable output values as H and use F to indicate the controller fuzzy output.

We recall that a fundamental property of the RAGE method is that the defuzzified value y^* is always one of the elements that have non-zero probability. The RAGE output is not an weighed average and as such it can easily provide a framework for implementing defuzzification in restricted environments.

A natural extension of the RAGE method appears to be one in which we combine via a conjunction the sets F and H and then use this new set to generate the probabilities to be used in the RAGE defuzzification procedure. Care must be taken in executing the conjunction. If we take

$$G = H \cap F$$

we are obtaining G as a desire to get a solution that is allowable, non restricted, and recommended by the controller. Then of course

$$p_i = \frac{G(x_i)}{\sum_j G(x_j)}.$$

Since $p_i \neq 0$ if $G(x_i) \neq 0$ we only get solutions that satisfy both.

While this simple aggregation at first seems appearing it has one dramatic drawback. Assume H and F are completely conflicting then $G = \Phi$. In this case we would get $p_i = 1/n$ for all x_i in the space. This means that we *lost* the restriction information.

Any aggregation procedure for combining F and H must keep the priority of the H. That is, it must not allow under any circumstances the defuzzified value to be in the forbidden region. In [5] Yager has introduced a new aggregation operator called the *nonmonotonic conjunction* which has the required priority condition.

Assume H and F are two fuzzy subsets of X. Let $G = \eta(H, F)$ where

$$G(x) = H(x) \wedge (F(x) \vee (1 - Poss(F|H))).$$

We recall

$$Poss(F|H) = Max_x[F(x) \wedge H(x)].$$

We note that if F and H have at least one element in common with membership grade one, $Poss(F|H) = 1$, then $G(x) = H(x) \wedge F(x)$, which is the usual intersection, $G = H \cap F$. At the other extreme if $F \cap H = \Phi$, then $Poss[F|H] = 0$ and $G(x) = H(x)$. Thus when the suggestion of the controller is in complete conflict with the restriction we use the restriction. We notice in this case of complete conflict if H is a crisp subset then $p(x) = \dfrac{1}{Card(H)}$ for all elements $x \in H$ and $p(x) = 0$ for all others.

We note that we can provide a more general formulation for the nonmonotonic intersection as

$$G(x) = H(x) \wedge S(F(x), 1 - Poss(F|H))$$

where S is any t-conorm [6, 7].

Another formulation for the nonmonotonic conjunction is

$$G(x) = H(x) \wedge (F(x))^{Poss(F|H)}.$$

This formulation suggests an interesting connection between this operation and the BADD transformation described earlier. We can consider

$$G = H \cap E$$

where

$$E(x) = (F(x))^{Poss(F|H)}$$

We see that we transform F where our confidence depends upon the compatibility of F and H. Thus $Poss(F|H)$ can be seen as our measure of confidence.

Using these ideas the RAGE method of defuzzification under restriction consists of the following process:

    (1) Aggregate the allowable region with the controller suggestion

$$G = \eta(H, F)$$

    (2) Normalize G to get the probabilities of the elements in the output space

$$P(x_i) = \frac{G(x_i)}{\sum_j G(x_j)}.$$

(3) Use $P(x_i)$ to randomly select an element.

Example: Assume X = {1, 2, 3, 4, 5, 6, 7, 8, 9}.

Case 1. Let $F = \{\frac{0}{1}, \frac{0}{2}, \frac{1}{3}, \frac{1}{4}, \frac{1}{5}, \frac{1}{6}, \frac{1}{7}, \frac{1}{8}, \frac{0}{9}, \frac{0}{10}\}$

$H = \{\frac{1}{1}, \frac{1}{2}, \frac{1}{3}, \frac{0}{4}, \frac{0}{5}, \frac{0}{6}, \frac{0}{7}, \frac{1}{8}, \frac{1}{9}, \frac{1}{10}\}$

Here Poss[F/H] = 1 and therefore $G(x) = H(x) \wedge F(x)$ hence

$G = \{\frac{0}{1}, \frac{0}{2}, \frac{1}{3}, \frac{0}{4}, \frac{0}{5}, \frac{0}{6}, \frac{0}{7}, \frac{1}{8}, \frac{0}{9}, \frac{0}{10}\}.$

From this we get $p_i = 0.5$ for i = 3, 8 and $p_i = 0$ for all others. In this case we would randomly select between 3 and 8. Thus if we randomly generate a number r in the unit interval we select our defuzzified value x* as follows:

$x^* = 3$ if $r \in [0, 0.5)$

$x^* = 8$ if $r \in [0.5, 1]$

Case 2. Let $F = \{\frac{0}{1}, \frac{0}{2}, \frac{1}{3}, \frac{1}{4}, \frac{1}{5}, \frac{1}{6}, \frac{1}{7}, \frac{1}{8}, \frac{0}{9}, \frac{0}{10}\}$

$H = \{\frac{1}{1}, \frac{1}{2}, \frac{0}{3}, \frac{0}{4}, \frac{0}{5}, \frac{0}{6}, \frac{0}{7}, \frac{0}{8}, \frac{1}{9}, \frac{1}{10}\}$

Here Poss[F/H] = 0 and therefore $G(x) = H(x)$ hence $p_i = .25$ for i = 1, 2, 9, 10 and $p_i = 0$ for all others. If we randomly generate a number r in the unit interval we select our defuzzified value x* as follows:

$x^* = 1$ if $r \in [0, .25)$

$x^* = 2$ if $r \in [.25, .5)$

$x^* = 9$ if $r \in [.5, .75)$

$x^* = 10$ if $r \in [.75, .1]$

Case 3a. $F = \{\frac{0}{1}, \frac{0}{2}, \frac{0}{3}, \frac{0}{4}, \frac{.4}{5}, \frac{.6}{6}, \frac{.8}{7}, \frac{1}{8}, \frac{1}{9}, \frac{1}{10}\}$

$H = \{\frac{1}{1}, \frac{1}{2}, \frac{1}{3}, \frac{.9}{4}, \frac{.6}{5}, \frac{.2}{6}, \frac{0}{7}, \frac{0}{8}, \frac{0}{9}, \frac{0}{10}\}$

Here Poss[F/H] = .4 and therefore $G(x) = H(x) \wedge (F(x) \vee 0.6)$ hence

$G = \{\frac{.6}{1}, \frac{.6}{2}, \frac{.6}{3}, \frac{.6}{4}, \frac{.6}{5}, \frac{.2}{6}, \frac{0}{7}, \frac{0}{8}, \frac{0}{9}, \frac{0}{10}\}$

$p_i = 0.1875$ for i = 1,2,3,4, 5

$p_i = 0.0625$ for i = 6

$p_i = 0$ for i = 7, 8, 9, 10.

If we randomly generate a number r in the unit interval we select our defuzzified value x* as follows:

$x^* = 1$ if $r \in [0, 0.1875)$

$x^* = 2$ if $r \in [0.1875, 0.375)$

$x^* = 3$ if $r \in [0.375, 0.5625)$

$x^* = 4$ if $r \in [0.5625, 0.75)$

$x^* = 5$ if $r \in [0.75, 0.9375)$

$x^* = 6$ if $r \in [0.9375, 1]$

Case 3b: In this case we use $S(a, b) = (a + b) \wedge 1$ instead of $S(a, b) = a \vee b$ as in case 3a thus we get $G(x) = H(x) \wedge (1 - Poss[F/H] + F(x))$ and since $Poss[F/H] = 0.4$ then $G(x) = H(x) \wedge (.6 + F(x))$. Thus $\{\frac{.6}{1}, \frac{.6}{2}, \frac{.6}{3}, \frac{.6}{4}, \frac{.4}{5}, \frac{.2}{6}, \frac{0}{7}, \frac{0}{8}, \frac{0}{9}, \frac{0}{10}\}$ and therefore

$p_i = 0.2$ for $i = 1, 2, 3, 4$

$p_i = 0.1333$ for $i = 5$

$p_i = 0.06666$ for $i = 6$

$p_i = 0$ for $i = 7, 8, 9, 10$.

If we randomly generate a number r in the unit interval we select our defuzzified value x^* as follows:

$x^* = 1$ if $r \in [0, .2)$
$x^* = 2$ if $r \in [.2, .4)$
$x^* = 3$ if $r \in [.4, .6)$
$x^* = 4$ if $r \in [.6, .8)$
$x^* = 5$ if $r \in [.8, .9333)$
$x^* = 6$ if $r \in [.9333, 1]$

5. Defuzzification by Constrained Optimization

The common methods used for defuzzification, COA and MOM, are based finding the _expected value_ of a probability distribution over the output base set. The probability distribution used is obtained from the fuzzy output of the controller. As described by Filev & Yager [1] the difference between these two methods, COA and MOM, lies in the procedure used to obtain the probability distributions from the fuzzy sets. In [1] it is shown that with the aid of the BADD transformation we can obtain COA and MOM as special cases by simply adjusting one parameter denoted α.

With this understanding we can view these methods as the same type, finding the defuzzified value as an expected value. It should be strongly emphasized that the process of obtaining the expected value can be seen as an optimization problem. In particular we see that the problem is to find d such that $E_x(x-d)^2$ is minimized, E_x is the expected value operator, $E_x(x-d)^2 = \sum_i (x_i - d)^2 p_i$. Since

$$E_x\{(x - d)^2\} = E_x\{x^2\} + (E_x\{x\} - d)^2 - (E_x\{x\})^2$$

$E_x\{(x - d)^2\}$ achieves its minimum for $d = E_x\{x\} = \sum_i x_i \, p_i$.

We shall now consider the extension of the expected value based method to the problem of defuzzification under constraints. We shall again assume that the fuzzy set F is the output of the fuzzy controller, it is a measure of the appropriateness of an element x as a solution, and that the fuzzy set H is the allowable region. We define $G = F \cap H$. We see G satisfies the criteria of being a good solution, F, and an allowable solution, H.

We shall now formulate the problem of defuzzification under restriction as a constrained nonlinear programming problem. As an optimization criterion we shall consider the requirement of minimization of the mean square of the error between the values of universe of discourse of G and the defuzzified value d_g. For the most generality we assume that the probabilities of the elements of the base set, the p_i's, defined by the BADD transformation [1] from the set G,

$$p_i = \frac{g_i^\alpha}{\sum_{j=1}^{n} g_j^\alpha} \quad , \quad i=(1, n).$$

We recall that the probabilities defined above satisfy both a monotonicity and identity conditions (Yager & Filev 1991):

$$g_i > g_j \implies p_i \geq p_j$$

$$g_i = g_j \implies p_i = p_j.$$

The probability distributions generated above for different $\alpha \geq 0$, are related to the membership function G. We select α to indicate the type of defuzzification we are using.

Then the criterion for a selected α has the form:

$$\text{MIN:} \quad \sum_{i=1}^{n} (x_i - d_g)^2 \frac{g_i^\alpha}{\sum_{j=1}^{n} g_j^\alpha} \qquad \qquad \mathbf{I}$$

The problem is to find d_g to minimize equation **I**. We note that if $\alpha = 1$ we are using a COA type method and if α is infinity then we are using a MOM type method.

On must also include a constraint on this problem. The constraint we need is one which assures us the the defuzzified value lies in the allowable region. We shall accomplish this by requiring that $H(d_g)$ is larger then some threshold value, t_g.

Then defuzzification problem under allowable region restrictions is representable as the following constrained nonlinear programming problem:

Find the value dg that Minimizes **I**

Subject to

$$H(d_g) \geq t_g \qquad \qquad \text{II.}$$

for a given $\alpha \geq 0$ and $t_g \in (0, 1]$. We note that α is fixed, the choice of α depends on issues discussed in [2]. The choice of t_g is fixed and it depends on how we define what it means to be contained in a fuzzy set. If H is crisp then any $t_g > 0$ works the same.

Note that for $t_g = 0$ the constrained optimization problem becomes unconstrained and has a global solution, defined by the expected value of the universe of discourse over the BADD transformation induced probability distribution, $p_i = \frac{g_i^\alpha}{\sum_{j=1}^{n} g_j^\alpha}$,

$$d_g = \sum_{i=1}^{n} x_i \frac{g_i^\alpha}{\sum_{j=1}^{n} g_j^\alpha}.$$

We now deal with the algorithmic realization of the solution to the constrained nonlinear programming problem

We see that the objective function can be expressed as follows

$$\sum_{i=1}^{n} (x_i - d_g)^2 \frac{g_i^\alpha}{\sum_{j=1}^{n} g_j^\alpha} = \sum_{i=1}^{n} x_i^2 \frac{g_i^\alpha}{\sum_{j=1}^{n} g_j^\alpha} - (\sum_{i=1}^{n} x_i \frac{g_i^\alpha}{\sum_{j=1}^{n} g_j^\alpha})^2 + (\sum_{i=1}^{n} x_i \frac{g_i^\alpha}{\sum_{j=1}^{n} g_j^\alpha} - d_g)^2$$

The first two terms are positive and functionally independent on d_g. The third term is also positive but depends on d_g, thus its minimum determines the minimum of the objective function. For an unconstrained d_g, the problem without condition II, the minimum coincides with d_g being the generalized defuzzified value via BADD:

$$d_g^* = \sum_{i=1}^{n} x_i \frac{g_i^\alpha}{\sum_{j=1}^{n} g_j^\alpha}$$

For the constrained defuzzification problem the minimum is obtained for those d_g, that satisfy II and minimize $(\sum_{i=1}^{n} x_i \frac{g_i^\alpha}{\sum_{j=1}^{n} g_j^\alpha} - d_g)^2$.

Equivalently we can use as our objective function

$$|\sum_{i=1}^{n} x_i \frac{g_i^\alpha}{\sum_{j=1}^{n} g_j^\alpha} - d_g| \qquad\qquad \textbf{III.}$$

Thus the solution of the constrained linear programming problem is obtained as those feasible d_g, satisfying II, that are at a **minimal** distance from the generalized BADD defuzzified value(unconstrained solution) d_g^*:

This formulation of the solution of the restricted defuzzification problem has the advantage that it doesn't require the membership function $G(x)$ to have an analytical expression. It is enough to select only those x_i, characterized with membership grade equal or greater than the threshold t_g and to pick up the one, that is closest to the generalized BADD defuzzified value for the selected α.

The simplified solution of the constrained nonlinear programming problem is summarized in the following algorithm.

Algorithm 1 (*Solution for fixed α and t_g*).

1. Calculate:

$$d_g^* = \sum_{i=1}^{n} x_i \frac{g_i^\alpha}{\sum_{j=1}^{n} g_j^\alpha}$$

2. Select all x_i, $i = (1, n)$ with membership grade $H(x_i) \geq t_g$ this is the t_g-level set of H which we denote as H_{t_g}

3. Find x^* such that: $|d_g^* - x^*| = Min_{x \in H_{t_g}} |d_g^* - x|$

4. Set defuzzified value to x*: $d_g = x^*$

The algorithm works for fixed α and t_g. This means that the threshold has to be specified. For $t_g = 0$ defuzzified problem becomes unconstrained. The predetermined value of α defines the type of defuzzified value.

Example. Let consider fuzzy set F with membership function F(x), defined as follows:

F(x) = (0/1, .3/2, .5/3, .9/4, 1/5, 1/6, 1/7, .9/8, .8/9, .9/10, 1/11, 1/12, 1/13, 1/14, 1/15, .9/16, 8/17, .8/18, .5/19, 0/20)

and fuzzy set of allowable restrictions H with membership function H(x):

H(x) = (0/1, .2/2, .4/3, .6/4, .8/5, .8/6, 0/7, 0/8, .8/9, .7/10, 0/11, 0/12, .8/13, .8/14, .8/15,   .8/16, .7/17, .6/18, .1/19, 0/20)

The intersection of H and F gives us

G(x) = (0/1, .2/2, .4/3, .6/4, .8/5, .8/6, 0/7, 0/8, .8/9, .7/10, 0/11, 0/12, .8/13, .8/14, .8/15, .8/16, .7/17, .6/18, .1/19, 0/20)

The generalized defuzzified value of fuzzy set G, without constraints on the membership grade, i.e. $t_g = 0$ is for different α as follows:

α	0	1	30
d_g^*	10.78	11.01	11.15

These values of d_g^* are not from the set of allowable values.

If $t_g = 0$ we obtain $d_g = d_g^*$. For $t_g = 0.2$, the candidates for d_g according Algorithm 1 are the following values of x_i (with membership grade $H(x_i) \geq 0.2$):

2, 3, 4, 5, 6, 9, 10, 13, 14, 15, 16, 17, 18

Because the distance $|d_g^* - 10|$ is minimal for $\alpha = 0$; $\alpha = 1$; $\alpha = 30$, then

$$d_g = 10$$

is the arithmetic mean - like, the COA - like and the MOM - like defuzzified value of fuzzy set G.

If $t_g = 0.8$, the candidates for d_g are the following values of x_i (with membership $H(x_i)$ ≥ 0.8): {5, 6, 9, 13, 14, 15, 16}. Since for $\alpha = 0$ $d_g^* = 10.78$ the closest element for $\alpha = 0$ is 9, thus the *arithmetic mean-like* defuzzified value of fuzzy set G is $d_g = 9$.

For $\alpha = 1$ $d_g^* = 11.01$ the closest of the candidates for $\alpha = 1$ is 11 and since for $\alpha = 30$ $d_g^* = 11.15$ then again 11 is the closest of the candidates. Thus the *COA-like* and the *MOM-like* defuzzified values of fuzzy set G coincide at $d_g = 11$.

The result of defuzzification is depicted on Fig.3.

6. Conclusion

Extensions of the defuzzification procedure useful in the presence of restrictions were presented in this paper. Two alternative approaches to defuzzification under restriction were proposed. The first approach is based upon a selection using the performance of a random experiment, RAGE defuzzification, it also makes use of a nonmonotonic aggregation of the allowable and suggested solution sets. The second approach involves the modeling of the defuzzification process as a constrained nonlinear programming problem. An efficient algorithm, simplifying the solution of the nonlinear programming problem was proposed.

7. References

[1]. Filev, D. and Yager, R. R., "A generalized defuzzification method under BAD distributions," International Journal of Intelligent Systems 6, 687-697, 1991.

[2]. Yager, R. R. and Filev, D. P., "On the issue of defuzzification and selection based on a fuzzy set," Fuzzy Sets and Systems 55, 255-272, 1993.

[3]. Yager, R. R., "On the use of combinability functions for intelligent defuzzification," Proceedings Joint Fourth IEEE Conference on Fuzzy Systems and IFES, Yokohoma, (To Appear).

[4]. Pfluger, N., Yen, J. and Langari, R., "A defuzzification strategy for a fuzzy logic controller employing prohibitive information in command formulation," Proceedings First IEEE Conference on Fuzzy Systems, 1991.

[5]. Yager, R. R., "Nonmonotonic set theoretic operations," Fuzzy Sets and Systems 42, 173-190, 1991.

[6]. Alsina, C., Trillas, E. and Valverde, L., "On some logical connectives for fuzzy set theory," J. Math Anal. & Appl. 93, 15-26, 1983.

[7]. Dubois, D. and Prade, H., "A review of fuzzy sets aggregation connectives," Information Sciences 36, 85 - 121, 1985.

ON QUALITY DEFUZZIFICATION

Theory and an Application Example

H. HELLENDOORN AND C. THOMAS
Siemens AG, Dept. ZFE T SN 4
D-81730 Munich

Abstract. We describe six important defuzzification methods and their respective merits and shortcomings, dependent on the rules, domains, *etc.* Furthermore, we give an alternative approach for the case where the output fuzzys sets have different shapes. Finally, we give an example.

KEYWORDS: Fuzzy Set Theory, Fuzzy Control, Defuzzification.

1. Several Defuzzification Methods

As the topic 'Fuzzy-Control' is a field of research which is in discussion since only about ten years, the use of terms describing the various basics of a fuzzy system is unfortunately still ambiguous. In order to facilitate the understanding of the following investigations, a definition of the defuzzification methods under concern in this paper will be given. The resulting behavior of fuzzy controllers using any of these defuzzification methods will be discussed in the following chapters. Our investigation covers six defuzzification methods: Center-of-Area/Gravity defuzzification, Center-of-Sums defuzzification, Center-of-Largest-Area defuzzification, First-of-Maxima defuzzification, Middle-of-Maxima defuzzification, and Height defuzzification

Before we start to give an overview of these defuzzification methods, we will introduce some symbols that are needed to calculate a crisp defuzzification value. A linguistic variable is defined by $\langle X, \mathcal{L}X, \mathcal{X}, M_X \rangle$. Here X denotes the symbolic name of a linguistic variable, e.g. *age, temperature, error, change of error*, etc. $\mathcal{L}X$ is the set of *linguistic values* that X can take on. A linguistic value denotes a symbol for a particular property of X. In the case of the linguistic variable *temperature* T we have $\mathcal{L}T = \{cold, cool, comfortable, warm, hot\}$. We denote an arbitrary element of $\mathcal{L}X$ by LX. $\mathcal{X}$ is the actual physical domain over which the linguistic variable X takes its quantitative (crisp) values. In the case of the linguistic

167

Z. Bien and K. C. Min (eds.),
Fuzzy Logic and its Applications, Information Sciences, and Intelligent Systems, 167–176.
© 1995 *Kluwer Academic Publishers.*

variable *temperature* it can be the interval $[-10°C, 35°C]$. M_X is a semantic function which gives a 'meaning' (interpretation) of a linguistic value in terms of the quantitative elements of $\mathcal{X}$, i.e.

$$M_X : LX \to \widetilde{LX}, \tag{1}$$

where $\widetilde{LX}$ is a denotation for a fuzzy set defined over $\mathcal{X}$. Instead of $\widetilde{LX}$ we will also use μ_{LX}, i.e. the membership function without an argument.

We can now define a set of m rules as

if e_1 is $LE^{(k)}$ and ... and e_n is $LE_n^{(k)}$ then u is $LU^{(k)}$.

The firing of these rules with physical, crisp input values $e_1^*, \dots, e_n^*$ will either result in m clipped fuzzy sets denoted by $\widetilde{CLU}^{(1)}, \dots \widetilde{CLU}^{(m)}$, or m scaled fuzzy sets denoted by $\widetilde{SLU}^{(1)}, \dots \widetilde{SLU}^{(m)}$. In the formulas given below it does not make any difference whether one uses clipped or scaled fuzzy sets. We will use CLU as the general term. The *union* of these fuzzy sets will be denoted by $\tilde{U}$ or μ_U, $\tilde{U} = \bigcup_{k=1}^m \widetilde{CLU}^{(k)}$; the actual crisp defuzzification value we will denote by u^*.

The *area* of the fuzzy set $\tilde{U}$ is defined as

$$\text{area}(\tilde{U}) = \int_{\mathcal{U}} \mu_U(u)\, du, \tag{2}$$

where $\int$ is the mathematical integral. The *height* of $\widetilde{CLU}^{(k)}$ is equal to the degree of match of the k-th rule antecedent, and will be denoted by f_k. The *peak value* of $\widetilde{CLU}^{(k)}$ is equal to the peak value of its unclipped version $\widetilde{LU}^{(k)}$ or $\mu_{LU}^{(k)}$. If $\widetilde{LU}^{(k)}$ is a triangular membership function, then its peak value is that domain element on $\mathcal{U}$ which has degree of membership 1.

Center-of-Area/Gravity

This method (in the literature also referred to as Center-of-Gravity method) is the best well known defuzzification method. This method determines the center of the area below the combined membership function $\tilde{U}$. In the continuous case the defuzzified value u^* of $\tilde{U}$ is given by

$$u^*_{\text{CoA}} = \frac{\int_{\mathcal{U}} u \cdot \mu_U(u)\, du}{\int_{\mathcal{U}} \mu_U(u)\, du} = \frac{\int_{\mathcal{U}} u \cdot \max_k \mu_{CLU^{(k)}}(u)\, du}{\int_{\mathcal{U}} \max_k \mu_{CLU^{(k)}}(u)\, du}, \tag{3}$$

where $\int$ is the mathematical integral. In the discrete case ($\mathcal{U} = \{u_1, \dots, u_\ell\}$), the integral $\int_{\mathcal{U}} \mu_U(u)\, du$ is replaced by the sum $\sum_{i=1}^{\ell} \mu_U(u_i)$, etc.

Center-of-Sums

A similar but faster defuzzification method is Center-of-Sums. The motivation for using this method is to avoid the computation of $\tilde{U}$. The idea is to consider the contribution of the area of each $\widetilde{CLU}^{(k)}$ individually. Overlapping areas, if such exist, are reflected more than once by this method. The Center-of-Sums defuzzification method is formally given by

$$u_{\text{CoS}}^* = \frac{\sum\limits_{i=1}^{\ell} u_i \cdot \sum\limits_{k=1}^{n} \mu_{CLU^{(k)}}(u_i)}{\sum\limits_{i=1}^{\ell} \sum\limits_{k=1}^{n} \mu_{CLU^{(k)}}(u_i)} \quad \text{or} \quad u^* = \frac{\int_{\mathcal{U}} u \cdot \sum\limits_{k=1}^{n} \mu_{CLU^{(k)}}(u)\, du}{\int_{\mathcal{U}} \sum\limits_{k=1}^{n} \mu_{CLU^{(k)}}(u)\, du}, \quad (4)$$

for the discrete and continuous case respectively.

Height

Height defuzzification takes the peak value of each $\widetilde{CLU}^{(k)}$ and builds the weighted (with respect to the height f_k of $\widetilde{CLU}^{(k)}$) sum of these peak values. Thus neither the support or shape of $\widetilde{CLU}^{(k)}$ play a role in the computation of u^*. The Height method is both a very simple and very quick method. Let $c^{(k)}$ be the peak value of $\widetilde{LU}$ and f_k the height of $\widetilde{CLU}^{(k)}$. Then the Height defuzzification method in a system of m rules gives

$$u_{\text{H}}^* = \left(\sum_{k=1}^{m} c^{(k)} \cdot f_k \right) \Big/ \sum_{k=1}^{n} f_k. \quad (5)$$

Center-of-Largest-Area

When the overall output fuzzy set $\tilde{U}$ is non-convex, i.e., consists of at least two convex fuzzy subsets, then the Center-of-Largest-Area method determines the convex fuzzy subset with the largest area and defines the crisp output value u^* to be the Center-of-Area of this particular fuzzy subset. It is difficult to represent this defuzzification method formally, because this involves finding the convex fuzzy subsets, then computing the areas, etc.

First-of-Maxima

First-of-Maxima uses $\tilde{U}$ and takes the smallest value of the domain $\mathcal{U}$ with maximal membership degree in $\tilde{U}$. This is realized formally in three steps. Let $\text{hgt}(U) = \sup_{u \in \mathcal{U}} \mu_U(u)$ be the highest membership degree of $\tilde{U}$, and let $\{u \in \mathcal{U} \mid \mu_U(u) = \text{hgt}(U)\}$ be the set of domain elements with degree of membership equal to $\text{hgt}(U)$. Then u^* is given by

$$u_{\text{FM}}^* = \inf_{u \in \mathcal{U}} \{u \in \mathcal{U} \mid \mu_U(u) = \text{hgt}(U)\}. \quad (6)$$

The alternative of this method is called the Last-of-Maxima, where infimum is substituted by extremum.

Middle-of-Maxima

Middle-of-Maxima is very similar to First-of-Maxima or Last-of-Maxima. Instead of determining u^* to be the first or last from all values where $\mathcal{U}$ has maximal membership degree, this method takes the average of these two values. Formally,

$$u^*_{\text{MoM}} = \left(\inf_{u \in \mathcal{U}} \{ u \in \mathcal{X} \mid \mu_U(u) = \text{hgt}(U) \} + \sup_{u \in \mathcal{U}} \{ u \in \mathcal{U} \mid \mu_U(u) = \text{hgt}(U) \} \right) / 2 \tag{7}$$

2. Comparison and Evaluation of Defuzzification Methods

We will discuss the advantages and disadvantages of defuzzification methods. But before we can do this, we have to develop criteria an ideal defuzzification method should satisfy. Now it must be stated in advance that none of our defuzzification methods satisfies all criteria listed below, i.e., one has to weight these criteria for the application under concern in order to be able to make the right choice of the defuzzification method.

Some criteria for defuzzification methods are

1. *Continuity:* A small change in the input of the fuzzy controller should not result in a large change in the output. For example, in the case of a two-input, one-output fuzzy controller, when two inputs $(e_1^*, \dot{e}_1^*)$ and $(e_2^*, \dot{e}_2^*)$ differ slightly, then the corresponding output values u_1^* and u_2^* should differ slightly too.
2. *Disambiguity:* A defuzzification method is disambiguous if the algorithm to find u^* is well-defined. This criterion is not satisfied by Center-of-Largest-Area if there are two equally large areas.
3. *Plausibility:* Every defuzzified control output has a horizontal component $u^* \in \mathcal{U}$, and a vertical component $\mu_U(u^*) \in [0,1]$. We define u^* to be plausible if it lies approximately in the middle of the support of $\widetilde{U}$ and has a high degree of membership in $\widetilde{U}$.
4. *Computational complexity:* This criterion is particularly important in practical applications of fuzzy controller. The Height method, together with the Middle- and First-of-Maxima are fast methods, whereas the Center-of-Area method is slower. The computational complexity of Center-of-Sums depends on the shape of the output membership functions and whether max-min composition based inference or scaled inference is chosen.

A fifth criterion is the so-called weighting of the output fuzzy sets which constitutes the difference between the Center-of-Area and Center-of-Sums method. This criterion is handled separately from the former four criteria, because it is hard to say whether it is to be preferred or not. We consider it as a positive property.

5. *Weight counting:* A defuzzification method is weight counting if it sums up the overlapping parts in the overall ouput fuzzy set $\tilde{U}$. Center-of-Sums and Height defuzzification are weight counting. Center-of-Area, for example, uses solely $\tilde{U}$ and therefore is not weight counting.

Table 1 gives an overview of these defuzzification methods and their performance with respect to these five criteria.

3. Defuzzification in Unbalanced Output Domains

Chen and Hsu (Chen, 1991) discuss a very interesting problem in fuzzy logic with asymmetric output domains, i.e., output domains with fuzzy sets of different shape and support. Now let us consider such a domain in a rather extreme case, e.g. there are two rules firing, expressed by

if $\langle$antecedent$_i\rangle$ *then* u is S_i, $\qquad i = 1, 2.$

where $\mu_{S_i}(x) = \Lambda(x; s_{i\ell}, s_{im}, s_{ir})$, i.e. S_i has a triangular membership function where $s_{i\ell}$ and s_{ir} have membership degree 0 and s_{im} has membership degree 1. The consequent of the first rule is a fuzzy set with a rather small support ($d_1 = s_{1r} - s_{1\ell}$ is small), whereas the second rule has a consequent with a very big support ($d_2 = s_{2r} - s_{2\ell}$ is big). This means that the first rule states that the output value is most certainly equal to s_{1m} and can have small deviations. The location of the output value of the second rule can not be given exactly, it can change rather much. This means that the first rule is a very good one and the second is a very bad one. However, when Center-of-Area, Center-of-Sums or Center-of-Largest-Area defuzzification is used, the crisp output value is overshadowed by the vague output value and will even not lie in the support interval of the first output fuzzy set, which is an unacceptable result. It is clear that in a situation like this the height method does not perform well too, because the it assumes that the support is the same for all the output fuzzy sets. The method proposed by Chen and Hsu (Chen, 1991) to solve this problem is some kind of normalizing the output fuzzy sets. They not normalize in advance, because in that case all the intersecting areas would change, but they do it afterwards. The main disadvantage of this method is that it ignores the fact that rules with 'crisper' outputs are *more important* than those with 'fuzzier' outputs.

Suppose now that there is a rule base with n rules

if $\langle$antecedent$_i\rangle$ *then* u is S_i, $\qquad i = 1, \ldots, n,$

where $\mu_{S_i}(x) = \Lambda(x; s_{i\ell}, s_{im}, s_{ir})$ like above. Let furthermore d_i be the length of the interval $[s_{i\ell}, s_{ir}]$, $d_i = s_{ir} - s_{i\ell}, i = 1, \ldots, n$. These values can be calculated in advance. Furthermore, let f_i $(i = 1, \ldots, n)$ be the value with which rule R_i fires, e.g. $f_1 = 0.83$ and $f_2 = 0.17$. So $\tilde{S}_i$ is clipped at f_i. We call f_i the clipping-value. Then we can use the quotient $w_i = f_i/d_i, i = 1, \ldots, n$, as the weight that has to be attached to each clipped output fuzzy set, i.e., w_i is proportional to f_i and inversely proportional to d_i: w_i is maximal if the support of the output set is small and the clipping value is high, like the first rule in Section 3; w_i is minimal if the support of the output set is large and the clipping value is low, like the second rule in Section 3. So w_i can be considered as a quality measure of the clipped output fuzzy set. If w_i is high, then the information of the clipped output set is high, if w_i is low, then this information is low.

The crisp defuzzified value u^* of this set of rules for particular input values is then given by

$$u^* = \left(\sum_{i=1}^{n} s_{im} \cdot w_i \right) \bigg/ \sum_{i=1}^{n} w_i. \tag{8}$$

We will call this defuzzification method the *Quality Method.*

As an example, suppose there are two rules R_1 and R_2 with conclusions $\mu_{S_1}(u) = \Lambda(u; 7, 8.5, 10)$ and $\mu_{S_2}(u) = \Lambda(u; 2, 5.5, 9)$; and suppose they fire with values $f_1 = 0.8$ and $f_2 = 0.2$. Then $w_1 = 0.8/3 \approx 0.267$ and $w_2 = 0.2/7 \approx 0.029$, i.e. the result of rule 1 is weighted approximately ten times more important as the result of rule 2. The defuzzified output value u^* is

$$u^* = \frac{s_{1m} \cdot w_1 + s_{2m} \cdot w_2}{w_1 + w_2} = \frac{5.5 \cdot \frac{0.2}{7} + 8.5 \cdot \frac{0.8}{3}}{\frac{0.2}{7} + \frac{0.8}{3}} \approx 8.2. \tag{9}$$

This means that a value very close to the peak value 8.5 of S_1 results.

An alternative approach is to additionally weight the supports of the output fuzzy sets with the following formula

$$u^* = \left(\sum_{i=1}^{n} \frac{s_{im} \cdot f_i}{d_i^{\xi}} \right) \bigg/ \left(\sum_{i=1}^{n} \frac{f_i}{d_i^{\xi}} \right), \qquad \xi \geq 0. \tag{10}$$

For $\xi = 1$ this results in Eq. (8); for $\xi = 0$ this is the Height defuzzification method, the supports of the output fuzzy sets are not taken into account; for $0 < \xi < 1$ one gets an output value that lies between the Height method and the Quality method; and for $\xi > 1$ one obtains output values that are even more influenced by the rule with the highest weight w_i. We will call this defuzzification method the *ξ-Quality Method.*

Another alternative approach is to use a convex sum of the Center-of-Area method and the Quality method. Let u_1^* be the defuzzified output

TABLE 1. A comparison of different defuzzification methods

	CoA	CoS	MoM	FoM	HM	CLA	QM
Continuity	++	++	--	--	--	0	++
Disambiguity	++	++	0	++	++	--	++
Plausibility	0	+	0	0	+	++	+
Comp. complexity	--	0	+	+	++	--	++
Weight counting	--	++	--	--	0	--	++
Quality regarding	--	--	0	0	-	--	++

value obtained by the Center-of-Area method and let u_2^* be the defuzzified output value obtained by the Quality method, then the convex sum of these two is given by

$$u^* = \gamma \cdot u_1^* + (1 - \gamma) \cdot u_2^*, \qquad 0 \leq \gamma \leq 1. \tag{11}$$

For $\gamma = 1$ this results in the Center-of-Area method, for $\gamma = 0$ this results in the Quality method. For values of γ between 0 and 1, u^* lies between these two methods.

Until now we have only considered membership functions with straight lines, but what happens when there are curved lines too? If we rewrite Eq. (8) into

$$u^* = \frac{\sum_{i=1}^{n} s_{im} \cdot w_i}{\sum_{i=1}^{n} w_i} = \frac{\sum_{i=1}^{n} \dfrac{s_{im} \cdot f_i^2}{\text{area}(S_i)}}{\sum_{i=1}^{n} \dfrac{f_i^2}{\text{area}(S_i)}} \tag{12}$$

where area(S_i) is the area under the original symmetric output fuzzy set $\tilde{S}_i$, i.e., let μ_{S_i} be the membership function of $\tilde{S}_i$, then

$$\text{area}(S_i) = \int_{\mathcal{X}} \mu_{S_i}(u) \mathrm{d}u, \tag{13}$$

which is equal to $\frac{1}{2} f_i d_i$ in the triangular case.

Table 1 shows the advantages of the ξ-Quality method in comparison to the six defuzzification methods given before. 'Quality regarding' is added as a criterion whether the defuzzification method takes into account unbalanced domains.

4. Description of the Example Scenario

In this section we will consider the problem of 'Overcrowding Control'. The task is to prevent a specified area, e.g. the platform of a subway station, from being overcrowded. To perform this task, one might want to use multiple sources of information provided by different types of sensors: in our example we could use optical information provided by a static camera and acoustical information provided by a set of microphones. These two sources of information lead to different estimates of the overcrowding level. The estimates can be uncertain, contradictory or even partially wrong. But the various types of sensor data can be fused to get a more reliable estimate of the overcrowding level on the platform. This estimate can be used for the subsequent decision process, which e.g. might cause the entry of the platform to be closed for a while.

This approach leads to a fuzzy system where the output domain of overcrowding-level can not be covered with a single set of membership functions. Instead it is necessary to describe the output domain by two separate sets of membership functions. From iterated experiments one could come to the conclusion (with respect to the applied algorithms) that information gained by the camera is more reliable than information gained by the microphone. Although the representation of uncertainty, vagueness or precision in some systems is not a very easy one there is a very simple and intuitive way to represent this reliability of information in a fuzzy system.

Let us assume two rules in our Data Fusion Knowledge Base:

Rule 1: IF Noise IS very high THEN OvercrowdingLevel IS high
Rule 2: IF Movement IS high THEN OvercrowdingLevel IS high

To express the higher precision of optical source of information (or the applied algorithms), one would probably use two membership functions for high and label them high_opt and high_acc (see Fig. 1). However the fuzzy system should behave in such a way that linguistic values which show less variance have more influence on the output value than others.

Let us now assume the following situation. The result of the decision making process might look like shown in Fig. 2, which means that the acoustical information leads to the estimate of the overcrowding to be high with an degree of 0.2 and the optical information estimates on very high with a degree of 0.8. Applying the center of gravity defuzzification would lead to a final result like shown in Fig. 2. This is due to the fact that Center of Gravity takes into account both the area below and the center of the cut fuzzy sets but does not consider the shape. In the situation described in Fig. 2 there are two reasons why the area of high_acc should nearly be neglected. First, the belief value of high_acc is only 0.2 whereas the belief value of high_opt is 0.8. Secondly, the shape of the original membership function high_acc expresses a high degree of vagueness whereas high_opt represents quite precise knowledge.

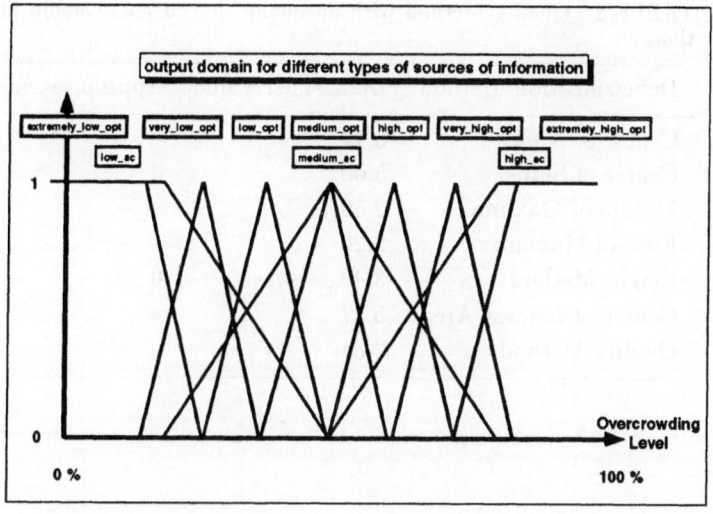

Figure 1. Representation of more and less precise knowledge in a Fuzzy System

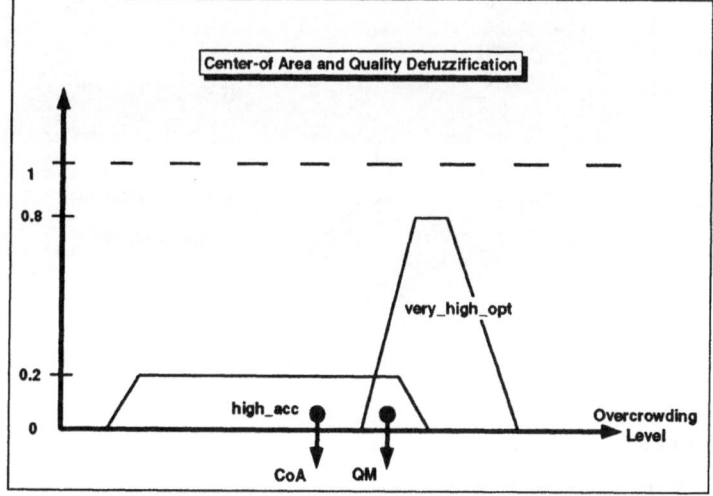

Figure 2. Center of Area and Quality defuzzification with unequally shaped membership functions

Obviously Center of Gravity does not lead to the result that we would like to have. One expects the system to provide as a final result of the defuzzification process a value as shown is Table 2 and Fig. 3. The quality defuzzification performs exactly this.

TABLE 2. Quality Method with unequally shaped membership functions

Defuzzification Method	Defuzzified Value	Appropriateness
Center of Area	5.17	−
Center of Sums	5.00	0
Middle of Maxima	2.50	− −
First of Maxima	2.20	− −
Height Method	3.40	0
Center of Largest Area	5.17	−
Quality Method	2.81	+ +

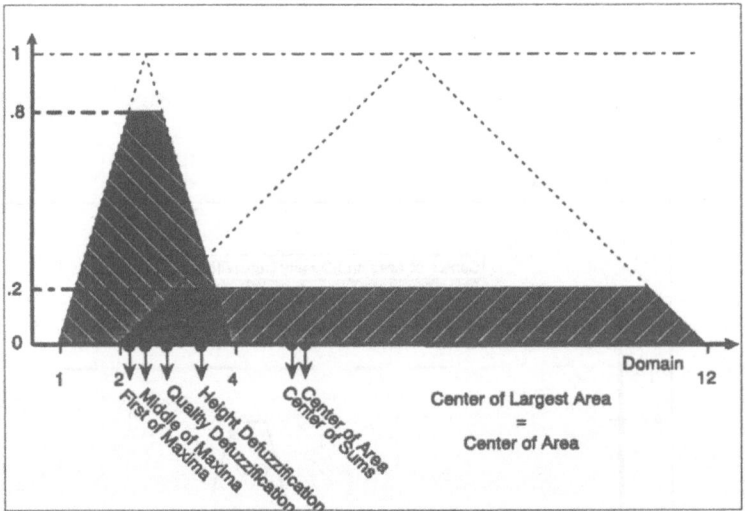

Figure 3. Results of different defuzzification strategies

References

Chen, Y.-Y. & Hsu, S.-T. I., *Rules Aggregations in Fuzzy Ruled-Based Systems*, Internal Report National Taiwan University, 1991.

Hellendoorn, H., "Fuzzy Logic and Fuzzy Control." In: René van der Vleuten et al. (eds.), *Clear Applications of Fuzzy Logic*, (Proceedings IEEE-Symposium Delft, 17 Oct. 1991) Delft, pp. 57–82.

Zadeh, L.A., "Fuzzy Sets," *Information and Control*, 8(1965)338–353.

A FUZZY NEURAL NETWORK: STRUCTURE AND LEARNING

M. FIGUEIREDO & F. GOMIDE
UNICAMP / Fee / Dca
CP. 6101
13081-970 - Campinas - SP - Brazil
mauri / gomide@dca.fee.unicamp.br

W. PEDRYCZ
University of Manitoba / Dece
R3T 2N2 - Winnipeg - Manitoba - Canada
pedrycz@eeserv.ee.umanitoba.ca

1. Introduction

A promising approach to get both the benefits of neural networks and fuzzy logic systems and to solve their respective problems is to combine them into an integrated system such that we can bring the learning and computational power of neural networks into the fuzzy logic systems, and the representation and reasoning capabilities of fuzzy logic systems into the neural networks. For system modelling and control purposes their combination should provide an approach where structured knowledge of complex ill-defined systems is processed in a qualitative way, allowing reasoning and consideration of essential a priori information and performance criteria. Learning features should provide training procedures for synthesis, design, and implementation. Systems that combine neural network with fuzzy logic are called neurofuzzy systems.

The fusion of fuzzy logic with neural network is, since in its early stages, concerned with the developments of multi-input, multi-output neuron models [4]. More recently, the compositional operator and fuzzy relations were used in a neurofuzzy structure proposed by Pedrycz [6]. Pedrycz's network represents knowledge at a very aggregated level. The inputs and outputs are fuzzy sets and the coded knowledge can not be extracted in fuzzy if-then rule format. Lin and Lee [5] describe a fuzzy neural network structure which is based on the possibilistic inference method rather than compositional one. The fuzzy predicate membership functions and the inference operators are represented as node parameters. This characteristic is not appropriate to be handled by neurocomputing. However, fuzzy rules are easily identified in the network structure. Gomide and Rocha [2] and Keller et al. [3] describe compositional based fuzzy neural network and possibilistic based fuzzy neural network, respectively.

In this work, a fuzzy neural network structure is proposed, exploring a particular, simplified approach for fuzzy reasoning. Given a set of fuzzy if-then rules, the network

Z. Bien and K. C. Min (eds.),
Fuzzy Logic and its Applications, Information Sciences, and Intelligent Systems, 177–186.

topology easily allows their encoding and processing. Otherwise, rules can be discovered if a set of input-output data is provided. For this task, a learning procedure for a particular class of membership functions is provided. Due to its structure, initial knowledge can easily be embedded in the network, speeding up learning and rules tuning.

2. Fuzzy Rules Based Neural Network

The structure of the system under discussion will be centered around a set of if-then conditional statements (rules) given as follows:

input premise: X_1 is A_1 and ... X_M is A_M.

rule 1: If X_1 is A_1^1 and ... X_M is A_M^1 then y is g^1.

....

rule N: If X_1 is A_1^N and ... X_M is A_M^N then y is g^N.

_____

consequence: y is g.

where: X_j is a fuzzy variable, A_j and A_j^i are the corresponding fuzzy sets, while "y" is a real variable, and g^j constants defined in the output space. All the universes are assumed to be discrete. Note that the essence of the above system is to provide approximation of "y" via a series of fixed (quantization) values. To simplify the notation we denote by z_k the grade of membership of x_k in Z, that is

$$Z(x_k) = z_k;$$

with x_k being a numerical value in the input space. The numerical consequence y is determined via a sequence of the three reasoning stages:

1) Matching: For each rule "i" and each antecedent "j" we compute the possibility measure P_j^i for fuzzy sets A_j and A_j^i. P_j^i is given by:

$$P_j^i = \max_k \{\min (a_{jk}, a_{jk}^i)\}, \text{where this maximum is taken over all k.} \quad (2.1)$$

2) Antecedent Aggregation: For each rule "i" its activation level is computed as intersection of its subconditions,:

$$H^i = \min_j \{P_j^i\}, j = 1, ... , M. \quad (2.2)$$

3) Rule aggregation: the overall numerical output is computed as a weighted sum:

$$y = \frac{\sum\limits_{i=1}^{N} H^i g^i}{\sum\limits_{i=1}^{N} H^i}. \tag{2.3}$$

In (2.1) and (2.2) the maximum and minimum operators are particular examples of T-norms and S-norms; obviously their use will give rise to much more general and flexible constructs.

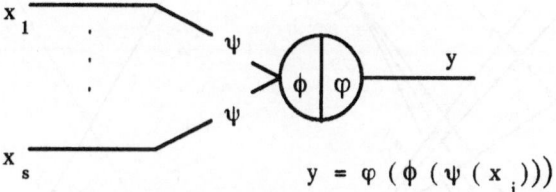

$$y = \varphi \, (\, \phi \, (\, \psi \, (\, x_i)))$$

Figure 2.1: The neuron model: x_i's are inputs, y is the output, ψ and ϕ are the synaptic operator and the input aggregation operator respectively, and φ is the decoding function.

In this work we borrow some recent results from the neurophysiology. Usually, the synaptic operator and the input aggregation operator are important neuron model specifications, see Fig. 2.1. The biological model suggests that these operators may be implemented using T-norms or S-norms [7].

The proposed fuzzy neural network (see Fig. 2.2) is a feedforward architecture with five layer of neurons.

The first layer is divided into groups of neurons. Each group corresponds to a single fuzzy variable standing in the antecedent of this rule. This implies M groups of neurons situated in this layer. The neurons in each of those groups are utilised to represent a discrete universe of discourse. More precisely, the neuron receives an input signal, transforms and transmits it to the second layer. In particular, a_{jk} will be used to denote a signal transmitted by the k-th neuron placed in the j-th group (universe). The output y is generically given by:

$$y = \varphi(\, u_j \,).$$

In this work we assume that the transformation operation of the input neuron applies to a pointwise numerical information (singleton). For an interval x_k such that $x_k = [x_I, x_F)$, the transformation performed by the neuron representing the k-th interval is given by, (see also Fig. 2.3);

The second layer comprises N groups (the number of rules). For each group there are M neurons. This layer accomplishes the first stage of inference namely, matching.

180

The j-th neuron of i-th group computers P_j^i. The k-th neuron of the j-th group of the first layer connects with it through the synapse whose weight is a_{jk}^i. The synaptic operator is taken as the minimum and the input aggregation operator is implemented as the maximum.

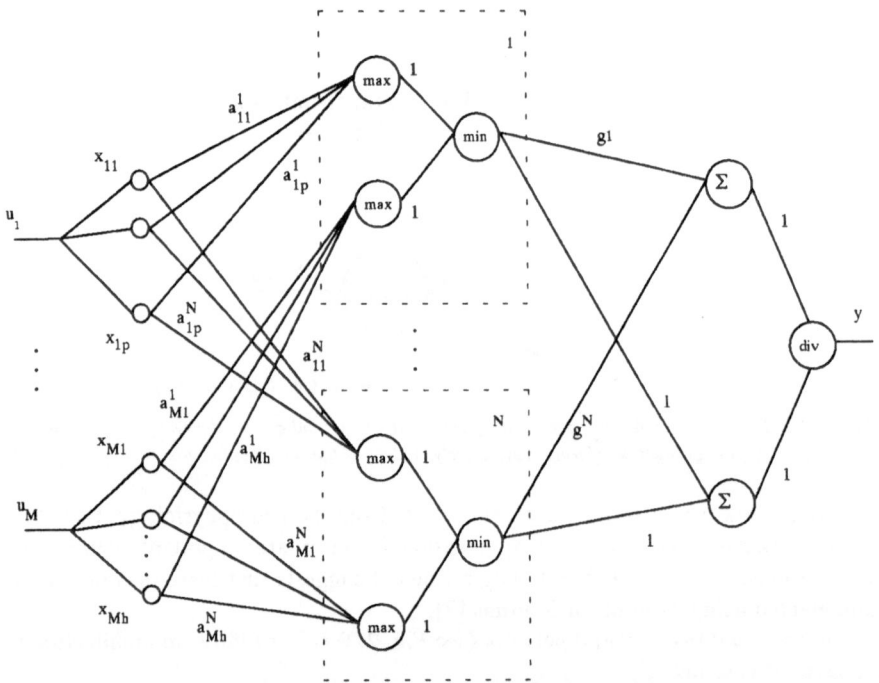

Figure 2.2: The fuzzy neural network.

For each group "j" of the second layer, a neuron in the third layer determines the degree of aggregation of the antecedents, see (2.2). The synapses do not modify the signal (we may assume that the synaptic weight is fixed at one and the synaptic operator is the algebraic product). Gomide and Rocha [1] describe a neuron model which provides the maximum and minimum operators (max and min neurons). These are the neurons which comprise the second and third layers

Each i-th neuron "i" in the third layer links with the two neurons of the fourth layer. The aggregation operator associated with these neurons is the algebraic sum. One of them connects with all the neurons in previous layer through the synapses whose weights are equal to g^j. The synaptic processing modulates the signal according the algebraic product. Its output constitutes the numerator of (2.3). The other neuron also connects with all the neurons in the previous layer. The signal H^i is received there without any modification, The output of it is just the denominator of (2.3).

The last layer consists of a single neuron which role is to compute the quotient of these two signals.

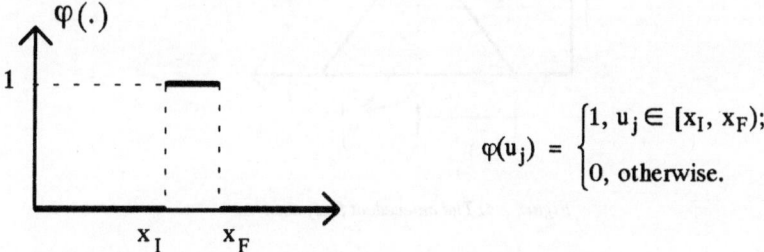

$$\varphi(u_j) = \begin{cases} 1, u_j \in [x_I, x_F); \\ 0, \text{ otherwise.} \end{cases}$$

Figure 2.3: Input neuron decoding function.

3. Learning Method

The learning is carried out in a supervised mode. We assume that a data set of input-output pairs $\{(u^1, y_d^1),..., (u^S, y_d^S)\}$ where $u^r = (u_1^r,..., u_M^r)^T$ is given. To start learning, additional information such as the fuzzy partitions of input space and the shape of the fuzzy sets A_j^i in the antecedents should be provided. To simplify the presentation, the fuzzy set shapes are assumed to be isosceles triangles. Let us remind that these triangular fuzzy sets should "cover" all input domain. Each combination of fuzzy sets in the partition gives rise to one fuzzy rule, with the consequent set as the center of domain. The first phase constructs the network and imbeds the initial knowledge about the structure e.g., the necessary fuzzy rules and preliminary antecedent fuzzy sets.

In the next phase, the parametric learning of the antecedent fuzzy sets of as well as the numerical values of consequents g_i are performed according to:

Min $(Q(y^r, y_d^r))$

$c_j^i, b_j^i,$ and $g_i.$

subject to: $\hfill (2.4)$

$$A_j^i(x_{jk}) = \begin{cases} 1 - \dfrac{|x_{jk} - c_j^i|}{b_j^i}, \text{ if } (c_j^i - b_j^i) \le x_{jk} \le (c_j^i + b_j^i); \\ 0, \text{ otherwise;} \end{cases}$$

where c_j^i and b_j^i are nodal values and bounds of the triangular fuzzy numbers, Fig. 2.4.

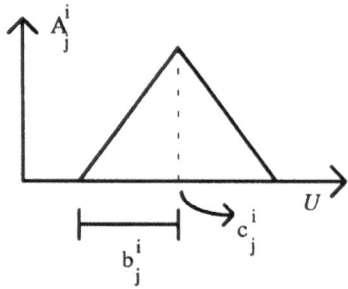

Figure 2.4: The antecedent fuzzy sets.

The performance index is a standard sum of squared errors (MSE),

$$Q(y^r, y_d^r) = \sum_{r=1}^{S} \frac{(y^r - y_d^r)^2}{2}.$$

(2.5)

Usually, due to a particular fuzzification strategy, the fuzzy set A_j is a singleton, and consequently, the possibility measure (2.1) is:

$$P_j^i = A_j^i(u_j^s) = \begin{cases} 1 - \dfrac{|x_{jk} - c_j^i|}{b_j^i}, & \text{if } (c_j^i - b_j^i) \leq x_{jk} \leq (c_j^i + b_j^i) \text{ and} \\ & u_j^s \text{ is a element of } x_{jk}; \\ 0, & \text{otherwise.} \end{cases}$$

(2.6)

By substituting P_j^i, H^i and g^i expressions [(2.2), (2.3) and (2.6)], the performance index Q(.) can be made more explicit. Considering z as being a generic parameter of the membership functions, the gradient descent method gives rise to the following adjustment formula,

$$z(t+1) = z(t) - \alpha_z \frac{\partial Q(z)}{\partial z}$$

(2.7)

Following the calculus suggested by Pedrycz [9], the derivatives of the min and max operations are defined as:

$$\frac{d \min(x,a)}{dx} = \begin{cases} 1, & x \leq a; \\ 0, & \text{otherwise.} \end{cases} \qquad \frac{d \max(x,a)}{dx} = \begin{cases} 1, & x \geq a; \\ 0, & \text{otherwise.} \end{cases}$$

The derivative of Q with respect to the quantization values g^k is computed accordingly,

$$\frac{\partial Q}{\partial g^k} = \sum_{r=1}^{S}\left(y^r - y_d^r\right)\frac{H^k g^k}{\sum\limits_{i=1}^{N} H^i} \tag{2.8}$$

For the parameters of the triangles numbers c_j^i and b_j^i the respective formulas can be derived in an analogous way. Hence,

$$\frac{\partial Q(z)}{\partial z} = \sum_{r=1}^{S}\left(y^r - y_d^r\right)\frac{\partial y^s}{\partial H^i}\frac{\partial H^i}{\partial P_j^i}\frac{\partial P_j^i}{\partial z} \tag{2.9}$$

Overall, the learning algorithm can be summarised by the following steps:

Begin

Initialise the parameters maintaining the basic requirements;
set iter = 0
Repeat

Update the parameters g^k, c_j^i, and b_j^i by computing the adjustments: $\Delta z = -\partial Q(z) / \partial z$.

iter = iter +1

Until $Q(y^r, y_d^r) \le \varepsilon$ or iter $\ge$ itermax.
End.

4. Simulation Results

We next present two experiments. In the first, the network learns from the data generated by the nonlinear function $f(.): [0,1] \times [0,1] \rightarrow [0,1]$:

$$f(x_1, x_2) = x_1 + x_2 - 2x_2 x_1.$$

We consider 25 rules. The fuzzy partitions consist of five fuzzy sets. The center positions and base lengths of the isosceles triangles were set in such way that the rules cover all the domain $[0,1] \times [0,1]$. The consequent part were set in the domain center. The Fig. 4.1 shows three of the initial rules induced by the initialisation step and

184

introduced in the network. The Fig. 4.2 shows the corresponding rules after learning. For all input $\mathbf{u}^r$ the errors ($y^r - y_d^r$) were less than 0.1. Fig 4.3 shows the learning steps evolution.

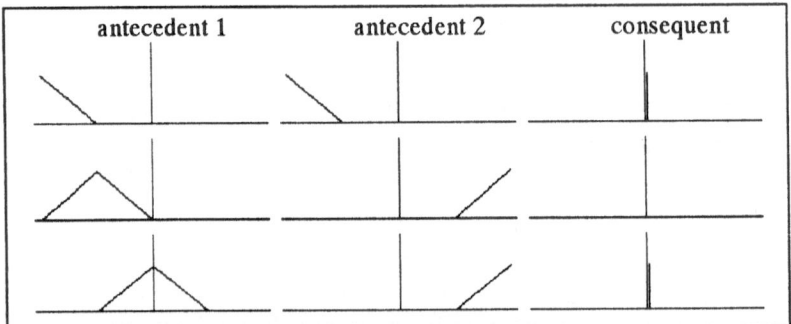

Figure 4.1: Example 1: three initial rules.

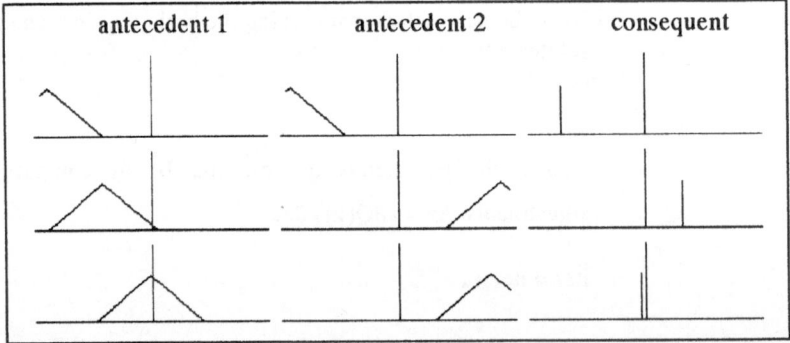

Figure 4.2: Example 1: corresponding rules after learning.

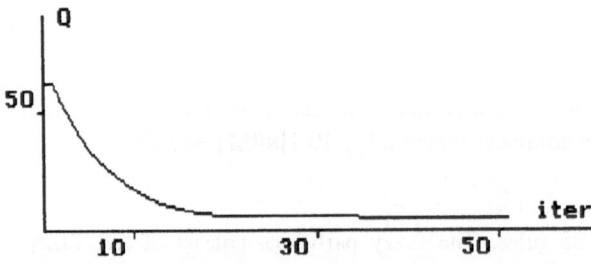

Figure 4.3: Example 1: Performance index evolution during learning.

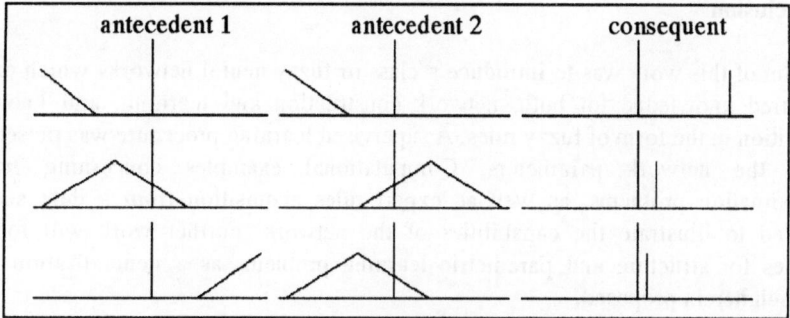

Figure 4.4: Example 2: three rules provided by an expert.

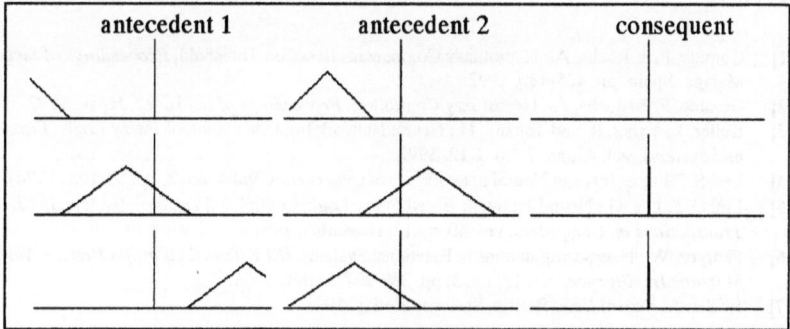

Figure 4.5: Example 2: Rules introduced into the network structure.

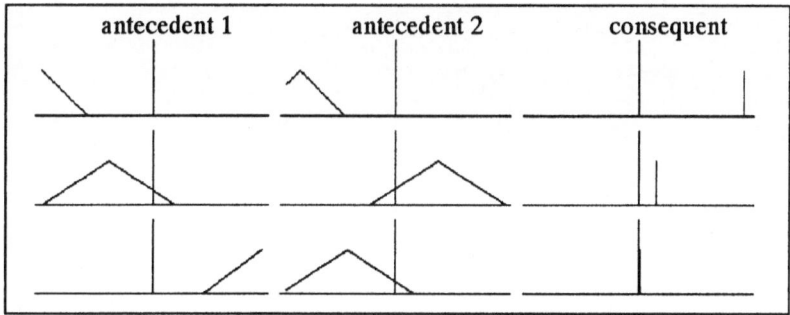

Figure 4.6: Example 2: the same three rules, extracted after learning.

In the next example the network learns rules provided by an expert. Sixteen rules are considered. The domains of the two antecedents and of the consequent are [-12,12]. The Fig. 4.4 shows three correct rules. The initial rules generated and the learned rules obtained are illustrated in Fig. 4.5 and 4.6, respectively.

5-Conclusion

The aim of this work was to introduce a class or fuzzy neural networks which exploits structured knowledge for both, network construction and learning, and knowledge acquisition in the form of fuzzy rules. A supervised learning procedure was presented to adjust the network parameters. Computational examples concerning function approximation problems, as well as expert rules acquisition from a data set were provided to illustrate the capabilities of the network. Further work will focus on schemes for structure and parametric learning problems as a generalisation of the approach herein proposed.

6. References

[1]   Gomide, F. & Rocha, A.: Neurofuzzy Components Based on Threshold, *Proceedings of Sicica'92*, Malaga, Spain, pp. 425-430, 1992.

[2]   Gomide, F. & Rocha, A.: Neurofuzzy Controllers, *Proceedings of Iizuka-92*, Japan, 1992.

[3]   Keller, J., Yager, R. and Tahami, H.: Neural Network Implementation of Fuzzy Logic, *Fuzzy Sets and Systems*, vol. 45, no. 2, pp. 1-12, 1992.

[4]   Lee, S.: "Fuzzy sets and Neural network", *J. of Cybernetics*, vol.4, no. 2, pp. 83-103 (1974).

[5]   Lin, C. & Lee, G.: Neural-Network-Based Fuzzy Logic Control and Decision System, *IEEE Transactions on Computers*, vol. 40, no.12, December, 1991.

[6]   Pedrycz, W.: Neurocomputations in Relational Systems, *IEEE Transactions on Pattern Analysis and Machine Intelligence*, vol. 13, no. 3, pp. 289-297, March, 1991.

[7]   Rocha, A. *Neural Nets,* Berlim, Springer-Verlag, 1992.

Acknowledgements: The first author acknowledges the support of FAPESP 93/3034-8. The second is grateful for CNPq grant #300729/86-3 and ESPRIT ECLA 005 Project.

NONLINEAR FUNCTION APPROXIMATION BASED ON NEURO-FUZZY INTERPOLATORS

IL HONG SUH and TAE WON KIM
Intelligent Control and Robotics Lab.,
Dept. of Electronics Eng., Hanyang Univ.,
Daehak-dong, Ansan-si,
Kyungki-do 425-791, KOREA

ABSTRACT

In this paper, a fuzzy-neural interpolating network is proposed to efficiently approximate a nonlinear function, where a linear scalar function is employed as an activation function of the network. Specifically, basis functions are first constructed by Fuzzy Membership Function-based Neural Network (FMFNN). And the fuzzy similarity, which is defined as the degree of matching between actual output value and the output of each basis function, is used to determine initial weightings of the proposed network. Then the weightings are updated in such a way that square of the error is minimized. To show the capability of function approximation of the proposed fuzzy-neural interpolating network, a numerical example is illustrated.

1. Introduction

It is well known that the nonlinear function approximation can be often solved by finding a set of coefficients for a finite number of fixed nonlinear basis functions (Sanger, 1991). The Radial Basis Function (RBF) network can offer approximation capabilities similar to those of the two-layer neural network, provided that the hidden layer of the RBF network is fixed appropriately (Powell, 1987). However, the performance of the RBF network critically depends upon the chosen centers of the basis functions (Chen *et. al.*, 1991). To overcome such difficulties in choosing centers of RBF, the FMFNN was proposed in (Suh and Kim, 1992, 1994), where it was shown that the structure of the RBF networks could be similar to that of fuzzy logic when employing the additive combination technique for the inference (Pedrycz, 1989) and the FMFNN could play a role of a basis function. And a simple interpolating network was proposed to reduce difficulties in determining fuzzy rules and membership functions, when a large number of input variable are necessary for the function approximation (Suh and Kim, 1994). However, since this interpolating

Z. Bien and K. C. Min (eds.),
Fuzzy Logic and its Applications, Information Sciences, and Intelligent Systems, 187–195.
© 1995 *Kluwer Academic Publishers.*

network has only one similarity node to learn function values, it might be difficult to accurately approximate a complex function.

In this paper, a modified version of the fuzzy-neural interpolating network in Suh and Kim (1994) is proposed to efficiently approximate a nonlinear function by employing a linear scalar function. Specifically, basis functions are first constructed by Fuzzy Membership Function based Neural Networks (FMFNN). And the fuzzy similarity, which is defined as the degree of matching between actual output value and the output of each basis function, is employed to determine how much each FMFNN should contribute to the approximation of a function. A fuzzy-neural interpolating network is then proposed by combining the FMFNN and the fuzzy similarity.

2. Fuzzy Membership Function-based Neural Network

Consider the following fuzzy relations :

$$R^i : \text{If } y_1 \text{ is } A_{i1}, y_2 \text{ is } A_{i2}, \text{, and } y_p \text{ is } A_{ip}, \text{ then } u \text{ is } B_i, \quad i = 1, 2, \ldots, q. \quad (1)$$

Here, y_i, for $i = 1, 2, \ldots, p$, is the input variable and u is the output fuzzy variable fuzzified with a singleton membership function. A_{ij} and B_i for $i = 1, 2, \ldots, q$ and $j = 1, 2, \ldots, p$ are input and output linguistic (fuzzy-set) values, respectively. And let $\mu_{ij}^A(y_j)$ and $\mu_i^B(u)$ be the membership functions for A_{ij} and B_i, respectively. If we let $\mu_i^B(u)$ be a normal singleton located at $u = \lambda_i$ for each i, and apply the centroidal defuzzification technique to $\mu^B(u)$, then $\mu_i^B(u)$ becomes $\mu_i^B(\lambda_i)$. And thus, regardless of types of inference, the scalar output u can be obtained by

$$u = \sum_{i=1}^{q} \lambda_i \frac{\Phi_i(y_1^0, y_2^0, \cdots, y_p^0)}{\sum_{k=1}^{q} \Phi_k(y_1^0, y_2^0, \cdots, y_p^0)} = \sum_{i=1}^{q} \lambda_i \tilde{\Phi}_i(\underline{y}^0), \quad (2)$$

where $\Phi_i(y_1^0, y_2^0, \cdots, y_p^0)$, $\underline{y}^0$ and $\tilde{\Phi}_i(\underline{y}^0)$ are defined as

$$\Phi_i(y_1^0, y_2^0, \cdots, y_p^0) \overset{\Delta}{=} \min\left\{ \mu_{ij}^A(y_j^0) | j = 1, 2, \cdots, p \right\}, \quad (3\text{-}1)$$

$$\underline{y}^0 \overset{\Delta}{=} (y_1^0, y_2^0, \cdots, y_p^0), \quad (3\text{-}2)$$

and

$$\tilde{\Phi}_i(\underline{y}^0) \stackrel{\Delta}{=} \frac{\Phi_i(y_1^0, y_2^0, \cdots, y_p^0)}{\sum\limits_{k=1}^{q} \Phi_k(y_1^0, y_2^0, \cdots, y_p^0)}. \tag{3-3}$$

Eq.1 can play a role of approximating a function as the RBF network can do [4, 5, 6]. Thus we called Eq.1 as *"Fuzzy Membership Function based Neural Network (FMFNN)"*, where λ_i's are the neural weights to be trained by using the linear least square method. In applying Eq.1 to a function approximation, as in (Kim, 1994), we put a linear scalar function $g : R \rightarrow R$ given by

$$g(u) = k_1 u, \tag{4}$$

at the output node together with the scaling factor K to effectively account for the maximum magnitude of the function output. In Eq.4, k_1 is a constant implying the slope of output node function. For the function approximation, let $f(\underline{y})$ be a scalar function to be approximated, and let $u(\underline{y})$ be an approximation of $f(\underline{y})$. Then $u(\underline{y})$ can be represented as

$$u(\underline{y}) = K g\left(\sum_{i=1}^{q} \lambda_i \tilde{\Phi}_i(\underline{y})\right), \tag{5}$$

if the FMFNN is utilized for the function approximation.

When the error function is given by

$$E = \tfrac{1}{2}\left(f(\underline{y}) - u(\underline{y})\right)^2, \tag{6}$$

weight changes $\Delta\lambda_i$'s could be chosen to be proportional to $-\partial E/\partial \lambda_i$, i.e.,

$$\Delta\lambda_i = \eta_1 k_1 K\left(f(\underline{y}) - u(\underline{y})\right)\tilde{\Phi}_i(\underline{y}), \tag{7}$$

where η_1 is learning-rate parameter. Thus the learning rule for adapting weight can be given as

$$\lambda_i^{t+1} = \lambda_i^t + \eta_1 k_1 K\left(f(\underline{y}) - u(\underline{y})\right)\tilde{\Phi}_i(\underline{y}), \tag{8}$$

where t is an integer implying the number of learning trials. The schematic diagram of the FMFNN with p inputs and a scalar output is depicted in Fig.1.

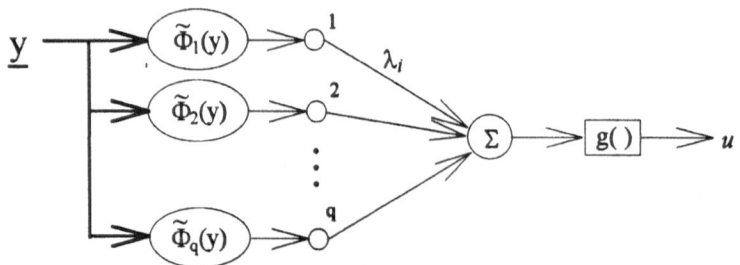

Fig.1. Schematic diagram of an FMFNN

It is remarked that the initial weight values are known to play important roles of obtaining the global minimum in most of neural networks (Rumelhart *et. at.*, 1986). In this respect, the heuristic choice of initial values considering fuzzy rules for the FMFNN can provide better performances than the random choice of the initial weight values for the RBF network, since the weights of the FMFNN have clear meanings as locations of the singleton fuzzy membership functions for the 'THEN' parts of the fuzzy rules, but the weights of the RBF networks have no physical meanings. It is also remarked that in some real applications, the number of membership functions and their centers seem to be well-selected by carefully observing data structures rather than the arbitrary selection mechanism usually employed in the RBF network.

It is further remarked that Wang and Mendal (1992) proposed the Fuzzy Basis Function (FBF) for function approximation, which is similar to the FMFNN. However, in their approach, there were no learning trials, which implies that FBF was not employed as the type of a neural network. Specifically, two arbitrary sets of initial FBF were first constructed by using input-output data pairs and linguistic IF-THEN rules. And then significant FBF's among initial FBF's were selected based on their error reduction ratio. Thus, the performance of their algorithm might be critically determined by initial choices of basis functions, their membership functions, and the number of basis functions to be finally fixed, which may require many a trial and error to achieve satisfactory performances. Compared to their approach, in the FMFNN, membership functions of THEN part in fuzzy IF-THEN rules are adjusted by the gradient descent method as usual in the RBF and the BPNN. And there is no need to determine many a initial basis functions as well as the number of basis function to be finally fixed, owing to the learning capability of the FMFNN. Especially, when a large number of fuzzy variables are necessary for the function approximation, the proposed fuzzy-neural interpolating network can be employed to make fuzzy rules be simple, but there is no such a scheme in Wang and Mendal (1992). In such view points, our approach can be considered as a better solution of fuzzy-neural fusion rather than the approach in Wang and Mendal (1992).

3. A Fuzzy-Neural Interpolating Network

When comparing the FMFNN with the RBF network, we could observe that the FMFNN might be considered as an effective fusion of the RBF network and fuzzy reasoning technique, since the network effectively reflect human expert's experiences and has a good learning capability (Suh and Kim, 1992, 1994a, 1994b). However, one might have difficulties in determining fuzzy rules and membership functions, when a large number of input variables are necessary for the function approximation.

To cope with such difficulties, a fuzzy-neural interpolation network is here proposed. To be specific, $f(y_1, y_2, \cdots, y_p)$ be a function to be approximated and be represented by M representative functions $H_i(y_1, y_2, \cdots, y_{p-1}) \overset{\Delta}{=} f(y_1, y_2, \cdots, y_{p-1}, y_{p,i}^*)$, for $i = 1, 2, \ldots, M$. $H_i(y_1, y_2, \cdots, y_{p-1})$ can be obtained by assigning a constant value $y_{p,i}^*$ to a variable of the function $f(y_1, y_2, \cdots, y_p)$. Let $H_i(y_1, y_2, \cdots, y_{p-1})$, for $i = 1, 2, \ldots, M$, be approximated as $\tilde{H}_i(y_1, y_2, \cdots, y_{p-1})$ by utilizing M FMFNN's. Then for a given input $(y_1, y_2, \cdots, y_p)$, if $\hat{y}_p$ is different from $y_{p,i}^*$ for all i, the output value of $f(y_1, y_2, \cdots, y_p)$ needs to be estimated by interpolating $\tilde{H}_i(y_1, y_2, \cdots, y_{p-1})$, for $i = 1, 2, \ldots, M$. For this, we employ fuzzy rules which inform how much $\hat{y}_p$ is similar to each $y_{p,i}^*$, $i = 1, 2, \ldots, M$. To be specific, let S_i be the *fuzzy similarity* between $\hat{y}_p$ and $y_{p,i}^*$, and n_s be the number of fuzzy rules for S_i, and let $\pi_{ij} : R \to [0,1]$ for $i = 1, 2, \ldots, M$, and $j = 1, 2, \ldots, n_s$ be the membership function for the j-th linguistic value of $\left| \hat{y}_p - y_{p,i}^* \right|$. Also let γ_{ij} be the location of singleton membership function for the j-th linguistic value of S_i. Then S_i can be represented as

$$S_i = \sum_{j=1}^{n_s} \gamma_{ij} \pi_{ij}(\hat{y}_p). \tag{9}$$

Now for a fuzzy-neural interpolation, a linear scalar function given by

$$\hat{g}(x) \overset{\Delta}{=} k_2 x \tag{10}$$

is used as an output node function of each S_i. Since the larger $\hat{g}(S_i)$ is, the more $H_i(y_1, y_2, \cdots, y_{p-1})$ are contributed to finding values of $f(\underline{y})$, it may be reasonable that $f(\underline{y})$ can be found by

$$f(\underline{y}) = Kg\left(\sum_{i=1}^{M} \hat{g}(S_i)\widetilde{H}_i(y_1, y_2, \cdots, y_{p-1})\right) \tag{11}$$

where $g(\bullet)$ and $\hat{g}(\bullet)$, respectively, are the scalar functions defined as in Eq.4 and Eq.10. The schematic diagram of the FMFNN incorporating the fuzzy-neural interpolating network is depicted in Fig.2. It is remarked that since n_s in Eq.9 was given as unity in [4], the result of function approximation was not satisfactory. This implies that there were no rooms for learning other function values except only a pre-learned datum.

Note that since only M representative functions are available, at most M function values can be approximated. Thus to cover the whole input space, we need to divide the input space by M subspaces. By choosing the normalized performance index J_i for the i-th subspace B_i as

$$J_i \stackrel{\Delta}{=} \sum_{\underline{y} \in B_i} \tfrac{1}{2}\left(u(\underline{y}) - f(\underline{y})\right)^2 \Big/ \sum_{\underline{y} \in B_i} f(\underline{y}), \tag{12}$$

and by applying the gradient descent method to minimize J_i, our updating rule for the weight γ_{ij} in Eq.9 can be obtained. Specifically, the derivative of the performance index J_i with respect to the weight γ_{ij} can be obtained as follows;

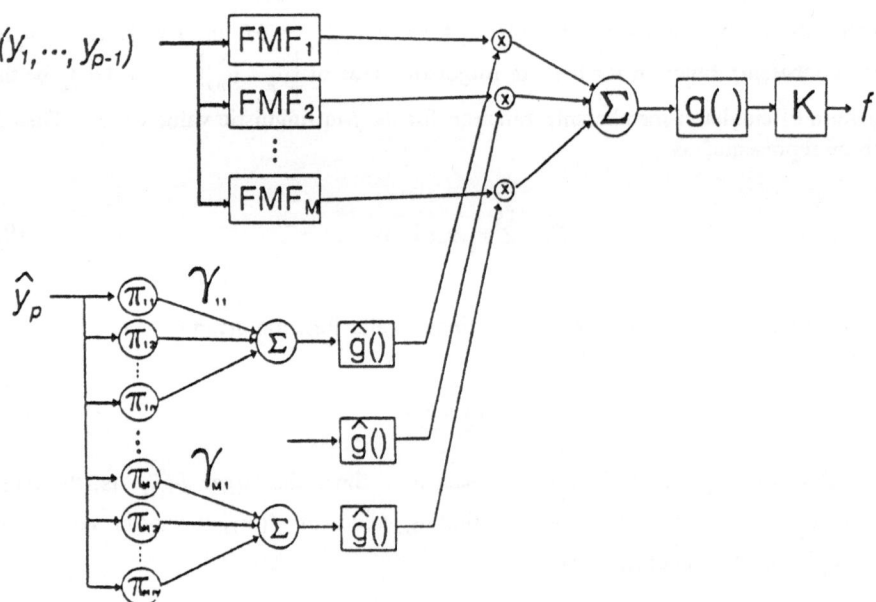

Fig. 2. Schematic diagram of the fuzzy-neural interpolating network

$$\frac{\partial J_i}{\partial \gamma_{ij}} = k_2 \left[\sum_{\underline{y} \in B_i} \left(u(\underline{y}) - f(\underline{y}) \right) \right] \left[\sum_{\underline{y} \in B_i} \tilde{H}_i \pi_{ij}(\underline{y}) \right] \left[\sum_{\underline{y} \in B_i} f(\underline{y}) \right] \Big/ \left[\sum_{\underline{y} \in B_i} f(\underline{y}) \right]^2$$

$$= k_2 \left[\sum_{\underline{y} \in B_i} \left(u(\underline{y}) - f(\underline{y}) \right) \right] \left[\sum_{\underline{y} \in B_i} \tilde{H}_i \pi_{ij}(\underline{y}) \right] \Big/ \left[\sum_{\underline{y} \in B_i} f(\underline{y}) \right] \qquad (13)$$

Thus, the learning rule for adapting weights γ_{ij} of the fuzzy-neural interpolating network can be given as

$$\gamma_{ij}^{t+1} = \gamma_{ij}^{t} + k_2 \left[\sum_{\underline{y} \in B_i} \left(u(\underline{y}) - f(\underline{y}) \right) \right] \left[\sum_{\underline{y} \in B_i} \tilde{H}_i \pi_{ij}(\underline{y}) \right] \Big/ \left[\sum_{\underline{y} \in B_i} f(\underline{y}) \right] \qquad (14)$$

where η_2 is the learning-rate parameter given as a positive constant not greater than or equal to unity.

4. A Numerical Example

To show the capability of the function approximation of the FMFNN incorporating the proposed fuzzy-neural interpolating network, a simulation is performed with a function known as the Mexican hat, *sombrero*, given by

$$f(x,y) = \begin{cases} \dfrac{40 \sin\left(\sqrt{x^2 + y^2} / 35\right)}{\sqrt{x^2 + y^2} / 35}, & \text{for } x \neq 0 \text{ and } y \neq 0, \\ \\ 40\pi, & \text{for } x = y = 0, \end{cases} \qquad (15)$$

which is depicted in Fig.3. The input and output universes of discourse are given as $x \in [-120, 120]$, $y \in [-120, 120]$ and $f(x,y) \in [-27.298, 125.6637]$. Assume that 169 input-output relations are available and these are divided into 13 subspaces. Then 13 FMFNN's are assigned in such a way that each FMFNN learns input-output mapping. The fuzzy rules for designing the i-th FMFNN can be generated by observing the training data. It is remarked that the whole input space is divided into 13 subspaces and 13 fuzzy rules are required for each subspace.

After completely training 13 FMFNN's, the proposed fuzzy-neural interpolating network with the learning rule in Eq.14 is applied to improve the degree of the function approximation. To verify the capability of approximation of our FMFNN incorporating the fuzzy-neural interpolating network, the approximated function is retrieved with 49 × 49 segmented input data. It may be observed from Fig.4 that the FMFNN incorporating the fuzzy-neural interpolating network can be used as a function approximator, but the network could not be completely reproduce the given function due to insufficient training data. In general, if the number of subspace is sufficiently large, the given function is expected to be satisfactorily approximated by the proposed FMF networks incorporating the fuzzy-neural interpolating network.

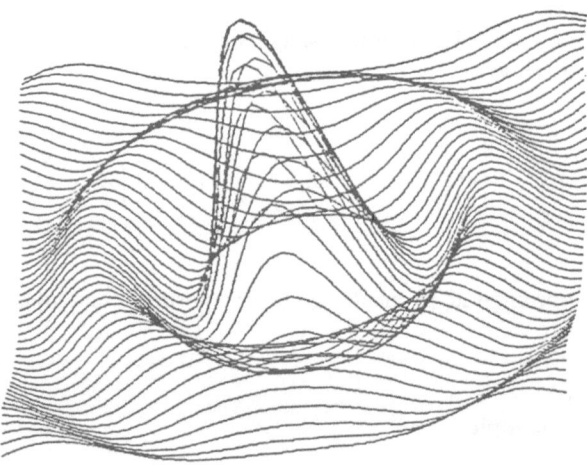

Fig. 3. Graphical representation of the function in Eq.15

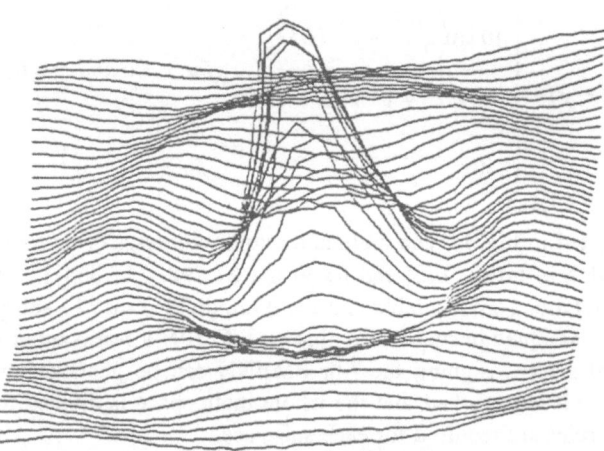

Fig. 4. Graphical representation of the function in Eq.15 approximated by 13 FMFNN incorporating the fuzzy-neural interpolating network

5. Concluding Remarks

It was shown that a fuzzy-neural interpolating network could be designed by employing both the fuzzy similarity and the FMFNN as a basis function. Simulation results showed that the performance of function approximation was satisfactory by incorporating the FMFNN with the fuzzy-neural interpolating network. It is remarked that since the structure of the proposed networks could effectively reflect the expert's knowledge, the proposed network is expected to show several desirable performances such as training simplicity, fast convergency, design simplicity, fast learning speed, and no computational complexity when retrieving.

REFERENCES

Chen, S., Cowan, C. F. N. and Grant, P. M. (1991) Orthogonal least square learning algorithm for radial basis function networks, *IEEE Trans. on Neural Networks* 2(2), 302-309.

Kim, T. W. (1994) *A Study on Fuzzy-Neural Network-based Visual Servoing of a Robot Manipulator* (in Korean), Ph.D. thesis, Hanyang Univ., Seoul, Korea.

Pedrycz, W. (1989) *Fuzzy control and Fuzzy systems*, Research Studies Press LTD.

Powell, M. J. D. (1987) Radial basis function approximations to polynomials, *Proc. 12th Biennial Numerical Analysis Conf. (Dundee)*, 223-241.

Rumelhart, D. E., Hinton, G. E. and Williams, R. J. (1986) Learning internal representations by error propagation, *Parallel Distributed Processing*, Ch.8, MIT Press, Cambridge, 318-336.

Sanger, T. D. (1991) A Tree-Structured Adaptive Network for Function Approximation in High-Dimensional spaces, *IEEE Trans. on Neural Networks* 2(2), 285-293.

Suh, I. H. and Kim, T. W. (1992) Nonlinear Function Approximation by the Fuzzy Membership Function based Neural Networks, *Proc. of ConFuSE '92 (Seoul, Korea)*, 153-156.

Suh, I. H. and Kim, T. W. (1994) Visual Servoing of Robot Manipulators by Fuzzy Membership Function Based Neural Networks, in K. Hashimoto (ed.), *Visual Servoing - Automatic Control of Mechanical Systems with Visual Sensors*, World Scientific Publishing Co., 285-315.

Wang, L. X. and Mendel, J. M. (1992) Fuzzy Basis Functions, Universal Approximation, and Orthogonal Least-Squares Learning, *IEEE Trans. on Neural Networks* 3(5), 807-814.

SEMANTICALLY VALID OPTIMIZATION OF FUZZY MODELS

W. PEDRYCZ

Dept. of Electrical & Computer Engineering
University of Manitoba
Winnipeg, Manitoba, Canada R3T 5V6

AND

J. VALENTE DE OLIVEIRA

INESC – Research Group on Control of Dynamic Systems
Apartado 13069, 1000 Lisboa, Portugal

1. Introduction

Fuzzy models are ubiquitous in both theoretical developments and applications of fuzzy set theory. They become particularly crucial when it comes to an extensive and thorough "what-if" type of analysis.

Concisely we can consider fuzzy models as mathematical entities expressing relationships between linguistic labels (fuzzy sets). In the presence of real-valued data these entities are extended with the well-known input and output interfaces linking the (fuzzy) procesing module with the environment. This define the overall architecture of the fuzzy model. The numeric input/output mapping accuracy of this type of models is often seen as the holly figure of merit to be considered by the optimization method. However, following the acceptance of the fuzzy model definition the general modelling methodology is only valid up to a certain extent. To take full advantage of the richness of the fuzzy model concept another crucial issue should be observed: Semantics. The fuzzy model should be passible of *linguistic interpretation*. Thus optimization should be realized at the different levels of the model. Furthermore, each module of the overall architecture can be optimized separately or simultaneously with the other modules, and subject to various semantically oriented criteria. These aspects give rise to a visible diversity of optimization techniques to be design.

Z. Bien and K. C. Min (eds.),
Fuzzy Logic and its Applications, Information Sciences, and Intelligent Systems, 197–206.
© *1995 Kluwer Academic Publishers.*

The aim of this paper is to describe the optimization tasks emerging at the different functional blocks of a fuzzy model, and the optimization criteria employed there. Detailed numerical examples are included.

2. Design of input and output interfaces

The essential role of the input and output interfaces is to transfer an information coming from or being sent out to the environment in which the modelling takes place. The transformation of the external form of information into an internal format usable at the processing level of the fuzzy model is carried out through a matching procedure between the external data and the so-called referential fuzzy sets. The matching procedure is usualy hinged on the extensive usage of the possibility measures (Zadeh, 1978; Dubois and Prade, 1988).

Although the roles of the input and output interfaces are different, both of them share the same basic framework i.e., they are both based on a class of referential fuzzy sets representing linguistic terms, providing thus a semantic description for the external variables. For a class of fuzzy sets to be interpreted as linguistic terms, the membership functions should satisfy the so-called requirements of semantic integrity.

2.1. THE REQUIREMENTS OF SEMANTIC INTEGRITY

The list of requirements that are deemed crucial to the effective development of a fuzzy system includes, cf. (Valente de Oliveira, 1993c):

   I: A justifiable number of elements: the number of linguistic terms should be compatible with the number of conceptual entities a human being can efficiently memorize and utilize in his inference activities. Therefore this number should not exceed the well-known limit of 7 ± 2 distinct terms;

  II: Natural zero positioning: whenever required, one of the linguistic terms should represent the "around zero" conceptual entity, i.e., its membership function should be convex, normal, and centered at zero;

 III: Normalization: since each linguistic term has a clear semantic meaning, then at least one numerical datum of the universe of discourse should yield full matching with each term. In addition, normalized allows one to use the full range of the unit fuzzy cube.

  IV: Coverage: the entire universe of discourse should be "covered" by linguistic terms, meaning that any numerical datum should be matched to a nonzero degree with at least one referential fuzzy set;

   V: Focus of attention (or distinguishability): each linguistic term should have a clear semantic meaning, while the corresponding referential fuzzy sets should clearly define a certain range in the universe of discourse. This implies a clear distinction of membership functions;

While the requirements I), II), and III) can be satisfyed by selecting an acceptable number of normal reference fuzzy sets in the preliminar phase of the modelling process, the requirements IV) and V) need to be monitorized during the optimization process. This will keep the overall model semantically valid.

2.2. SEMANTICALLY VALID OPTIMIZATION OF INTERFACES

For assuring coverage we will use the concept of optimal interfaces. This concept, first suggested in (Valente de Oliveira, 1993a), entails an equivalence between the numeric and linguistic information formats, i.e., it states that an error free conversion should exist when the values of a given numeric variable are successively transformed by an input and output variable. More specifically,

$$\forall_{x \in X} \; x \; = \; \mathcal{F}^{-1}(\mathcal{F}(x))$$
$$= \; \mathcal{F}^{-1}(\mathbf{x}) \tag{1}$$

where $\mathbf{x} \in [0,1]^n$ stands for a vector of matching values (degrees of membership) between the numeric datum x and the n linguistic labels. Specially, $\mathcal{F}$ will be used to denote the numeric-to-linguistic transformation implemented by the input interface. Similarly $\mathcal{F}^{-1}$ will be used to denote the linguistic-to-numeric conversion provided by the output interface.

The basic motivation beyond this criterion is to preserve information when data is coverted from one format to another. One way to assure this is to apply the inverse conversion to the converted data and ensure that the results equals the original data.

Criterion (1) can be specified depending upon the accepted linguistic-to-numeric transformation. A typical example is the following member of the family of averaging transformations:

$$\hat{x} \; = \; \mathcal{F}^{-1}(\mathbf{x})$$
$$= \; \frac{\sum_{i=1}^{n} \mu_i(x)\bar{a}_i}{\sum_{i=1}^{n} \mu_i(x)} \tag{2}$$

where $\bar{a}_i$ denotes the modal value of the i-th membership function μ_i. Considering a set of numerical data $\{x_1, \ldots, x_N\}$, this gives rise to an optimization task driven by the following performance index:

$$J_1 = \frac{1}{2} \sum_{k=1}^{N} (x_k - \hat{x}_k)^2 \tag{3}$$

where $\hat{x}_k$ results from converting x_k using (2).

More generally, the interface optimization problem can be stated as follows (Valente de Oliveira, 1993c):

Minimize Q with respect to the parameters of the membership functions and subject to a distinguishability constraint.

In general, distinguishability is guaranteed if each and every internal representation of data satisfies:

$$\forall_{x \in X} M_p(\mathcal{F}(x)) \leq 1 \tag{4}$$

where M_p is the sigma-count of a fuzzy set (Kosko, 1992). A trivial observation is that an interface formed by equality distributed triangular membership functions successively overlapping at 0.5 of membership, constitute a viable solution to the above problem. Nevertheless the overall performance of a fuzzy model with this type of interfaces might not be satisfactory.

3. Processing module

The processing module implements the mapping $[0,1]^n \to [0,1]^m$, where n and m are the number of linguistic terms at the input and output interfaces, respectively. In other words, it performs computations at the level of linguistic terms (fuzzy sets). The construction of the processing module calls for the determination of the logical relationships between these terms. We will be concerned with the s-t composition operators as the basic computational means used in the formation of this module. These operators constitute a generalization of the two well-known operations on fuzzy sets, namely the max-min composition. More formally, let X and Y be the input and the output fuzzy set of the processing module, respectively. The s-t composition of X and R, where $R : \mathbf{X} \times \mathbf{Y} \to [0,1]$ reads as:

$$Y = X \bullet R \tag{5}$$

that coordinatewise translates into

$$y_j = S_{i=1}^n(x_i \mathrm{\ t\ } r_{ij}); \quad j = 1,2,\ldots,m \tag{6}$$

the symbols S and t denoting s-norm and t-norm, respectively. One can look at (6) as examples of logic-based neurons where the overall aggregation can be interpreted as an extended AND, cf. (Pedrycz, 1993). The optimization of the processing module is accomplished through a sequence of adjustments in the entries of the fuzzy relation R. The optimization process is basically a procedure of supervised learning. This procedure requires that the values of

R are modified by changing its individual entries according to the gradient of a given performance index, J.

$$\Delta R = -\alpha \frac{\partial J}{\partial R} \tag{7}$$

where $\alpha \in [0, 1]$ is the learning step parameter. The detailed formulas can be derived once the triangular norms (both s and t) have been specified. For more computational details the reader is referred to (Pedrycz, 1991; Pedrycz, 1993; Valente de Oliveira, 1993b).

4. Optimization scenarios in fuzzy models

For assessing the effectiveness of the proposed semantic requirements, the following optimization scenarios are worth analyzing:

A: Optimization of input and output interfaces in advance and separately from the processing module. Their parameters are tuned to meet the semantic requirements.

B: This scenario is similar to that described in A: the parameters of the interfaces are tuned in advance but after that they are allowed to be optimized even further during the optimization phase of the processing module without any further semantic concerns.

C: A simultaneous optimization of the interfaces along with the processing module without semantic concerns. This option is quite frequently utilized in the existing literature;

D: A global optimization of all the modules of the model (as in scenario C) but now with the parameters of interfaces subject to the constraints of semantic integrity.

4.1. EXAMPLE 1

This example illustrates the importance of the pre/post-processing stages occuring at the interface level of the fuzzy model. It is shown that a non-linear processing at these stages (being implied by nonlinear membership functions) improves the model's performance.

Consider the static nonlinearity (sigmoide function) $f(x) = \frac{1}{1+e^{-x}}$. For training purposes, a set of $N = 50$ non-uniformly distributed data points were collected, as presented in Fig. 1

For the interfaces, two types of membership functions (piece-wise linear and non-linear) are tested, resulting in two distinct models. Hereafter the models are named a T-model (the model exploiting triangular membership functions), and G-model (the model with Gaussian membership functions). Three membership functions were assumed at each interface (input and

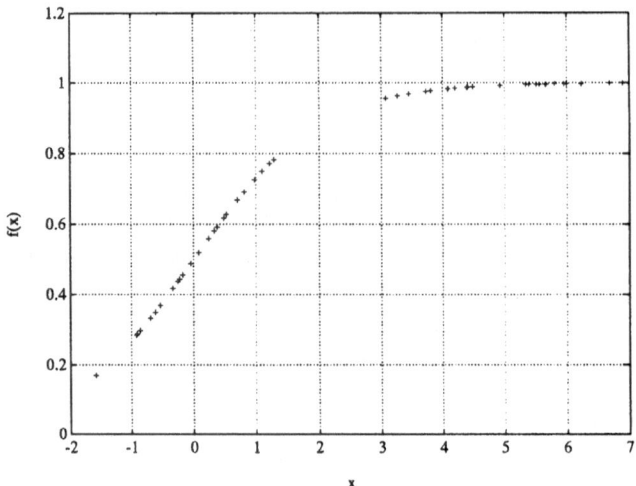

Figure 1. Training data set for Example 1.

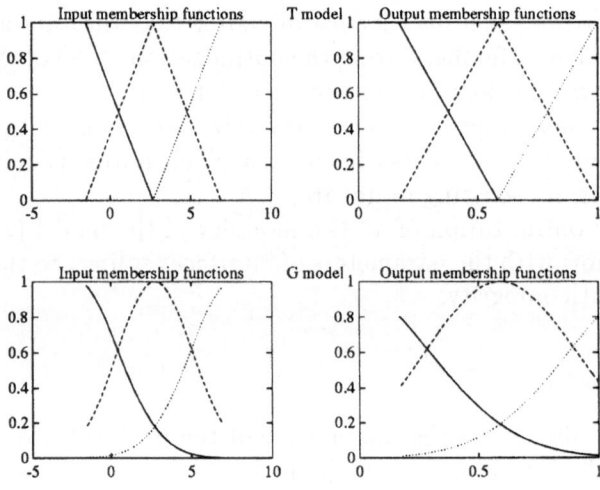

Figure 2. Three optimized membership functions for both models.

output) for both models. Fig. 2 shows the optimized membership functions for both of them.

The processing part of the fuzzy model uses the t-s composition applied to X and R, $Y = X \bullet R$, with $n = m = 3$, with the t-norm defined as the product, and the s-norm taken as the probabilistic sum in (6). The overall

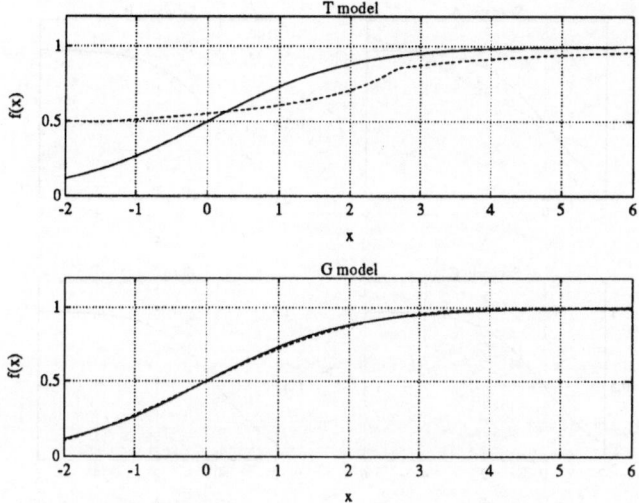

Figure 3. Cross validation results from both models using 3 linguistic terms: dashed line – fuzzy model; solid line – data.

performance index is:

$$J = \frac{1}{2} \sum_{i=1}^{N} (y_i - f(x_i))^2 \tag{8}$$

where y_i the (numeric) output of model when the i-th input, x_i, is presented.

The optimization policy being used correspond to scenario D, that is the simultaneous optimization of the processing module and the interface subject to the semantic requirements. Optimization was carried out until reaching a minimum value of the performance index.

For assessing the performance of the optimized models, a sequence of 121 testing inputs have been applied to both of them. The results of this cross-validation are presented in Fig. 3.

This figure indicates a good performance for the G-model while a relatively weak performance is observed for the T model. Increasing the number of referential fuzzy sets (linguistic labels) either at the input or output interface, is not sufficient to improve this comparatively weak performance of model T.

4.2. EXAMPLE 2

This example compares the performance of the fuzzy models optimized according to four different optimizing scenarios. The comparison considers the performance of the optimized models, and the semantic meaning of the linguistic terms at their interfaces. For assessing these features, a fuzzy

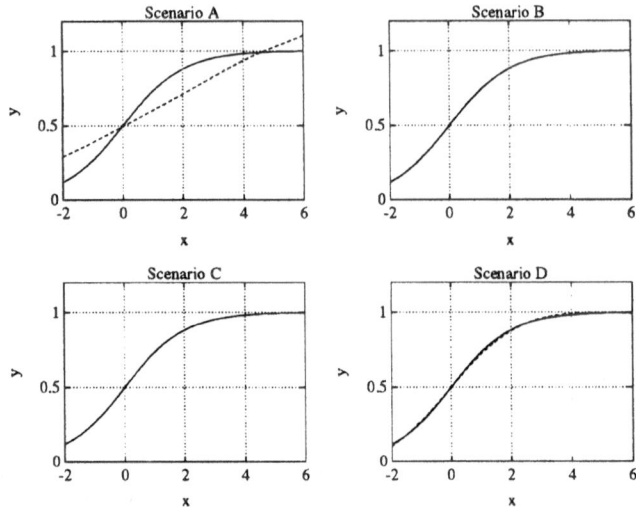

Figure 4. Cross validation results for models optimized according to the different sc-narios: dashed line – fuzzy model; solid line – data

model is built according to each scenario and then validated. Both the training and the testing data sets used now are the same as in Example 1.

For each model, three linguistic terms (represented by Gausssian membership functions) were assumed for both input and output interfaces. As in Example 1, the processing module is described by (5) with the same triangular norms. All scenarios assume the same overall performance index (8). As seen in Fig. 4 a good cross validation performance is obtained for all the models except for that derived with scenario A, i.e., optimization of both input and output interfaces in advance, and separetely from the processing module.

The reason for this poor performance is that, in this case, the processing subsystem (processing part) has failed to provide the required mapping $[0, 1]^3 \rightarrow [0, 1]^3$ whose domain and co-domain were defined *a priori* by the input and output interfaces. To overcome the weak "capacity" of the simple processing subsystem used, one should select those optimal interfaces that also optimize the overall model performance (i.e. to optimize the model subject to the constraint imposed by the optimal interfaces). This is what scenario D is aimed at.

Looking for the optimized membership functions for both input and output interfaces in figures 5 and 6, one can notice that scenario A and D produce membership functions whose semantic meaning can be easily assessed by a human being. For instance, the membership functions can be interpreted as the linguistic terms *small*, *medium*, and *large*. By the oppo-

205

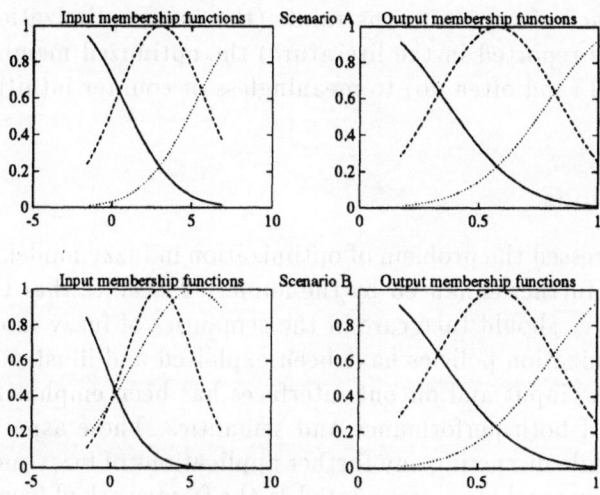

Figure 5. Final membership functions obtained from scenarios A and B, for the input and output interfaces.

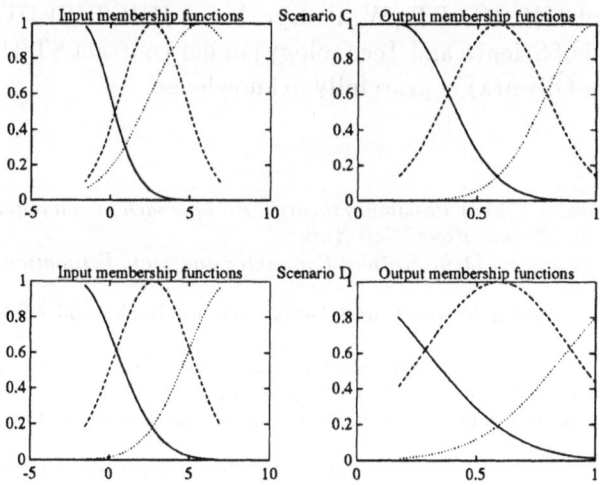

Figure 6. Final membership functions obtained from scenarios C and D, for the input and output interfaces.

site, scenarios B and C have generated membership functions for the input interface that can be more dificult to interpret as the dotted and dashed membership functions are far from providing a clear distinction for the respective linguistic entities that they are aiming at representing. Notice that, in absence of semantic constraints (the usual optimization process of fuzzy systems reported in the literature) the optimized membership functions can lead (and often do) to meaningless or counter-intuitive linguistic terms.

5. Conclusions

We have addressed the problem of optimization in fuzzy models. It has been revealed and further enhanced by the numerical studies that the optimization procedures should take care of the semantics of fuzzy models. Several specific optimization policies have been exploited and illustrated. The crucial role of the input and output interfaces has been emphasized from the standpoints of both performance and semantics. These aspects should be also raised while discussing any further applications of fuzzy models such as prediction or control tasks formulated in the framework of fuzzy modelling.

Acknowledgements

Support from the Natural Sciences and Engineering Research Council of Canada and MICRONET (W. Pedrycz) and JNICT/FEDER (The Portuguese Board of Science and Technology) under contract STRDA/C/TIT/94/9: (J. Valente de Oliveira) is gratefully acknowledge.

References

Dubois, D., Prade, H. (1988) *Possibility theory - An approach to computerized processing of uncertainty*. Plenum Press, New York.

Eyckoff, P. (1974) *System Identification: Parameter and State Estimation*. J. Wiley, London.

Kosko, B. (1992), *Neural Networks and Fuzzy Systems*. Englewood Cliffs, NJ: Prentice Hall.

Pedrycz, W. (1991) Neurocomputing in relational systems, *IEEE Trans. on Pattern Analysis and Machine Intelligence*, **Vol 13**, no. 3, pp. 289-297.

Pedrycz, W. (1993) Fuzzy neural networks and neurocomputations, *Fuzzy Sets and Systems*, **Vol 56**, pp. 1-28.

Pedrycz, W, Valente de Oliveira, J. (1993a) An alternative architecture for application driven fuzzy systems, *Proc. 5th IFSA World Congress*, Seoul, pp. 985-988.

Pedrycz, W., Valente de Oliveira, J. (1993b) Optimization of fuzzy relational models, *Proc. 5th IFSA World Congress*, Seoul, pp. 1187-1190.

Valente de Oliveira, J. (1993a) On optimal fuzzy systems I/O interfaces, *Proc. of the Second IEEE International Conference on Fuzzy Systems*, San Francisco, CA, pp. 851-856.

Valente de Oliveira, J. (1993b) Neuron inspired learning rules for fuzzy relational structures, *Fuzzy Sets and Systems*, **Vol. 57** pp. 41-53.

Valente de Oliveira, J. (1993c – accepted for publication) A design methodology for fuzzy system interfaces, *IEEE Trans. on Fuzzy Systems*.

Zadeh, L.A. (1978) Fuzzy sets as a basis for a theory of possibility, *Fuzzy Sets and Systems*, **Vol. 1**, pp. 3-28.

HULL FORM GENERATION BY USING TSK FUZZY MODEL

Y.S. Lee[*], S.J. Jeong[*], S.Y. KIM[*], and G. KANG[**]
*Dept. of Naval Architecture, Pusan National Univ.,
Jangjeun-Dong, Dongrae-Gu, Pusan, KOREA
**Dept. of Electronics Eng., National Fisheries Univ. of Pusan
Daeyeun-Dong, Nam-Gu, Pusan, KOREA

1. Introduction

A relatively accurate decision of a hull form is essential to estimate a main engine horse power and a ship cost in the initial design stage. Therefore, many designers have elaborated to express a hull form in mathematical ways. Recently, due to the rapid development of computer technology, new methods of describing the hull form have been developed[1].

There are several hull form generation methods that can be used in the initial process of ship design. For example, the standard series approach, lines distortion from parent lines, and form parameter design method can be illustrated[2][3][4].

Though the standard series approach and lines distortion from parent lines have been widely used due to their simplicity, they have a limitation in the sense that they can only generate a hull form, similar to the standard hull form, within the restricted distortion range. Thus, if the distortion range is widened, it leads to degeneration of a whole hull form. The form parameter design method expresses an aimed hull form in a mathematical way by combining form paramters that define the hull form. It is quite useful in calculating geometrical and numerical properties. It can also generate any kind of hull form which satisfies designer's demand. However, the form parameter design method can be affected by the way how the form parameters are combined. For example, if the hull form is expressed in a polynomial, which requires relatively simple calculations, problems may arise such as an increase of an order of polynomial and oscillation of a curve when the number of form parameters increases. Moreover it is impossible to express the hull form

Z. Bien and K. C. Min (eds.),
Fuzzy Logic and its Applications, Information Sciences, and Intelligent Systems, 207–214.
© 1995 Kluwer Academic Publishers.

which has vertical tangent vector. When B-spline is used in expressing the equation by combining form parameters, there are several merits such as a guarantee of the enough continuity, permission of local control, the possibility of expressing discontinuities and straight curve and easiness of anticipating a shape. But it requires many trial and errors to determine form parameters for a new hull form. Thus it accompanies many difficulties to use when the designer is not equipped with the experienced knowledge. In addition, form parameters are not general kinds of design paramter but confined form definition paramters having only geometrical characteristics like position, slope, curvature, area, moment, and centroid. Because the hull form generation from form paramters has a chance to bring about an unexpected performance problem and a design fault which are not within the design constraints, it definitely requires overall consideration about design conditions, design parameters and a hull form. Since ship design is a complex process and it needs many feedbacks at design stage, it has been difficult to generate a hull form that satisfies all the requirements when we use above mentioned methods.

On the contrary, an application of a fuzzy model to hull form generation makes it possible to express not only a general design kowledge concerned with a ship design but also an experienced kowledge which can be only transferred verbally. Traditionally the hull form has been determinded so far by producing and considering basic curves of SAC, DWL, and keel profile. But with hull form generation by using TSK fuzzy model[5], the hull form can be generated from SAC only. The Fig. 1 shows the difference between the two approaches. This paper discusses the hull form generation from TSK fuzzy model constructed with actual ship data.

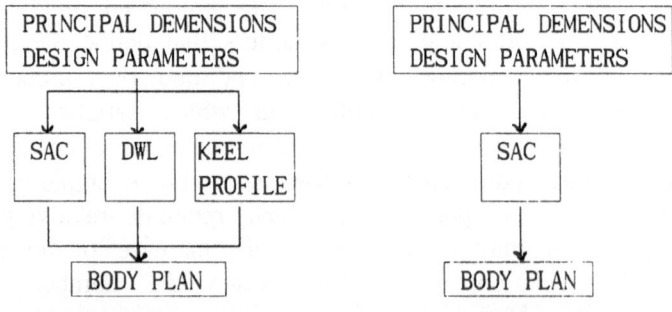

(a)general design method   (b) fuzzy model method

Fig. 1 Comparison of hull form generation procedures

2. TSK Fuzzy Model

A TSK(Takagi-Sugeno-Kang) fuzzy model is composed of the following fuzzy implications.

$$L^i: \text{If } z_1 \text{ is } A_1^i, \quad z_2 \text{ is } A_2^i, \cdots, z_m \text{ is } A_m^i \quad (1)$$
$$\text{then } y^i = c_0^i + c_1^i x_1 + c_2^i x_2 + \cdots + c_m^i x_m$$

where, z_j is the premise variable, A_j^i is a fuzzy set, x_j is a consequent variable, and y^i is the output form of the i-th fuzzy implication.

Given an input $(z_1^0, z_2^0, \ldots, z_m^0)$, the output y is inferred by taking the weighted average of the y^i's.

$$y = \sum_{i=1}^{n} W^i y^i / \sum_{i=1}^{n} W^i \quad (2)$$

where, n is the number of fuzzy rules, and the weigt W^i implies the overall truth value of the premise of the i-th rule for the input

$$W^i = \prod_{j=1}^{m} A_j^i(z_j^0) \quad (3)$$

where, $A_j^i(z_j^0)$ is the membership value of z_j^0 in A_j^i.

A TSK fuzzy implication represents a linear input-output relation in a fuzzy subspace specified at the premise part. The TSK fuzzy models can represent complicate nonlinear systems precisely with a few implicatiuons. The algorithm identifing a TSK fuzzy model had been suggested by Sugeno and Kang[6].

Here we use the input-output data of actual ship in a fuzzy model identification, which consists of two parts :structure identification and parameter identification.

3. TSK Fuzzy Modelling of SAC and Body Plan

Fuzzy modeling means a process that expresses global relationships among parameters in a system in mathematical models. Though a statistical method has been used for this mathematical modeling, it has been still difficult to find out a global function of complex nonlinear systems. Fuzzy modeling used in this paper puts its basis on finding a set of local input-output relations describing a system. This is meant to analyze a nonlinear structure of a system by input-output data analysis and to regard the structure as a kind of superposition form made of several linear relation structures. This view is adequate to establish relationships among design conditions, principal dimensions, and hull forms.

In this paper, two-step fuzzy modeling is considered. The first step is the fuzzy modeling process to get SAC from design condition and principle dimensions. The second is a fuzzy modeling process to generate a body plan from SAC. The method suggested by Sugeno and Kang[7] is used to identify the TSK fuzzy model.

3.1 TSK Fuzzy Modeling of SAC

SAC has the greatest influence on determining a hull form in an initial designing process because it describes the most basic properties of a hull form. It is a reasonable procedure to decide SAC from estimated parameters, but it is also very difficult since the complex relationship between parameters and SAC has to be considered overall. To solve the problem, the relationship between parameters and SAC is described as a TSK fuzzy model. In the TSK fuzzy modeling, it is assumed that principal dimensions and design parameters are given. L/B, B/D, C_b, Vs, LCB, C_p-Curve, F_n, C_m, are used as input data for model identification. The definition and the range of each design parameters is given below.

L/B : Length between Perpendiculars(m) / Moulded Breadth(m)
B/D : Moulded Breadth(m) / Design Draft(m)
V_s(kt) : Design Speed (des)
C_b : Block Coefficient
C_m : Midship Coefficient
LCB(%) : Longitudinal Center of Buoyancy
F_n : Froude Number

The range of the each design parameter
$$5.5 < L/B < 8.63 \qquad 0.97 < C_m < 0.99$$
$$2.39 < B/D < 4.31 \qquad -2.08 < LCB < 3.64$$
$$14.0 < V_s < 24.6 \qquad 0.13 < F_n < 0.25$$
$$0.53 < C_b < 0.83$$

As defined so far, each parameter is usually known as a crucial element in determining SAC, and it is selected synthetically with a view of a designer. Through the fuzzy model identification, we decide which parts of SAC these parameters influence, and which parameters mostly affect in determining SAC. As the result of the fuzzy model identification, the premise variable of SAC fuzzy model is the longitudinal axis X, and consequent variables are X, C_b, LCB and F_n. The resultant fuzzy models are given in Fig. 2.

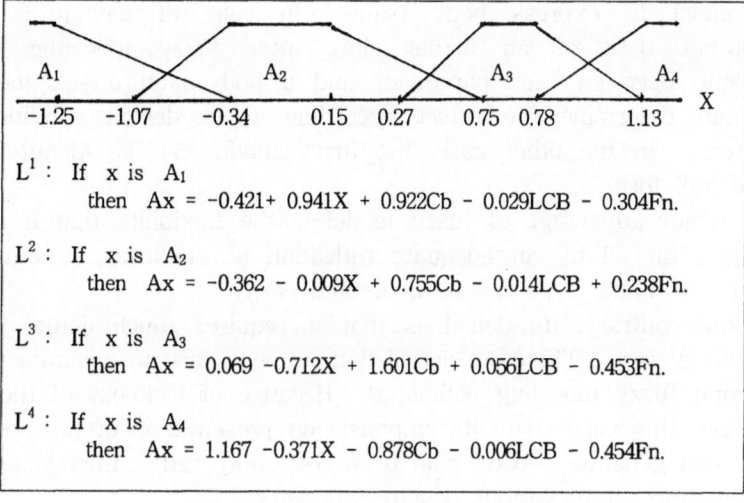

Fig. 2 The TSK fuzzy model of SAC

The SAC inferred from the TSK fuzzy model of Fig. 2 is shown in Fig. 3. Considering the SAC characteristics, a value greater than 1 is defined as 1 and a value less than 0 is defined as 0.

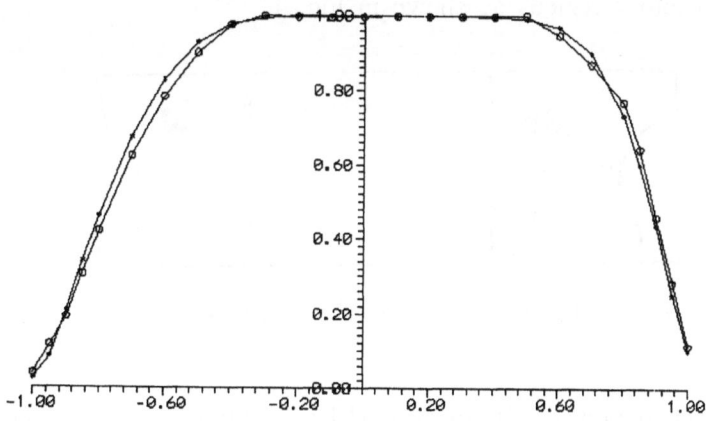

Fig. 3 SAC inferred by the fuzzy model of SAC
(C_b = 0.8081, LCB = 2.431, Fn = 0.162)

3.2 TSK Fuzzy Modeling of Body Plan

The fuzzy models can be categorized into two groups when we make fuzzy model to express body plan. In case of having a lot of input-output data of an actual ship, after fuzzy modeling of the relationship between each parameter and a body plan offset, the fuzzy model can determine the offset according to a design condition and parameters. In the other case, the fuzzy model can be identified in a confined hull form.

The major advantage of fuzzy model is the flexibility that if a fuzzy model is achieved by an adequate reflection of nonlinear structure, any hull form of similar type can be generated easily.

On the contrary, its defect is that it requires much actual data to model the system. Therefore actual data is necessary to generate the hull form using fuzzy modeling technique. Because of lackness of the actual offset data, this paper puts its emphasis on presentation of the possibility that we can generate a body plan from SAC only rather than general hull form generation from various design parameters.

The relationship between SAC and body plan offset is constructed into fuzzy model and actual ship data such as X(section), each sectional area Ax of SAC, Z(depth), Y(breadth) are used. There are some problems in expressing a stern and a bow which vary abruptly with Ax variation in spite of the advantage of fuzzy concept. Therefore, emphasising the variation of a stern and a bow which have much variation, the total area is divided into 4 regions as shown in Fig. 4

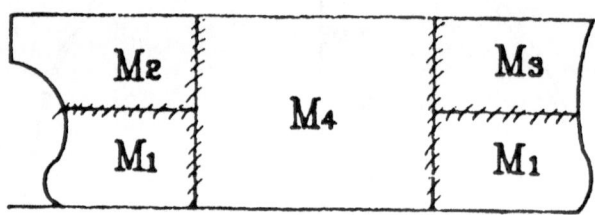

Fig. 4 The divided sections of body plan data.

For each region a TSK fuzzy model is identified. Each model consists of 4 fuzzy implications. Premise variables are Ax and Z, and consequent variables are X, Ax and Z. One of fuzzy models of body plan is shown in Fig. 5

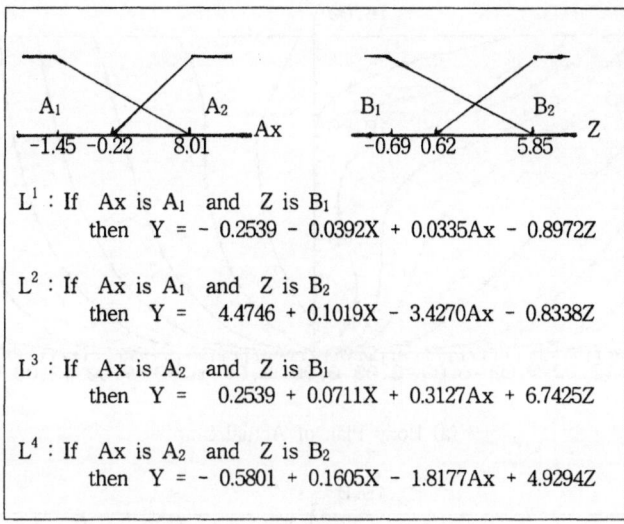

L¹ : If Ax is A₁ and Z is B₁
        then   Y = - 0.2539 - 0.0392X + 0.0335Ax - 0.8972Z

L² : If Ax is A₁ and Z is B₂
        then   Y =   4.4746 + 0.1019X - 3.4270Ax - 0.8338Z

L³ : If Ax is A₂ and Z is B₁
        then   Y =   0.2539 + 0.0711X + 0.3127Ax + 6.7425Z

L⁴ : If Ax is A₂ and Z is B₂
        then   Y = - 0.5801 + 0.1605X - 1.8177Ax + 4.9294Z

Fig. 5 The TSK fuzzy model of body plan

For generation of body plan, the sectional area Ax is inferred from the SAC fuzzy model by using X, C_b, LCB, and F_n., and the offset Y is inferred from each body plan fuzzy model.

Since each sectional data has been superposed in the modeling process, relatively good results can be obtained with the few data. But the body plan obtained from 4 fuzzy models tends to be unnatural at the boundaries of 4 regions.

To conclude, as shown in Fig.6 for comparision, a new hull form relatively similar to an actual ship data is generated.

4. Conclusion

It is possible to construct fuzzy models of SAC and body plan by using the TSK fuzzy model, which express the relationship among the design parameters, SAC, and offsets.

The suggested TSK fuzzy models can generate the SAC which satisfies the design condition and the design parameters

By using TSK fuzzy model, It is possible even for an inexperienced desginer to generate a hull form wihtout DWL, Keel profile to satisfy design parameters and design conditions only from the SAC.

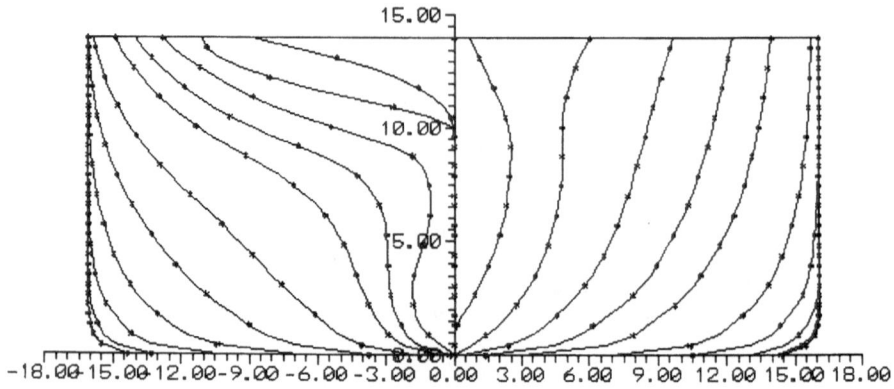

(a) Body Plan of Actual Ship

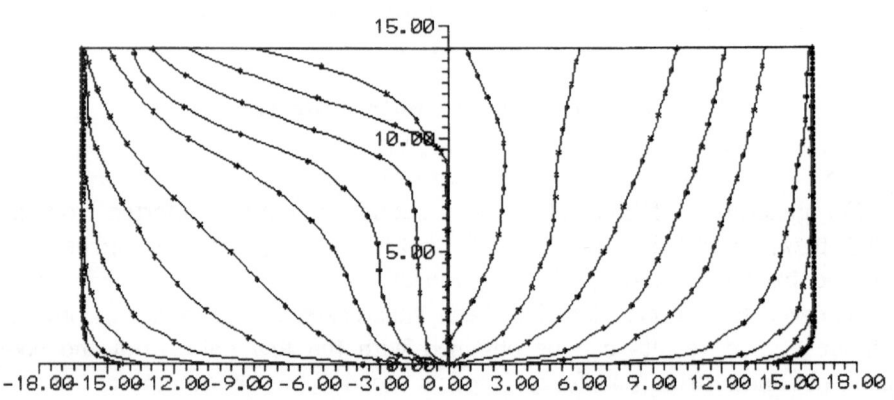

(b) Body Plan Inferred from TSK Fuzzy Model
Fig. 6 Body plan (C_b = 0.8081, LCB = 2.431, Fn = 0.162)

References

[1] H.Nowacki, G.Creutz, F.C. Munchmeyer, "Ship Lines Creation by Computer-Objectives, Methods and Results", The Society of Naval Architects and Marine Engineers, 1977.

[2] Creutz, G., "Curve and Surface Design from Form parameters by means of B-splines"(In German), Doctoral Thesis, University of Berlin, 1977.

[3] A.M.Reed and H.Nowacki, "Interactive Creation of Fair Ship Lines", Journal of Ship Research, Vol.18, pp.96-112, 1974.

[4] H.Lackenby, "On the Systematic Geometrical Variation of Ship Forms", Transactions INA, Vol. 92, pp.289-316, 1950.

[5] T.Takagi and M.Sugeno, "Fuzzy Implication of Systems and Its Applications to Modeling and control", IEEE Trans. Systems, Man and Cybernetics, 15-1, pp.116-132, 1985.

[6] M.Sugeno and G.T.Kang, "Structure Identification of Fuzzy Model", Fuzzy Sets and Systems Vol. 28, pp.15-33, 1988.

INTERPOLATIVE REASONING FOR COMPUTATIONALLY EFFICIENT OPTIMAL MULTISTAGE FUZZY CONTROL

JANUSZ KACPRZYK
Systems Research Institute
Polish Academy of Sciences
ul. Newelska 6
01-447 Warsaw, Poland
E-mail: `kacprzyk@ibspan.waw.pl`

Abstract. Fuzzy optimal multistage control is considered meant as to find an optimal sequence of controls best satisfying fuzzy constraints on the controls applied and fuzzy goals on the states (outputs) attained, with a fuzzy system under control. Control with a fixed and specified, implicitly specified, fuzzy, and infinite termination time is discussed. For computational efficiency, a small number of reference fuzzy states and controls is assumed by which all fuzzy controls and states are approximated. Optimal control policies relating optimal reference fuzzy controls to current reference fuzzy states are determined as a fuzzy relation which is used, via the compositional rule of inference, to derive optimal controls. This requires a large number of overlapping reference fuzzy controls and states implying a low computational efficiency. To overcome this, a small number of nonoverlapping reference fuzzy states and controls is assumed, and then interpolative reasoning is used to infer an optimal fuzzy control for a current fuzzy state.

Key words: fuzzy control, fuzzy optimal control, fuzzy dynamic system, fuzzy inference, analogical reasoning, interpolative reasoning

1. Introduction

As opposed to usually assumed approaches to fuzzy control, a different one, more consistent with the spirit of control, has been advocated in Kacprzyk's

215

Z. Bien and K. C. Min (eds.),
Fuzzy Logic and its Applications, Information Sciences, and Intelligent Systems, 215–223.
© 1995 *Kluwer Academic Publishers.*

(1983a) book: the temporal evolution of a fuzzy system under control, the fuzzy constraints and the fuzzy goals are assumed known, and an optimal sequence of controls is sought. The operator's knowledge concerns how the system behaves (and not how to control the process as in traditional approaches), and the fuzzy controls and states represent requirements on how control is to proceed. A best (optimal) control is determined by an algorithm.

Kacprzyk's (1983a) book appeared in a "bad" time, long before the recent eruption of interest in fuzzy control. However, it may lead to a *new generation* of fuzzy control (or better *optimal multistage fuzzy control*) (cf. Kacprzyk, 1992).

The problem considered in Kacprzyk (1983a), which is also the point of departure for our present discussion, is as follows. First, $\mathcal{U} = \{C_1, \ldots, C_n\}$ is a set of *fuzzy controls* defined in U, $\mathcal{X} = \{S_1, \ldots, S_m\}$ is a set of *fuzzy states* (outputs) in X, the *system under control* is governed by $X_{t+1} = F(X_t, U_t)$, where $X_t, X_{t+1} \in \mathcal{X}$ are fuzzy states at time (control stage) t and $t+1$, and $U_t \in \mathcal{U}$ is a fuzzy control at t, $t = 0, 1, \ldots, N-1$; N is the *termination time* (planning horizon). Next, $\mu_{\overline{C}^t}(U_t)$ is a *fuzzy constraint* on U_t, $\mu_{\overline{G}^{t+1}}(X_{t+1})$ is a *fuzzy goal* on X_{t+1}, and $\mu_D(.\,|\,.)$ is a *fuzzy decision*; $\overline{G}^{t+1}$ and $\overline{C}^t$ account for the fuzziness of X_{t+1} and U_t as, e.g., $\mu_{\overline{G}^{t+1}}(X_t) = 1 - \mathrm{d}(X_{t+1}, G^{t+1})$, where $\mathrm{d}(.,.)$ is some distance; an similarly for $\overline{C}^t$.

We seek an *optimal sequence of fuzzy controls* $U_0^*, \ldots, U_{N-1}^*$ such that

$$
\begin{aligned}
\mu_D(U_0^*, \ldots, U_{N-1}^* \mid X_0) &= \\
&= \max_{U_0, \ldots, U_{N-1}} \mu_D(U_0, \ldots, U_{N-1} \mid X_0) = \\
&= \max_{U_0, \ldots, U_{N-1}} \bigwedge_{t=0}^{N-1} (\mu_{\overline{C}^t}(U_t) \wedge \mu_{\overline{G}^{t+1}}(X_{t+1}))
\end{aligned}
\tag{1}
$$

Problem (1) leads to various problem classes which may be differentiated due to (cf. Kacprzyk, 1983a): (1) **the termination time:** (a) fixed and specified in advance, (b) implicitly given (by entering a termination set of states), (c) fuzzy, and (d) infinite; (2) **the system under control:** (a) deterministic, (b) stochastic, and (c) fuzzy. The cases of all termination types, and with a fuzzy system under control will be discussed here.

In all these problems the numerical efficiency requires a finite and relatively small number of fuzzy states and controls. Since in general this number is very high (theoretically infinite), we assume some reference fuzzy states and controls by which all fuzzy states and controls are to be approximated in the course of the algorithms. Then, we derive optimal policies relating optimal reference fuzzy controls to reference fuzzy states. These policies are represented by fuzzy relations which are in turn used, via the compositional rule of inference, to determine an optimal fuzzy control (not necessarily reference) for a current fuzzy state (not necessarily reference). For this procedure to give meaningful results, there should be a relatively large number of overlapping reference fuzzy states and controls. This is however harmful to the numerical efficiency of the optimal control algorithms used! We assume therefore a small number of nonoverlapping reference fuzzy states and controls, and then use an interpolative (analogical) reasoning scheme to infer an optimal fuzzy control for a current fuzzy state.

2. Control with a fixed and specified termination time

The problem considered, (1), can be solved by dynamic programming (Baldwin and Pilsworth, 1982) which boils down to seeking $U_0^*, \ldots, U_{N-1}^*$ such that

$$
\mu_D(U_0^*, \ldots, U_{N-1}^* \mid X_0) =
$$
$$
\doteq \bigwedge_{t=1}^{N} (\max_{U_{t-1}}((\max(\mu_{U_{t-1}}(u_{t-1}) \wedge \mu_{C^{t-1}}(u_{t-1})) \wedge
$$
$$
\wedge \max_{X_t}(\max_{x_t}(\mu_{X_t}(x_t \wedge \mu_{G^t}(x_t))))) \tag{2}
$$

whose corresponding dynamic programming recurrence equations are

$$
\begin{cases}
\mu_{\tilde{G}^N}(X_N) = \max_{x_N}(\mu_{X_N}(x_N \wedge \mu_{G^N}(x_N)) \\
\mu_{\tilde{G}^{N-i}}(X_{N-i}) = \max_{U_{N-i}}(\max_{u_{N-i}}(\mu_{U_{N-i}}(u_{N-i}) \wedge \\
\quad \wedge \mu_{C^{N-i}}(u_{N-i})) \wedge \mu_{\tilde{G}^{N-i+1}}(X_{N-i+1})) \\
\mu_{X_{N-i+1}}(x_{N-i+1}) = \max_{x_{N-i}}(\max_{u_{N-i}}(\mu_{U_{N-i}}(u_{N-i}) \wedge \\
\quad \wedge \mu_{X_{N-i+1}}(x_{N-i+1} \mid x_{N-i}, u_{N-i})) \wedge \mu_{X_{N-i}}(x_{N-i})) \\
i = 1, \ldots, N
\end{cases} \tag{3}
$$

This set of recurrence equations may be solved but $\mu_{\tilde{G}^t}(X_t)$'s are to be specified for all X_t whose number may be huge (and similarly $\max_{U_{t-1}}$ is to proceed over a large set of U_{t-1}'s). A natural approach (cf. Kacprzyk and Staniewski, 1982) is to use some prespecified *reference fuzzy states* and *reference fuzzy controls*, denoted $\overline{U}_t \in \overline{U} = \{\overline{C}_1, \ldots, \overline{C}_l\} \subseteq U$ and $\overline{X}_{t+1} \in \overline{X} = \{\overline{S}_1, \ldots, \overline{S}_r\} \subseteq X$, and to express all U_{t-1}'s and X_t's by their closest reference counterparts (fuzzy matching!). Such an approach is used to solve (3), and will be used throughout this paper.

The solution of (3), an *optimal control policy*, a_t^*, such that $\overline{U}_t^* = a_t^*(\overline{X}_t)$, $t = 0, 1, \ldots, N - 1$ – is represented by "IF $\overline{X}_t = \overline{S}_k$ THEN $\overline{U}_t = \overline{C}_w$", $k = 1, \ldots, r$, equated with a fuzzy relation R in $X \times U$. Thus, for a current X_t the U_t^* sought is determined by the *compositional rule of inference* $U_t^* = X_t \circ R$.

Unfortunately, this dynamic programming scheme is not efficient (the infamous curse of dimensionality!). Moreover, for the compositional rule of inference to work properly there should be a high number of overlapping fuzzy controls and states. This is however harmful to the computational efficiency of dynamic programming! This contradiction will be resolved by assuming a small number of nonoverlapping reference fuzzy states and controls, and then using interpolative reasoning.

Simpler and more efficient than dynamic programming is an earlier branch-and-bound approach by Kacprzyk (1978a, 1979): we seek $\overline{U}_0^*, \ldots,$ $\overline{U}_{N-1}^*$ such that

$$
\mu_D(\overline{U}_0^*, \ldots, \overline{U}_{N-1}^* \mid \overline{X}_0) =
$$
$$
= \max_{\overline{U}_0, \ldots, \overline{U}_{N-1}} (\mu_{\overline{C}^0}(\overline{U}_0) \wedge \mu_{\overline{G}^1}(\overline{X}_1) \wedge \ldots
$$
$$
\ldots \wedge \mu_{\overline{C}^{N-1}}(\overline{U}_{N-1}) \wedge \mu_{\overline{G}^N}(\overline{X}_N)) \tag{4}
$$

The solution of problem (4) is based on the following property of "$\wedge$" (and of many other t-norms too): if $\mu_D(\overline{U}_0, \ldots, \overline{U}_k \mid X_0) = \wedge_{t=0}^{k-1}(\mu_{\overline{C}^t}(\overline{U}_t)$ $\wedge \mu_{\overline{G}^{t+1}}(\overline{X}_{t+1})$, then $N - 1 > k > l \Rightarrow \mu_D(U_0, \ldots, U_{N-1} \mid X_0) \leq \mu_D(U_0, \ldots,$ $U_k \mid X_0) \leq \mu_D(U_0, \ldots, U_l \mid X_0)$. That is, by "adding" next controls we cannot increase the value of $\mu_D(. \mid .)$. The branching is via $\overline{U}_t$'s and the bounding is via $\mu_D(\overline{U}_0, \ldots, U_t \mid \overline{X}_0)$'s.

An efficient branching requires a small number of reference fuzzy controls, $\overline{C}_1, \ldots, \overline{C}_w$. As a solution we obtain optimal control policies $\overline{U}_t^* = a^*(\overline{X}_t)$ which are determined as a fuzzy relation, and for a current (not necessarily reference) fuzzy state the optimal control is determined via the compositional rule of inference. And again, there should be sufficiently many, overlapping reference fuzzy states and controls to obtain meaningful results. For numerical efficiency we should however have as few as possible, nonoverlapping reference fuzzy states and controls. Then, interpolative reasoning can be used again.

3. Control with an implicit termination time

The termination time N is now when the system enters for the first time a terminating set of states, $\mathcal{X}^T \subset \mathcal{X}$. Suppose that $\overline{\mathcal{X}} = \{\overline{S}_1, \ldots, \overline{S}_p, \overline{S}_{p+1}, \ldots, \overline{S}_r\}$ and $\overline{\mathcal{X}}^T = \{\overline{S}_{p+1}, \ldots, \overline{S}_r\}$.

We seek $\overline{U}_0^*, \ldots, \overline{U}_{K-1}^*$ such that

$$
\mu_D(\overline{U}_0^*, \ldots, \overline{U}_{K-1}^* \mid \overline{X}_0) =
$$
$$
= \max_{\overline{U}_0, \ldots, \overline{U}_{K-1}} (\mu_{\overline{C}}(\overline{U}_0 \mid \overline{X}_0) \wedge \ldots \wedge \mu_{\overline{C}}(\overline{U}_{K-1} \mid \overline{X}_{K-1}) \wedge \mu_{\overline{G}}(\overline{X}_K)) \tag{5}
$$

where $\overline{X}_K \in \mathcal{X}^T$ and $\overline{X}_{K-1} \in \mathcal{X} - \mathcal{X}^T$. Using the approximation by reference fuzzy controls and states, (5) may be solved using: an iterative approach, a graph-theoretic approach, and a branch-and-bound approach (cf. Kacprzyk, 1983a). The first is related to the case of an infinite termination time (cf. Section 5), the second is not operational, and the last is simple and efficient – analogous to that discussed in Section 2.

The model with an implicit termination time may be very useful, in particular for optimizing the first part of the trajectory until a stable operation when the termination time is evidently unknown in advance.

4. Control with a fuzzy termination time

The idea of a fuzzy termination time, exemplified by *a couple of, some, about ten,* ... control stages, appeared in Fung and Fu (1977) and Kacprzyk (1977, 1978a, c). It is given as a fuzzy set, W, in $V = \{1, \ldots, K, K + 1, \ldots, N\}$; $\mu_W(t) \in [0, 1]$ represents how "good" t is as the termination time. The process should terminate at some $M \in \operatorname{supp} V = \{t \in T : \mu_W(t) > 0\}$, and we assume that $M \in \{K, K+1, \ldots, N\}$.

We seek an optimal termination time M^* and $\overline{U}_0^*, \ldots, \overline{U}_{M^*}^*$ such that

$$
\mu_D(\overline{U}_0^*, \ldots, \overline{U}_{M^*}^* \mid X_0) = \max_{M, \overline{U}_0, \ldots, \overline{U}_{M-1}} (\mu_{\overline{C}^0}(\overline{U}_0) \wedge \ldots
$$

$$
\ldots \wedge \mu_{\overline{C}^{M-1}}(\overline{U}_{M-1}) \wedge (\mu_W(M)\, \mu_{\overline{G}^M}(\overline{X}_M))) \tag{6}
$$

For solving (6), Kacprzyk's (1977, 1978a, c) dynamic programming approach similar to (2), and Kacprzyk's (1978b, 1979) branch-and-bound scheme (cf. Section 3) may be used.

In the former case, we devise two sets of recurrence equations:

$$
\begin{cases}
\overline{\mu}_{\tilde{G}M}(\overline{X}_M, M) = \max_{x_M}(\mu_{X_M}(x_M) \wedge \mu_W(M)\, \mu_{\overline{G}^M}(\overline{X}_M)) \\
\overline{\mu}_{\tilde{G}M-i}(\overline{X}_{M-i}, M) = \\
\quad = \max_{U_{M-i}}(\max_{u_{M-i}}(\mu_{\overline{U}_{M-i}}(u_{M-i}) \wedge \mu_{\overline{C}^{M-i}}(\overline{U}_{M-i})) \wedge \\
\quad \wedge \overline{\mu}_{\tilde{G}M-i+1}(\overline{X}_{M-i+1}, M)) \\
\mu_{\overline{X}_{M-i+1}}(x_{M-i+1}) = \max_{x_{M-i}}(\max_{u_{M-i}}(\mu_{U_{M-i}}(u_{M-i}) \wedge \\
\quad \wedge \mu_{\overline{X}_{M-i+1}}(x_{M-i+1} \mid x_{N-i}, u_{N-i})) \wedge \mu_{\overline{X}_{M-i}}(x_{M-i})) \\
i = 1, \ldots, M - K + 1; M = K, \ldots, N
\end{cases} \tag{7}
$$

$$
\mu_{\overline{G}^{K-1}}(\overline{X}_{K-1}) = \max_M \overline{\mu}_{\tilde{G}^{K-1}}(\overline{X}_{K-1}, M) \tag{8}
$$

$$
\begin{cases}
\mu_{\overline{G}^{K-i-1}}(\overline{X}_{K-i-1}) = \max_{\overline{U}_{K-i-1}}(\max_{u_{K-i-1}}(\mu_{\overline{U}_{K-i-1}}(u_{K-i-1}) \wedge \\
\quad \wedge \mu_{\overline{C}^{K-i-1}}(\overline{U}_{K-i-1})) \wedge \mu_{\overline{G}^{K-i}}(\overline{X}_{K-i})) \\
\mu_{\overline{X}_{K-i}}(x_{K-i}) \max_{x_{K-i-1}}(\max_{u_{K-i-1}}(\mu_{\overline{U}_{K-i-1}}(u_{K-i-1}) \wedge \\
\quad \wedge \mu_{\overline{X}_{K-i}}(x_{K-i} \mid x_{K-i-1}, u_{K-i-1})) \wedge \mu_{\overline{X}_{K-i-1}}(x_{K-i-1})) \\
i = 1, \ldots, K - 1
\end{cases} \tag{9}
$$

We obtain optimal control policies $\overline{U}_t^* = a_t^*(\overline{X}_t)$, and all the previous remarks on the number of reference fuzzy states and controls and on interpolative reasoning are also valid here.

5. Control with an infinite termination time

When the process is to proceed over a long time and is low-varying, with the goal just to maintain some (stable) conditions, it may be expedient to assume an infinite termination time and use some specific apparatus proposed first by Kacprzyk and Staniewski (1982, 1983) (see also Kacprzyk, 1983a).

We seek an *optimal stationary policy* $a_\infty^* : X \longrightarrow U$ such that

$$\mu_D(a_\infty^* \mid X_0) =$$
$$= \max_{a_\infty} \mu_D(a_\infty \mid X_0) = \lim_{N \to \infty} \bigwedge_{t=0}^{N} b^t(\mu_{\overline{C}}(a \mid X_t) \wedge \mu_{\overline{G}}(X_{t+1})) \quad (10)$$

where $b > 1$ is a *discount factor* expressing a higher importance of earlier control stages.

A policy iteration algorithm was given by Kacprzyk and Staniewski (1982, 1983) to solve problem (10) which, by the approximation by reference fuzzy states and controls, was transformed into one with an auxiliary finite state deterministic system. Thus, if $A(.)$ means this approximation, the auxiliary deterministic system representing the fuzzy system under control is given by $\overline{X}_{t+1} = A(F(\overline{X}_t, \overline{U}_t)), t = 0, 1, \ldots$, and, e.g., $\mu_{\overline{C}}(\overline{U}_t \mid \overline{X}_t) = e((\overline{C}(\overline{X}_t), \overline{U}_t)$ and $\mu_{\overline{G}}(\overline{X}_t) = e(G, \overline{X}_t)$, where $e(.,.)$ is a degree of equality.

We seek an optimal stationary policy $a_\infty^* : \overline{X} \longrightarrow \overline{U}$ such that

$$\mu_D(a_\infty^*) = \max_{a_\infty} \lim_{N \to \infty} \bigwedge_{t=0}^{N} b^t(\mu_{\overline{C}}(a \mid \overline{X}_t) \wedge \mu_{\overline{G}}(\overline{X}_{t+1})) \quad (11)$$

and notice that an optimal reference stationary policy found above relates an optimal reference fuzzy control to a reference fuzzy state.

An a_∞^* solving (11) is determined by a policy iteration type algorithm (Kacprzyk and Staniewski, 1982, 1983) whose essence is:

Step 1 Choose an arbitrary $a_\infty = (a, a, \ldots)$ (relating an optimal reference fuzzy control to a reference fuzzy state!).

Step 2 Solve in $\mu_D(a_\infty \mid \overline{S}_i)$, $i = 1, \ldots, r$:

$$\mu_D(a_\infty \mid \overline{S}_i) = \mu_{\overline{C}}(a(\overline{S}_i) \mid \overline{S}_i) \wedge$$
$$\wedge \mu_{\overline{G}}(A(F(\overline{S}_i, a(\overline{S}_i)))) \wedge b\mu_D(a_\infty \mid A(F(\overline{S}_i, a(\overline{S}_i)))) \quad (12)$$

Step 3 Improve a_∞, i.e. find a $z^* : \overline{X} \longrightarrow \overline{U}$ maximizing $\mu_{\overline{C}}(z(\overline{S}_i) \mid \overline{S}_i) \wedge \mu_{\overline{G}}(A(F(\overline{S}_i, z(\overline{S}_i)))) \wedge b\mu_D(a_\infty \mid A(F(\overline{S}_i, a(\overline{S}_i))))$.

Step 4 If z^* found in Step 3 is the same as the previous one, it is an optimal reference stationary policy sought. Otherwise, assume $a_\infty = z^*$ and return to Step 2.

We obtain (in a finite number of steps!) an a_∞^*, i.e. for the reference fuzzy states only. Needless to say that the number of them (and of reference fuzzy controls) has a decisive impact on the efficiency of the algorithm. Thus, as in the previous sections, we assume a small number of reference fuzzy states and controls, solve the problem, and then use interpolative resoning.

6. Interpolative reasoning in the derivation of optimal controls

In all the above cases there is a conflict between a large number of reference fuzzy controls and states required for obtaining meaningful results via the compositional rule of inference, and a small number of them required for the efficiency.

Since for real problems the efficiency may be decisive, we may be forced to assume a small number of nonoverlapping reference fuzzy controls and states, and arrive at the situation: we obtain an optimal control policy a_t^* stating "IF $\overline{X}_t = \overline{S}_1$ THEN $\overline{U}_t^* = \overline{C}_{t1}$ ELSE ...ELSE IF $\overline{X}_t = \overline{S}_i$ THEN $\overline{U}_t^* = \overline{C}_{ti}$ ELSE IF $\overline{X}_t = \overline{S}_{i+1}$ THEN $\overline{U}_t^* = \overline{C}_{t(i+1)}$ ELSE ...ELSE IF $\overline{X}_t = \overline{S}_r$ THEN $\overline{U}_t^* = \overline{C}_{tr}$". The problem is the implementation of a_t^*. Suppose we wish to determine U_t^* for a current X_t, not e reference one. Let X_t be a fuzzy number between the two reference fuzzy states $\overline{S}_i$ and $\overline{S}_{i+1}$. We seek therefore U_t^* corresponding to X_t via this a_t^*. Notice that since X_t is not a reference one, U_t^* will not be a reference one either.

The determination of U_t^* is meant here, assuming a representation by triangular fuzzy numbers, as the determination of the mean value and width. It is reasonable to require U_t^* to be similar (close) to one of these optimal $\overline{C}_{ti}$ and $\overline{C}_{t(i+1)}$ corresponding to $\overline{S}_i$ and $\overline{S}_{i+1}$.

The idea of our approach is as follows. The first problem is to determine the mean value of the fuzzy optimal control sought. We apply here Kóczy and Hirota's (1992) approach whose essence may be expressed as

$$d(\overline{S}_i, X_t)/d(X_t, \overline{S}_{i+1}) = d(\overline{C}_i, U_t^*)/d(U_t^*, \overline{C}_{t(i+1)}) \tag{13}$$

where $d(.,.)$ is a distance. The sense of (13) is that the relative position of U_t^* with respect to its closest reference counterparts should be the same as that concerning X_t and its reference counterparts.

The second problem is to determine the width of U_t^*. The reasoning is that the lower the number of reference fuzzy states and controls, the less precise is the available information. Hence, the fuzzier (of a larger width)

U_t^* should be. For instance, we can use a formula

$$\overline{w}(U_t^*) = \frac{1}{5}[\overline{w}(\overline{S}_i) + \overline{w}(\overline{S}_{i+1}) + \overline{w}(\overline{X}_t) + \overline{w}(\overline{C}_{ti}) + \overline{w}(\overline{C}_{t(i+1)})] \qquad (14)$$

where $\overline{w}(.)$ is a relative width (related to the universe of discourse of the fuzzy states and controls, and the simplest arithmetics mean (14) can be replaced by another formula expressing the above rationale.

The approach sketched above works very well, and is relatively easy to implement.

7. Concluding remarks

Models of optimal multistage fuzzy control were discussed. To efficiently solve them, a small number of reference fuzzy states and controls was assumed, and optimal control policies were derived. To make it possible to use these policies not necessarily for the reference fuzzy states and controls, an interpolative reasoning scheme was proposed which determines the position of a (nonreference) fuzzy optimal control sought and its width (fuzziness).

8. References

Baldwin J.F. and B.W. Pilsworth (1982) Dynamic programming for fuzzy systems with fuzzy environment. *Journal of Mathematical Analysis and Applications* **85** 1–23.

Fung L.W. and K.S. Fu (1977) Characterization of a class of fuzzy optimal control problems. In M.M. Gupta, G.N. Saridis and B.R. Gaines (Eds.): *Fuzzy Automata and Decision Processes*, New York: North-Holland, pp. 209–219.

Kacprzyk J. (1977) Control of a non-fuzzy system in a fuzzy envirnment with fuzzy termination time. *Systems Science* **3** 320–331.

Kacprzyk J. (1978a) Control of a stochastic system in a fuzzy environment with fuzzy termination time. *Systems Science* **4** 291–300.

Kacprzyk J. (1978b) A branch-and-bound algorithm for the multistage control of a nonfuzzy system in a fuzzy environment. *Control and Cybernetics* **7** 51–64.

Kacprzyk J. (1978c) Decision-making in a fuzzy environment with fuzzy termination time. *Fuzzy Sets and Systems* **1** 169–179.

Kacprzyk J. (1979) A branch-and-bound algorithm for the multistage control of a fuzzy system in a fuzzy environment. *Kybernetes* **8** 139–147.

Kacprzyk J. (1983a) *Multistage Decision-Making under Fuzziness*. Cologne: Verlag TÜV Rheinland.

Kacprzyk J. (1983b) A generalization of fuzzy multistage decision making and control via linguistic quantifiers. *International Journal of Control* **38** 1249–1270.

Kacprzyk J. (1992) Fuzzy optimal control revisited: toward a new generation of fuzzy control. *Proceeings of 2nd International Conference on Fuzzy Logic and Neural Networks (Iizuka '92)*, Vol. 2, pp. 429–432.

Kacprzyk J. and C. Iwański (1987) Generalization of discounted multistage decision making and control through fuzzy linguistic quantifiers: an attempt to introduce commonsense knowledge. *International Journal of Control* **45** 1909–1930.

Kacprzyk J. and P. Staniewski (1982) Long term inventory policy making through fuzzy decision making models. *Fuzzy Sets and Systems* **8** 117–132.

Kacprzyk J. and P. Staniewski (1983) Control of a deterministic system in a fuzzy environment over infinite planning horizon. *Fuzzy Sets and Systems* **10** 291–298.

Kóczy L. and K. Hirota (1992) Analogical fuzzy reasoning and gradual inference rules. *Proceedings of 2nd International Conference on Fuzzy Logic and Neural Networks (Iizuka '92)*, Vol. 1, pp. 329–332.

FUZZY PETRI NETS AND THEIR APPLICATIONS TO FUZZY REASONING SYSTEMS CONTROL

Tadashi MATSUMOTO, Atsushi SAKAGUCHI, and Kohkichi TSUJI

Department of Electrical and Electronics Engineering
Faculty of Engineering, Fukui University, Fukui 910, JAPAN

1. INTRODUCTION

A fuzzy control has a distinguished feature that the control is capable of incorporating expert's empirical control rules using his linguistic description. [4] ∼ [11] But, automated extraction or identification of the fuzzy if - then rules is one of big issues in a fuzzy control. Several researchers have done on the identification capability as well as the learning capability of neural networks for their applications to fuzzy control. [7] ∼ [10]

However, another model for a fuzzy control has been desired because the neural network model has the linkage explosion and the computation of exponentials for each linkage of neurons. [4], [5]

In this paper, first, the fuzzy Petri net inference mechanism with learning function is proposed by using the extended fuzzy Petri nets. The new controller can automatically identify the if - then rules and tune the membership functions by utilizing expert's control data. Identification capability of the new fuzzy controller is also examined using numerical data. Secondly, an adaptive control system with this new inference engine is proposed. It is shown that this system can be controlled adaptively under the big parameter change.

2. FUZZY PETRI NETS AND FUZZY INFFRENCE MECHANISM

2.1. Definitions and Notations for the Extended Fuzzy Petri nets[1]∼[5].

First, let us define the ordinary Petri nets[1].

[Definition 1] Ordinary Petri Nets[1],[2]

The structure of an ordinary Petri net can be defined as a directed bipartite graph with two disjoint sets of nodes, S and T, called a set S of places (symbol:○) and a set T of transitions (symbol:|). Assume that cardinality of S is n. Marking $M \in N^n$ is defined as a non-negative integral vector whose component $M(p)$ is a number of tokens on the place p at a marking M. For a subset U of S, the symbol $^\bullet U$ denotes the set of all transitions t's such that there exists an arc from t to $p \in U$. The symbol $^\bullet U$ is called the set of input transitions of U. Similarly, $U^\bullet$ denotes the set of all transitions t's such that there exists an arc from $p \in U$ to t and is called the set of output transitions of U. For subset Q of T, the set of input places $^\bullet Q$ and the set of output places $Q^\bullet$ of Q are similarly defined. For a transition t, t is said to be firable at M iff $M(p) > 0$ for each $p \in {^\bullet t}$. When the firable transition t at M fires, the tokens are moved as follows and the new resulting marking $M'(p) \geq 0$ is defined as

$$M'(p) = M(p) + 1 : p \in t^\bullet \cap p \notin {^\bullet t}$$
$$M'(p) = M(p) - 1 : p \in {^\bullet t} \cap p \notin t^\bullet \qquad (1)$$
$$M'(p) = M(p) : \text{otherwise.}$$

Next, we review the new extended fuzzy Petri nats[4][5]. The definition of ordinary Petri nets is included in that of the extended fuzzy Petri nets. For modeling of vague discrete events, let us define the extended fuzzy Petri nets by changing the definition of ordinary Petri nets.

225

Z. Bien and K. C. Min (eds.),
Fuzzy Logic and its Applications, Information Sciences, and Intelligent Systems, 225–234.

[Definition 2] Extended fuzzy Petri nets (EFPNs) [5]

- Place: The same definition and symbol of a place as those in ordinary Petri nets are used.

- Transition: Symbol | :Free firing transition, but this firing is selective.

- Token: Let the symbol $\bullet^W$ be the token with the weight w which is any positive real number. Let us define the virtual token when W equals zero and denote it $\bigcirc = \bullet^\circ$. The virtual token is useful to distinguish the following two cases:

① The premise of if-then rule is denied, then the place has no token.

② The if-then rule has not worked at that time, then the place has no token.

The virtual token implies the case ①.

- Firing rule: By a firing of firable transition t at M, the token is moved and the new resulted marking $M'(p) \geq 0$ is defined as follows, where $a = \overset{min}{\underset{i}{}} M(p_i)$ and $p_i \in {}^\bullet t$.

$M'(p) = M(p) + a : p \in t^\bullet \cap p \notin {}^\bullet t$.

$M'(p) = M(p) - a : p \in {}^\bullet t \cap p \notin t^\bullet$. $\hspace{3cm}$ (2)

$M'(p) = M(p) : \text{otherwise}$.

- Arc: There exists no weight on it. See also Remark 1. ∎

[Definition 3] The max.transition

Let the max.transition be the transition t which transfers the maximum token-weight of all the input places p's $\in {}^\bullet t$ into all the output places p''s $\in t^\bullet$. ∎

However, if we use the virtual tokens, some inhibitors ($-\circ$), and the regular transitions which obey the firing rule of Definition 2, we can represent the max.transition by using EFPNs.

[Remark 1]

Although it is defined that an arc has no weight in the extended fuzzy Petri nets as in Definition 2, it is simple to use the weight on the arc (t, p) as shown in §2·2. However, the weight on the arc is transformed into the net of Difinition 2 because of the firing rule of the regular transitions. ∎

Hereafter, let us denote the extented fuzzy Petri nets EFPNs for simiplicity.

2.2. Fuzzy Inference Mechanism using EFPNs[5]

Let us consider the if-then rule: If X is A, then Y is C, where X and Y are fuzzy variables. First,let us explain how to model the discrete membership functions of both the premise and the consequence by using examples.

If X has the discrete membership functions of Fig.1(a), then the EFPN model for Fig.1(a) is obtained as in Fig.1(b), where " ↑ " implies " ↑↓ ", i.e., the double arc. When the input for X is 1.0, then the weights for P and Z is 0.5 and 0.3, respectivey.

If $Y = C$ has the discrete membership function of Fig.2(a) after Mamdani's min-max opetrations, then we have the EFPN model of Fig.2(b). When the resulted weights for P, Z, and N is 0.8, 0.2, and 0.0, we have the value of 0.8, 0.5, 0.2, 0.2, and 0.0 on the discrete universe 2, 1, 0, -1, and -2, respectively.

Secondly, let us consider the fuzzy inference mechanism of two input variables, X_1 and X_2, and one output variable, Y, by using Mamdani's method, where each variable has three linguistic labels, P, Z, and N and the decision rules are shown in Fig. 3(a). Then, we have nine if-then rules:

If X_1 is A_i and X_2 is B_i, then Y is C_i, $i = 1, 2, \cdots, 9$. Moreover, the resultant fuzzy number from the above each rule is determined by the max. operation (i.e., the max. transition).

Therefore, we have the EFPN model shown in Fig.3(b) for the above fuzzy inference mechanism.

If we adopt gravity calculation as a defuzzification method, the output Y^* of the fuzzy inference mechanism is determined by $Y^* = \int Y C^*(Y) dy / \int C^*(Y) dy$. where $C^*(Y) = C_1(Y) \vee$

$C_2(Y) \vee \cdots \vee C_9(Y)$ and $\vee$ denotes maximum. Note that we can use the resultant fuzzy numbers, P, Z, and N, of Fig.3(b) for the above defuzzification.

From Fig.3, we can easily understand each reasoning process and can do easily and effectively the knowledge acquisition because the extended fuzzy Petri nets EFPNs avoid the linkage explosion of the usual neural networks by linking only conditions that combine into rules, whereas neural networks link every neuron cell in a layer with every other neuron in adjacent layers. Further, the outputs of EFPN transitions are extermely simple and efficient to compute, while the output functions of the usual neural networks involve the computation of exponentials for each linkage of neurons[4],[5].

3. AUTOMATED EXTRACTION OF FUZZY IF-THEN RULES

3.1. Automated Rule-Extraction Method

The automated extraction or identification of the fuzzy if-then rules is one of big issues in a fuzzy control or a fuzzy inference engine.

In this subsection, let us show an automated rule-extraction method for the EFPN model fuzzy inference engine.

(1) The setting of the membership functions for the premise of a rule:

The membership functions for the premise of a rule is fixed such that each membership function has the equi-width and that it is regularly spaced on the universe, where each fuzzy gain is adjusted to cover all the input data.

(2) Initialization of the consequence of a rule:

All the membership functions of the consquence are set zero at the biginning.

(3) Automated extraction of the membership functions for the consequence of a rule:

The consequence of an if-then rule is characterized by the weights of arcs from the transitions, of which input places imply the resultant conditions of each rule, to the places which imply the discrete points on the universe. Then, we can generate the membership functions of the consequence adjusting the above weights to the expert's input data. Let X_s and Y_s be the input-output pair at time t and let the token weight on the place which implies the adaptability of the premise of the i-th rule be $O_i(s)$. Then, the token weight of the consequence at time $t + 1$, $W_{ij}(t+1)$, is determined by the next equation:

$$W_{ij}(t + 1) = W_{ij}(t) - (Y_s - Y^{**}) \cdot (Y^{**} - j) \cdot O_i(s) \cdot LF. \qquad (3)$$

where j is the discrete value on the universe, $W_{ij}(0) = 0$ .and $LF = 0.001$ is the learning factor.

(4) The performance evaluation of a rule:

Measure the infered output to the expert's input data by using the new extracted rules and calculate the error between the infered output and the expert's output. If the error is too big, adjust the interval of the premise and extract again the membership functions of the consequence.

3.2. Performance Evaluation

In order to check the usefulness of the learning fuzzy inference engine, we use the following nonlinear system with three inputs (X_1, X_2, X_3) and an output (Y), which was used to verify the learning capability in Refs.[7],[8],[11]:

$$Y = (1 + X_1^{0.5} + X_2^{-1} + X_3^{-1.5})^2 \qquad (4)$$

Table 1 is the obtained results about evaluation, where X_4 is a dummy variable. In Table 1, E_1 is the average error for identification data (No.1~No.20) and E_2 is the average error for evaluation data (No.21~No.40). In our experiment, E_1 is very good compared with others [7], [8], [11], but E_2 is not so good. As a whole, we can say that our fuzzy engine has the same identification capability as that in [7], [8], [11].

4. SIMULATION

In this section, the simulated results for the controlled object of the first-order lagging system with dead time are given. The control system used is Fig.6 without the output trajectry estimator, where the learning fuzzy controller has two inputs (e, Δe) and an output (Δu) with seven premise membership functions.

We adopted the input and output data of PI control as the expert's data for automated extraction of rules, where $K_p = 4.2$, $K_I = 0.12$, and the sampling period $\tau = 0.2$. The number of input and output data for identification is 250 from $t = 0.0$ to $t = 50.0$ (sec). Moreover, the fuzzy gains are $g_e = 0.02$, $g_{\Delta e} = 0.5$, and $g_{\Delta u} = 1.7$, where e, Δe, and Δu is the error, the change of error, and the change of manipulated variable, respectively.

Table 2 shows the obtained fuzzy rule table in which the number implies each membership function. In this simulation, 35 rules out of 49 are automatically generated.

One of controlled results under the automatically extracted rules is shown in Fig.4 together with PI control, which are step responses of a feedback control system which incorporates the above fuzzy engine and a controlled object of a first order lagging system with a dead time : $G(s) = \dfrac{e^{-2s}}{1 + 20s}$.

5. CONTROL SYSTEM WITH FUZZY PETRI NET MODEL INFERECE ENGINE

Let us consider the adaptive modification of if-then rules to overcome the system paramater change or the disturbance by using the knowledge about the output trajectory which implies a kind of the higher rank knowledges than the expert's operating knowledges. This problem is another important issue in a fuzzy control[6]. Fig.6 is the blockdiagram of the fuzzy control system, where the Fuzzy controller in Section 4 is used, and the output trajectory estimation and the performance evaluation are as follows.

5.1. Output Trajectory Estimator

The fuzzy if-then rule for the output trajectory estimator has two inputs (e and Δe) and an output (the change of error at the next time : Δne) and the range of the universe of both the premise and the consequence is from -9 to +9 and it is discretized as the integer. The number of rules is $9 \times 9 = 81$. The estimation process is as follows.

(1) Collect the operator's input and output data and determine each fuzzy gain for e, Δe, and Δne.

(2) The initial learning or identification:

Extract the knowledge for each fuzzy if-then rule for the output trajectry estimator from the operator's input and output data by using the same automated rule-extraction method as that in Section 3.

(3) On line adaptation:

Each consequence of rules of the fuzzy controller is modified such that $\Delta u(t_2 - 1)$ equals

$D\_\Delta u(t_2 - l)$ by using the method of §3.1 three times and the next equation:

$$D\_\Delta u(t_2 - l) = \frac{D\_\Delta e(t_2) - \Delta e(t_1)}{\Delta e(t_2) - \Delta e(t_1)} \cdot \Delta u(t_2 - l). \tag{5}$$

where $\Delta e(t_1)$, $\Delta e(t_2)$, and $D\_\Delta e(t_2)$ are defined in Fig.5 and $\Delta u(t_2 - l)$ is the change of manipulating variable at time $t_2 - l$.

Each fuzzy gain for e, Δe, and Δne is determined as follows: $g_e = 0.1$, $g_{\Delta e} = 2.0$, and $g_{\Delta ne} = 40.0$. Note that l is empirically chosen as eleven.

The idenification of the 9×9 rules for the output trajectory estimator is done as the same way as that in Section 4 and 48 rules out of 81 are automatically generated, where the maximum learning repetitions is 2000 times.

5.2. Simulation for Fuzzy Control System

In order to check the usefulness for the system parameter change in Fig.6, we simulate the next case, where the parameter is changed at time $t = 20$(sec).

$$G(s) = \frac{e^{-2s}}{1 + 20s} \rightarrow G(s) = \frac{e^{-2s}}{1 + 10s}$$

The simulated result is shown in Fig.7 together with PI control results.

Fig.8 is the simulated result with the disturbance $d = 80$ in the input of controlled system of Fig.6 at time $t = 25$(sec), where $G = \dfrac{e^{-2s}}{1 + 20s}$. We could recognize the usefulness of adaptation for the disturbance of being less than or equal to 80, but for more than $d > 80$, it was difficult to control the disturbance because the knowledge of the output trajectory estimator was not enough to control those bigger disturbances. If we identify directly the knowledge of the estimator from the desired output trajectory instead of the operator's input and output data, it seems that we can improve the adaptivity for bigger disturbance.

6. CONCLUSIONS

In this paper, we have proposed a learning fuzzy inference mechanism, or a fuzzy controller, by using the extented fuzzy Petri nets. This fuzzy engine could acquisite easily and effectively the knowledge for each if-then rule from the expert's operating input and output data.

For demonstrating the capability of the new fuzzy engine, we have also simulated step responses of a feedback control system which incorporated the fuzzy engine and a controlled object of a first order lagging system with a dead time. Moreover, we have applied this new engine to a control system with parameter change and disturbance. Then, we have comfirmed that the system could be controlled adaptively under rather big parameter change and disturbance.

Therefore we can conclude that the fuzzy Petri net inference engine has some superiorities to neural networks; the net structure as well as computation is simpler and each reasoning process can be easily understood. Then the knowledge acquisition can also be done easily and effectively.

References

[1] J.L.Peterson : "Petri Net Theory and the Modelling of Systems". Prentice-Hall(1981)

[2] T.Murata : "Petri nets : properties, analysis and applications". Proc.of the IEEE, Vol.77, No.4, pp.541-580 (1989).

[3] K.Jensen : "Coloured Petri Nets and the Invariant-Method". Theoret.Comput. Sci.. Vol.14,pp.317-336 (1985).

[4] C.G.Looney : "Fuzzy Petri Nets for Rule Based Decisionmaking". IEEE Trans.on Systems,Man.and Cibernetics.Vol.18,pp.178-183 (1988).

[5] T.Tsuji and T.Matsumoto : "Extended Petri Net Models for Neural Networks and Fuzzy Inference Engines".Proc. of ISCAS'90.pp.2670-2673 (1990).

[6] T.Terano,K.Asai.and M.Sugeno : "Fuzzy Systems Theory and Its Applications". Ohm-Sha(1987).

[7] I.Hayashi and H.Takagi : "Formulation of Fuzzy Reasoning by Neural Network".Proc.of 4th Fuzzy System Symposium. pp.55-60.Tokyo.May.pp.30-31(1988).

[8] S.Horikawa,T.Furuhashi. S.Okuma, and Y.Uchikawa : "A Learing Fuzzy Controller Using a Neural Network". Proc. of SICE. Vol.27. No.2. pp.208-215(1991).

[9] Y.Hayashi and M.Nakai : "Automated Extraction of Fuzzy IF-THEN Rules Using Neural Networks", Trans. IEE Japan. Vol.110-c. No.3. pp.198-206(1990).

[10] T.Yamaguchi,N.Imazaki. and K.Haruki : "A Reasoning and Learining Method for Fuzzy Rules with Associative Memory". ibid.. pp.207-214(1990).

[11] G.T.Kang and M.Sugeno : "Fuzzy Modelling". Proc.of SICE. Vol.26. No.6.pp.106-108(1987).

[12] Y.Deng and S.K.Chang : "A G-Net Model for Knowledge Representation and Reasoning". IEEE Trans. Knowledge and Data Engeneering. Vol.2. No.3. pp.295-310(1990).

[13] S.M.Chen, J.S.Ke. and J.F.Chang : "Knowledge Representation Using Fuzzy Petri Nets". ibid., pp.313-319(1990).

[14] T.Matsumoto, A.Sakaguchi. and K.Tsuji : "Automated Extraction of Fuzzy If-Then Rules Using Fuzzy Petri Nets and its Application to Adaptive System Control". Procs. of Korea-Japan Joint Conf. on Fuzzy Systems and Engeneering. pp.37-40(1992).

[15] M.G.Chun and Z.Bien : "Expert's Fuzzy Knowledge Representation and Inference Methods via Fuzzy Petri Nets". ibid.. pp.64-67(1992).

Table 1 Precision of identified model

	Input variables	E_1 (%)	E_2 (%)
Fuzzy Controller using EFPN	x_1,x_2,x_3,x_4	$0.11*10^{-3}$	19.13
	x_1,x_2,x_3	$0.38*10^{-3}$	5.41
Fuzzy Controller using neural network[8]	x_1,x_2,x_3,x_4	0.07	14.69
	x_1,x_2,x_3	0.60	2.41
Fuzzy Controller using neural network[7]	x_1,x_2,x_3	0.47	4.79
Sugeno's fuzzy model I [6,11]	x_1,x_2,x_3	1.5	2.1
Sugeno's fuzzy model II [6,11]	x_1,x_2,x_3	1.1	3.6

Table 2 Identified control rules

Δe \ e	PB	PM	PS	ZO	NS	NM	NB
PB	—	—	—	—	—	—	—
PM	—	⑥	⑫	⑱	㉔	㉚	—
PS	①	⑦	⑬	⑲	㉕	㉛	—
ZO	②	⑧	⑭	⑳	㉖	㉜	—
NS	③	⑨	⑮	㉑	㉗	㉝	—
NM	④	⑩	⑯	㉒	㉘	㉞	—
NB	⑤	⑪	⑰	㉓	㉙	㉟	—

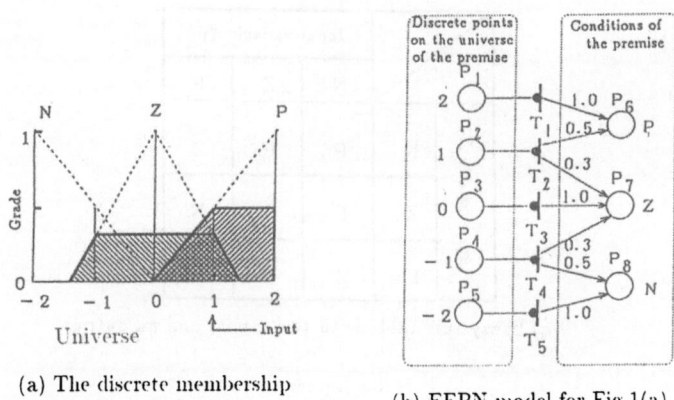

(a) The discrete membership functions for the premise.

(b) EFPN model for Fig.1(a).

Fig.1 EFPN model for the discrete membership functions for the premise.

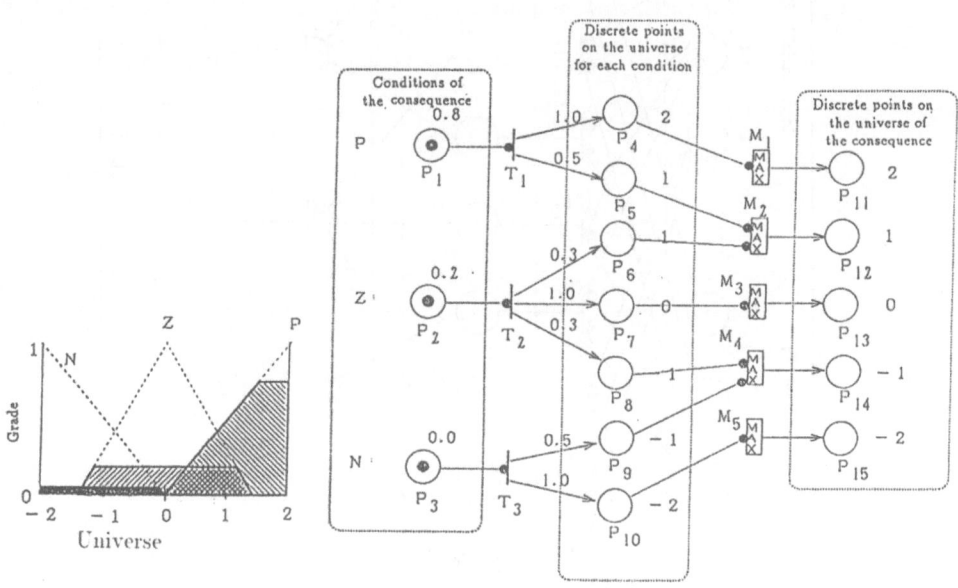

(a) The resutaulant discrete membership functions of the consequence

(b) EFPN model for Fig.2(a).

Fig.2 EFPN model for the discrete membership functions for the consequence.

		Input variable X_1		
		N	Z	P
Input variable X_2	N	P	P	Z
	Z	P	Z	N
	P	Z	N	N

(a) Fuzzy rule table with two inputs and an output.

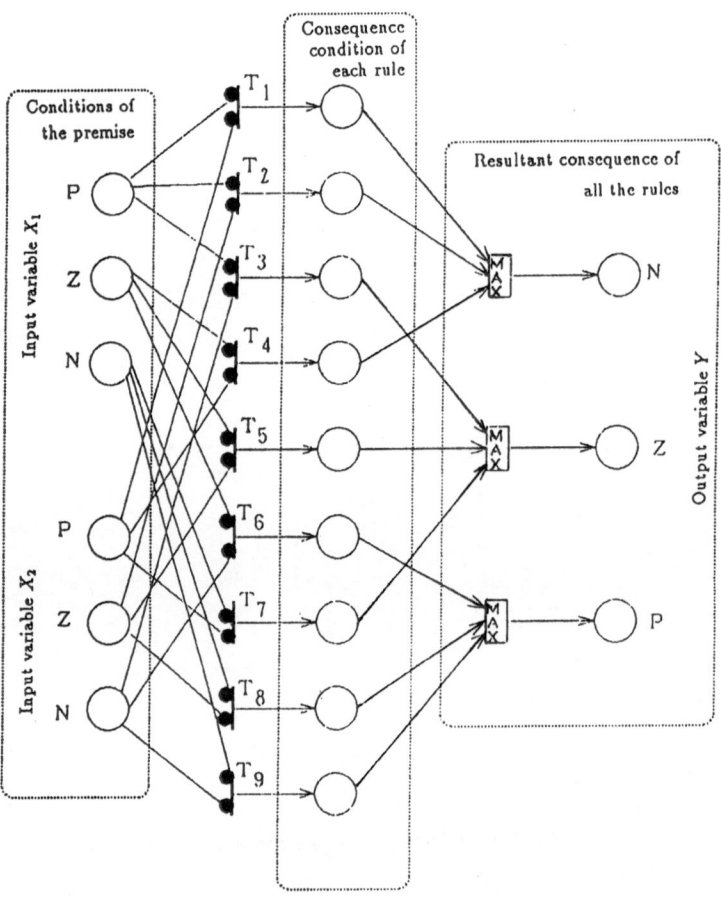

(b) EFPN model for Fig.3(a).

Fig.3 EFPN model fuzzy inference engine: each variable has three lingstic labels.

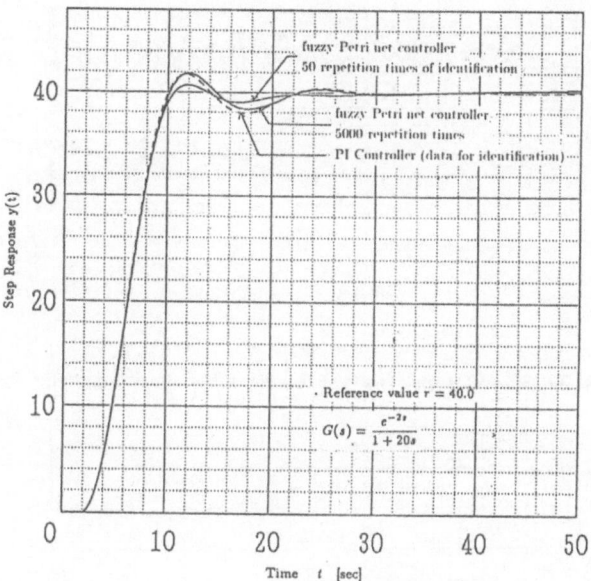

Fig.4 Step responses of a control system with PI controller or fuzzy Petri net controller

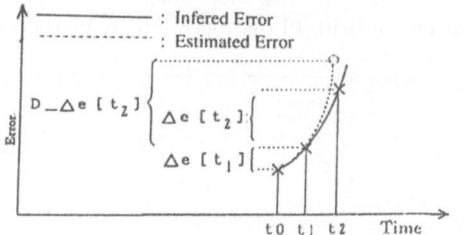

Fig 5 Illustration of eq.(5).

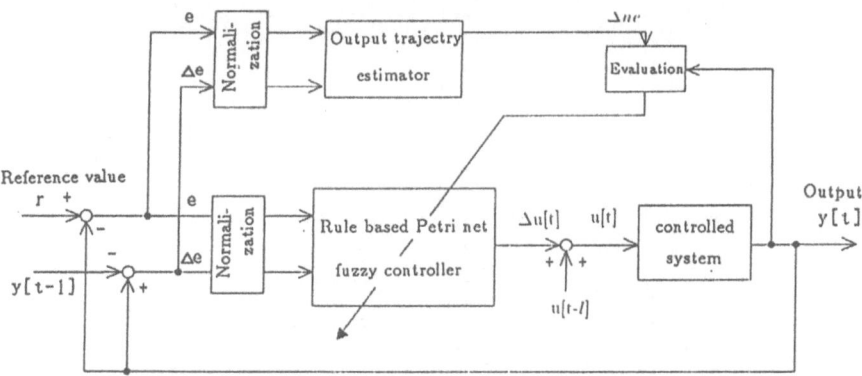

Fig. 6 System configuration of an adaptive control system.

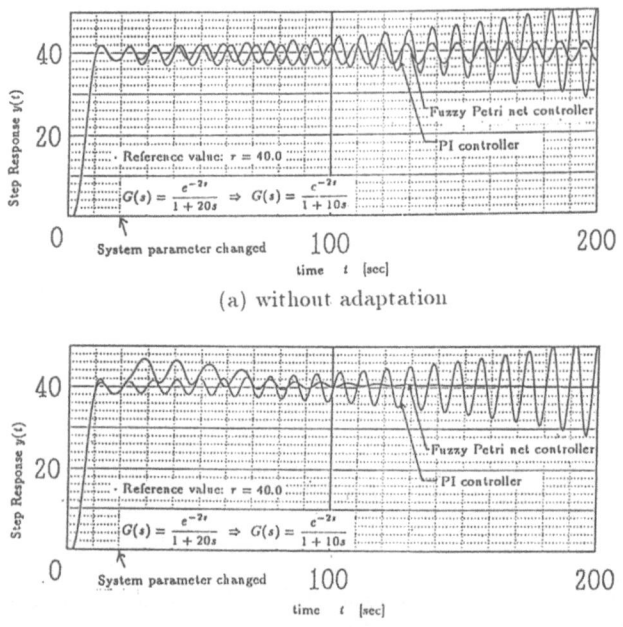

(a) without adaptation

(b) with adaptation

Fig.7 Experimented results with PI controller of fuzzy Petri net controller.

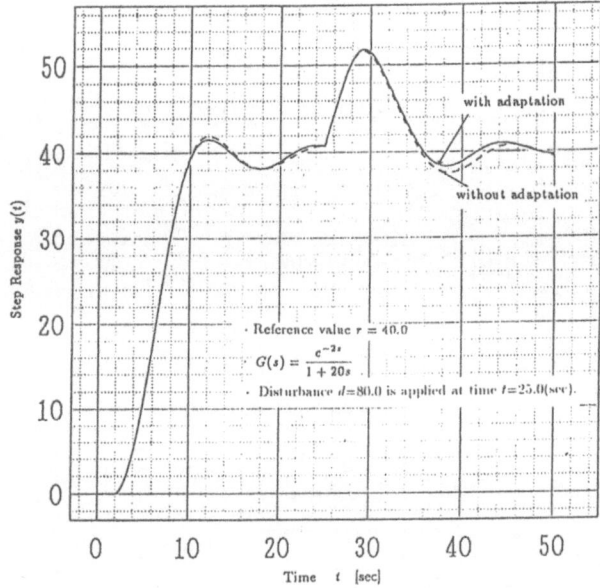

Fig.8 Step responses of a control system with or without fuzzy adaptation under disturbance.

Simulation Study on Self-learning Fuzzy Control of Carbon Monoxide Concentration

Kazuo TANAKA

Department of Mechanical Systems Engineering

Kanazawa University

2-40-20 Kodatsuno Kanazawa 920 Japan

TEL & FAX +81-762-34-4736

email tanaka@kicews2.ms.t.kanazawa-u.ac.jp

1. Introduction

CO (carbon monoxide) is one of important factors in air pollution problems. It is known that CO concentration system has

(1)high non-linearity, and

(2)many predictor variables.

On the other hand, it has been reported that fuzzy modeling techniques are useful for identification of complex systems [2,3,4,5].

In a previous paper [1], we have identified a fuzzy prediction model for CO concentration in the air at a traffic intersection point of a large city of Japan. Moreover, we have reported that the identified fuzzy model is very useful for predicting CO concentration.

In this paper, we simulate self-learning control systems of CO concentration using the identified fuzzy model. The purpose of this control is to keep CO concentration at a constant level. We adaptively adjust controller parameters by introducing Widrow-Hoff

235

Z. Bien and K. C. Min (eds.),
Fuzzy Logic and its Applications, Information Sciences, and Intelligent Systems, 235–244.
© 1995 *Kluwer Academic Publishers*.

learning rule since dynamics of the real CO concentration system changes gradually over a long period of time.

2. Fuzzy Modeling of CO Concentration

A fuzzy model, proposed by Takagi and Sugeno [3], is described by fuzzy IF-THEN rules which locally represent linear input-output relations of a system. This fuzzy model is of the following form:

Rule i : IF x_1 is A_{i1} and $\cdots$ and x_n is A_{in} THEN $y_i = c_{i0} + c_{i1}x_1 + \cdots + c_{in}x_n$ (1)

where $i=1, 2, \cdots, r$. r is the number of IF-THEN rules, y_i is the output from the i-th IF-THEN rule, and A_{ij} is a fuzzy set.

Given an input $(x_1, x_2, \cdots, x_n)$, the final output of the fuzzy model is inferred by as follows [8]:

$$y = \sum_{i=1}^{r} w_i y_i,$$
(2)

where y_i is calculated for the input by the consequent equation of the i-th implication, and the weight w_i implies the overall truth value of the premise of the i-th implication for the input calculated as

$$w_i = \prod_{k=1}^{n} A_{ik}(x_k) = \prod_{k=1}^{n} \exp(-\frac{(x_k - d_{ik})^2}{b_{ik}}),$$
(3)

where d_{ik} and b_{ik} are parameters of the membership functions.

Next, we explain outline of identification algorithm of a fuzzy model proposed by Tanaka and Sano [2]. Our method is a simplified version of a fuzzy modeling method proposed by Sugeno and Kang [4, 5].

Fig.1 shows the identification algorithm. The identification procedure is classified into three steps:

(Step 1)choice of the premise structure and the consequent structure;

(Step 2)identification of the parameters of the structure determined in (Step 1);

(Step 3)verification of the premise structure and the consequent structure.

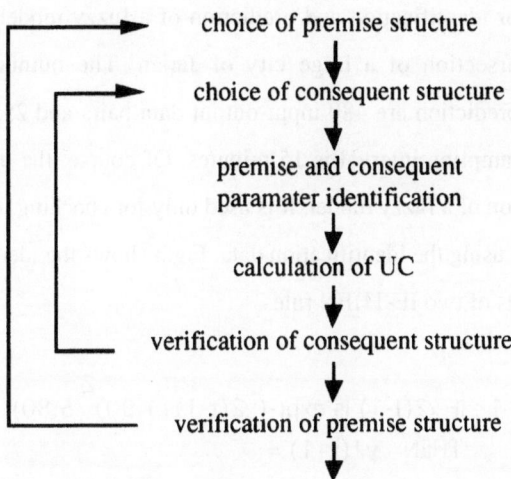

Fig.1 Identification algorithm.

In a previous paper [1], we have applied this fuzzy modeling method to identification of a fuzzy prediction model for CO concentration in the air at a traffic intersection point of a large city of Japan. Inputs and an output of a fuzzy model are shown in Fig.2. x_1 is the wind velocity, x_2 is the volume of traffic, x_3 is the temperature, x_4 is the amount of sunshine, and y is the CO concentration. For each variable, we perform normalization so that the mean and the variance of normalized variables equal 0 and 1, respectively. In other words, we transform the distribution to $N(0,1)$.

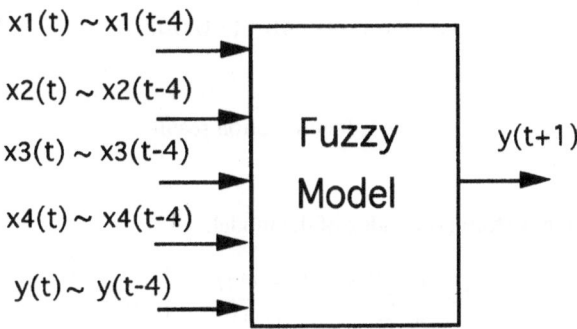

Fig.2 Inputs and an output of the CO concentration model.

238

The data used for identification and prediction of a fuzzy model are collected at the busiest traffic intersection of a large city of Japan. The number of data used for identification and prediction are 480 input-output data pairs and 253 input-output pairs, respectively. The sampling interval is 15 minutes. Of course, the prediction data is not used for identification of a fuzzy model. It is used only for checking the validity of a fuzzy model identified by using the identification data. Fig.3 shows the identification result. The fuzzy model consists of two IF-THEN rules.

Rule 1 : IF x2(t-1) is exp(-(x2(t-1)+1.90)^2/5.80)
 THEN   y1(t+1) =
 -0.008x1(t-2)+0.026x1(t-1)-0.032x1(t)
 -0.205x2(t-1)+0.249x2(t)
 -0.138x4(t-4)-0.181x4(t-3)+0.335x4(t-2)
 +0.088x4(t-1)-0.112x4(t)
 +0.090y(t-4)-0.106y(t-3)+0.011y(t-2)
 -0.460y(t-1)+1.356y(t)-0.018

Rule 2 : IF x2(t-1) is exp(-(x2(t-1)-2.06)^2/5.72)
 THEN   y2(t+1) =
 -0.013x1(t-2)+0.059x1(t-1)-0.032x1(t)
 -0.497x2(t-1)+0.827x2(t)
 +0.006x4(t-4)+0.066x4(t-3)-0.130x4(t-2)
 +0.095x4(t-1)-0.036x4(t)
 -0.009y(t-4)+0.007y(t-3)-0.037y(t-2)
 -0.103y(t-1)+0.748y(t)-0.002

Fig.3 Identification result

Eq.(4) shows the performance index of the model.

$$J = \frac{1}{m}\sum_{t=0}^{m-1}\left|\frac{y(t+1)-y*(t+1)}{y(t+1)}\right| \times 100, \qquad (4)$$

where m is the number of input-output data. y*(t+1) and y(t+1) are the outputs of the identified fuzzy model and the real system at time instant t+1, respectively. y*(t+1) and y(t+1) are raw data and are not normalized.

Table 1 shows the values of performance index for a linear model and the fuzzy model. J_1 and J_2 are the values of performance index for the identification data and the prediction data, respectively. The performance index J_2 of the fuzzy model is superior to that of linear model. This means that the CO concentration system is essentially non-linear. Table 1 shows the usefulness of the identified fuzzy model.

Table 1 Performances of models

	Linear model	Fuzzy model
J_1	5.7	4.8
J_2	11.8	5.9

3. Self-learning Controls

We simulate two self-learning control systems of keeping CO concentration at a constant level using Widrow-Hoff learning rule:linear control and fuzzy control. Fig.4 shows the self-learning control system, where $x_2(t)$ is the manipulated variable, $y(t)$ is the controlled variable and r is the setpoint of CO concentration.

In this simulation,

(1)with respect to wind velocity x_1, temperature x_3 and amount of sunshine x_4, we use real values of the prediction data,

(2)we use the identified fuzzy model as a controlled object.

Linear controller is constructed as follows.

$$\Delta x_2(t) = a_1 z_1(t) + a_2 z_2(t),$$
$$x_2(t) = x_2(t-1) + \Delta x_2(t),$$

where

$$z_1(t)=r-y(t) \text{ and } z_2(t)=z_1(t)-z_1(t-1)$$

and a_i (i=1, 2) is a parameter of the controller. As mentioned above, we adaptively adjust controller parameters by using Widrow-Hoff learning rule. The idea which adaptively optimizes parameters of fuzzy controller using Widrow-Hoff learning rule was first introduced by Ichihashi [8].

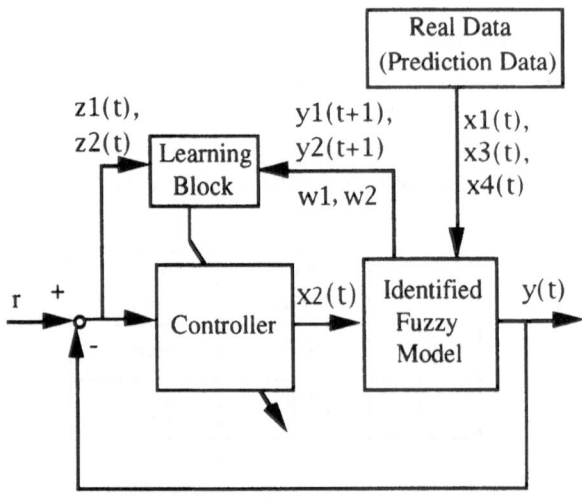

Fig.4 Self-learning control system.

Let us consider the following performance function.

$$J = \frac{1}{2}(r - y(t+1))^2 \qquad (5)$$

By partially differentiating J with respect to each controller parameter a_i, we obtain

$$\frac{\partial J}{\partial a_i} = -\{r - \sum_{j=1}^{2} w_j(t)y_j(t+1)\} \cdot z_i(t) \cdot \sum_{j=1}^{2} w_j(t)p_j \qquad (6)$$

where $w_j(t)$ is the membership value of j-th rule of fuzzy model at time instant t and p_j is a consequent parameter of $x_2(t)$, that is, $p_1 = 0.249$ and $p_2 = 0.827$. We can successively adjust controller parameters using Eq.(7).

$$a_i^{NEW} = a_i^{OLD} + \varepsilon_i \{r - \sum_{j=1}^{2} w_j(t)y_j(t+1)\} \cdot z_i(t) \cdot \sum_{j=1}^{2} w_j(t)p_j \qquad (7)$$

where ε_i is a learning factor and $\varepsilon_i = 0.003$ (i=1, 2). $w_j(t)$ and $y_j(t+1)$ can be calculated by the identified fuzzy model shown in Fig.3.

On the other hand, fuzzy controller is constructed as follows.

Rule 1:IF $x_2(t-1)$ is A_1 THEN

$$\Delta x_{12}(t) = a_{11} z_1(t) + a_{12} z_2(t),$$

Rule 2:IF $x_2(t-1)$ is A_2 THEN

$$\Delta x_{22}(t) = a_{21} z_1(t) + a_{22} z_2(t),$$

where A_1 and A_2 are the same fuzzy sets as in the identified fuzzy model shown in Fig.3, that is,

$$A_1(x_2(t-1)) = \exp(-\frac{(x_2(t-1)+1.90)^2}{5.80}),$$

$$A_2(x_2(t-1)) = \exp(-\frac{(x_2(t-1)-2.06)^2}{5.72}).$$

The final output of the fuzzy controller is calculated as follows.

$$\Delta x_2(t) = w_1(t)\Delta x_{12}(t) + w_2(t)\Delta x_{22}(t),$$

$$x_2(t) = x_2(t-1) + \Delta x_2(t),$$

where $w_1(t)$ and $w_2(t)$ are the membership values of A_1 and A_2, that is, $w_1 = A_1(x_2(t-1))$, $w_2 = A_2(x_2(t-1))$. The learning law of the fuzzy controller can be derived in the same way as the linear controller.

$$a_{ik}^{NEW} = a_{ik}^{OLD} + \varepsilon_{ik} \{r - \sum_{j=1}^{2} w_j(t)y_j(t+1)\} \cdot w_k(t)z_i(t) \cdot \sum_{j=1}^{2} w_j(t)p_j \qquad (8)$$

where $\varepsilon_{ik} = 0.003$ (i, k=1, 2).

Fig.5 shows relations between the number of learning and summation of squared error (SE), where

$$SE = \sum_{t=1}^{m-1} (r - y(t))^2, \qquad (9)$$

and m=253. It is found from Fig.5 that the learning efficiency of the fuzzy controller is superior to that of the linear controller. We can point out that the self-learning fuzzy controller effectively compensate non-linearity of the controlled object.

Fig.6 ~ Fig.8 show simulation results of self-learning fuzzy control system. The setpoint r is set as follows. If time of a day is from 8 o'clock to 20 o'clock, then r=30, else r=5. It is found from these figures that the self-learning fuzzy control of CO concentration are effectively realized.

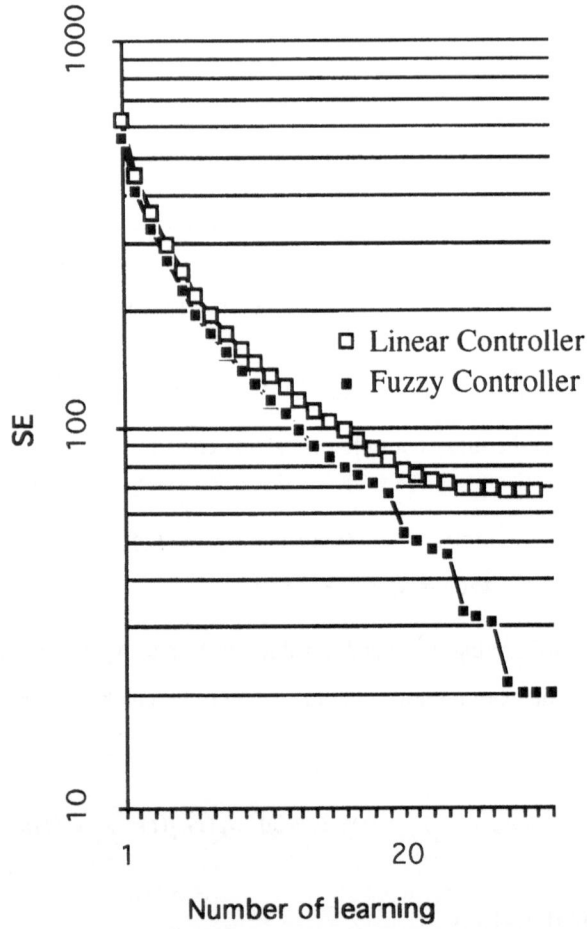

Fig.5 Learning process.

4. Conclusion

We have simulated two self-learning control systems for an identified fuzzy prediction model of CO concentration by using Widrow-Hoff learning rule. The purpose of this control is to keep CO concentration at a constant level. It has been assumed in this simulation that the identified fuzzy model perfectly represents real CO concentration system. Therefore, we should investigate robustness of this control system for parameter perturbation of the CO concentration model. This point is discussed in [10].

References

[1]K.Tanaka, M.Sano and H.Watanabe:Identification and Analysis of Fuzzy Model for Air Pollution, Proceedings of IEEE International Conference on IECON'92, Vol.3, pp.1431-1436 (1992).

[2]K.Tanaka, M.Sano and K.Suzuki:A Simplified Method on Fuzzy Identification Algorithm and Its Applications to Modeling of a Municipal Refuse Incinerator, Trans. Soci. Instrum. Contr. Eng., Vol. 28, No.11, pp.1355-1363 (1992) (in Japanese).

[3]T.Takagi and M.Sugeno:Fuzzy identification of systems and its applications to modeling and control, IEEE Trans. SMC 15, pp.116-132 (1985).

[4]M.Sugeno and G.T.Kang:Fuzzy modeling and control of multilayer incinerator, FUZZY SETS AND SYSTEMS 18, pp.329-346 (1986).

[5]M.Sugeno and G.T.Kang:Structure identification of fuzzy model, FUZZY SETS AND SYSTEMS 28, pp.15-33 (1988).

[6]A.G.Ivakhnenko at el.:Principle versions of the minimum bias criterion for a model and an investigation of their noise immunity, Soviet Automat. Control 11, pp.27-45 (1978).

[7]K.Tanaka and M.Sano:Frequency Shaping for Fuzzy Control Systems with Unknown Nonlinear Plants by Neural Network, FUZZY SETS AND SYSTEMS (to appear).

[8]H.Ichihashi and T.Watanabe:Learning Control by Fuzzy Models Using a Simplified Fuzzy Reasoning, vol.2, no.3, pp.429-437 (1990) (in Japanese).

[9]H.Ichihashi : Iteraive Fuzzy Modeling and a Hierarchical Network, proceedings of IFSA'91, pp.49-52 (1991).

[10]K.Tanaka, M.Sano and H.Watanabe:Modeling and Control of Carbon Monoxide Concentration using a Neuro-Fuzzy Technique, IEEE Transactions on Fuzzy Systems (to appear).

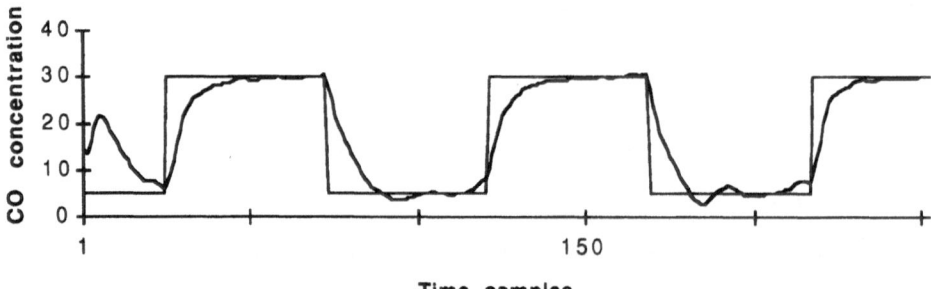

Fig.6 Control result (number of learning : 1).

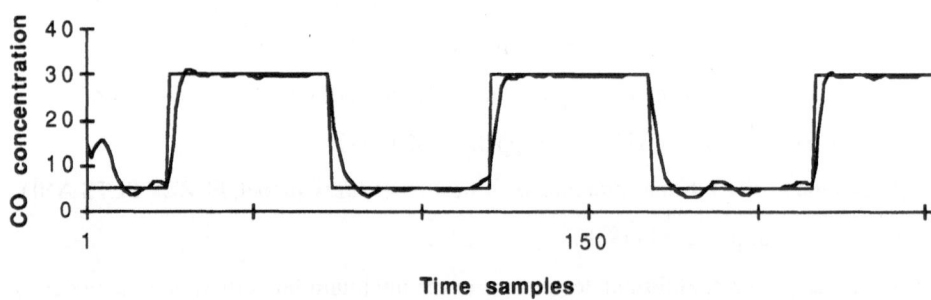

Fig.7 Control result (number of learning : 10).

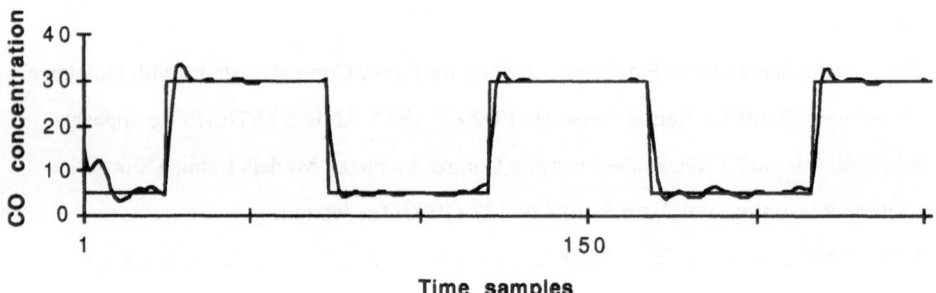

Fig.8 Control result (number of learning : 20).

FLC DESIGN WITH MULTIPLE CONTROL OBJECTIVES AND APPLICATION TO OVERHEAD CRANE CONTROL

T. LIM AND Z. BIEN

Dept. of EE, KAIST

373-1 Kusongdong, Yusung-gu, Taejon 305-701 KOREA

1. Introduction

The fuzzy logic control can be viewed, in a certain sense, as a means to model a human operator in human-in-loop system[1]. However, when the system has many control objectives to satisfy and/or many variables to refer to, it is often very difficult to make the rule base for the FLC(Fuzzy Logic Controller). The control of the overhead crane can be an example(See Figure 1)[2][7][8][9]. The control objectives of the overhead crane are two fold; one is positioning the trolley as fast as possible and the other is reducing the swing of the load as small as possible. The operators of the overhead crane in the Pohang Iron & Steel Company in Korea are controlling the crane so as to satisfy both of the control objectives as much as possible. However, when we had interviews with the operators to make the rule base for the FLC, we found that the operators described their behaviors in terms of fuzzy If-Then rules referring to only one of the two control objectives. That is, they could explain their behaviors of positioning the trolley only without referring to reducing the swing and could explain their behaviors of reducing the swing only without considering the positioning of the trolley. Nevertheless, the operators were claiming that they can exert both actions of positioning the trolley and reducing the swing at the same time. For this experience, we have found that, when there are uncertainties in making the rule base for controlling a plant with many control objectives to satisfy, we may obtain separate groups of the rules satisfying partial control objectives each instead of the rules satisfying all the control objectives simultaneously.

In this paper, we study the problem of designing FLC by using the obtained multiple groups of the rules satisfying the partial control objectives.

Z. Bien and K. C. Min (eds.),
Fuzzy Logic and its Applications, Information Sciences, and Intelligent Systems, 245–253.
© 1995 *Kluwer Academic Publishers.*

In order to deal with the problem, there are available several methods in the literature[2][3]. One of the methods is to discard and/or modify the obtained groups of the rules based on heuristic knowledge so that the final rule base can satisfy all the control objectives. However, it can be often a tedious and difficult task when there are many rules and/or a prior knowledge is weak. Also we can design an FLC for each partial objectives and simply add all the outputs of the multiple FLCs to make the final input for the plant. However, note that the operators usually change the priorities of the control objectives based on the state of the plant. Such operator's action may not be taken into consideration if we use the latter method.

In this paper, the following two schemes are proposed to deal with the problem. The first scheme is to make multiple FLCs by using the obtained groups of the rules and then combine the output of each FLC in terms of the weights which may vary according to the state of the plant. The second scheme is to apply all the obtained groups of rules in the same rule base and extract the core information by suitable fuzzy inference.

2. Multi-Objective FLC Design by using the variable weights

In this section, we propose the Multi-Objective FLC(MOFLC) to deal with the situation discussed in the introductory section. The structure is shown in Figure 2. The MOFLC consists of a supervisory controller and sub-controllers. Each subcontroller is a usual FLC in which the corresponding group of the rules are used as a rule base, and the supervisory controller is a kind of SOC(Self Organizing Controller) to yield the weights for the subcontrollers as the output. The output of the MOFLC, and at the same time, the input of the plant is the weighted sum of all the outputs of the subcontrollers as

$$u(t) = \sum_{k=1}^{M} w_k(t)u_k(t) \tag{1}$$

where M is the total number of the subcontrollers, $u(t)$ is the input of the plant, $u_k(t)$ is the output of the kth subcontroller, and $w_k(t)$ is the weight of the kth subcontroller at time t.

The weight decision part in the supervisory controller utilizes the states of the plant as the input, and generates the final output of the supervisory controller. The weight decision part is made up with the rule base and the inference engine. The rule base is expressed in the following linguistic description.

If x is $L(x)$ Then w_1 is $L(w_1)$, w_2 is $L(w_2)$,..., w_M is $L(w_M)$.

Here, x is the state of the plant and $L(x)$, $L(w_k)$ are linguistic labels of the states and the weights, respectively. Since it is difficult to determine the membership functions of the rule base, some learning mechanism is introduced to the MOFLC. We adopt the concept of SOC[4] for the learning part to modify the membership functions. The SOC consists of the performance table part, the model part and the rule modification algorithm part as shown in Figure 2. The function of each block is as follows. The performance table part evaluates the control result in terms of the control objectives, while the model part converts the evaluation value of the performance table part into the membership functions to be modified, and the rule modification algorithm part updates the membership functions.

The learning procedure proceeds as the following four steps. First, the performance table part evaluates the control result according to the satisfaction degree of each control objective. The satisfaction degree of the control objective is defined as the degree satisfying the control objective on the basis of the designer's selected criterion. The satisfaction degree of kth control objective is expressed as $P_k(k=1,2,...,M)$, where P_k is in $[0,1]$. Secondly, the following condition is checked;

$$\min_{k=1,..,M} P_k \geq \alpha, \tag{2}$$

where α is a value of the designer's choice($\alpha \in [0,1]$). If the condition (2) is satisfied, i.e. the current MOFLC achieves the overall goal, the procedure is stopped. If not, the next step is continued. Thirdly, the model part determines the membership functions to be modified. The membership functions to modify are the membership functions of the consequents in the rule base – the membership functions of weights. Let the shape of the membership functions be isosceles triangle, λ_{kj} be the central value of the jth membership functions of $w_k(j=1,2,...J)$. The mean of the satisfaction degrees, P_{mean}, is calculated as

$$P_{mean} = \frac{\sum_{k=1}^{M} P_k}{M} \tag{3}$$

P_{mean} offers the reference value for determining which of the control objectives is accounted of. So, if $P_k \leq P_{mean}$, the λ_{kj} moves to the larger part. In the same manner, if $P_k \geq P_{mean}$, λ_{kj} moves to the smaller part. δ_k as the change value of λ_{kj} is determined by the equation (4).

$$\delta_k = \eta \frac{P_{mean} - P_k}{\sum_{l=1}^{M} |P_{mean} - P_l|} \tag{4}$$

Here, η is the learning gain, i.e. the increasing step size. Finally the rule modification algorithm part utilizes δ_k as the input and updates the central values of the membership functions of w_k as

$$\lambda_{kj,new} = \lambda_{kj} + \delta_k \quad \text{where } j = 1,2,...,J, \tag{5}$$

where $\lambda_{kj,new}$ is the update value of λ_{kj}. It can make up for the small satisfaction degee by increasing the central values of the corresponding membership functions. The MOFLC controls the plant repetitively by modifying the membership functions of the weight decision rule base if the condition (2) is not satisfied.

3. Design scheme by using MDIM

As an another approach to deal with the situation where independent groups of the rules considering partial objectives are given, we propose that all the obtained groups of the rules be applied to the same rule base at the same time. This concept is shown in Figure 3. The group of the rules only for reducing the swing of the overhead crane and the group of the rules only for positioning the trolley are applied to the same rule base for the inference engine. However, in this case, the rule from one group of rules and the rule from the other group of rules can be inconsistent with each other, and conventional inference engine scheme[6] may render undesirable effect in dealing with the situation. To be more specific, the following two rules with the membership functions in Figure 4 should be inferred at the same time.

R1: If (angle is PS) and (change of angle is ZE) Then (output is AZ).
R2: If (pos. is ZE) and (change of pos. is NS) Then (output is PVS).

R1 is derived from the rules for reducing the swing, and R2 is derived from the rules for positioning the trolley. The rule R1 says that the output should be AZ(*almost zero*) for the anti-swing objective when the angle is PS(*positive small*) and the change of angle is ZE(*zero*). The rule R2 says that any output value near the value of zero is possible for the objective for positioning when the position error is ZE and the change of position error is NS(*negative small*). In the above case, it is our common understanding that both objectives are well satisfied if the output is PVS(*positive very small*). However, when the well known direct method of Mamdani's min-max inference method[5] is employed, the output value of the FLC is dominantly influenced by the fat shape fuzzy set AZ(*almost zero*). This is because conventional FLC has the phenomenon that rules having fatter shape consequent fuzzy sets are more influential.

To cope with the difficulty, we have proposed a new type of inference method, MDIM(Minimum Distance Inference Method)[2][7]. We consider the problem of Generalized Modus Ponens(GMP)[6] with multiple implicants as follows:

Implicants:

R_1: IF $(x_1$ is $\tilde{A}_{11})$ and $\cdot\cdot$ $(x_m$ is $\tilde{A}_{1m})$ THEN (Y is $\tilde{O}_1)$
R_2: IF $(x_1$ is $\tilde{A}_{21})$ and $\cdot\cdot$ $(x_m$ is $\tilde{A}_{2m})$ THEN (Y is $\tilde{O}_2)$

$\cdots$

$\cdots$

R_n: IF $(x_1$ is $\tilde{A}_{n1})$ and $\cdot\cdot$ $(x_m$ is $\tilde{A}_{nm})$ THEN (Y is $\tilde{O}_n)$

Observation: x_1 is u_1, x_2 is u_2, $\cdots x_m$ is u_m.

Conclusion: Y is $\tilde{c}_f$.

Here, $x_1 \sim x_m$ are m inputs and Y is the output. The problem is to find the centroid of the conclusion fuzzy set $\tilde{c}_f$ when the rules are given as above and the inputs of FLC $x_1 \sim x_m$ are given as $u_1 \sim u_m$, respectively. For slim shaped consequent fuzzy sets to be more emphasized, we define the measure of certainty. Let the width of a fuzzy set $W(\tilde{A})$: $F_X \to R^1$ be defined as:

$$W(\tilde{A}) = \begin{cases} 0 & \text{if max } \mu_{\tilde{A}}(x) = 0 \\ \frac{\int_X \mu_{|\tilde{A}|}(x)dx}{max\{\mu_{\tilde{A}}(x)\}} & \text{otherwise} \end{cases} \quad (6)$$

and let the width-certainty C_w: $R^1 \to R^1$ be defined as:

$$C_w(x) = \begin{cases} 1 & \text{for } 0 \leq x < p_1 \\ \frac{p_2-x}{p_2-p_1} & \text{for } p_1 \leq x < p_2 \\ 0 & \text{for } p_2 \leq x, \end{cases} \quad (7)$$

where p_1 and p_2 are user's choice positive constants satisfying $p_1 < p_2$. And, let the height-certainty C_h: $F_X \to R^1$ be defined as:

$$C_h(\tilde{A}) = max\{\mu_A(x)\} \quad \text{for } x \in X \quad (8)$$

The measure of certainty $M_c(\tilde{A})$ is defined as:

$$M_c(\tilde{A}) = min\{C_w(W(\tilde{A})), C_h(\tilde{A})\} \quad (9)$$

Further, we define the distance $\rho_f(\tilde{A}, \tilde{B})$ between two fuzzy sets as:

$$\rho_f(\tilde{A}, \tilde{B}) = |p_{\tilde{A}} - p_{\tilde{B}}| \quad (10)$$

where p_A and p_B are the centroid of the membership functions of $\tilde{A}$ and $\tilde{B}$, respectively. Finally we define compatibility of the i-th rule as

$$W_i = min\{\mu_{\tilde{A}_{i1}}(u_1), \mu_{\tilde{A}_{i2}}(u_2), \cdots, \mu_{\tilde{A}_{im}}(u_m)\} \quad \text{for } i = 1, 2, \cdots, n. \quad (11)$$

From the above definitions, we propose that the conclusion Y of the above GMP be the fuzzy set $\tilde{c}_f$ minimizing

$$J = \sum_{i=1}^{n} W_i \cdot M_c(\tilde{O}_i) \cdot \rho_f{}^2(\tilde{O}_i, \tilde{c}_f), \quad (12)$$

where n is the number of rules. By a simple calculation, the centroid of the final conclusion $\tilde{c}_f$ can be written as

$$p_{\tilde{c}_f} = \sum_{i=1}^{n} p_{\tilde{O}_i} \cdot W_i \cdot M_c(\tilde{O}_i) / \sum_{i=1}^{n} W_i \cdot M_c(\tilde{O}_i). \quad (13)$$

This is the final output of the proposed FLC.

4. Simulation Result

For computer simulation, we use the following model as the dynamics of the overhead crane:

$$\ddot{x} = \frac{f}{M}$$

$$\ddot{\theta} = \frac{-g\sin\theta + \ddot{x}\cos\theta}{l}$$

where f is the input force (N), x is the trolley position (m), θ is the angle of the load (rad), l is the length of the rope (m), M is the weight of the trolley (Kg), g is the gravity constant (m/sec^2).

In the simulation, it is assumed that $M = 1\ Kg$, $l = 1\ m$ and $g = 9.8$ m/sec^2. The first control objective in this simulation is to regulate θ from the initial condition of $\theta = \theta_0$ to zero with the tolerance of $0.05\ rad$. The second control objective is to regulate x from the initial condition of $x = x_0$ to zero with the tolerance of 5 cm.

The first approach is to use the MOFLC. We are given two independent groups of the rules based on the operator's knowledge. One group of the rules are designed for positioning the trolley and the other group of the rules are designed for reducing the swing, which are used for the rule base for the subcontrollers. The weight decision rule base and the membership functions are shown in Figure 5 and Figure 6, respectively. The control result with $\theta_0 = 0.7\ rad$ and $x_0 = 1.0\ m$ is shown in Figure 7. This result shows that the elapsed time satisfying the two control objectives is reduced owing to the learning.

The second approach is to apply the two groups of the rules in the same rule base with MDIM. The control result with $\theta_0 = 1.0 \ rad$ and $x_0 = 1.0$ m is shown in Figure 8. Note that the proposed inference scheme is more suitable to deal with inconsistent rules than well known min-max inference scheme[5][6].

Comparing the elapsed time to satisfy the two control objectives, we can find that of the MOFLC is shorter than that of the second approach. Therefore, we can say that the MOFLC deals with the given example better than the second approach using the simple addition of the rules.

5. Conclusion

We proposed two types of FLC which are suitable to deal with the situation where the multiple groups of the rules satisfying the partial control objectives are given. First, we design the MOFLC to combine outputs of the subcontrollers by varying the weights for the subcontrollers. Secondly, we apply all the groups of the rules in the same rule base by using MDIM as an inference scheme. The proposed two FLCs will be effectively applicable when there is uncertainty in extracting rules, in particular, when it is very difficult to obtain control rules for all the control objectives.

References

[1] L. Wang, J. M. Mendel.(1992) Generating Fuzzy Rules by Learning from Examples, *IEEE Trans, Syst., Man, Cybern.*, **Vol. 22 no. 6**, pp. 1414–1427

[2] W. Yu, and Z. Bien.(1994) Design of fuzzy logic controller with inconsistent rule base, *Journal of Intelligent and fuzzy systems*, **Vol. 2**, pp. 147–159

[3] K. Kim, and J. Kim.(1994) Multicriteria Fuzzy Control, *Journal of Intelligent and fuzzy systems*, **Vol. 2**, pp. 279–288

[4] W. Pedrycz. (1989) *Fuzzy Control and Fuzzy Systems*, Research Studies Press Ltd., Somerset England.

[5] E.H. Mamdani(1974) Applications of fuzzy algorithms for control of simple dynamic plant, *Proc. IEE*, **Vol. 121 no. 12**, pp. 1585–1588

[6] C.C. Lee.(1990) Fuzzy logic in control systems: fuzzy logic controller-part I II, *IEEE Trans, Syst., Man, Cybern.*,**Vol. 20 no. 2** pp. 404–435

[7] W. Yu, T. Lim, & Z. Bien.(1993) Fuzzy Logic Control of a Roof Crane with Conflicting Rules, *Proc. 5th IFSA Congress*, pp. 1370–1373

[8] M. Nakatsuyama, S. Wang, Y. Hasimoto, H. Kaminaga, B Song.(1994) Automatic Operation of an Overhead Crane Based on Fuzzy Algorithm with Matrix Representation, *J. of Japan Society for Fuzzy Theory and Systems*, **Vol. 6 no. 6** pp. 1211–1221

[9] S. Yamada, H. Fujikawa.(1989) Fuzzy Control of The Roof Crane, *IEEE Industrial Electronic Conference Proceedings*, pp. 709–714

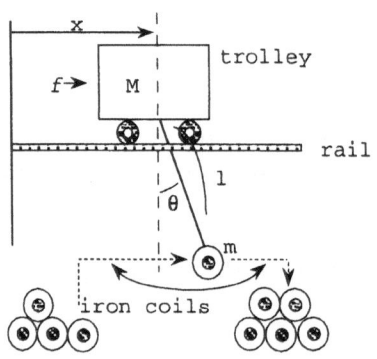

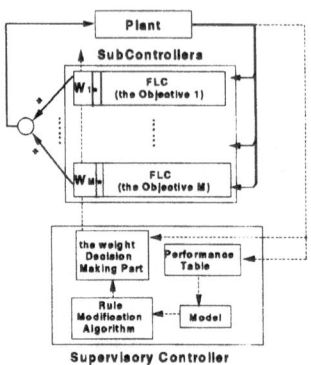

Fig 1. The Overhead Crane   Fig 2. The Structure of MOFLC

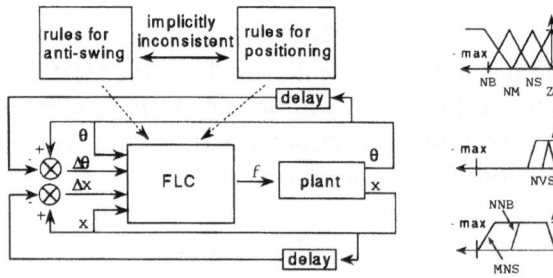

Fig 3. The Structure of FLC   Fig 4. The Membership func. in use(III)

Swing Angle

	NB	NS	ZE	PS	PB
NB	BG	BG	BG	SM	MD
NS	BG	BG	SM	MD	SM
ZE	BG	SM	MD	SM	BG
PS	SM	MD	SM	BG	BG
PB	MD	SM	BG	BG	BG

(Position)

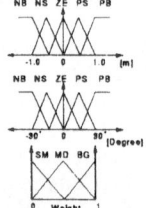

Fig 5. Weight dec. rule(pos)   Fig 6. Membership func. in use(II)

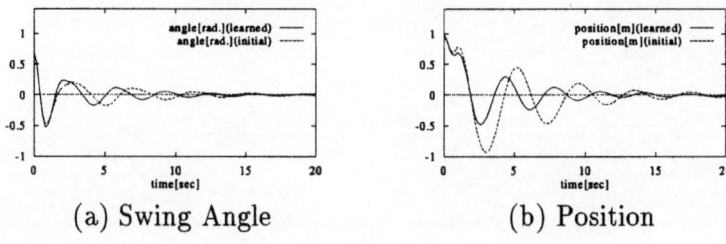

(a) Swing Angle   (b) Position

Fig 7. The result of MOFLC

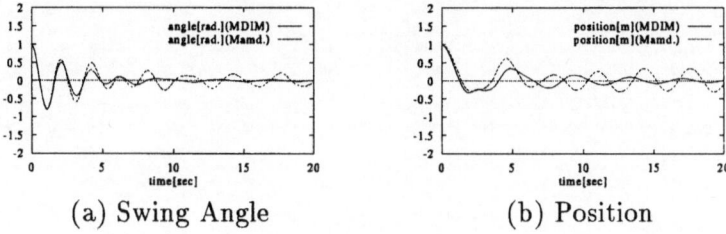

(a) Swing Angle   (b) Position

Fig 8. The result of MDIM

PERFORMANCE EVALUATION FOR FAULT DETECTION OF ANALOG ELECTRONIC CIRCUITS

Masaki Hashizume, Yoshihiro Iwata and Takeomi Tamesada
Faculty of Engineering, The University of Tokushima
Tokushima-shi, Tokushima, 770 JAPAN

Abstract: When an analog electronic circuit is tested, fault detection, which is to determine whether the circuit is faulty or not, is performed. In the first stage of the fault detection, performance evaluation is performed. That is, it is checked whether the performances can satisfy the requests. In this paper, a performance evaluation method for fault detection of analog circuits is proposed, which is based on a fuzzy reasoning technique. In our method, some evaluation mechanisms of an expert test engineer are defined by means of directed graphs. By using the graphs, performance evaluation results will be derived, which are satisfied by the expert, even if test engineers are novices. The effectiveness of our method is checked by some experiments for an operational amplifier circuit.

keywords: Fault Detection, Analog Circuit, Fuzzy Reasoning,
Performance Evaluation

1. Introduction

Fault detection is an operation to judge whether a circuit under test(CUT) is faulty or not. The fault detection is always performed when an electronic circuit is tested. According to the result of fault detection, which is either faulty or unfaulty, it is decided whether the circuit should be repaired or abandoned.

Fault detection of analog circuits is more difficult than logic circuits. A result of fault detection of a logic circuit can be obtained by checking whether an output logic value of the circuit is different from the one of the unfaulty circuits. If any outputs different from the unfaulty circuits are measured, it will be decided as a faulty one. On the other hand, in the case of an analog circuit, even if a circuit element is not damaged

255

Z. Bien and K. C. Min (eds.),
Fuzzy Logic and its Applications, Information Sciences, and Intelligent Systems, 245–253.
© 1995 *Kluwer Academic Publishers.*

fully and has a value, which is different from the one of the unfaulty circuits, the circuit will work well in many cases. Also, since there are some variances in the performances of unfaulty circuits, the expected outputs of unfaulty circuits can not be defined without any fuzziness. The difficulty of fault detection makes it more difficult to test analog electronic circuits, especially to develop automatic test equipment for them[1,2].

In order to obtain a result of fault detection, performance evaluation is performed. That is, it is checked whether each performance of the circuit is the same as the one of unfaulty circuits. From the results of performance evaluation, it is decided whether the circuit is faulty or not. The difficulty of performance evaluation leads to the difficulty of fault detection of analog circuits.

The difficulty in the performance evaluation is caused by the existence of some fuzziness in the performances of unfaulty circuits and the criterion of judgment for determining whether each performance of the CUT is the same as the unfaulty circuits. The fuzziness can be coped by means of the fuzzy set theory. Therefore, we attempted to develop a performance evaluation method based on a fuzzy reasoning technique[3].

Until now, analog circuits have been tested by a small number of expert test engineers. If the evaluation mechanism of an expert test engineer is implemented in a test system, novice test engineers can derive the same result as the expert engineer and productivity of analog circuit testing can be improved, especially at the field testing of analog circuits. Therefore, in this paper, we propose a new performance evaluation method so that the productivity of testing analog circuits can be improved.

A performance evaluation problem can be formalized as a subjective evaluation problem. Thus, it may be solved by using a fuzzy measure[4]. However, since there are many factors that affect on final results of performance evaluation, the method can not be applied to many performance evaluation problems of electronic circuits. Therefore, in this paper, we propose a performance evaluation method, in which two kinds of graphs are used for implementing evaluation mechanisms of an expert test engineer.

In section 2, our method is presented. Experimental evaluation results of our method are shown in section 3.

2. Performance Evaluation Method

2.1 PHILOSOPHY OF OUR APPROACH

Our goal is to develop a performance evaluation method, with which productivity of analog electronic circuit testing can be improved. Until now, analog circuits have been diagnosed by expert test engineers. Therefore, we attempt to develop a method, with which novice test engineers can derive an evaluation result satisfied by an expert test engineer.

In a test system based on our performance evaluation method, evaluation mechanisms of a human expert for a CUT are stored in advance. At the testing, the performances of the circuit, for example, the output voltage, the frequency response, etc., are measured and inputted to the test system by a test engineer. From them, the system generates a performance evaluation result satisfied by the expert test engineer. According to the evaluation result, the test engineer derives the result of fault detection. In this way, by using our test system, test engineers can obtain advice from our system instead of a human expert.

Generally, a result obtained by evaluating what degree the circuit is faulty is not complementary with the one obtained by evaluating what degree it is unfaulty. At the testing, if both results can be obtained, it is easy for test engineers to decide the result of fault detection. Therefore, μ_{Fi} and μ_{Ni}, which are the evaluation values of i-th evaluation factor obtained by evaluating what degree the i-th factor is unpreferable and preferable, respectively, are derived for all factors by our method. After that, both μ_{FT} and μ_{NT} as final evaluation results are derived from the μ_{Fi}'s and the μ_{Ni}'s, where μ_{FT} and μ_{NT} are the evaluation values obtained by evaluating what degree the circuit is faulty and is unfaulty, respectively.

2.2 EVALUATION GRAPH

There are many factors that have an effect on μ_{FT} and/or μ_{NT}. They are classified into two kinds of factors: the ones are the factors whose evaluation characteristics of an expert test engineer can be defined by means of membership functions; the others are the factors whose evaluation results can be obtained from the evaluation result of other factors. The former factors correspond to the ones concerned with the circuit specifications of a CUT. For example, in performance evaluation of a voltage amplifier circuit, an evaluation characteristic curve of the voltage gain can be defined with a membership function and the evaluation result can be obtained by using the function. On the other hand, an evaluation result of the frequency response can not be determined by using any membership functions. It can be obtained from evaluation results of the lower cut-off frequency(f_L) and the higher cut-off frequency(f_H).

The relations among factors can be defined by a directed graph as shown in Figure 1, which is referred to as an "evaluation graph" in this paper. In the evaluation graph, only keywords are described, whose meanings are shown in Tables 1.

If the expected functions of a CUT are not obtained fully, the circuit can be decided as a faulty one. Even if the functions are obtained fully, the circuit, whose good performances are not obtained, can not be decided as an unfaulty circuit. For example,

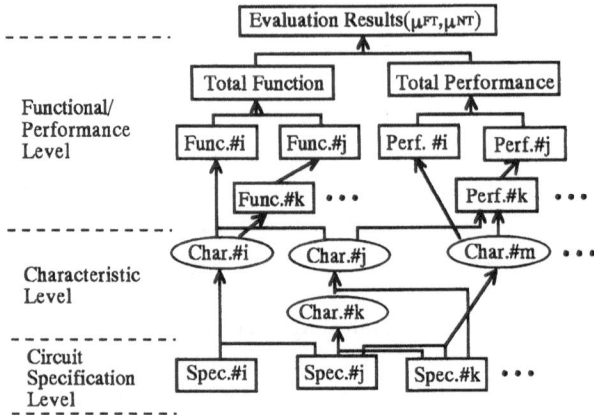

Figure 1 Evaluation graph.

TABLE 1. Abbreviations in Figure 1

Abbr.	Meaning for μ_{FT}	Meaning for μ_{FN}
Func.#i	The i-th function is abnormal.	The i-th function is obtained.
Perf. #i	The i-th performance is beyond the permissible range.	The required i-th performance is obtained
Char.#i	The i-th characteristic is abnormal.	The required i-th characteristic is obtained.
Spec.#i	The i-th specification does not satisfy the design request.	The i-th specification satisfies the request.

even if an input signal can be amplified by a voltage amplifier circuit, the circuit, whose power dissipation is extremely large, will not be decided as an unfaulty circuit. Thus, both μ_{FT} and μ_{NT} depend on whether the expected functions of the circuit are obtained, and/or whether the expected performances are obtained. That is, final evaluation results are determined from the evaluation results of the functions and the performances. Therefore, at the top level of our evaluation graph, the relation among them is defined as shown in Figure 1.

Some functions and performances of electronic circuits can be defined in detail by other evaluation factors. For example, the evaluation results of the frequency response of a voltage amplifier circuit are determined from the ones of f_L and f_H. In such a case, the reasoning relation is defined by using directed arcs.

The process to make the functions and the performances of a CUT in detail will be continued, until all the circuit specifications are connected to any factor. As the result, evaluation graphs of a CUT can be derived, which consist of three kinds of hierarchy level as shown in Figure 1. At the top level, evaluation factors on the functions and the performances are denoted. At the bottom level, the names of the circuit specifications, which can be measured by test engineers, are denoted. In the middle level, the names of characteristics, which are obtained by classifying some circuit specifications, are denoted.

In our method, two kinds of evaluation graphs are defined. One is for μ_{FT} and the other is for μ_{NT}. That is the reason why the evaluation mechanism for μ_{FT} is different from the one for μ_{NT} in many cases.

2.3 REASONING BASED ON EVALUATION GRAPHS

For each factor of the bottom hierarchy level in evaluation graphs, evaluation characteristic of an expert test engineer is defined by using a membership function.

When an evaluation result of a factor is obtained from other factors by using such an evaluation graph, it should be considered the differences of importance among the factors. For example, an evaluation result of the amplitude function of a voltage amplifier circuit has more effects on the final evaluation results than the power dissipation characteristic. Therefore, for each factor, a weight is defined in our method. The weight $w_i (0 \le w_i \le 1)$ of i-th factor is determined by considering the effect of only the factor with the effects of other factors neglected.

Our method can not always derive the same evaluation result as an expert test engineer, because our method is an approximate implementation of the evaluation mechanisms. Therefore, our method derives a stricter performance evaluation result of μ_{FT} than μ_{FN}, so that faulty circuits can not be decided as unfaulty ones. Thus, when μ_{Fi} is derived from other evaluation results, additional effect is considered. Since the effect depends on the degree of distinction among the factors and on the importance of the evaluation, both w_i and k_{ij} are used for determining the effect in our method. k_{ij} $(0 \le k_{ij} \le 1)$ is a coefficient, which is referred to as a "duplicate coefficient" in this paper, for expressing what extent i-th factor expresses the same characteristics as j-th factor, that is, what degree they resemble each other. For example, if the i-th factor is similar to the j-th factor at the degree of 0.2, k_{ij} is set to be 0.2. The k_{ij} means that 80% of the total characteristics expressed by the i-th factor is expressed by only the factor, and the rest is expressed by both the i-th factor and the j-th one. If the characteristic expressed by the i-th factor is independent of the j-th factor, k_{ij} is set to be 0.

In our method, it is assumed that additional effects are in proportion to the duplication coefficients. Therefore, when an evaluation result of a factor C is obtained from evaluation results of factors A and B, the evaluation result is derived by using (1).

$$\mu_{FC} = \begin{cases} w_A \cdot \mu_{FA} + (1-k_{AB}) \cdot (1-w_A \mu_{FA}) \cdot w_B \cdot \mu_{FB}, & \text{if } w_A \cdot \mu_{FA} \geq w_B \cdot \mu_{FB} \\ w_B \mu_{FB} + (1-k_{AB}) \cdot (1-w_B \mu_{FB}) \cdot w_A \mu_{FA}, & \text{if } w_A \cdot \mu_{FA} < w_B \cdot \mu_{FB} \end{cases} \qquad (1)$$

where

μ_{FA} and μ_{FB} are evaluation values of factors A and B, w_A and w_B are the weights of A and B, respectively, and k_{AB} ($0 \leq k_{AB} \leq 1$) is a duplicate coefficient between A and B.

As shown in (1), if $w_A \cdot \mu_{FA} \geq w_B \cdot \mu_{FB}$, μ_{FC} is determined as $w_A \cdot \mu_{FA} + (1-w_A \cdot \mu_{FA}) \cdot \beta$, where β is a coefficient for defining an additional effect generated by the factor B, since μ_{FC} should be larger than $w_A \cdot \mu_{FA}$ and the additional effect should be within $1 - w_A \cdot \mu_{FA}$. In our method, β is defined as the percentage of the evaluation values of the factor C based on the factor B in the sum of the maximum evaluation values of all the factors except for the factor A. At the derivation of β, all effects of the factor A are removed. Therefore, since the similarity between the factors A and B is defined with k_{AB}, β is $(1-k_{AB}) \cdot (w_B \cdot \mu_{FB})/1$.

When μ_{Fi} is derived from Np factors(Np>2), the following procedure is used after specifying all duplicate coefficients between any two factors.

[Derivation procedure for μ_{Fi}]
[1]$S=\{x_1, x_2, \ldots, x_{Np}\}$, $S_0 = \emptyset$ and $\beta=0$.
[2]For each element in S, calculate $w_{xi} \cdot \mu_{Fxi}$ and select a factor having the maximal evaluation value in S as P.
[3]Extract P from S and store it into S_0.
[4]Continue the following until S is empty.
　　(1)Calculate the evaluation value for each factor in S, in which effects of all the factors in S_0 are removed.
　　(2)Select a factor having the maximum evaluation value in S as Px.
　　(3)Divide the evaluation value of Px by Np-1 and add the obtained value to β.
　　(4)Extract Px from S and store it into S_0.
[5]Calculate μ_{Fi} by $\mu_{Fi} = w_P \cdot \mu_{FP} + (1-w_P \cdot \mu_{FP}) \cdot \beta$.

The similarity among evaluation factors can be illustrated by using a Venn diagram, in which duplicate coefficients are denoted. For example, when an evaluation result of a factor D is derived from factors A, B and C, the similarity among factors A, B and C can be defined as shown in Figure 2. In Figure 2, if $w_A \cdot \mu_{FA} \geq w_B \cdot \mu_{FB} \geq w_B \cdot \mu_{FC}$ and $(1 - k_{AB}) \cdot w_{FB} \cdot \mu_{FB} \geq (1 - k_{CA}) \cdot w_{FC} \cdot \mu_{FC}$, μ_{FD} is obtained from (2) and (3) by means of the above procedure.

$$\mu_{FD} = w_A \cdot \mu_{FA} + (1 - w_A \cdot \mu_{FA}) \cdot \beta \tag{2}$$

$$\beta = \{(1 - k_{AB}) w_B \cdot \mu_{FB} + (1 - k_{BC} - k_{CA} + k_{AB} \cdot k_{BC} \cdot k_{CA}) \cdot w_C \cdot \mu_{FC}\}/2 \tag{3}$$

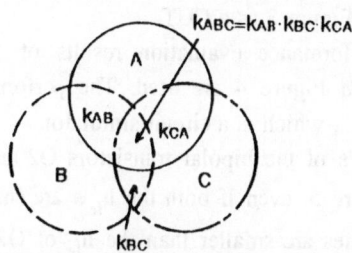

Figure 2 Similarity among evaluation factors A, B and C.

In many cases, μ_{FT} can be determined by an evaluation result of only a factor. For example, when μ_{Fi} of the voltage gain is extremely small in performance evaluation of a voltage amplifier circuit, the circuit can be decided as a faulty one from only the evaluation result. Therefore, in our method, α_{Fi}, which is the upper bound of μ_{Fi}, is defined for each factor, and if $\mu_{Fi} \geq \alpha_{Fi}$ is satisfied for an evaluation factor, μ_{FT} is set to be 1. If all μ_{Fi}'s are less than the upper bounds for all factors, μ_{FT} is derived by using the evaluation graph.

For μ_{NT} and μ_{Ni}, any additional effects are not considered in our method. They are calculated by the same method as in [5], according to an evaluation graph of μ_{NT}. For each factor, α_{Nj}, which is the lower bound of μ_{Nj}, is defined. If it is satisfied that $\mu_{Nj} \geq \alpha_{Nj}$ for all N_p factors, μ_{Ni} of i-th factor, which is to be derived from the evaluation values of the N_p factors, is calculated with (4). Otherwise, μ_{Ni} is set to μ_{Nj} having the minimum value in the N_p factors as in (4).

$$\mu_{Ni} = \begin{cases} \left(\sum_{j=1}^{N} w_j \mu_{Nj}\right) \Big/ \sum_{j=1}^{N} w_j\,, & \text{if } \mu_{Nj} \geq \alpha_{Nj} \text{ for all } j \\ min(\mu_{Nj}) & \text{Otherwise} \end{cases} \tag{4}$$

3. Performance Evaluation of Amplifier Circuits

In [3], we evaluate performances of a small size of amplifier circuit. From the experiments in [3], it is shown that our method can derive performance evaluation results, which can be satisfied by expert test engineers. Furthermore, the experiments show that our method can derive results that are more accepted by expert test engineers than a conventional method based on λ-fuzzy measure[4] and the Choquet integration.

In this paper, we show performance evaluation results of a more complex amplifier circuit than the circuit in [3]. In these experiments, we use an operational amplifier circuit (μPC741) shown in Figure 3 as a CUT.

In order to derive performance evaluation results of the circuit in Figure 3, evaluation graphs shown in Figure 4 are used. The performances of the circuit are obtained by using "PSPICE", which is a circuit simulator.

In our experiments, h_{fe}'s of the bipolar transistors Q2 and/or Q16 in Figure 3 are changed. As shown in Figure 5, even if both the h_{fe}'s are changed simultaneously, the performance evaluation values are smaller than the h_{fe} of Q2. That is the reason why the h_{fe}'s of Q2 and Q16 affect on the amplitude characteristic complementally. From Figure 5, it is found that our method can derive performance evaluation results expressing the complementary relation.

In performance evaluations of the circuit in Figure 3, there are many factors that affect on the final evaluation results. Thus, it is difficult to define the evaluation mechanism of a human expert preciously, if the method based on λ-fuzzy measure is used. On the other hand, our method can define the evaluation mechanism easier than the method, since an approximate reasoning technique is used in our method. Moreover, our method can derive evaluation results satisfied by human experts. Therefore, our method is more suitable than the method based on the fuzzy measure, especially for performance evaluations of larger size of analog circuits.

Also, these experiments of the circuit in Figure 3 and the results in [3] show that by using our method, if test engineers are novices, they can derive the same evaluation result as the one derived by an expert test engineer. Thus, it is expected that productivity of testing analog circuits can be improved by using our method.

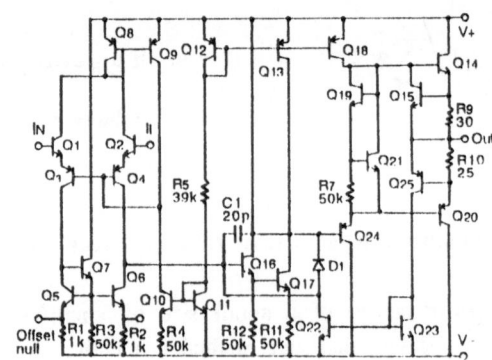

Fig.3 Operational amplifier circuit.

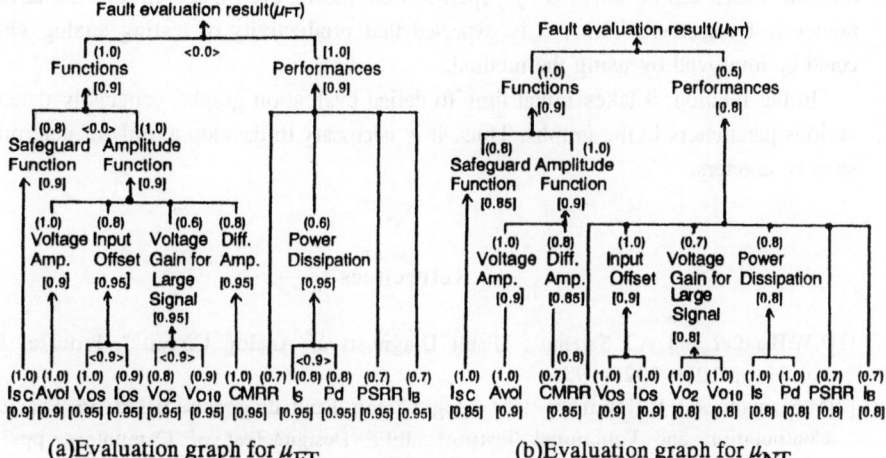

(a)Evaluation graph for μ_{FT} (b)Evaluation graph for μ_{NT}

Figure 4 Performance evaluation graphs for operational amplifier circuit.
(The numbers in (), [] and <> are w_i, α_{Fi} or α_{Ni} and k_{ij}, respectively.)

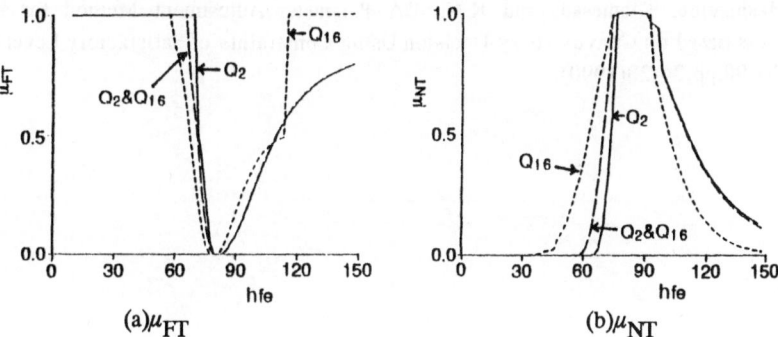

(a)μ_{FT} (b)μ_{NT}

Figure 5 Performance evaluation results of the circuit in Figure 3.

4. Conclusion

A performance evaluation method is presented in this paper, in which evaluation mechanisms of an expert test engineer are defined by means of directed graphs. Also, the method is applied to performance evaluation problems of an operational amplifier circuit. It is convinced from the results that evaluation results will be obtained by the method, which can be satisfied by expert test engineers, even if the circuits are tested by novice test engineers. Thus, it is expected that productivity of testing analog circuits could be improved by using the method.

In the method, it takes much time to define evaluation graphs, especially determine various parameters in the graphs. Thus, it is necessary to develop a tool for determining such parameters.

References

[1]J.W.Bandler and A.E.Salama : "Fault Diagnosis of Analog Circuits", Proc. of IEEE, Vol.73, pp.1279-1325(1985).

[2]M.Slamani and B.Kaminska : "Analog Circuit Fault diagnosis Based on Sensitivity Computation and Functional Testing", IEEE Design&Test of Computers, pp.30-39 (March, 1992).

[3]M.Hashizume, Y.Iwata and T.Tamesada: "Fault Evaluation Based on Fuzzy Logic for Analog Electronic Circuits", Proc. of Fifth IFSA World Congress, pp.1402-1405(1993).

[4]K.Ishii and M.Sugeno: "A Model of Human Evaluation Process Using fuzzy Measure", International Journal of Man-Machine Studies, No.22,pp.19-38(1985).

[5]M.Hashizume, T.Tamesada and K.Nii: "A Parameter Adjustment Method for Analog Circuits Based on Convex Fuzzy Decision Using Constraints of Satisfactory Level", IEEE ICCD-90,pp.24-28(1990).

SUBJECTIVE SYSTEM RELIABILITY ANALYSIS
AND MUTUAL AGREEMENT OF ITS RESULTS

T. ONISAWA
Institute of Engineering Mechanics
University of Tsukuba
1-1-1, Tennodai, Tsukuba 305, Japan

Abstract — This paper describes a model of the subjective reliability analysis, which uses a fuzzy set, natural language expressions and parameterized operations of fuzzy sets. The model has a problem of many different analysis results being obtained dependent on analysts since the subjective reliability analysis reflects analysts subjectivity. As one of the solutions of the problem two kinds of mutual agreements are considered. One is the intersection and the union of the fuzzy sets obtained by the analysis. The other is the weighted average of the fuzzy sets. This paper also gives these interpretations from the viewpoint of a system reliability analysis. Finally examples of these considerations are shown.

1. Introduction

A disastrous earthquake struck Kobe in Japan on January 17, 1995. Many concrete buildings as well as woodern frame houses collapsed. Some expressways also collapsed. These buildings and expressways had satisfied design standards that were determined for a less severe earthquake than this Kobe earthquake. Strong earthquakes have occurred in Japan since the standards were determined. So design standards have been revised. However the collapsed buildings and expressways have not been reinforced to stand up to a more severe earthquake yet.

How are design standards determined? They are not absolute and not unchangeable. Kobe city authorities determined the standards for a less severe earthquake than this Kobe earthquake since no strong earthquake had struck a Kobe area. Usually final standards may be determined based on experts engineering and experienced judgements. In case of Kobe city the standards might be also dependent on their judgements. Subjectivity, i.e., fuzziness, is inherent in their judgements.

The same thing can be said about a system safety and reliability analysis. It is not likely that enough amount of immediate data can be collected to estimate a failure probability and an error probability. The estimation of the failure probability and the error probability cannot help being dependent on experts engineering judgements. And even if the probability of a system accident is obtained by the analysis, the probability is very very small from the viewpoint of our daily life. Safety evaluation of the analyzed system is also closely related to experts subjectivity. Experts subjectivity plays an important role in the system reliability analysis. The necessity of fuzzy sets theory has been recognized

Z. Bien and K. C. Min (eds.),
Fuzzy Logic and its Applications, Information Sciences, and Intelligent Systems, 265–274.
© 1995 *Kluwer Academic Publishers.*

in this area and its applications to this area have increased recently[1][2].

First of all this paper mentions a model of subjective system reliability analysis[3]. The model uses a fuzzy set, natural language expressions and parameterized fuzzy sets operations. This model reflects analysts subjectivity more than our previous model[4]. However for the reason that the analysis results depend on analysts subjectivity, the subjective reliability analysis has a problem of various analysis results being obtained. In this paper as one of the solutions of the problem two kinds of mutual agreements are considered and discussed from the viewpoint of a system reliability analysis.

2. A Model of Subjective Reliability Analysis

2.1. FAILURE POSSIBILITY

The model uses a fuzzy set, called a failure possibility, on the unit interval with the membership function

$$F(x) = \frac{1}{1+20 \times |x-x_0|^m},$$ (1)

where x_0 and m are parameters.

The failure possibility is considered as a subjective unreliability measure. The parameter x_0 gives the maximal grade of the membership and the parameter m is related to fuzziness.

TABLE 1 Expressions of reliability estimate
and parameter x_0

Class	Expressions of reliability estimate	Parameter x_0 (Representative value)
	(System, subsystem, instrument or human operator has)	
1	no reliability	—
2	very low reliability	0.9 — 1.0(0.95)
3	low reliability	0.7 — 0.9(0.8)
4	rather low reliability	0.55 — 0.7(0.625)
5	standard reliability	0.45 — 0.55(0.5)
6	rather high reliability	0.3 — 0.45(0.375)
7	high reliability	0.2 — 0.3(0.25)
8	quite high reliability	0.1 — 0.2(0.15)
9	extremely high reliability	0.05 — 0.1(0.075)
	(Accident, failure or human error is)	
10	next to impossible	0.0 — 0.05(0.025)
11	impossible	—

Natural language expressions of the estimation of reliability are more commensurate with the case in which we have not enough amount of data to estimate the failure and the error probabilities since fuzziness inherent in reliability estimate can be covered well by natural language. In this model the estimation of reliability is expressed in the form of reliability estimate and its fuzziness. The parameter x_0 corresponds to the natural language expressions of reliability estimate shown in Table 1. The failure possibility in the class 1 has the membership function $F(1) = 1(F(x)=0;x\neq1)$ and that in the class 11 has the membership function $F(0) = 1(F(x)=0;x\neq0)$. The parameter m corresponds to the expressions of fuzziness of reliability estimate shown in Table 2.

TABLE 2 Expressions of fuzziness and parameter m

Class	Expressions of fuzziness	Parameter m
1	low	2.0
2	medium	2.5
3	rather high	3.0
4	high	3.5

2.2. FUZZY SETS OPERATIONS

The present reliability analysis applies a fault tree analysis that uses the failure possibility instead of the failure and the error probabilities. Then basic operations in the analysis are an **and** and an **or** operations of failure possibilities.

The parameterized operations (2) and (3) are used as the **and** and the **or** operations, respectively, where the extension principle[5] is used.

$$H(x,\ y) = \frac{1}{1+\left(\left((1-x)/x\right)^{1/nH}+\left((1-y)/y\right)^{1/nH}\right)^{nH}},\qquad (2)$$

where $0<x,y\leq1$, $H(x,0)=H(0,y)=0$, and nH is a non-negative parameter.

$$G(x,\ y) = \frac{\left((x/(1-x))^{nG}+(y/(1-y))^{nG}\right)^{1/nG}}{1+\left((x/(1-x))^{nG}+(y/(1-y))^{nG}\right)^{1/nG}},\qquad (3)$$

where $0\leq x,y<1$, $G(x,1)=G(1,y)=1$, and nG is a non-negative parameter.

The operations H and G are so called Dombi t-norm and Dombi t-conorm, respectively[6]. The operation H can cover a min operation through a drastic product. The operation G can cover a drastic sum through a max operation. From the viewpoint of a reliability analysis the operations H and G can cover the range of reliability estimate from the most optimistic estimate through the most pessimistic one[3]. The parameterized operations H and G can reflect analysts subjectivity. The larger the parameters nH and nG are, the more optimistic the reliability estimate is.

2.3. DEPENDENCE

2.3.1 Dependence Model

Let us consider a parallel system with dependence as depicted in Figure 1. The failure of subsystem A has an influence on that of subsystem B. Let F_A and F_B be failure possibilities of subsystems A and B, respectively, and let R be a fuzzy set representing the dependence level. The following two cases are considered in the dependence model. The extension principle is used in the following operations (4) through (7).

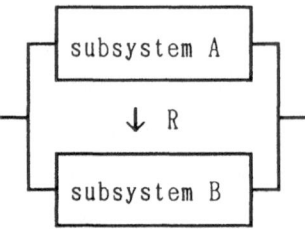

Figure 1. Parallel system with dependence

One is the case in which the failure of subsystem A has an influence on that of subsystem B. In this case the failure possibility of the system F_B' is estimated by

$$F_B' = H(F_A, R). \tag{4}$$

The other is the case in which the failure of subsystem A has no influence on that of subsystem B. As far as the dependence is not complete, the failure of subsystem A does not always have an influence on that of subsystem B. In this case the portion of the failure possibility, which has no influence on the failure of subsystem B, is estimated by

$$F_A = G(F_A', F_B'), \tag{5}$$

where F_A' is the portion.

The operation (5) implies that the failure of subsystem A has an influence on that of subsystem B or not. In this case the failure possibility of the system is estimated by

$$F' = H(F_A', F_B). \tag{6}$$

The operation (6) implies that the failure of the system occurs when the failure of subsystem A and the failure of subsystem B occur without influence.

Considering the above two cases the failure possibility of the whole system is estimated by

$$F = G(F_B', F'). \tag{7}$$

2.3.2 *Natural Language Expressions of Dependence Level*

The information of the event dependence is usually provided by the experts, who are familiar with the analyzed system or task, based on their engineering and experienced judgements. The estimation of the dependence level is expressed by natural language better than by numerical values since we have not enough amount of data to estimate the dependence information by numerical values. In this model it is expressed in the form of the dependence level estimate and its fuzziness. Five kinds of expressions are used as the dependence level estimate and four kinds of expressions are used as its fuzziness. Table 3 shows them. Their meanings are expressed by a fuzzy set with the membership function (8).

$$R(r) = \frac{1}{1+20\times|r-r_0|^{mr}},$$

(8)

where r_0 and mr are parameters.

The parameters r_0 and mr correspond to the expressions of the dependence level estimate and to the expressions of its fuzziness, respectively. The fuzzy sets in the level 1 and in the level 5 have the same membership functions as the failure possibilities in the class 1 and in the class 11, respectively.

TABLE 3 Expressions of dependence level estimate
and its fuzziness

Level	Expressions of dependence level estimate	
1	complete dependence	
2	high dependence	
3	medium dependence	
4	low dependence	
5	zero dependence	

Class	Expressions of fuzziness	Parameter mr
1	low	2.0
2	medium	2.5
3	rather high	3.0
4	high	3.5

2.4. NATURAL LANGUAGE EXPRESSIONS OF ANALYSIS RESULT

In this model the analysis result is expressed by natural language. Let F_R be the failure possibility obtained by the analysis and F_S be the failure possibility with the membership function (1). The distance between F_R and F_S is defined by

$$d = \int_0^1 \sqrt{(x_{1R}(\alpha)-x_{1S}(\alpha))^2+(x_{2R}(\alpha)-x_{2S}(\alpha))^2}\,d\alpha$$

(9)

where α-cuts[5] of F_R and F_S are defined by $(F_R)_\alpha= (x_{1R}(\alpha),\ x_{2R}(\alpha))$ and $(F_S)_\alpha=(x_{1S}(\alpha),\ x_{2S}(\alpha))$, respectively.

The parameters x_0 and m of F_S are selected so as to minimize the distance, and natural language expressions are selected from Tables 1 and 2.

3. Determinations of Parameters

3.1. PARAMETERS nH AND nG

Let us consider a parallel system and a series system. Each system is assumed to consist of two independent subsystems. Let reliability estimates of subsystems and their fuzziness be *standard reliability* and *medium*, respectively. Parameters nH and nG are determined by analysts based on their reliability estimates of the parallel system and the series system, respectively. For example, when an analyst estimates reliability of the parallel system and that of the series system to be $high(x_0=0.2)$ and $standard(x_0=0.55)$, respectively, the estimates lead to his determination of the parameters $nH=2.5$ and $nG=2.5$.

3.2. PARAMETERS r_0 AND mr

The parameter mr is selected from Table 3, i.e., natural language expressions of fuzziness of the dependence level estimate.

The parameter r_0 is determined by the following way. Let reliability estimates of the subsystems in Figure 1 and their fuzziness be *standard reliability* and *medium*, respectively. When the dependence level is estimated to be *complete*, reliability of the system and its fuzziness are estimated to be *standard* and *medium*, respectively. On the other hand when the dependence level is estimated to be *zero*, reliability of the system is estimated by the operation H. Let the parameter x_0 of the failure possibility F_S be x_{0i}, where F_S with x_{0i} is the estimated failure possibility of the system with the dependence level $i(i = 1$; *complete, i = 2 ; high, i = 3 ; medium, i = 4 ; low, i = 5 ; zero*). And let the parameter r_0 of the fuzzy set $R(r)$ representing the dependence level i be $r_{0i}(i = 2, 3, 4)$. The parameters $r_{0i}(i = 2, 3, 4)$ are determined so as to satisfy

$$(x_{0i}-x_{05}):(x_{01}-x_{0i}) = \begin{cases} 3:1 & (i = 2) \\ 1:1 & (i = 3). \\ 1:3 & (i = 4) \end{cases} \qquad (10)$$

4. Example

Figure 2 shows an example of the analyzed fault tree. Let us assume that two analysts analyze this tree and that reliability estimate of each basic event and

TABLE 4 Dependence estimate and parameters nH and nG

(Dependence level, Fuzziness)		(nH, nG)
first	(medium, medium)	(2.05, 1.0)
second	(low, medium)	(2.5, 2.5)

its fuzziness are *standard reliability* and *medium*, respectively. It is also assumed

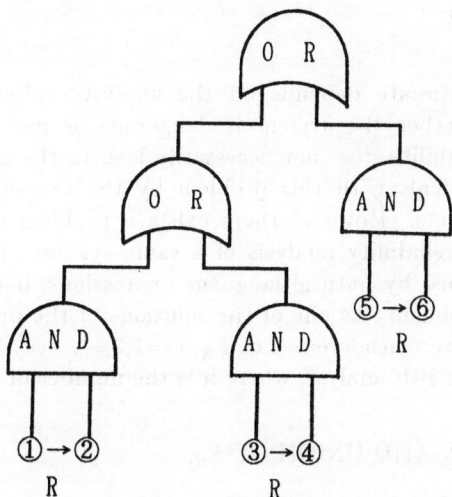

Figure 2. Example of fault tree

that the parameters nH and nG, and the dependence estimates are different between two analysts as shown in Table 4. Table 4 implies that the first analyst's reliability estimates of the parallel system and the series system are *high*(x_0=0.25) and *rather low*(x_0=0.63), respectively. On the other hand the second analyst's estimates of them are *high*(x_0=0.2) and *standard*(x_0=0.55), respectively. It is found that the first analyst is more pessimistic than the second one.

Figure 3 shows their analysis results. The first analyst estimates system reliability to be *rather low* and its fuzziness to be *medium*(Figure 3(1)). On the other hand the second analyst estimates system reliability to be *rather high* and its fuzziness to be *medium*(Figure 3(2)). It is found that the results reflect analysts subjectivity.

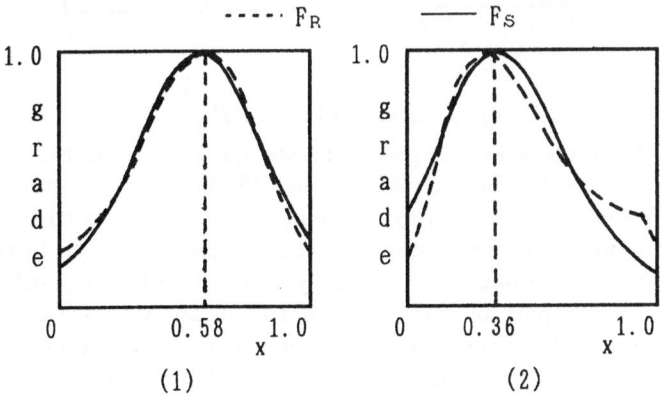

Figure 3. Analysis results

5. Mutual Agreement

System reliability estimate depends on the analysts subjectivity toward the analyzed system whether the system is dangerous or not. Same estimate of each basic event reliability does not necessarily lead to the same analysis result. The present model deals with this problem by the use of the parameterized operations of fuzzy sets. However there exists a problem that various results are obtained in the reliability analysis of a same system. Even if the analysis results can be classified by natural language expressions, it is necessary to consider further this problem. As one of the solutions of the problem two kinds of mutual agreements are considered. Let F_{Rj} $(j=1,2,---,n)$ be the failure possibility obtained by the j-th analyst, where n is the number of analysts.

5.1. INTERSECTION AND UNION OF F_{Rj}

The intersection and the union of F_{Rj} are defined by $F_I = \cap_j F_{Rj}$ and $F_U = \cup_j F_{Rj}$. A normalized F_I is defined by

$$F_{NI} = F_I / MF_I, \tag{11}$$

where MF_I is the maximal grade of F_I.

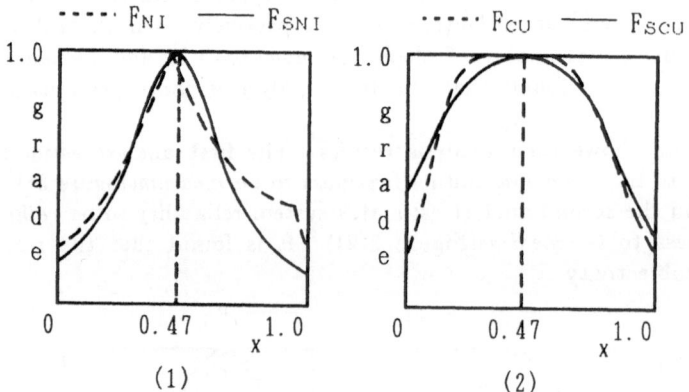

Figure 4. Intersection and union of F_{Rj}

Let F_{CU} be the result of the convexity operation[5] on F_U. And let F_{SNI} and F_{SCU} be fuzzy sets which minimize the distance (9) between F_S and F_{NI}, and that between F_S and F_{CU}, respectively. F_{SNI} and F_{SCU} are expressed by natural language in the form of reliability estimate and its fuzziness. F_{SNI} is interpreted as the least part of the result which all analysts agree with one another since F_{SNI} includes only the part of the results which all analysts agree with one another. Then the small maximal grade of F_I means the small grade of mutual agreement among the analysts. On the other hand F_{SCU} is interpreted as the largest part of the result which all analysts agree with one another since F_{SCU} includes all results obtained by analysts. Then high fuzziness of F_{SCU} means large difference of reliability estimates among the analysts.

The failure possibility in Figure 4(1) is the intersection of F_{Rj} in the example, where $MF_I = 0.75$. This result implies that *system reliability is standard and its fuzziness is low*. The failure possibility in Figure 4(2) is the union of F_{Rj}. This implies that *system reliability is standard and its fuzziness is high*. The results show rather large difference of reliability estimates between two analysts.

5.2. WEIGHTED AVERAGE OF F_{Rj}

Let W be the weight that means analyst's belief for the result. Five kinds of expressions about the belief are considered as shown in Table 5 and their meanings are expressed by fuzzy sets with the membership function (12).

$$W(w) = \frac{1}{1 + 20 \times |w - w_0|^2},\tag{12}$$

where w_0 is a parameter and its numerical value is also shown in Table 5.

TABLE 5 Expressions of belief and parameter w_0

Expressions of belief	Parameter w_0
(Grade of belief is)	
very high	0.9
high	0.75
medium	0.5
low	0.25
very low	0.1

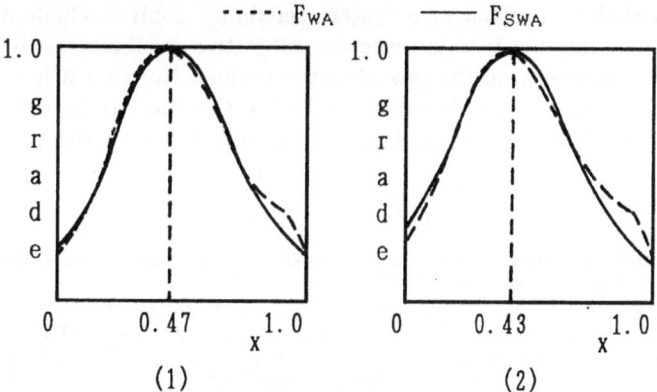

Figure 5. Weighted average of F_{Rj}

The weighted average of F_{Rj} is defined by

$$F_{WA} = \frac{\Sigma_j \ F_{Rj} \times W_j}{\Sigma_j \ W_j}\tag{13}$$

where W_j is the j-th analyst's weight and the extension principle is used.

Let F_{SWA} be the failure possibility that minimizes the distance between F_S and F_{WA} in the sense of Equation (9). F_{SWA} is expressed by natural language.

Figure 5(1) shows the weighted average of F_{Rj} in the example, where the grade of both analyst's belief is *very high*. In this case system reliability is expressed by natural language such as *system reliability is standard and its fuzziness is medium*. Figure 5(2) shows the case in which the grade of the first analyst's belief is *medium* and that of the second one is *very high*. In this case system reliability is expressed by natural language such as *system reliability is rather high and its fuzziness is medium*.

6. Concluding Remarks

This paper mentions a model of subjective reliability analysis. This model analyzes system reliability by the use of the failure possibility and natural language expressions. This model also considers subjectivity in fuzzy sets operations as well as in reliability estimate and dependence level estimate. The parameterized fuzzy sets operations can reflect analysts subjectivity toward the analyzed system whether the system is dangerous or not. However there exists a problem that the difference of subjectivity leads to various analysis results. Even if analysis results can be classified by natural language expressions, it is necessary to consider this problem. This paper considers two kinds of mutual agreements based on the results of the subjective analysis. One is the intersection and the union of the failure possibilities obtained by the analysis. The intersection and the union are interpreted as the least part of the result and the largest one, respectively, which all analysts agree with one another. The other is the weighted average of the failure possibilities. The weighted means the grade of analysts belief toward the result.

The probabilistic method of a system reliability analysis eliminates subjectivity from the analysis. In this sense the subjective reliability analysis may be against the consideration in the probabilistic method. However it is necessary to consider the subjectivity openly in the analysis rather than to hide it behind the numerical value. This paper considers system reliability from this point of view.

References

[1] T. Onisawa and J. Kacprzyk eds.: *Reliability and safety analyses under fuzziness*, Physica-Verlag, A Springer-Verlag Company, Heidelberg(1995).

[2] T. Onisawa ed.: Special issues of applications of fuzzy sets theory to reliability analysis and risk analysis, *Journal of Japan Society for Fuzzy Theory and Systems*, Vol.5, No. 5(1993)(in Japanese).

[3] T. Onisawa: A model of subjective reliability analysis, *Proc. of Second IEEE International Conference on Fuzzy Systems, San Francisco*, Vol.2, pp.756-761(1993).

[4] T. Onisawa: Use of natural language in system reliability analysis, in R. Lowen and M. Roubens(eds.), *Fuzzy Logic, State of the Art*, Kluwer Academic Publishers, Dordrecht, pp.517-529(1993).

[5] D. Dubois and H. Prade: *Fuzzy sets and systems: Theory and applications*, Academic Press, New York, pp.19-20, pp.36-37, pp.25-26(1980).

[6] M. Mizumoto: Pictorial representations of fuzzy connectives, Part I: Cases of t-norms and t-conorms, and averaging operators, *Fuzzy Sets and Systems*, Vol. 31, No. 2, pp. 217-242(1989).

FUZZY DECISION-MAKING APPLICATIONS IN NUCLEAR SCIENCE

D. RUAN

Nuclear Research Centre (SCK•CEN)
Boeretang 200, 2400 Mol, Belgium

Abstract.

Fuzzy set theory has been extensively researched in various fields of engineering. In nuclear science, a significant influence of fuzzy sets can be noticed. However, applications of fuzzy set theory to nuclear engineering is novel. In this paper, we start with a basic statement of the decision-making process based on fuzzy set theory, and then apply it to nuclear science with some practical applications (a fuzzy decision making in an accidental release to the atmosphere as well as in a problem of land suitability classification). We believe that the use of fuzzy set theory in nuclear science has potential advantages, and the fuzzy approach represents the available information in a meaning tractable way, and supplies the decision maker with more analytical information on ranges of sensitivity of decisions, rather than for obtaining a ranking of feasible actions.

1. Introduction

Progress in science and technology has made our modern society very complex, and, as a consequence, decision processes have become increasingly vague and hard to analyse. In almost all engineering activity, one is faced with decision-making. By decision-making, we cite the original statement of Bellman and Zadeh [1] about the role of fuzzy sets in decision analysis:

Much of the decision-making in the real world takes place in an environment in which the goals, the constraints and the consequences of possible actions are not known precisely. To deal quantitatively with imprecision, we usually employ the concepts and techniques of probability theory and, more particularly, the tools provided by decision theory, control theory and

Z. Bien and K. C. Min (eds.),
Fuzzy Logic and its Applications, Information Sciences, and Intelligent Systems, 275–284.
© 1995 *Kluwer Academic Publishers.*

information theory. In so doing, we are tacitly accepting the premise that imprecision -whatever its nature- can be equalled with randomness. This, in our view, is a questionable assumption. Specifically, our contention is that there is a need for differentiation between *randomness* and *fuzziness*, with the latter being a major source of imprecision in many decision processes. By *fuzziness*, we mean a type of imprecision which is associated with fuzzy sets [15], that is, classes in which there is no sharp transition from membership to nonmembership.

The term *decision* has been used with many different meanings and in many disciplines. Zimmermann [16] has specified clearly what will be meant by "decision", "decision model", "decision theory," and "decision technology" or "decision analysis," for a number of purposes, and in many different areas of life. In order to facilitate our task, it is, therefore, necessary to distinguish between definitions used in the scientific area and those interpolated in the technology areas. It is important to indicate that the main goals of these areas are different. While the main purpose of a scientific discipline is to generate knowledge and to come closer to the truth without making any value statement, technologies normally try to generate tools for solving problems better and very often by either accepting or building on given value schemes.

The search for better tools for decision analysis is an unending one. The rapidly growing number of applications of fuzzy set theory suggests that it represents a natural development in the field of nuclear science of our understanding of how humans can reason effectively within a vague and fuzzy environment. The significant influence of this new theory in this area becomes clear from a glance through INIS (International Nuclear Information Systems). With respect to various disciplines of nuclear applications, a lot of work has been done on the investigations of the potential of fuzzy sets and related approaches for the field of nuclear science and continue to develop. The best well known work in this area is particularly topical, as it deals with the Chernobyl accident in which fuzzy human reliability analyses in man-machine systems have been considered [5, 6]. The ideas developed by the above work could be useful in a broader context, and it is worthwhile to see how the relevant concepts are related to those of fuzzy set theory. Moreover, fuzzy set theory can take into account the imprecision of the factors affecting decision making in nuclear science. Hitherto known studies constitutes merely the initial investigation of a fuzzy-logical qualification of the uncertainty in risk and reliability assessment that arises as a consequence of the vagueness attached to expert judgement and seems incomplete. An answer to the question "how safe is safe enough"? for example, may be cast ideally in a fuzzy logic context since safety is by no means a crisp concept. Hence, the potential contribution of fuzzy set theory techniques towards the issues of safety criteria and regulatory decision making is significant. The use of fuzzy set theory in the decision process has been investigated by many researchers. As reported recently [4], the main advantage of using fuzzy set theory has been in overcoming the difficulties of decision making in a fuzzy situation represented by ill-defined terms. The inherent imprecision of such terms makes crisp ranking very difficult and application of statistical decision theory doubtful. The situation can be handled by the analyst by ranking these quantities verbal, which is the normal behaviour of human beings to account for inherent imprecision. Verbal ranking is then represented by fuzzy sets. The final ranking of alternatives from best to worst can be obtained using fuzzy operations.

Present investigation first starts with a basic statement of the decision-making process formulated by Bellman and Zadeh (1970), and then places it in the general setting of fuzzy sets, applying them further into a decision aiding system in an accidental release to the atmosphere as well as in a problem of land suitability classification.

2. Decision making in fuzzy environments

We formulate the problem of decision making in terms of fuzzy sets. The literature is quite extensive [1, 14, 2, 16, 7]. Sometimes the decision-making process is quite straight-forward, but there are also situations that give one pause in the presence of divers and contradictory objectives(constraints and goals). For instance, in the decision problem [7]:

Find a control ensuring *high precision* and *low energy consumption*.

It is obvious that the goal *high precision* and constraint *low energy consumption* are essentially at odds. High precision requires a control action with high energy consumption. Thus, satisfying the first objective violates the second one. The basic idea of a model suggested by Bellman and Zadeh for decision making in a fuzzy environment argues as follows:

The fuzzy objective function is characterized by its membership function and so are the constraints. Since we want to satisfy (optimize) the objective function as well as the constraints, a decision in a fuzzy environment is defined by analogy to nonfuzzy environments as the selection of activities which simultaneously satisfy objective function(s) and constraints. According to the above definition and assuming that the constraints are noninteractive, the logic "and" corresponds to the intersection. The decision in a fuzzy environment can therefore be viewed as the intersection of fuzzy constraints and fuzzy objective function(s). The relationship between constraints and objective functions in a fuzzy environment are therefore fully symmetric, that is, there is no longer a difference between the former and the latter.

More formally, let X denote a space in which all goals and constraints are defined. All are represented in terms of fuzzy sets. Thus for 'n' goals $G_1, G_2, \ldots, G_n$ and 'm' constraints $C_1, C_2, \ldots, C_m$:

$$G_1, G_2, \ldots, G_n; C_1, C_2, \ldots, C_m : X \to [0,1] \tag{1}$$

We can see that a decision to be made is also a fuzzy set D, $D : X \to [0,1]$, so that it results from all the objectives (1). Therefore:

$$D = f(G_1, G_2, \ldots, G_n, C_1, C_2, \ldots, C_m) \tag{2}$$

where 'f' expresses ties with the fuzzy set of decision D. 'f' would result from a translation of this statement: the decision must result from a satisfaction of *all* objectives, i.e., *all* the goals and constraints.

If we use the original notion of Bellman and Zadeh, we have:

$$D = G_1 \cap G_2 \cap \ldots \cap G_n \cap C_1 \cap C_2 \cap \ldots \cap C_m \tag{3}$$

where $\cap$ denotes an operation of intersection.
i.e.,

$$D(x) = T(G_1(x), G_2(x), \ldots, G_n(x), C_1(x), C_2(x), \ldots, C_m(x)), x \in X \tag{4}$$

where T is an extension of t-norm, for the related materials, the reader is referred to [8, 9, 10, 11]. In [1], T was specified as minimum, the concept is illustrated by the following example [16].

Example
Objective function x should be substantially larger than 10, and characterized by the membership function

$$\mu_G(x) = \begin{cases} 0, & x \le 10 \\ (1 + (x-10)^{-2})^{-1}, & x > 10 \end{cases} \tag{5}$$

Constraint x should be in the vicinity of 11 and characterized by the membership function

$$\mu_C(x) = (1 + (x-11)^4)^{-1} \tag{6}$$

The fuzzy set "decision" is then characterized by its membership function for all $x \in X$

$$\begin{aligned} \mu_D(x) &= min(\mu_G(x), \mu_C(x)) \\ &= \begin{cases} min\{(1 + (x-10)^{-2})^{-1}, (1 + (x-11)^4)^{-1}\}, & x > 10 \\ 0, & x \le 10 \end{cases} \end{aligned} \tag{7}$$

Some remarks from [1, 16]:
1. In the above example the min-operator was used on the basis of the

following argument: in the classical (crisp) choice model of a decision the (verbal) linkage between constraints and goals is usually "and". The only well-defined "and" is the "logical and" and this corresponds to the set-theoretic intersection. The model of the intersection of fuzzy sets might in certain contexts not be the min-operator but rather the product-operator or others.

2. Furthermore, the intersection might not even be the appropriate model of the "and" and it seems better to take the confluence of goals and constraints into account rather than of the intersection.

3. In defining a fuzzy decision, D, as the intersection-or more generally as the confluence- of the goals and constraints, we are tacitly assuming that all of the goals and constraints that enter into D are, in a sense, of equal importance. There are some situations, however, in which some of the goals and perhaps some of the constraints are of greater importance than others. In such cases, D might be expressed as a convex combination of weighting coefficients reflecting the relative importance of the consistent terms.

3. Decision aiding system in an accidental release to the atmosphere

The development of emergency response systems, which are able to support decision making in the event of a nuclear accident, is a relatively new area of R & D work, only a small number of systems are already in operation and, in general, they can only respond to a limited number of questions posed by a decision maker.

During the early stages of an accidental release of radioactive material to the atmosphere, the immediate aims of the off-site emergency management scheme are twofold: firstly, to determine the extent of any contamination occurring close to the site (i.e., out of a few km) for purposes of protecting the local public; secondly, to provide early estimates of the source term and hence permit consequences farther afield to be assessed. The source term module has the task to supply the system with all information concerning the release of radionuclides. However, the exact reconstruction of the source term is up to now an unresolved problem. Several mathematical methods are under development for atmospheric dispersion models. Some of them already give results with acceptable accuracy in special but not too complex situations. A direct application of the methods used for probabilities assessments of accident consequence is not suitable, since time consuming calculations are not feasible during an emergency and the nature of the uncertainties in a large fraction of the input information and model predictions differs fundamentally. Therefore, we seek another approach based on fuzzy sets and decision theory to treat uncertain information (incomplete

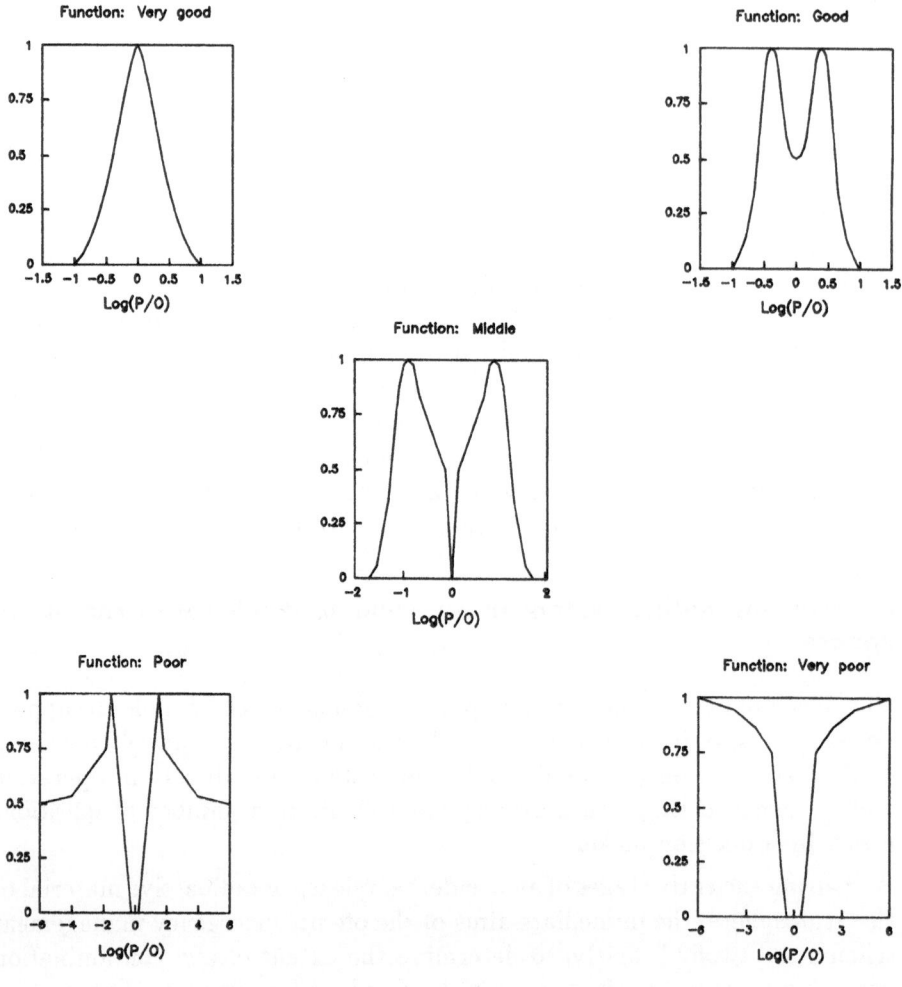

Figure 1. Five membership functions for the system.

and inexact). For example, the ambiguous problem of the classification with linguistic values such as very good, good, middle, bad and very bad according to the ratio of predictions and observation data can be treated with membership functions 1. Also a simple fuzzy algorithm is designed for the system to provide a reasonable solution for the practical users. The global view of the approach is depicted in Figure 2 which could propose different solutions to the user (visualisation of the situation) from the view of practice.

For further details, the reader is referred to the joint work [12]. The basic

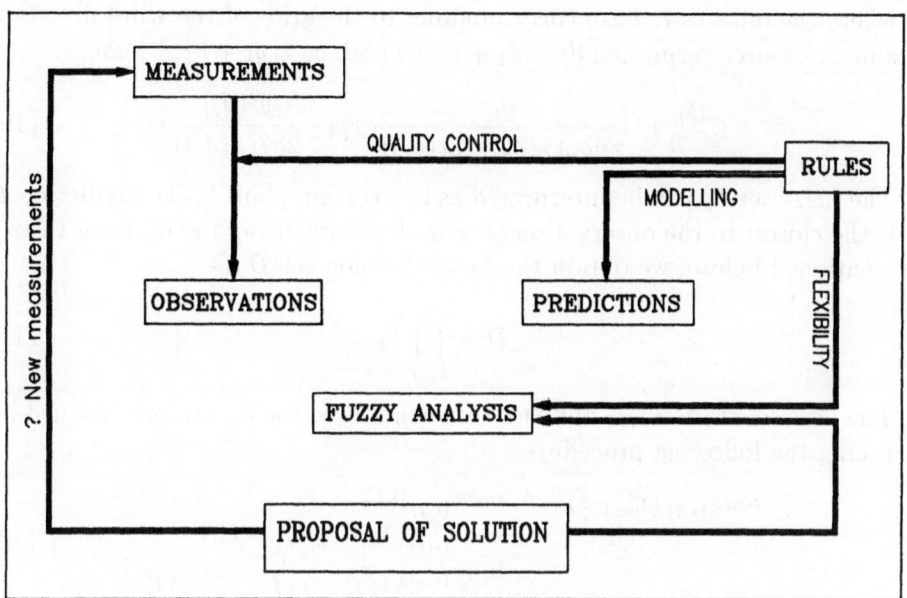

Figure 2. Fuzzy system shows a global diagnosis of the situation.

idea of the system is built on the following simple mathematical model.

Mathematical Model:

$$
\begin{aligned}
C &= \frac{Q}{\pi u \sigma_y(x)\sigma_z(x)} exp[-\frac{y^2}{2\sigma_y^2(x)}] \\
x &= x(D) \\
y &= y(D)
\end{aligned}
\tag{8}
$$

In the above equation, C is the concentration of radioactive material, which is dependent of an unknown source term Q and an unknown wind direction D (x and y are coordinates depending on the wind direction D, and the other physical parameters are given in this study case). In a real situation, we can on one hand obtain by means of observation (O) measured concentration data denoted as $C_m^{(k)}$ for certain points ($k \in \{1,2,\ldots,n\}$), and on the other hand we can calculate prediction (P) denoted as $C_{(i,j)}^{(k)}$ by the model if the source term Q and the wind direction D given for the same points as used in the observation. For each point k, we define a fuzzy set F_k as

$$
F_k = \frac{1}{1 + |C^{(k)}{}_m - C^{(k)}{}_{(i,j)}|}
\tag{9}
$$

where the indexes i, j are corresponding to the grids of the wind direction and the source term, i.e., $d_i = d_1 + (i-1)\Delta d$, $q_j = q_1 + (j-1)\Delta q$.

$$C_{(i,j)}^{(k)} = \frac{q_j}{\pi u \sigma_y(x(d_i)) \sigma_z(x(d_i))} exp\left(-\frac{y^2(d_i)}{2\sigma_y^2(x(d_i))}\right) \tag{10}$$

The fuzzy set F_k can be interpreted as in a certain point k, the prediction P is the closest to the observation O. For all points, using the decision theory mentioned before, we obtain the fuzzy decision set D as

$$D = \bigcap_{k=1}^{n} F_k \tag{11}$$

The best source term q_0 and the best wind direction d_0 can be obtained by taking the following procedure:

$$\begin{aligned} max_{(i,j)} \cap_{k=1}^{n} F_k &= max_{(i,j)} \Pi F_k \\ &= max_{(i,j)} \{F_1 \cdot F_2 \cdot \ldots \cdot F_n\} \\ &= max_{(i,j)} \{\frac{1}{1+f_{ij}^1} \cdot \frac{1}{1+f_{ij}^2} \cdots \frac{1}{1+f_{ij}^n}\} \end{aligned} \tag{12}$$

where $f_{ij}^k = |C^{(k)}{}_m - C^{(k)}{}_{(i,j)}|, k = 1, 2, \ldots, n$

Remarks from this research work:
1. The intersection operator here is used product-operator, which is better than min-operator after having compared the results for the two operators.
2. Nonfuzzy values of Q and D are obtained by max-procedure. If there are more than one maximum value, we will treat their average value as the representative one, i.e., a mean-maxima approach for the actural use in this resaerch work.
3. From a practical point of view, we can set the finite grids for wind direction since we more-or-less know the range of it. But we do not have any idea of the source term. If we set a very large range of it, then we face the time-consuming problem for this simulation. For that reason, we use the concept of ratios of $\frac{P}{O}$ (prediction over observation), the situation for which all these ratios (for all measurement points) are as close to 1 as possible is being searched with respect to the five different membership functions. By this way, we obtain an approximate value of source, therefore a more-or-less correct range of it. Several examples have been tested and the method proposed here seems acceptable from the point of view of the practice [12].

4. Land suitability classification

Land evaluation is concerned with the assessment of land performance for specified land utilization purposes. Such evaluation is essential in the process of land use planning, because it may guide decisions on land utilization in such a way that the resources of the environment are optimally used and that a sustained land management is achieved.

Land suitability classification is an approach in land evaluation that concerns the appraisal and grouping of specific areas of land in terms of their suitability for defined uses. FAO [3] proposed a general classification for the land suitability in which two suitability orders were discerned: suitable (S) and unsuitable (N). The order S was subdivided into a very suitable (S_1), moderately suitable (S_2) and marginally suitable class (S_3). The order N was subdivided into a currently unsuitable (N_1) and permanently unsuitable class (N_2). Although FAO defined principles for evaluation, no specific methodology was suggested to achieve the classification. In recent years, a number of methodologies have been developed under the FAO framework. However, these methods are based on either land characteristics or land qualities and result in a qualitative evaluation. The suitability classes are defined as discrete groupings, separated by strict class definitions or fixed class limits. Land units that have a degree of suitability somewhat intermediate between classes can however only be classified in one single suitability class. A new methodology is developed, based on fuzzy set theory, which gives additional information on the neighbouring suitability class.

The land suitability classification using fuzzy set theory is performed in four successive steps. These steps involve the determination of the membership functions (for each characteristic and for each suitability class membership functions express the degree to which a value of a land characteristic belongs to a suitability class), the determination of the membership values (for a given land unit, the membership values for the different land characteristics and suitability classes are subsequent arranged in a fuzzy relation from land characteristics to suitability classes), the establishment of a weight matrix (land characteristics have a relative importance with regard to different objects under consideration) and the calculation of the evaluation matrix (the suitability class for the considered land unit coincides with the element of the matrix that has the highest value). For the related details, the reader is referred to the joint research work [13].

5. Acknowledgements

The author wishes to thank Prof. Kerre for his valuable advice for the early version of this paper.

References

1. Bellman, R.E. and Zadeh, L.A. (1970) Decision-making in a fuzzy environment, *Manag. Sci.* **17**, pp. 141–164.

2. Dubois, D. and Prade, H. (1980) *Fuzzy sets and systems: theory and applications.* Academic Press, New York.

3. FAO (1976) A framework for land evaluation, *FAO Soils Bulletin* **32**, FAO, Rome.

4. Gençay, S. (1991) Site selection for nuclear plants using fuzzy decision analysis, *Kerntechnik* **56 (5)**, pp. 320–327.

5. Onisawa, T. (1988) An approach to human reliability in man-machine systems using error possibility, *Fuzzy Sets and Systems* **27**, pp. 87–103.

6. Onisawa, T. and Nishiwaki, Y. (1988) Fuzzy human reliability and analysis on the Chernobyl accident, *Fuzzy Sets and Systems* **28**, pp. 115–127.

7. Pedrycz, W. (1989) *Fuzzy control and fuzzy systems.* Research Studies Press Ltd, Somerset, England; John Wiley & Sons Inc.

8. Ruan, D. (1990) A critical study of widely used fuzzy implication operators and their influence on the inference rules in fuzzy expert systems, PhD thesis, University of Gent, Belgium.

9. Ruan, D. (1991) Critical study of fuzzy implication operators and their influence on approximate reasoning, In E.E. Kerre (Ed.), *Introduction to the basic principles of fuzzy set theory and some of its applications,* Communication and Cognition, Gent, Belgium, pp. 214–251.

10. Ruan, D. and Kerre, E.E. (1993) Fuzzy implication operators and generalized method of cases, *Fuzzy Sets and Systems* **54**, pp. 23–37.

11. Ruan, D. and Kerre, E.E. (1993) On the extension of the compositional rule of inference, *Int. J. of Intelligent Systems* **8**, pp. 807–817.

12. Sohier, A., Van Camp, M., Ruan, D. and Govaerts, P. (1993) Methods for radiological assessment in the near-field during the early phase of an accidental release of radioactive material using an incomplete data base, *Radiation Protection Dosimetry* **50 (2–4)**, pp. 321–325.

13. Tang, H. and Ruan, D. (1992) Land use suitability assessment for Irrigated maize based on fuzzy set theory, in *Int. Conf. Advances in planning, design and management of irrigation systems as related to sustainable land use,* Leuven, Belgium, 14–17 September 1992. pp. 597–605.

14. Yager, R.R. (1977) Multiple objective decision-making using fuzzy sets, *Int. J. Man-Mach. Stud.* **9**, pp. 375–382.

15. Zadeh, L.A. (1965) Fuzzy sets, *Information and Control* **8**, pp. 338–353.

16. Zimmermann, H.-J. (1987) *Fuzzy Sets, Decision making, and Expert Systems.* Kluwer Academic Publishers, Boston.

Chapter 3.

MATHEMATICAL FOUNDATIONS

CONDITIONAL EVENTS AND FUZZY CONDITIONAL EVENTS VIEWED FROM A PRODUCT PROBABILITY SPACE PERSPECTIVE

I.R. GOODMAN

CODE 422, SEASIDE
NCCOSC RDTE DIV (NRaD)
SAN DIEGO, CA 92152-7463

1. Introduction

This paper first provides a brief review of the product space approach to conditional event algebra and the one-point random set coverage function representation of fuzzy sets followed by a natural extension to a fuzzy set structure.

Until recently, no systematic way existed for the modeling and evaluation of logical compounds of conditional -- or if-then -- statements such as

$$s = \text{"(if b, then a) and (if d, then c)"}, \tag{1}$$

when antecedents b and d do not coincide -- for which the individual conditional statements are compatible with probability, i.e., the first conditional statement yielding the (conditional) probability evaluation

$$P(\text{if b, then a}) = P(a|b), \qquad P(\text{if d, then c}) = P(c|d) \tag{2}$$

assuming $P(b) > 0$, where

a = "the enemy will most likely consider areas A or B" ,
b = "the temperature is well over 45 degrees F in the morning" ,
c = "the enemy will probably go into areas A or C (in the morning)",
d = "the temperature is below, perhaps, 60 degrees F". (3)

It should be noted in conjunction with this, if one attempted to employ the well-established classical logic interpretation of conditioning via the material conditional operator $\Rightarrow$, where typically

$$b \Rightarrow a = b' \vee a = b' \vee ab, \tag{4}$$

then one has the corresponding probability evaluation

$$P(b \Rightarrow a) = 1 - P(b) + P(ab) = P(a|b) + (P(b') \cdot P(a'|b)) > P(a|b), \tag{5}$$

in general, unless $P(b) = 1$ or $P(a'b) = 0$, obviously showing a discrepency with eq.(2). On the other hand, the material conditional interpretation of conditional statements does yield the completely computable evaluation for s in eq.(1) as

$$P(s) = P((b \Rightarrow a) \cdot (d \Rightarrow c)) = P(b'd' \vee abd' \vee cdb' \vee abcd) = P(b'd') + P(abd') + P(cdb') + P(abcd). \tag{6}$$

Z. Bien and K. C. Min (eds.),
Fuzzy Logic and its Applications, Information Sciences, and Intelligent Systems, 287–296.
© 1995 *Kluwer Academic Publishers*.

The above illustrates the gap that exists between the classical logic approach which produces computable logical combinations of arbitrary conditionals which are not compatible with probability (in the sense of eq.(2)) vs. the usual development of conditional probability for which no standard calculus of logical operators exist for combining conditionals compatible with probability. However, over the past several years this difference has now begun to be filled with the development of "conditional event " algebras, whereby conditional expressions as "if b, then a" can be consistently interpreted as objects (or even events) , denoted from now on for convenience as (a|b), in a space with well-defined operators extending the usual boolean / logical ones (such as negation ( )', conjunction · , disjunction v) such that eq.(2) holds. (See, e.g. [1].) Although a number of conditional event algebras have been proposed in direct response to the above issue, a number of other difficulties arose. These included the fact that none of the algebras possessed a boolean structure, thus not allowing many standard laws of probability to hold for conditionals. In addition, problems occurred regarding consistency of independence, higher order conditioning, and compatibility with the usual conditioning of random variables. To this end, the role of conditional event algebra was re-examined and the product space -based conditional event algebra -- denoted here as PS -- originally proposed by Van Fraasen in 1976, but not further developed [2], was independently rediscovered and found to address affirmatively all of the above-mentioned issues. Indeed, it was also shown that this conditional event algebra can be essentially characterized as the unique boolean conditional event algebra which satisfies modus ponens and a weak form of exportation -- properties satisfied by all previously proposed conditional event algebras, as well as by the material conditional:

$$P^\wedge((a|b)\cdot b) = P(ab), \qquad P^\wedge((a|b) \mid c)) = P(a|b) , \qquad (7)$$

for all events a,b in the given boolean algebra of unconditional events, where $P(b) > 0$, c is naturally identified with $(c|\Omega)$, $P^\wedge$ denotes the countable product probability extension of probability measure P to the product space setting; similarly for $P^{\wedge\wedge}$ [3].

Although PS requires an infinite construction, computation-wise, it requires actually only a finite number of operations for implementation of any of its logical operators. However, the number of such operations required for implementation of conjunctions or disjunctions is significantly larger than the corresponding ones needed by the other previously-proposed conditional event algebras. For example, for trinary conjunction, for the previously proposed Goodman-Nguyen-Walker conditional event algebra [1], the Adams-Calabrese conditional event algebra [4], PS [3], respectively, one has

$$P^\wedge((a|b)\cdot(c|d)\cdot(e|f)) = P(abcdef \mid a'b \lor c'd \lor e'f \lor abcd); \qquad (8)$$

$P^\wedge((a|b)\cdot(c|d)\cdot(e|f)) = P(abcdef \lor abefd' \lor cdefb' \lor abcdf \lor cdb'f \lor abd'f \lor efb'd' \mid b \lor d)$
$=P(abcdef)+P(abefd')+P(cdefb')+P(abcdf)+P(cdb'f)+P(abd'f)+P(efb'd')] /P(b\lor d\lor f) ; \qquad (9)$

$P^\wedge((a|b)\cdot(c|d)\cdot(e|f)) = [P(abcdef)+P(abefd')P(c|d) +P(cdefb')P(a|b)+P(abcdf)P(e|f)$
$+P(cdb'f)P^\wedge((a|b)\cdot(e|f))+P(abd'f)P^\wedge((c|d)\cdot(e|f))+P(efb'd')P^\wedge((a|b)\cdot(c|d))] / P(b\lor d\lor f) , \qquad (10)$

where typically the binary conjunction for PS is given as
$$P^\wedge((a|b)\cdot(c|d)) = [P(abcd) + P(abd')P(c|d) + P(cdb')P(a|b)] / P(b \lor d). \qquad (11)$$

Thus, the more complex, but theoretically superior, PS will be only be consdered.

2. Modeling Linguistic Information Via Fuzzy Sets and Random Sets

Fuzzy sets and fuzzy logic, introduced by Zadeh in 1965 have become a basic tool in the modeling of linguistic expressions by use of appropriately chosen logical and functional compounds of formal membership relations between elements and their possible attributes. For example, the linguistic expression representing expression a in eq.(3), instead of being considered an ordinary (unconditiopnal) event in an appropriate boolean algebra, can also be represented in fuzzy logic as $h(f_1(x_1),f_2(x_2))$, where $f_j:D_j \to [0,1]$ is the fuzzy set membership function, identified as usual with its membership function over some natural domain D, with values in the unit interval [0,1], not just at the endpoints 0,1, as ordinary , or crisp, set membership functions are required to have. (Thus, note that all ordinary sets $a_j \subseteq D_j$) have membership functions $\phi(a_j):D_j \to [0,1]$ which also can be considered as special cases of fuzzy set membership functions.) $f_j(x_j)$ represent the degree to which elements x_j in D_j satisfy f_j, j=1,2. Here, f_1 represents the compound attribute consisting of temperature being well-over 45 degrees F (potentially) implying the breakthrough area for enemy is most likely A or B, while x_1 represents the ordered pair element (actual temperature, actual enemy breakthrough area), with D_1 being the natural domain of possible values x_1 can take on. this is here, some reasonable range of temperatures, say [20 degrees F,100 degrees F] × (cartesian product with) the set of possible enemy breakthrough areas such as {A,B,C, D}. Similar remarks hold for f_2 , x_2 , D_2 . Finally, $h:[0,1]^2 \to [0,1]$ is a fuzzy set representing the logical (boolean) connector for disjunction extended to fuzzy logic. Unlike the classical logic situation where unique interpretations exist for disjunction and conjunction, h can be interpreted several consistent ways. Often, h is given as the function max (arithmetic maximum) , or probsum (the DeMorgan transform of prod, the arithmetic product), or as a function chosen from a general class of functions which abstracts the basic characteristics of max and probsum- such as the class of *t-conorms* (i.e., DeMorgan transforms of t-norms, see below), or the different, but overlapping, class of *co-copulas* (similarly, DeMorgan transforms of copulas - see below). In a related vein, fuzzy conjunction is often interpreted by min (arithmetic function minimum) or prod , or by use of functions from a more general class of functions which abstract the basic characteristics of min and prod- such as the class of *t-norms* or the distinct, but overlapping class of *copulas*. (The latter are characterized as all joint distribution functions all of whose marginals correspond to uniformly distributed random variables over the unit interval - see [5] for further details on t-norms, t-conorms, copulas, co-copulas.) On the other hand, fuzzy negation is most often simply interpreted by the simple form 1-(.) and a DeMorgan relation is often assumed to hold between g and h.

Ever since Zadeh, introduced fuzzy set theory, there has been interest in establishing connections with probability theory, especially when both linguistic and probabilistic expressions are present as inputs to data fusion. A number of individuals, including Orlov, Höhle, and Goodman [6], concluded independently that there was a natural relation between fuzzy set descriptions - and hence linguistic descriptions - and probability ones, via use of the *one-point coverage functions* of appropriately chosen *random sets*. That is, consider $f:D \to [0,1]$, any given fuzzy set, . Then, one can find a random set - i.e., a random quantity, which instead of being point-valued as the familiar typical random variable is, is now set-valued - say, S(f) over D (or, equivalently, S(f) is a random subset of D) - so that

$$P(x \text{ in } S(f)) = f(x), \quad \text{all } x \text{ in } D. \tag{12}$$

In fact, one such *one-point coverage equivalent random set* S(f) to f, among infinitely many possible in general satisfying eq.(12), is given by the one point coverage equivalent *canonical nested random set*

$$S(f) = f^{-1}[U,1], \tag{13}$$

where U is any random variable over the unit interval which is uniformly distributed (i.e., induces lebesgue measure) and f^{-1} is the functional inverse of f. In fact, it can be shown (see the recent paper [7] for details) that the entire solution class S(f) of distribution- distinct random sets S(f) satisfying (2.1), when D is finite, is completely determined by the class of all copulas g, relative to index set D as $g:[0,1]^D \rightarrow [0,1]$, together with the fixed family of cumulative distribution functions (cdfs) $(G_{f(x)})_{x \text{ in } D}$, where each $G_{f(x)}$ corresponds to the two-point mass or dirac distribution assigning probability f(x) to 1 and 1-f(x) to 0. Each S(f) is then explicitly constructed by first obtaining the cdf H of its ordinary (random) set membership function, $\phi(S(f)):D \rightarrow \{0,1\}$, a zero-one-valued random variable, as the composition $H = g_o[(G_{f(x)})_x \text{ in } D]: \{0,1\}^D \rightarrow [0,1]$. Then, we recover the distribution of S(f) via the usual relations and straightforward combinatorics ([7], Appendix) as:

$$P(S(f)=a) = P(\phi(S)=\phi(a)) = \int_{t \text{ in } \phi(a)} dH(t) = \sum_{\emptyset \subseteq K \subseteq a} (-1)^{card(K)+1} g[(1-f(x))_x \text{ in } D] , \tag{14}$$

the case for the canonical nested random set given in eq.(13) corresponding to g = min. When g = prod, the random set S(f) corresponds to $\phi(S(f)) = (\phi(S(f))(x))_x \text{ in } D$ being a family of mutually statistically independent zero-one-valued random variables and is seen to be the maximum entropy one point coverage equivalent random set to f [8]. Finally, note the natural compatibility condition: when $f=\phi(a)$, for $a \subseteq D$, any ordinary set,

$$S(\phi(a)) = a . \tag{15}$$

Next, consider the problem of carrying over this correspondence between fuzzy sets and random sets via the one-point coverage probabilities, as given in eq.(12) to logical operators acting among fuzzy sets and ordinary (boolean) logical operators acting among random sets. For example, eq.(1) is a conjunction of conditional forms. Hence, it is natural to seek the relations ([6], Sect. 4.2,5.2)

$$P((x_1 \text{ in } S(f_1)) \& (x_2 \text{ in } S(f_2))) = h(f_1(x_1),f_2(x_2)); \quad P(x_j \text{ in } S(f_j)) = f_j(x_j), \text{all } x_j \text{ in } D_j \tag{16}.$$

More generally, when a linguistic expression can be analyzed formally as a compound of disjunctions h of conjunctions g of fuzzy sets $f_{ij}:D_{ij} \rightarrow [0,1]$, we seek analogous to (16), random sets $S(f_{ij})$ such that, denoting the choice of fuzzy conjunction operation by h and using obvious notation for multiple arguments, the following *homomorphic-like relations* hold for all finite index sets M,N, all x_{ij} in D_{ij} ,

$$P(\underset{i \text{ in } N}{\text{ or }} \underset{j \text{ in } M}{\&} (x_{ij} \text{ in } S(f_{ij}))) = h(g(f_{ij}(x_{ij});j \text{ in } M);i \text{ in } N), \quad P(x_{ij} \text{ in } S(f_{ij})) = f_{ij}(x_{ij}), \tag{17}$$

Similar comments hold for the dual of the above: conjunctions of disjunctions.

Recently, the issue of the non-uniqueness of choice of fuzzy operators for such arbitrary finite combinations has been resolved :

Theorem 1 (Goodman [7]) Under assumptions of continuity, DeMorgan and additional mild constraints, including finiteness assumptions for M,N, and the D_{ij}, the only copula h, cocopula g, and $S(f_{ij})$ satisfying the homomorphic-like relation in eq.(17) for all possible choices of f_{ij} and finite D_{ij}, are: g=min, h =max, or g=prod, h = probsum, or (g,h) = any *ordinal sum of* (prod,probsum) [a certain affine-like transforms of (prod,probsum) relative to the unit interval [6], Sect. 2.3.6)] , while the $S(f_{ij})$ as marginal random variables can be arbitrary one-point coverage equivalent random sets to the f_{ij}, subject to joint distributional constraints imposed by g,h.   ♦

We also remark that when only (repeated) conjunctions or only disjunctions are utilized in a linguistic phrase, Theorem 1 is not applicable and essentially any joint distributional constraint can hold among the one-point coverage equivalent random sets representing the fuzzy sets. In applying the above one-point coverage equivalent random sets and their homomorphic-like relations to given fuzzy logic interpretations of linguistic-based information, we note that basically, while we may choose definite fuzzy logic connectors - such as guided by Theorem 1, or other prior considerations, in general, we *will not* choose any specific one-point coverage equivalent random set from the class of all possible ones, for each given fuzzy set, unless the information is mixed with stochastic descriptions. All conclusions and combining procedures will be essentially independent of the choice of particular one-point coverage equivalent random sets from each class $S(f)$.

3.   Extension of PS to a Fuzzy Set Structure

We have seen in Section 1 a basic motivation for considering conditional event algebra in general, utilizing the example in eqs. (1)-(3). In order to accomplish this, the example was interpreted purely stochastically, i.e., the relevant components a,b,c,d, were all assumed to be ordinary unconditional events which lie in a common sample space (or boolean or sigma-algebra) B, so that ordinary probability evaluations P are valid over B. In turn, with this paradigm, the general problem of treating conditionals was briefly considered and PS, considered as the best candidate, at least from a theoretical foundations basis, despite its relatively longer implementations. However, as pointed out in Section 2, the expression in eqs.(1), could well be considered not a classical logical combination of standard events or sets, but rather a formal logical combination of linguistic based entities , the latter naturally interpreted as fuzzy sets (or ordinary sets, if actually warranted), while the former as fuzzy logic operators. Thus, in that case, the symbolization of eq.(1) remains valid,

$$s = (a|b) \cdot (c|d) , \tag{18}$$

but now with the more-detailed interpretations:
$$a = most(break(A) \vee break(B)); b = tempabov(45); c = prob(break(A) \vee break(C)); d = tempbel(60) \tag{19}$$
where for simplicity, temporal and other considerations are kept to a minimum and where most, prob, break, tempabov(45), tempbel(60), all represent "most likely", "probably", "enemy breaks through at", "temperature is well over 45", "temperature is below 60", respectively, all fuzzy sets with appropriately chosen domains : most:$D_1 \rightarrow$ [0,1], prob:$D_2 \rightarrow$ [0,1],break:$D_3 \rightarrow$ [0,1], tempabov:$D_4 \rightarrow$ [0,1], tempbel:$D_5 \rightarrow$ [0,1].

For example, D_1 is most appropriately chosen as the unit interval [0,1], since it quantifies all possible membership function values and most can be chosen as some continuous nondecreasing function with critical values

$$most(0) = 0 , \quad most(1/2) \text{ close to 0}, \quad most(3/4) \text{ close to 1}, \quad most(1) = 1 . \tag{20}$$

Similar comments hold for $D_2 = [0,1]$ with prob having a form similar to most, but in general for arguments strictly inside $[0,1]$ somewhat higher than the first function, because of the linguistic weakness in emphasizing truth using "probably " compared to "most likely". On the other hand, we may well have $D_3 = \{A,B, C,D,E\}$, representing all the areas (possibly overlapping and appropriately discretized) where the enemy can break through with "break" taking various values at each argument, based on subject a priori knowledge, so that e.g. break(A) might be 1 , while break(B) might be 0.8, etc., and the sum of the evaluations (possibilities) need not equal unity. In fact, compatible with the usual results of fuzzy set theory and the one-pont random set representation of fuzzy sets (again, seek Section 2), it is easily seen that the one point coverage function of any given random subset of D_3 itself need not be a probability function and can add up to strictly less than unity (in which case the null event $\varnothing$ is being assigned some non-zero probability), equal to unity (in which case the one-point coverage funstion or fuzzy set membership function is also formally a probability function), or strictly greater than unity (due to probabilities of overlaps). tempabov(45) has domain D_4, which could reasonably be (depending on the location, possible extremes of temperatures, and accuracy of informer in meaning "below") as interval [40, 110], with the membership function here being monotone increasing, close to unity at 45, etc. Then, expression "a" can be considered as a compound of "most " applied to the fuzzy logic disjunction (see Section 2 again) of break evaluated separately at two arguments, i.e., a fuzzy set membership function itself, identified as, say $a:D_a \to [0,1$; similarly, for b: $D_b \to [0,1]$, $c:D_c \to [0,1]$, $d:D_d \to [0,1]$ where

$$D_a =(\text{def}) D_3 \times D_3 , \quad D_b =(\text{def}) D_4, \quad D_c =(\text{def}) D_3 \times D_3 , \quad D_d =(\text{def}) D_5 \quad (21)$$

Returning to the modeling of the overall sentence s in eqs.(1), (3), we must be able to interpret also the individual conditional expressions (a|b), (c|d) and combine them by an appropriate choice of conjunction operator $\cdot$. Thus, in effect, we need to be able to define a *conditional fuzzy set, which we note, on one hand, must reduce to the membership function of a conditional events as prtesented in the previous section , when the antecedent and consequent membership functions represent those of ordinary sets or events, and on the other hand, must reduce to the consequent fuzzy set, when the antecedent is unity.* It is for these reasons that the now accepted forms in fuzzy set theory for forming conditional fuzzy sets are not satisfactory - especially the proposals to use the convenient material conditional fuzzy set counterpart (see the discussion in [9]). In fact, in [10] it was shown that conditional fuzzy sets satisfying these requirements could be defined and we present here, for completeness, a brief summary of, and some new insights into, these results.

The reasoning is essentially as follows:From Section 2, we have the correspondences:

Diagram 1. One-Point Coverage Probability Correspondences Between Fuzzy Sets and Random Sets

fuzzy sets $\qquad \to (\cdot) \to \qquad$ random sets

one-to (infinitely, in general) many

via the one-point coverage probabilities

$a:D_a \to [0,1] \qquad\qquad\qquad\qquad\qquad\qquad S(a):\Omega \to P (D_a)$

$\vdots \qquad\qquad\qquad\qquad\qquad\qquad\qquad\qquad \vdots$

$d:D_d \to [0,1] \qquad\qquad\qquad\qquad\qquad\qquad S(d):\Omega \to P (D_d),$

$f(x) = P(x \text{ in } S(f)) ,$ all x in $D_f,$ f=a,b,c,d,

where (Ω, C, P) is some appropriately chosen probability space, $P(\ )$ denotes power class. Then, it is natural that the formal conditioning operator $(.|..)$ in eq.(1) preserves the one-point coverage probability relations in Diagram 1, compatible with the development of conditional events and conditional event algebra as in the previous section. Hence, we have, analogous to the construction of random variables:

Diagram 2. One Point Coverage Probability Correspondences Between
Conditional Fuzzy Sets and Conditional Random Sets

$$(a|b): (D_a \times D_b)_0 \to [0,1] \to (.|..) \to (S(a) \times S(b) \mid D_a \times S(b)): \Omega_0 \to (P(D_a \times D_b))_0$$

where the subscript o indicates the product space construction and use of PS. Specifically, the one-point coverage relation here yields the definition

$$(a|b)(z_0) = (\text{def}) \; P_0(z_0 \text{ in } (S(a) \times S(b) \mid D_a \times S(b))) , \qquad (22)$$

for all $z_0 = (x_j, y_j)_{j=1,2,...}$ in $(D_a \times D_b)_0$, i.e., x_j in D_a, y_j in D_b, all j . It follows from the basic product property of P_o , the form of PS conditonal events (see again [2]) with a replaced by $S(a)$, b by $S(b)$, etc.), and the one-point coverage probability relationships between $S(a)$ and a and $S(b)$ and b,

$$(a|b)(z_0) = P_0(z_0 \text{ in } \overset{+\infty}{\underset{j=0}{v}} ((D_a \times S(b)')^j \times (S(a) \times S(b)))) = \qquad (23)$$

$$\overset{+\infty}{\underset{j=0}{\Sigma}} \; \overset{j}{\underset{i=1}{\Pi}} \; (P(y_i \text{ in } S(b)')P((x_{j+1} \text{ in } S(a))\&(y_{j+1} \text{ in } S(b))) = \overset{+\infty}{\underset{j=0}{\Sigma}} \; \overset{j}{\underset{i=1}{\Pi}}(1-b(y_i))q(a,b;g)(x_{j+1},y_{j+1}).$$

Here, the product factor is taken to be vacuous when $j = 0$ and indicates the usual membership function of ordinary sets c by $\phi(c)$, and hence similar notation for random sets, noting that $\phi(S(a))$ and $\phi(S(b))$ are now zero-one-valued random variables and where g denotes that copula determining the joint distribution between $S(a)$ and $S(b)$ (again, see Section 2). Finally,

$$q(a,b;g)(x_{j+1},y_{j+1}) = (\text{def}) \; P((x_{j+1} \text{ in } S(a)) \& (y_{j+1} \text{in } S(b)))$$
$$= P((\phi(S(a))(x_{j+1}) = 1) \& (\phi(S(b))(y_{j+1}) = 1)). \qquad (24)$$

Note also the easily derived recursive form for (23), analogous to that for ordinary PS conditionals [2], is

$$(a|b)(z_0) = q(a,b;g)(x_1,y_1) + (1-b(y_1))(a|b)(z_{0,1}) , \qquad (25)$$

where $z_{0,2}$ is z_0 with the first term removed: $z_{0,2} = (\text{def}) \; (x_j,y_j)_{j=2,3,...}$. Also, note by a simple argument, it can be seen that necessarily, slightly abusing notation,

$$(a|b) = (q(a,b:g) \mid b). \qquad (26)$$

The further evaluation of q in eq.(24) - and in fact logical combinations of conditional fuzzy sets - depends on the following lemma

Lemma 1. (See also [10].) Let (Ω, B, P) be a given probability space and $(V_j)_{j=1,...,n}$ a collection of zero-one-valued random variables $V_j: \Omega \to \{0,1\}$ whose joint distribution (by Sklar's Theorem [5]) is determined by copula $g:[0,1]^n \to [0,1]$:

$$P(\underset{j=1}{\overset{n}{\&}}(V_j =1)) = \underset{\varnothing \neq K \subset \{1,..,n\}(proper)}{\sum} (-1)^{card(K)+1} r(g)(K) = \underset{\varnothing \neq K \subseteq \{1,..,n\}}{\sum} (-1)^{card(K)+1} r(g)(K) + \underset{j=1}{\overset{n}{\sum}} P(V_j =1), \quad (27)$$

where

$$r(g)(K) =(def)\ 1- P(\underset{j\ in\ K}{\&}(V_j = 0)\ ) = 1 - g(P(V_j = 0);\ all\ j\ in\ K)), \quad (28)$$

by restricting g to K-arguments and noting that $r(g)(K)$ is the (DeMorgan) cocopula corresponding to g evaluated at $(P(V_j = 1))_j$ in K. (See also Section 2.)

Proof : Follows from straightforeward combinatorics. ◆

Corollary 1 Let (Ω, B, P) be a given probability space and $(S(f_j))_{j=1,..,n}$, a collection of random sets $S(f_j):\Omega \to P (D_{f_j})$ with copula g determining their joint distribution , each $S(f_j)$ one-point coverage equivalent to fuzzy set $f_j:D_{f_j} \to [0,1]$ as in eq.(2.1). Then,

$$q(f_j(x_j), j =1,..,n;\ g) =(def)\ P(\underset{j=1}{\overset{n}{\&}}(x_j\ in\ S(f_j))\ ) = P(\underset{j=1}{\overset{n}{\&}} (\phi(S(f_j))(x_j) = 1)\ )$$

$$= \underset{\varnothing \neq K \subseteq \{1,..,n\}}{\sum} (-1)^{card(K)+1} r(g)(K;f,x)) , \quad (29)$$

where here

$$r(g)(K,f,x) =(def)\ 1-g(1-f_j(x_j);\ all\ j\ in\ K) = g^*(f_j(x_j);all\ j\ in\ K), \quad (30)$$

g* being the (DeMorgan) co-copula corresponding to g. In particular, for n=2,

$$q(f_1(x_1),f_2(x_2);g) = f_1(x_1) + f_2(x_2) - g^*(f_1(x_1),f_2(x_2)). \quad (31)$$

Proof : The proof is self evident by use of Lemma 1, with $V_j = \phi(S(f_j))(x_j)$. ◆

Corollary 1 shows that $q(a,b,g)(x_{j+1},y_{j+1})$ in (eq.24) can be relatively easily evaluated, as well as its n-argument generalization. Most importantly, note that the definition in eq.(22) , via eq.(23), *does not depend upon any particular choice of one-point equivalent random set S(a) to a and S(b) to b*. Furthermore, relative to the remarks following eq.(21), all of the desired special case reductions to membership functions of conditonal ordinary events and to ordinary (i.e., unconditional) fuzzy sets, hold. Next, consider logical combinations of conditional fuzzy sets. In particular, consider the conjunction of (a|b) and (c|d), for a,b,c,d arbitrary fuzzy sets with the same notation as as in Diagram 1, but for simplicity, now assuming without loss of generality that all fuzzy sets here have the same domain D. Choosing any appropriate copula g to determine the jointness of the random sets involed, let, for any $x_0 = (x_1,x_2,...)$ in $D_0 = D \times D \times...,$

$$(32)$$

$$((a|b)\&(c|d))(x_0) =(def)\ P_0(x_0\ in\ ((S(a)\ |\ (S(b))\&(S(c)|S(d))) = P_0(x_0\ in\ [\alpha\ |\ S(b)\cup S(d)]),$$

using the same result for conjunctions of conditionals for PS as in eq.(11), where now

$$\alpha =(def)\ ((S(a)\cap S(b)\cap S(c)\cap S(d)) \times D_0) \cup ((S(a) \cap S(b) \cap S(d)') \times (S(c)|S(d))\ ) \cup$$
$$((S(c)\cap S(d)\cap S(b)') \times (S(a)|S(b))). \quad (33)$$

Hence, analogous to the evaluation in eq.(23), we have

$$((a|b)\&(c|d))(x_0) = \sum_{j=0}^{+\infty} \prod_{i=1}^{j} (1-s(b,d;g)(x_i)) \bullet q(a,b,c,d;g)(x_{j+1}) , \tag{34}$$

where from eq.(21),

$$\tag{35}$$

$$s(b,d;g)(x_i) =(def) \, P(x_i \text{ in } S(b) \cup S(d)) = b(x_i)+d(x_i)-q(b,d;g)(b(x_i),d(x_i))=g^*(b(x_i),d(x_i)),$$

and $q(a,b,c,d;g)(x_{j+1})$ is readily obtainable from eq.(29) (n=4, a=f_1,b=f_2,c=f_3,d=f_4).

Finally, applying eq.(26) to this case, we obtain the closure form

$$(a|b)\&(c|d) = (q(a,b,c,d;g) \mid g^*(b,d)) . \tag{36}$$

A similar argument applies to three or more conjunctions, analogous to the computations for PS conjunctions of conditional events discussed in the last section. Negations when handled analogous to eq.(32) - in the form of the probability of a point being covered by the negation of a conditional event - clearly produce the simple result that

$$(a|b)' = 1- (a|b) = (1-a \mid b). \tag{37}$$

Disjunctions, similarly defined, because PS is boolean, yield the DeMorgan transform of conjunctions in eq.(36):

$$(a|b) \vee (c|d) = 1- (q(1-a,b,1-c,d;g) \mid g^*(b,d)) = (1-q(1-a,b,1-c,d;g) \mid g^*(b,d)). \tag{38}$$

Thus, we see, that the resulting calculus of operations, though complicated, is not really much more involved than that for PS logical combining of conditional events. It should also be remarked that the logical combination of fuzzy conditional events with crisp unconditional events or crisp conditional events procedes just as for the more general logical combination of fuzzy conditional events, but with possible drastic simplifications, because of the consistency and reduction properties mentioned after eq.(31). For example, note that when $(a|b)$ represents an ordinary set, say Q, so that a = $\phi(Q)$, b=1, then eq.(36) simplifies, because now for all i,j

$$q(a,b,c,d;g)(x_{j+1}) = \begin{cases} q(c,d;g)(x_{j+1}), & \text{if } x_{j+1} \text{ in } Q \text{ ,} \\ \\ 0 & \text{, if } x_{j+1} \text{ not in } Q, \end{cases} \tag{39}$$

$$s(b,d:g)(x_i) \quad = \quad 1, \tag{40}$$

whence eq.(34) reduces to

$$(a|b) \& (c|d)(x_0) = q(c,d;g)(x_1). \tag{41}$$

Also, in view of the infinite series appearing in the basic conditional fuzzy set forms, it is of some interest, for evaluation purposes to restrict oneself to certain simplified values. In particular, note for the equal-argument case where for eq.(23), x_j =x, y_j = y, for all j, then the conditional fuzzy event evaluates to

$$(a|b)(z_0) = \sum_{j=0}^{+\infty} (1-b(y))^j \cdot q(a,b;g)(x,y) = q(a,b;g)(x,y) / b(y) , \qquad (42)$$

provided $b(y) > 0$. When also g is modular, such as $g = \min$, or prod, eq.(31) shows that $q = g$. These restricted forms were proposed by Goodman [9], among others, originally for the definition of fuzzy conditional events.

Lastly, note that a natural probability evaluation of any conditional fuzzy set is readily obtained by extending the well-known evaluation of any unconditional fuzzy sets such as $a:D_a \to [0,1]$ used above, as an expectation $E(a(X))$, where X is any random variable over D_a, noting if $a = f(Q)$, for Q crisp, then the standard result obtains

$$E(a(X)) = P(X \text{ in } Q). \qquad (43)$$

The basic extension of this to any nontrivial conditional fuzzy event $(a|b)$, for the simplified case as before where a,b both have common domain D in light of the definition in eq.(22), yields as a probability evaluation, $E((a|b)(X_0))$, where $X_0 = (X_j)_{j=1,2,..}$ is a stochastic process over D_0. Thus, eq.(23) yields

$$E((a|b)(X_0)) = \sum_{j=0}^{+\infty} E \left(\prod_{i=1}^{j} (1-b(X_i)) \cdot q(a,b;g)(X_{j+1}) \right) , \qquad (44)$$

but, unless independence and/or equal antecedent assumptions are made, eq.(4.4) will not simplify. The issue of probability assignment still remains an open one.

4. References

1. Goodman, I.R., Nguyen, H.T. & Walker, E.A., *Conditional Inference and Logic for Intelligent Systems: A Theory of Measure-Free Conditioning*, North-Holland, Amsterdam, 1991.

2. Goodman, I.R., "Toward a comprehensive theory of linguistic and probabilistic evidence:two new approaches to conditional event algebra", IEEE Trans. Systems, Man & Cybernetics (special issue on conditionals) 24(12), Dec., 1994, 1685-1698.

3. Van Fraasen, B., "Probabilities of Conditionals", in book *Foundations of Probability Theory, Statistical Inference, and Statistical Theories of Science* , 1(W.L. Harper & C.A. Hooker, eds.), D.Reidel, Dordrecht, Netherlands, 1976, pp. 261-308.

4. Calabrese, P.G., "A theory of conditional information with applications", *IEEE Transactions on Systems, Man & Cybernetics* (special issue on conditionals), 24(12), Dec., 1994, 1676-1684.

5. B. Schweizer, B. & Sklar, A., *Probabilistic Metric Spaces*, North-Holland, New York, 1983.

6. Goodman, I.R. & Nguyen, H.T., *Uncertainty Models for Knowledge-Based Systems*, North-Holland, New York, 1985.

7. Goodman, I.R., "A new characterization of fuzzy logic operators producing homomorphic-like relations with one-point coverages of random sets", in book *Advances in Fuzzy Theory & Technology*, 2 (P.P. Wang, ed.), Bookwright Press, Durham, NC., pp. 133-159.

8. Goodman, I.R., "Some new results concerning random sets and fuzzy sets", *Information Sciences*, 34, Nov., 1984, 93-113.

9. Goodman, I.R., "Algebraic and probabilistic bases for fuzzy sets and the development of fuzzy conditioning", in book *Conditional Logic in Expert Systems* (I.R. Goodman, M.M. Gupta, H.T. Nguyen & G.S. Rogers, eds.), North-Holland, Amsterdam, Netherlands, 1991, pp. 1-69.

10. Goodman, I.R., "Applications of product space algebra of conditional events and one- point random set representations of fuzzy sets to the development of conditional fuzzy sets", *Fuzzy Sets & Systems* (special issue, D. Ralescu, ed.), to appear.

L-FUZZY ULTRA-COMPACTIFICATION

RALPH C. STEINLAGE
Department of Mathematics
University of Dayton
Dayton, Ohio USA 45469-2316

1. Introduction

In this paper we construct for an L-fuzzy topological space (X, Δ) an extension which is ultra-compact. The technique used is a generalization of that used by Lowen, Steinlage, and Wuyts in [9] which in turn had its roots in Shanin's generalization of Wallman's construction as given in [14]. The construction of the extension is analogous to Shanin's generalization of Wallman's compactification for topological spaces in that a basis for the closed sets in the extension is explicitly defined by extending the closed sets in the underlying space X. In fact, if (X, Δ) is topologically generated, our construction coincides with Wallman's compactification. Thus the construction presented here is a good extension of Wallman's compactification to the fuzzy setting and even to the L-fuzzy setting. The construction given here applies to all L-fuzzy topological spaces. No separation properties are required and (X, Δ) need not be fully stratified. Thus our construction has advantages over previous fuzzy compactifications which apply to a smaller class of fuzzy spaces and sometimes only to topologically generated spaces.

2. Preliminaries

We assume familiarity with most of the concepts and notations in the realm of fuzzy topology. Nevertheless, we recall here some of those which are particularly needed in this paper. We shall work with L-valued fuzzy sets [4] where L is a *fuzzy lattice*: a complete, completely distributive lattice with an order reversing involution $a \rightarrow a^c$, and with least and greatest elements denoted 0 and 1 respectively. That is, a *fuzzy subset* of X is a function $\mu : X \rightarrow L$ where L is a fuzzy lattice. A collection Δ of fuzzy subsets of X is called an *L-fuzzy topology* (or simply a *fuzzy topology*) on X provided

Z. Bien and K. C. Min (eds.),
Fuzzy Logic and its Applications, Information Sciences, and Intelligent Systems, 297–304.
© 1995 *Kluwer Academic Publishers.*

a) $\bar{r} \in \Delta$ for $r = 0$ and $r = 1$ where $\bar{r}$ denotes the constant function with value r

b) if $\mu \in \Delta$ and $\nu \in \Delta$, then $\mu \wedge \nu \in \Delta$

c) if $\mu \in \Delta$ for all $\mu \in \mathfrak{G} \subset \Delta$, then $\vee \{\mu : \mu \in \mathfrak{G}\} \in \Delta$.

Note that we do not require an L-fuzzy topology to be fully stratified in the sense of Lowen and Wuyts [10]; i.e., we do not require that $\bar{r} \in \Delta$ for all $r \in L$. If Δ is an L-fuzzy topology on X then (X, Δ) is called an *L-fuzzy topological space* (L-fts). If Γ is a collection of fuzzy sets we denote by Γ^c the collection of all its *pseudo-complements*; i.e., $\Gamma^c = \{\mu^c : \mu \in \Gamma\}$.

An L-fuzzy topological space (Y, Γ) in which the L-fuzzy topological space (X, Δ) is embedded is called an *extension* of (X, Δ). A subset X of an L-fuzzy topological space (Y, Γ) is said to be *dense* in Y provided $\inf\{\mu(x) : x \in X\} = \inf\{\mu(x) : x \in Y\}$ for each $\mu \in \Gamma^c$. An L-fuzzy topological space (Y, Γ) is called a *compactification* of (X, Δ) if X is densely embedded in Y and Y is compact. For basic results on compactification in topological (as opposed to fuzzy topological) spaces we refer the reader to [13] and [14]. For discussions of compactness notions in fuzzy topological spaces, we refer the reader to [1], [2], [6], [7], [8], [9], [11], [15], and [17].

3. Ultra-Compactification of L-Fuzzy Topological Spaces

3.1. THE INITIAL TOPOLOGY FOR AN L-FUZZY TOPOLOGICAL SPACE

Definition: An element $p \in L$ is called

i) *prime* provided $p \geq a \wedge b$ iff $p \geq a$ or $p \geq b$

ii) *co-prime* provided $p \leq a \vee b$ iff $p \leq a$ or $p \leq b$.

Note that 1 and 0 are trivially prime and co-prime respectively and that p is prime iff p^c is co-prime.

Definition: The *initial topology* for an L-fuzzy topological space (X, Δ) is the ordinary topology $\iota(\Delta)$ on X generated by using

$$\mathfrak{D} := \{\mu^{-1}([\alpha, 1]) : \mu \in \Delta^c, \alpha \text{ co-prime}, \alpha \in L\} \cup \{\emptyset\}$$

as a subbasis for the collection of closed sets.

Note that $\mathfrak{D}$ contains X $(= \mu^{-1}[0, 1]$ for any $\mu)$.

Definition: The L-fuzzy topological space (X, Δ) is said to be *ultra-compact* provided $(X, \iota(\Delta))$ is compact.

3.2. CONSTRUCTING THE "NEW POINTS"

In this section we construct the pool of *new points* which will be appended or adjoined to an L-fuzzy topological space (X, Δ) in order to produce an ultra-compactification of that fuzzy topological space. Since 𝔇 is a subbasis for the closed sets in $\iota(\Delta)$, the collection $\mathfrak{D}(\Delta)$ of all finite unions of finite intersections of sets in 𝔇 is a basis for the closed sets of $\iota(\Delta)$ and is itself closed under the operations of taking finite unions and finite intersections. A subset $\mathfrak{H}$ of $\mathfrak{D}(\Delta)$ is called a $\mathfrak{D}(\Delta)$-*FIP-family* if it has the finite intersection property (FIP); it is called a *maximal* $\mathfrak{D}(\Delta)$-FIP-family if it is not properly contained in any other $\mathfrak{D}(\Delta)$-FIP-family; and it is called *vanishing* if it has an empty intersection. In [13] it was shown that a maximal $\mathfrak{D}(\Delta)$-FIP-family $\mathfrak{M}$ fulfills the *prime-property*; i.e., if $\mathfrak{M}$ is a maximal $\mathfrak{D}(\Delta)$-FIP-family and if $A \cup B \in \mathfrak{M}$ with A,B $\in \mathfrak{D}(\Delta)$ then either $A \in \mathfrak{M}$ or $B \in \mathfrak{M}$. Given an L-fuzzy topological space (X, Δ) and $\mathfrak{D}(\Delta)$ constructed as above we let V(X) denote the collection of all vanishing maximal $\mathfrak{D}(\Delta)$-FIP-families. V(X) is the pool of "new points" we wanted to describe in this section. Adjoining these new points to X, we obtain $X^* := X \cup V(X)$.

3.3. EXTENDING THE FUZZY TOPOLOGICAL STRUCTURE TO X^*

To extend the fuzzy topological structure of X to X^* we extend the closed fuzzy sets in X to X^* and then use these as a basis for the closed fuzzy sets in X^*; pseudo-complements of the closed fuzzy sets in X^* then form the open fuzzy sets in X^*. Given $\mu \in \Delta^C$ we define $\mu^*(x) = \mu(x)$ for all x $\in$ X; if $\mathfrak{M} \in X^* - X$, then $\mathfrak{M} \in V(X)$ and we define

$$\mu^*(\mathfrak{M}) = \sup\{\alpha \in L : \mu^{-1}[\alpha, 1] \in \mathfrak{M}\}.$$

We shall have occasion to use several other characterizations of $\mu^*(\mathfrak{M})$. Recall that
$$x \ll y \qquad \text{(read x is "way below" y)}$$
means that
for every directed set $D \subset L$, $y \leq \sup D \to x \leq z$ for some $z \in D$

([3] and [5]). We then observe that
   i)   every element of L is a sup of co-primes
and that
   ii) $\alpha = \sup\{\gamma : \gamma \ll \alpha\}$ for every $\alpha \in L$
([3], pp. 41, 72). In particular, $\gamma \ll \alpha$ implies $\gamma \leq \alpha$. It follows that
   iii)   every element is the sup of co-primes which are "way below" that element
and we can then establish the following alternate characterizations of $\mu^*(\mathfrak{M})$.

Lemma 1.

$$\mu^*(\mathfrak{M}) = \sup\{\alpha \in L : \mu^{-1}[\alpha, 1] \in \mathfrak{M}\}$$
$$\mu_*(\mathfrak{M}) = \sup\{\alpha \in L : \alpha \text{ co-prime and } \mu^{-1}[\alpha, 1] \in \mathfrak{M}\}$$
$$\mu^*(\mathfrak{M}) = \sup\{\alpha \in L : \alpha \ll \mu^*(\mathfrak{M})\}$$
$$\mu_*(\mathfrak{M}) = \sup\{\alpha \in L : \alpha \text{ co-prime and } \alpha \ll \mu^*(\mathfrak{M})\}$$

Lemma 2. For a fixed $\mathfrak{M} \in \mathfrak{D}(\Delta)$, the set $\{\alpha : \mu^{-1}[\alpha, 1] \in \mathfrak{M}\}$ is directed. Furthermore,

$$\mu^{-1}([\beta, 1]) \in \mathfrak{M} \text{ for all co-primes } \beta \ll \mu^*(\mathfrak{M}).$$

Proof. For any $\alpha, \beta \in L$ and $x \in X$, we note that $\mu(x) \geq \alpha \vee \beta$ if and only if $\mu(x) \geq \alpha$ and $\mu(x) \geq \alpha \vee \beta$. Thus

$$\mu^{-1}([\alpha \vee \beta, 1]) = \mu^{-1}([\alpha, 1]) \cap \mu^{-1}([\beta, 1]).$$

Since $\mathfrak{M}$ satisfies the FIP and is maximal in $\mathfrak{D}(\Delta)$ which is closed under finite intersections, this establishes that the set $\{\alpha : \mu^{-1}[\alpha, 1] \in \mathfrak{M}\}$ is directed (upward).

If $\beta \ll \mu^*(\mathfrak{M}) = \vee\{\alpha : \mu^{-1}[\alpha, 1] \in \mathfrak{M}\}$, then since this latter set is directed, it follows that $\beta \leq$ some α for which $\mu^{-1}[\alpha, 1] \in \mathfrak{M}$. Thus for this α, $\mu^{-1}[\beta, 1] \supset \mu^{-1}[\alpha, 1] \in \mathfrak{M}$. If β is a co-prime, it then follows that $\mu^{-1}[\beta, 1] \in \mathfrak{M}$ since $\mathfrak{M}$ is maximal.

Proposition For $\mu, \nu \in \Delta^c$ the following properties hold:

(1) $\bar{0}^* = \bar{0}$ and $\bar{1}^* = \bar{1}$

(2) $(\mu \vee \nu)^* = \mu^* \vee \nu^*$

(3) $(\mu \wedge \nu)^* = \mu^* \wedge \nu^*.$

Proof. Regarding (1), we note that $\mu^{-1}[\alpha, 1] \in \mathfrak{M}$ only for $\alpha = 0$ when $\mu = \bar{0}$ and that $\mu^{-1}[1, 1] = X$ when $\mu \equiv 1$. Clearly $X \in \mathfrak{M}$ since $\mathfrak{M}$ is maximal.

For (2) and (3) we first remark that if $\xi, \theta \in \Delta^c$ satisfy $\theta \leq \xi$, then $\theta^{-1}[\alpha, 1] \subset \xi^{-1}[\alpha, 1]$ so that if $\theta^{-1}[\alpha, 1] \in \mathfrak{M}$, then $\xi^{-1}[\alpha, 1] \in \mathfrak{M}$ by the maximality of $\mathfrak{M}$. Thus $\theta^* \leq \xi^*$ so we already have $\mu^* \vee \nu^* \leq (\mu \vee \nu)^*$ and $(\mu \wedge \nu)^* \leq \mu^* \wedge \nu^*$.

Next observe that if α is co-prime, then $\mu(x) \vee \nu(x) \geq \alpha$ if and only if $\mu(x) \geq \alpha$ or $\nu(x) \geq \alpha$. Thus, for α co-prime, $(\mu \vee \nu)^{-1}([\alpha, 1]) = \mu^{-1}([\alpha, 1]) \cup \nu^{-1}([\alpha, 1])$. Now let $\mathfrak{M} \in (X^* - X)$, let $\mu, \nu \in \Delta^c$, and let α be a co-prime such that $(\mu \vee \nu)^{-1}([\alpha, 1]) \in \mathfrak{M}$. Then, since $\mathfrak{M}$ has the prime property and $(\mu \vee \nu)^{-1}([\alpha, 1]) = \mu^{-1}([\alpha, 1]) \cup \nu^{-1}([\alpha, 1])$, either $\mu^{-1}([\alpha, 1]) \in \mathfrak{M}$ or

$$(\mu \vee \nu)^*(\mathfrak{M}) = \sup\{\alpha \text{ co-prime} : (\mu \vee \nu)^{-1}([\alpha, 1]) \in \mathfrak{M}\}$$

$$\leq (\sup\{\alpha \text{ co-prime} : \mu^{-1}([\alpha, 1]) \in \mathfrak{M}\})$$

$$\vee (\sup\{\alpha \text{ co-prime} : \nu^{-1}([\alpha, 1]) \in \mathfrak{M}\})$$

$$= \mu^*(\mathfrak{M}) \vee \nu^*(\mathfrak{M}).$$

This establishes (2).

Analogously if α, β, and γ are co-primes such that $\gamma \leq \alpha \wedge \beta$, $\mu^{-1}([\alpha, 1]) \in \mathfrak{M}$, and $\nu^{-1}([\beta, 1]) \in \mathfrak{M}$, then by the maximality of $\mathfrak{M}$ and the fact that $\mu^{-1}([\alpha, 1]) \cap \nu^{-1}([\beta, 1]) \subset (\mu \wedge \nu)^{-1}([\gamma, 1])$, we have $(\mu \wedge \nu)^{-1}([\gamma, 1]) \in \mathfrak{M}$. Thus

$$\mu^*(\mathfrak{M}) \wedge \nu^*(\mathfrak{M}) = (\sup\{\alpha \text{ co-prime} : \mu^{-1}([\alpha, 1]) \in \mathfrak{M}\})$$

$$\wedge (\sup\{\beta \text{ co-prime} : \nu^{-1}([\beta, 1]) \in \mathfrak{M}\})$$

$$= \sup\{\alpha \wedge \beta : \alpha, \beta \text{ both co-prime}, \mu^{-1}([\alpha, 1]) \in \mathfrak{M},$$

$$\text{and } \nu^{-1}([\beta, 1]) \in \mathfrak{M}\}$$

$$\leq \sup\{\gamma \in I_1 : (\mu \wedge \nu)^{-1}([\gamma, 1]) \in \mathfrak{M}\}$$

$$\text{(since each } \alpha \wedge \beta \text{ is a sup of co-primes)}$$

$$= (\mu \wedge \nu)^*(\mathfrak{M}).$$

This establishes (3).

As a consequence of this proposition the family $B^* = \{\mu^* : \mu \in \Delta^c\}$ is a basis for the closed fuzzy sets of some fuzzy topology on X^*, which we shall denote Δ^*.

3.4. ULTRA-COMPACTNESS

Theorem (X^*, Δ^*) is an ultra-compactification of (X, Δ).

Proof. Ultracompactness of X^* means that $\iota(\Delta^*)$ is compact as an ordinary topology. We shall use Alexander's Subbase Lemma with the subbasis

$$\mathfrak{D}^* := \{\mu^{*-1}[\alpha, 1] : \mu \in \Delta^c, \alpha \text{ co-prime}, \alpha \in L\} \cup \{\emptyset\}$$

for the closed sets in $\iota(\Delta^*)$. Using Lemmas 1 and 2, we obtain

$$\mu^{*-1}[\alpha, 1] = \mu^{-1}[\alpha, 1] \cup \{\mathfrak{M} \in V(X) : \mu^{-1}[\beta, 1] \in \mathfrak{M} \text{ for all co-primes } \beta \ll \alpha\}$$

for all $\mu \in \Delta^c$ and $\alpha \in L$. Let $(F_j^*)_{j \in J}$ be a subfamily of $\mathfrak{D}^*$ with $\bigcap_{j \in J} F_j^* = \emptyset$ and write $F_j^* := \mu_j^{*-1}[\alpha_j, 1]$. Then

$$F_j^* = P_j \cup Q_j^*$$

where $\qquad P_j \;\; = \;\; \mu_j^{-1}([\alpha_j, 1]) \qquad \subset X$

and $\qquad Q_j^* = \{\mathfrak{M} : \mu_j^{-1}[\beta, 1] \in \mathfrak{M} \; \forall \text{ co-primes } \beta \ll \alpha_j\} \;\; \subset V(X).$

Therefore, for all $K \subset J$, we have

$$\bigcap_{j \in K} F_j^* = (\bigcap_{j \in K} P_j) \cup (\bigcap_{j \in K} Q_j^*)$$

and in particular

$(*)$ $\qquad\qquad\qquad\qquad \bigcap_{j \in J} P_j = \bigcap_{j \in J} Q_j^* = \emptyset$

Now we consider three cases:

(1°) There is some finite collection $K \subset J$ with $\bigcap_{j \in K} P_j = \emptyset$ and there is some finite collection $L \subset J$ with $\bigcap_{j \in L} Q^*{}_j = \emptyset$.

Then $K \bigcup L \subset J$, $K \cup L$ is finite and $\bigcap_{j \in K \cup L} F_j^* = \emptyset.$

(2°) For every finite $K \subset J$, $\bigcap_{j \in K} P_j \neq \emptyset.$

This means that $(P_j)_{j \in J}$ is a $\mathfrak{D}^*(\Delta)$-FIP-family which thus is contained in a maximal $\mathfrak{D}^*(\Delta)$-FIP-family $\mathfrak{M}$ and which by $(*)$ is vanishing. For each $j \in J$ we now have

$$\mu_j^{-1}([\alpha_j, 1]) = P_j \in \mathfrak{M}$$

so that $\mu_j^*(\mathfrak{M}) \geq \alpha_j$ for each $j \in J$. This implies that $\mathfrak{M} \in \bigcap_{j \in J} Q_j^*$ which by $(*)$ is a contradiction.

(3°) For every finite $L \subset J$, $\bigcap_{j \in L} Q_j^* \neq \emptyset.$

Then if L is finite, $L \subset J$ and $\mathfrak{M} \in \bigcap_{j \in L} Q_j^*$ and if for each $j \in J$ we take a co-prime $\beta_j \ll \alpha_j$ it follows again from Lemma 2 that $\mu_j^{-1}([\beta_j, 1]) \in \mathfrak{M}$ for all $j \in L$ so that

$$\bigcap_{j \in L} \{\mu_j^{-1}([\beta_j, 1])\} \neq \emptyset$$

since $\mathfrak{M}$ satisfies the FIP. Since this holds for all finite $L \subset J$ and all choices of co-primes $\beta_j \ll \alpha_j$ it follows that the entire family

$$\{\mu_j^{-1}([\beta, 1]) : j \in J, \beta \text{ co-prime}, \beta \ll \alpha_j\}$$

is a $\mathfrak{D}^*(\Delta)$-FIP-family and therefore is contained in some maximal $\mathfrak{D}^*(\Delta)$-FIP-family $\mathfrak{M}^*$.

Now if $\mathfrak{M}^*$ is non-vanishing, there exists $x \in X$ such that $\mu_j(x) \geq \beta$ for all $j \in J$ and all co-primes $\beta \ll \alpha_j$; but then $\mu_j(x) \geq \alpha_j$ for each $j \in J$ and thus $x \in \underset{j \in J}{\cap} P_j$ which is a contradiction of $(*)$.

On the other hand, if $\mathfrak{M}^*$ is vanishing then it is a "new point" and $\mu_j^{-1}([\beta, 1]) \in \mathfrak{M}^*$ for all co-primes $\beta \ll \alpha_j$ and all $j \in J$. It then follows from Lemma 1 that $\mu_j(\mathfrak{M}^*) \geq \alpha_j$ so that $\mathfrak{M}^* \in Q_j^*$ for all $j \in J$ which is also a contradiction of $(*)$. This shows that of the three possible cases only the first one can occur. Alexander's Subbase Lemma then indicates that $\iota(\Delta^*)$ is compact.

3.5. DENSENESS

To establish that (X, Δ) is dense in (X^*, Δ^*), it suffices to show for each $\mu \in \Delta^c$ that

$$\wedge \{ \mu^*(x^*) : x^* \in X^*\} = \wedge \{\mu(x) : x \in X\}$$

since the collection $\{\mu^* : \mu \in \Delta^c\}$ forms a basis for the L-closed fuzzy sets in X^*.

Clearly,

$$\wedge \{ \mu^*(x^*) : x^* \in X^*\} = (\wedge \{\mu(x) : x \in X\}) \wedge (\wedge \{\mu^*(\mathfrak{M}) : \mathfrak{M} \in V(X)\})$$

so it suffices to show for each $\mu \in \Delta^c$ that

$$\wedge \{\mu^*(\mathfrak{M}) : \mathfrak{M} \in V(X)\} \geq \wedge \{\mu(x) : x \in X\}.$$

This can be accomplished by showing, for a fixed $\mu \in \Delta^c$, that $\mu^*(\mathfrak{M}) \geq \wedge \{\mu(x) : x \in X\}$ for each $\mathfrak{M} \in V(X)$. Let $\beta = \wedge \{\mu(x) : x \in X\}$. Then $\mu^{-1}[\beta, 1] = X \in \mathfrak{M}$ (by maximality of $\mathfrak{M}$) and

$$\mu^*(\mathfrak{M}) = \sup\{\alpha \in L : \mu^{-1}[\alpha, 1] \in \mathfrak{M}\} \geq \beta = \wedge \{\mu(x) : x \in X\}.$$

This completes the proof that (X, Δ) is dense in (X^*, Δ^*) so that (X^*, Δ^*) is indeed an ultra-compactification of (X, Δ).

4. Relation to Topological Wallman Compactification

The construction given here coincides with the Wallman ultra-compactification given by Lowen, Steinlage, and Wuyts [9] when L = [0, 1] and thus coincides with the usual Wallman compactification when X is a topological space. Because of the relationship of this compactification to the Wallman compactification for topological spaces and for [0, 1]-fuzzy topological spaces, we shall call it the *Wallman ultra-compactification* of (X, Δ).

REFERENCES

1.   Cerruti, U. (1981) The Stone-Cech compactification in the category of fuzzy topological spaces, *J. Fuzzy Sets and Systems* **6**, 197-204.

2.   Gantner, T.E., Steinlage, R.C., and Warren, R.H. (1978) Compactness in Fuzzy Topological Spaces, *J. Math. Anal. Appl.* **62**, 547-562.

3.   Gierz, G., et al. (1980) *A Compendium of Continuous Lattices*, Springer-Verlag, Berlin-Heidelberg-New York.

4.   Goguen, J.A. (1967) L-Fuzzy Sets, *J. Math. Anal. Appl.* **18**, 145-174.

5.   Johnstone, P.T. (1982) *Stone Spaces*, Cambridge University Press.

6.   Liu Ying-Ming and Luo Maokang (1986) Fuzzy Stone-Cech type compactifications, *Proc. Polish Symp. Interval & Fuzzy Mathematics*, 117-137.

7.   Lowen, R. (1978) A Comparison of Different Compactness Notions in Fuzzy Topological Spaces, *J. Math. Anal. Appl.* **64**, 446-454.

8.   Lowen, R. (1976) Fuzzy Topological Spaces and Fuzzy Compactness, *J. Math. Anal. Appl.* **56**, 621-633.

9.   Lowen, R., Steinlage, R.C., and Wuyts, P. (1992) Wallman Compactification in FTS, *Rocky Mountain J. Math.* **22**, 1435-1446.

10.   Lowen, R. and Wuyts, P. (1988) Concerning the Constants in Fuzzy Topology, *J. Math. Anal. Appl.* **129**, 256-268.

11.   Martin, H.W. (1980) A Stone-Cech ultrafuzzy compactification, *J. Math. Anal. Appl.* **73**, 453-456.

12.   McLean, R.G. and Warner, M.W. (to appear) Locale Theory and Fuzzy Topology.

13.   Nagata, J.I. (1968) *Modern General Topology*, North Holland.

14.   Shanin, N.A. (1943) On the theory of bicompact extensions of topological spaces, *Dokl. SSSR* **38**, 154-156.

15.   Warner, M.W. (to appear) On Compact Hausdorff L-Fuzzy Spaces.

16.   Wuyts, P. (1988) The R_0-property in fuzzy topological spaces, *Comm. IFSA Math.* **Chapter 2**, 36-40.

17.   Wuyts, P. and Lowen, R. (1983) Separation axioms in fuzzy topological spaces, fuzzy neighborhood spaces and fuzzy uniform spaces, *J. Math. Anal. Appl.* **93**, 27-41.

A CLASS OF WEAKLY INDUCED SPACES AND ITS APPLICATION TO THE THEORY OF EMBEDDING *

YING-MING LIU AND DE-XUE ZHANG

Institute of Mathematics, Sichuan Union University

Chengdu 610064, *P. R. China*

Introduction

Induced spaces play a rather important role in the study of L-fuzzy topology. The reason lies in that, at first, they are generated by topological spaces naturally, hence the rationality of the fuzzifications of the basic topological notions should be examined in these spaces; Secondly, the topologies of these spacees just consist of all the lower semicontinuous functions, so the study of them has a global analysis flavour in itself. Weakly induced spaces are a kind of spaces which are closely related to induced spaces and they have many good properties. The purpose of this paper is to find a class of weakly induced spaces which has close connection with L-fuzzy topological spaces (not only topological spaces). At first, we investigate weakly induced spaces from the point of view of category theory, precisely we prove that the category of weakly

---

* The authors acknowledge the support of NSF of China and the Science Fund of the state Education Commision of China.

Z. Bien and K. C. Min (eds.),
Fuzzy Logic and its Applications, Information Sciences, and Intelligent Systems, 305–313.
© 1995 *Kluwer Academic Publishers.*

induced spaces is a reflective and coreflecfive subcategory of L-fuzzy topological spaces, i. e. , for every L-fuzzy topological space, its reflection and coreflection in weakly induced spaces are constructed. We call the coreflection of an L-fuzzy topological space in weakly induced spaces its associate weakly induced topological space, and we devote section 2 to the study of these spaces. As an application, we obtain a short proof for the embedding theorem for sub-T_0 λ completely regular spaces in [10].

L denotes a completely distributive lattice with an order-reversing involution in this paper, and a topology δ always means an open topology.

1 Weakly induced spaces Associated with an L-fuzzy topological space

In this section, the reflection and coreflection of an L-fuzzy topological space in weakly induced spaces will be given, and some of the basic properties will be discussed.

1. 1 Definition   An L-fuzzy topological space (L^X, δ) is called weakly induced if for each $A \in \delta$, $a \in L$, $X \backslash A^{[a]} \in \delta$, where $A^{[a]} = \{x \in X | A(x) \leqslant a\}$. If δ is moreover stratified (δ contains all the constant mappings) then (L^X, δ) is called induced.

1. 2 Proposition   Let (L^X, δ) be an L-fuzzy topological space, $\beta \subset \delta$ is a subbasis, if for each $a \in L$, $A \in \beta$, $X \backslash A^{[a]} \in \delta$, then (L^X, δ) is weakly induced.

Proof   If β is also a basis, then the conclusion holds trivially, hence what remains is to prove that for each $A, B \in \beta$, $a \in L$, $X \backslash (A \wedge B)^{[a]} \in \delta$.

By complete distributivity of L, a can be represented as the intersection of a collection of prime elements $[1,9]$ $\{a_j\}_{j\in J}$ in L, hence

$$X\backslash(A \wedge B)^{[a]} = X\backslash \bigwedge_{j\in J} (A \wedge B)^{[a_j]}$$

$$= \bigvee_{j\in J} X\backslash(A \wedge B)^{[a_j]}$$

$$= \bigvee_{j\in J} (X\backslash A^{[a_j]}) \vee (X\backslash B^{[a_j]}) \in \delta \qquad \square$$

For every L-fuzzy topological space (L^X,δ), let $\iota_L^*(\delta)$ denote the collection of crisp open sets in δ (trivially $\iota_L^*(\delta)$ is a topology on X), and $\iota_L(\delta)$ denote the topology on X generated by $\{X\backslash U^{[a]}; a \in L, U \in \delta\}$ as a subbasis. $(X,\iota_L(X))$ is called the topological modification of (L^X,δ) in the literature.

1.3 Proposition $[6,9]$ (L^X,δ) is induced if and only if δ consists of all the lower semicontinuous functions from $(X,\iota_L^*(\delta))$ to L, thus δ is generated by $\iota_L^*(\delta)$; and in this case, $\iota_L^*(\delta) = \iota_L(\delta)$.

If (X,J) is atopological space, let $(L^X,\omega_L(J))$ denote the L-fuzzy topological space generated by J, i.e., $\omega_L(J)$ is the collection of the lower semicontinuous functions from (X,J) to L.

Let Top denote the category of topological spaces; L-fts, the category of L-fuzzy topological spaces and WI-fts, the category of weakly induced spaces. Obviously WI-fts is a full subcategory of L-fts, and we write $i: WI$-fts $\rightarrow L$-fts for the inclusion functor.

By proposition 1.3 it can be easily verified that Top is equivalent to the subcategory of L-fts consisting of induced spaces, and ω_L is an isomorphism functor. Moreover we have

1.4 Theorem (see $[8]$ for the case $L = [0,1]$) When restricted to stratified spaces, ι_L is the right adjoint of ω_L, and ι_L^* is the left adjoint of ω_L.

In the following theorem the right and left adjoint of the inclusion functor $i: WI$-fts $\rightarrow L$-fts will be given, thus WI-fts is a reflective and coreflective full subcategory of L-fts.

1.5 **Theorem** The inclusion functor $i: WI$-fts $\to L$-fts has both a right adjoint and a left adjoint, precisely.

(i) For every L-fuzzy topological space (L^X, δ), its WI-fts reflection is given by

$$(L^X, \delta) \xrightarrow{\ id_X\ } (L^X, \lambda(\delta))$$

where $\lambda(\delta) = \{U \in \delta: X \backslash U^{[a]} \in \delta$ for each $a \in L\}$. Obviously $\lambda(\delta)$ is the finest weakly induced topology contained in δ.

(ii) For every L-fuzzy topological space (L^X, δ), its WI-fts coreflection is given by

$$(L^X, \lambda(\delta)) \xrightarrow{\ id_X\ } (L^X, \delta)$$

where $\lambda(\delta)$ is generated as a subbasis by the union of δ and its topological modification $\iota_L(\delta)$. Trivially $\lambda(\delta)$ is the coarsest weakly induced topology containiag δ.

Proof Straightforward. $\square$

For every L-fuzzy topological space (L^X, δ), the weakly induced space $(L^X, \lambda(\delta))$ is called the associate weakly induced space of (L^X, δ), also denoted $\lambda(L^X, \delta)$.

1.6 **Corollary** (i) $\lambda(\prod_{\iota \in T} (L^{X_\iota}, \delta_\iota)) = \prod_{\iota \in T} (L^{X_\iota}, \lambda(\delta_\iota))$.

(ii) For every subset $Y \subset X$, $(L^X, \lambda(\delta))|L^Y = (L^Y, \lambda(\delta|L^Y))$.

2 Associate weakly induced spaces of completely regular spaces

The recent work [10] by Wang and Xu has discussed the refining of the canonical topology on the L-fuzzy unit interval $I(L)$. In fact, the refining is just the associate weakly induced space $\lambda(I(L))$ of $I(L)$. Using $\lambda(I(L))$ as model space, Wang and Xu have established a fuzzy embedding theory in the spirit of [4], which says that every sub-T_0

λ-completely regular space (defined below) is a subspace of some power of $\lambda(I(L))$. Since the idea and the techniques for the proof of this result in [10] is analogous to that in [4], so it is unavoidably lengthy. In this section, we will give a very simple proof of this result as an application of the functor λ in Theorem 1.5.

At first we give a brief description of the L-fuzzy unit interval $I(L)$.

2.1 Definition The L-fuzzy unit interval is defined to be the set $\{\lambda: [0,1] \to L \mid \lambda$ preserves unions$\}$; and the canonical fuzzy topology on $I(L)$ is generated by $\{L_t, R_t \mid t \in [0,1]\}$ as a subbasis, where for each $\mu \in I(L)$,

$$L_t(\mu) = \mu(t); \quad R_t(\mu) = \bigvee_{s>t} \mu'(s)$$

The equivalence of the definition of $I(L)$ given above and those in the literature [2,4,] can be verified easily by the reader, or see [11, 12] for details.

2.2 Lemma The L-fuzzy unit interval $I(L)$ forms a completely distributive lattice under the pointwise order, and the intersection of two elements μ_1 and μ_2 in $I(L)$ is given by $\mu_1 \wedge \mu_2(t) = \bigvee_{s<t} \mu_1(s) \wedge \mu_2(s)$ for each $t \in [0,1]$.

Proof A more general result is proved in [5,11,12]. □

2.3 Lemma $I(L)$ is a fuzzy topological lattice. that is to say, the binary union and intersection operations are both continuous with respect to the canonical topology on $I(L)$.

Proof We just prove the continuity of the binary intersection $\wedge$, the continuity of the binary union can be proved similarly.

Let $P_1, P_2: I(L) \times I(L) \to I(L)$ denote the projections on the first coordinate and the second coordinate respectively.

For $\lambda, \mu \in I(L)$, $t \in [0,1]$,

(i) $\bigwedge^{-1}(L_t)(\lambda,\mu) = L_t(\lambda \wedge \mu) = \lambda \wedge \mu(t)$

$$= \bigvee_{s<t} \lambda(s) \wedge \mu(s) = \bigvee_{s<t} \lambda(s) \wedge \bigvee_{s<t} \mu(s)$$

$$= P_1^{-1}(L_t)(\lambda,\mu) \wedge P_2^{-1}(L_t)(\lambda,\mu)$$

(ii) $\bigwedge^{-1}(R_t)(\lambda,\mu) = R_t(\lambda \wedge \mu) = \bigvee_{s>t} (\lambda \wedge \mu)'(s)$

$$= \bigvee_{s>t} (\bigvee_{r<s} \lambda(r) \wedge \mu(r))'$$

$$= \bigvee_{s>t} (\lambda(s) \wedge \mu(s))'$$

$$= \bigvee_{s>t} \lambda'(s) \vee \bigvee_{s>t} \mu'(s)$$

$$= P_1^{-1}(R_t)(\lambda,\mu) \vee P_2^{-1}(R_t)(\lambda,\mu)$$

hence $\bigwedge$ is continuous. $\square$

Note Lemma 2. 3 is implicit in [4].

2. 4 **Corollary** The associate weakly induced space $\lambda(I(L))$ of $I(L)$ is a fuzzy topological lattice.

Proof An immediate consequence of Corollary 1. 6 and Lemma 2. 3.

2. 5 **Proposition** The topological modification of the canonical topology on $I(L)$ coincides with the interval topology on $I(L)$ regarded as a complete lattice, hence it is compact Hausdorff.

Proof See [11,12]. $\square$

2. 6 **Definition** (L^X, δ) is called completely regular (λ-completely regular resp) if for every $U \in \delta$, there exiists a collection of fuzzy sets $\{W_t | t \in T\}$ such that $\bigvee_{t \in T} W_t = U$ and for each $t \in T$, there exists a continuous mapping $f_t: (L^X, \delta) \to I(L)(\lambda(I(L))$ resp) such that $W_t \leqslant f_t^{-1}(R_0') \leqslant f_t^{-1}(L_1) \leqslant U$.

2. 7 **Proposition** Let (L^X, δ) be an L-fuzzy topological space, $\beta \subset \delta$ a subbasis. If for each $U \in \beta$, there exists a collection of sets $\{W_t | t \in T\}$ with $\bigvee_{t \in T} W_t = U$ and for each $t \in T$, there exists a continuous function

$f_i: (L^X, \delta) \to I(L)(\lambda(I(L)))$ such that $W_i \leqslant f_i^{-1}(R'_0) \leqslant f_i^{-1}(L_1) \leqslant U$,

then (L^X, δ) is completely regular (λ-completely regular).

Proof The reader can deduce the conclusion easily from the fact that $I(L)(\lambda(I(L)))$ is a fuzzy topological lattice, or see [4] for the case of complete regularity. □

If δ is the fuzzy topology on L^X just consisting of all the constant functions, then if can be easily checked that (L^X, δ) is both completely regular and λ-completely regular (for λ-complete regularity, also see Proposition 2.9).

2.8 Corollary (i) Both complete regularity and λ-complete regularity are productive.

(ii) The stratification of a completely regular space (λ-completely regular space) is completely regular (λ-completely regular). □

2.9 Proposition (L^X, δ) is a weakly induced space, then (L^X, δ) is completely regular if and only if it is λ-completely regular.

Proof Sufficiency is trivial.

Necessity follows from the observation that a function $f: (L^X, \delta) \to I(L)$ is continuous if and only if it is continuous with respect to the associate weakly induced topology on $I(L)$, i.e., that of $\lambda(I(L))$. □

2.10 Corollary $\lambda(I(L))$ is λ-completely regular.

Proof By Proposition 2.9 it suffices to prove that $\lambda(I(L))$ is completely regular, and this follows from Proposition 2.7 and the fact that the topology on $\lambda(I(L))$ is generated as a subbasis by the union of the canonical fuzzy topology on $I(L)$ which is completely regular and its topological modification which is compact Hausdorff. □

2.11 Proposition If (L^X, δ) is λ-completely regular, then it is weakly induced.

Proof It suffices to show that for each $U \in \delta$, $a \in L$, $X \backslash U^{[a]} \in \delta$.

By λ-complete regularity of (L^X, δ), there exists a collection of fuzzy sets $\{W_t | t \in T\}$ with $\bigvee_{t \in T} W_t = U$ and for each $t \in T$, there exists a continuous function $f_t : (L^X, \delta) \to \lambda(I(L))$ such that $W_t \leqslant f_t^{-1}(R_0^t) \leqslant f_t^{-1}(L_1) \leqslant U$. Thus $W_t^{[a]} \geqslant (f_t^{-1}(L_1))^{[a]} = f_t^{-1}(L_1^{[a]}) \geqslant U^{[a]}$ for each $t \in T$. By the equality $U = \bigvee_{t \in T} W_t$, we have $\bigwedge_{t \in T} W_t^{[a]} = U^{[a]} = \bigwedge_{t \in T} f_t^{-1}(L_1^{[a]})$, the last term is closed since $\lambda(I(L))$ is weakly induced, hence $X \backslash U^{[a]} \in \delta$. $\qquad \square$

An L-fuzzy topological space (L^X, δ) is called sub-T_0 if for $x \neq y$ in X, there is $U \in \delta$ with $U(x) \neq U(y)$.

2.12 Theorem Every sub-T_0 λ-completely regular space is homeomorphic to a subspace of some power of $\lambda(I(L))$.

Proof Let (L^X, δ) be a sub-T_0 λ-completely regular space, then it is weakly induced and completely regular. Hence it is homeomorphic to a subspace of some power of $I(L)$, that is to say there is an embedding $i :$ $(L^X, \delta) \to I(L)^\Gamma$ for some index set Γ. Then it is easy to see $i : (L^X, \delta) \to$ $\lambda(I(L))^\Gamma$ is also an embedding by corollary 1.6 and the functoriality of λ in theorem 1.5. $\qquad \square$

2.13 Corollary The associate weakly induced space of a sub-T_0 completely regular space is λ-completely regular.

Refreences

[1] G. Gierz, et al., A compendium of continuous lattices, Springer-Verlag, 1980 Berlin.

[2] B. Hutton, Normality in fuzzy topological spaces, J. Math. Anal. Appl., 50(1975)74 — 79

[3] P. T. Johnstone, Stone Spaces, Camb. Univ. Press, 1982.

[4] Liu, Ying-ming, Pointwise characterization of complete regularity in fuzzy topological spaces and the embedding theorem, Scientia

Sinica, Ser. A, 25(1982), 675-82.

[5] Liu Ying-ming, He-Ming, Induced mappings on completely distributive lattices, Proc, 15th Inter Symp. on multple-Valued logic, 1985, 346-353.

[6] Liu, Ying-ming and Luo, Mao-kang, Induced spaces and fuzzy Stone-Cech compatifications, Scientia Sinica, Ser. A, 30(1987), 359-68.

[7] Liu, Ying-ming and Luo, Mao-kang, Lattice-valued Hohn Dieudonne-Tong insertion theorem and stratification structures, Top. Appl. , 45(1992), 173-88.

[8] E. Lowen, et al. , The categorical topology approach to fuzzy topology and fuzzy convergence, Fuzzy Sets and Systems, 40(1991), 347-73.

[9] Wang, Guo-jun, A Theory of L-fuzzy Topological Spaces, Shanxi Normal University Press, Xian, 1988.

[10] Wang, Guo-jun and Xu, Luo-shan, Lawson topology and the refining of the Hutton unit interval, Scientia Sinica, Ser. A, 35(1992), 705-712.

[11] Zhang, De-xue and Liu, Ying-ming, L-fuzzy modification of completely distributive lattices, Math. Nachr. 168(1994), 79-95.

[12] Zhang, De-Xue, The fuzzy unit interval and fuzzy Stone representation theorem. Thesis, Sichuan University, 1993.

STABLE FUZZY COMPACTNESS

Kyung Chan Min and Yong Bae Kim
Department of Mathematics
Yonsei University
Seoul 120-749, Korea

Abstract   Using prefilter we introduce a notion of stable fuzzy compactness. This new notion is characterized in terms of ultra $\bar{\alpha}$ filter which allows us to show the Tychonoff theorem in a usual way and compared with other notions of fuzzy compactness.

1. Introduction

In fuzzy topology, various notions of compactness have been studied using fuzzy-open sets, fuzzy nets, prefilters and functors between fuzzy topological spaces and topological spaces. However, so far the notion of ultra-prefilter is not much utilized in this direction [1,8]. Wang[11] defined the notion of N-compactness in terms of fuzzy α-net, α-filter and remoted neighborhoods and proved a Tychonoff theorem in terms of cluster points. In this paper we introduce a notion of $\bar{\alpha}$-filter and obtain a characterization of ultra $\bar{\alpha}$-filter. Using $\bar{\alpha}$-filters we define a notion of stable fuzzy compactness (S-compact) as a good extension and prove a Tychonoff theorem after characterizing the notion of S-compactness in terms of ultra $\bar{\alpha}$-filters. This characterization of ultra $\bar{\alpha}$-filter will allow us to show the Tychonoff theorem in a usual way. Moreover we compare this new notion with others on fuzzy compactness.

2. $\bar{\alpha}$-filters

In this section, we introduce a notion of $\bar{\alpha}$-filter and obtain characterizations of an ultra $\bar{\alpha}$-filter which is useful to prove a Tychonoff theorem.

Definition 2.1 [11]. Let A be a fuzzy set in X. We call $\sup\limits_{x \in X} A(x)$ the *height of* A and denote it by hgt(A). A prefilter $\mathcal{F}$ on X is called an α-*filter* if $\inf\limits_{F \in \mathcal{F}}$ hgt(F) = $\alpha (0 < \alpha \le 1)$.

Definition 2.2. An α-filter $\mathcal{F}$ on X is called an $\bar{\alpha}$-*filter* if $F_\alpha \ne \phi$ for every $F \in \mathcal{F}$, where F_α is the α-cut of F.

Definition 2.3. An $\bar{\alpha}$-filter $\mathcal{F}$ on X is called an *ultra $\bar{\alpha}$-filter* if there is no strictly finer $\bar{\alpha}$-filter than $\mathcal{F}$.

Z. Bien and K. C. Min (eds.),
Fuzzy Logic and its Applications, Information Sciences, and Intelligent Systems, 315–323.
© 1995 *Kluwer Academic Publishers.*

Definition 2.4. Let $\mathcal{F}$ be a prefilter on X and $S \subseteq X$. S is called α-*included in* $\mathcal{F}$ if every fuzzy set A in X with α-cut S is contained in $\mathcal{F}$.

Theorem 2.5. Let X be a set and $\mathcal{F}$ an $\bar{\alpha}$-filter on X. Then the following are equivalent.

(1) If $S \subseteq X$ then either S or S^c is α-included in $\mathcal{F}$.

(2) $\mathcal{F}$ is an ultra $\bar{\alpha}$-filter.

(3) If $A \in I^X$ and $A \notin \mathcal{F}$ then $A \cap F < \alpha$ for some $F \in \mathcal{F}$.

Proof. (1) $\Rightarrow$ (2) Suppose that $\mathcal{F}$ is not an ultra $\bar{\alpha}$-filter. Then there is an $\bar{\alpha}$-filter $\mathcal{G}$ such that $\mathcal{F} \subseteq \mathcal{G}$. Take $G \in \mathcal{G}$ such that $G \notin \mathcal{F}$ and put $S = G_\alpha$. By the condition, S^c is α-included in $\mathcal{F}$ since S is not α-included in $\mathcal{F}$. Put $B = 1_{S^c}$. Then $B_\alpha = S^c$ and $B \in \mathcal{F} \subseteq \mathcal{G}$. Since $G \in \mathcal{G}$, we have $B \cap G \in \mathcal{G}$. But $B \cap G < \alpha$ and this is impossible since $\mathcal{G}$ is an $\bar{\alpha}$-filter.

(2) $\Rightarrow$ (3) Suppose that $A \notin \mathcal{F}$ and $A \cap F < \alpha$ for any $F \in \mathcal{F}$. Then there is a crisp point $x^F \in X$ such that $(A \cap F)(x^F) \geq \alpha$. Hence the set $B = \{A \cap F | F \in \mathcal{F}\}$ is a prefilter base which generates an $\bar{\alpha}$-filter $\mathcal{G}$. Notice that $\mathcal{F} \subseteq \mathcal{G}$ and $A \in \mathcal{G}$. Therefore $\mathcal{F} \subseteq \mathcal{G}$ which contradicts that $\mathcal{F}$ is an ultra $\bar{\alpha}$-filter.

(3) $\Rightarrow$ (1) Let $S \subseteq X$ such that S and S^c are not α-included in $\mathcal{F}$. Then there are fuzzy sets A and B such that $A_\alpha = S$, $B_\alpha = S^c$ and $A, B \notin \mathcal{F}$. By the condition, $A \cap F_A < \alpha$ and $B \cap F_B < \alpha$ for some $F_A, F_B \in \mathcal{F}$. Hence $F_A \cap F_B < \alpha$ which contradicts that $\mathcal{F}$ is an $\bar{\alpha}$-filter.

Theorem 2.6. Let X and Y be sets and $f : X \to Y$ a map. If $\mathcal{F}$ is an α-filter (resp. $\bar{\alpha}$-filter, ultra $\bar{\alpha}$-filter) on X then $f(\mathcal{F})$ is an α-filter (resp. $\bar{\alpha}$-filter, ultra $\bar{\alpha}$-filter) on Y.

Proof. Recall that $f(\mathcal{F}) = \langle \{f(F) | F \in \mathcal{F}\} \rangle$ is a prefilter on Y.

Step 1: Let $\mathcal{F}$ be an α-filter on X. Notice that

$$\inf_{F' \in f(\mathcal{F})} \mathrm{hgt}(F') = \inf_{F \in \mathcal{F}} \mathrm{hgt}(f(F))$$

and

$$\mathrm{hgt}(f(F)) = \sup_{y \in Y} f(F)(y) = \sup_{y \in Y} \sup_{x \in f^{-1}(y)} F(x)$$
$$= \sup_{x \in X} F(x) = \mathrm{hgt}(F).$$

Hence we have $\inf_{F \in \mathcal{F}} \mathrm{hgt}(f(F)) = \inf_{F \in \mathcal{F}} \mathrm{hgt}(F) = \alpha$, and so $f(\mathcal{F})$ is an α-filter.

Step 2: Let $\mathcal{F}$ be an $\bar{\alpha}$-filter on X. For any $F' \in f(\mathcal{F})$, there is $F \in \mathcal{F}$ such that $f(F) \subseteq F'$. Since $F_\alpha = \phi$, $F(x) \geq \alpha$ for some $x \in X$. Put $y = f(x) \in Y$, then

$$f(F)(y) = \sup_{x' \in f^{-1}(y)} F(x') \geq F(x) \geq \alpha$$

and hence $F'_\alpha \supseteq f(F)_\alpha = \phi$. Therefore $f(\mathcal{F})$ is an $\bar{\alpha}$-filter.

Step 3: Let $\mathcal{F}$ be an ultra $\bar{\alpha}$-filter on X. Let $T \subseteq Y$ and let $S = f^{-1}(Y_0) \subseteq X$. Then either S or S^c is α-included in $\mathcal{F}$ by Theorem 2.5. Suppose S is α-included in $\mathcal{F}$. Let A be a fuzzy set in Y with α-cut T, and put $B = f^{-1}(A) = A \circ f$. Then

$$B_\alpha = B^{-1}[\alpha, 1] = f^{-1}(A^{-1}[\alpha, 1]) = f^{-1}(T) = S$$

hence $B \in \mathcal{F}$ i.e. $f(B) \in f(\mathcal{F})$. Therefore we have $A \in f(\mathcal{F})$ since $f(B) = f(f^{-1}(A)) \subseteq A$. This shows that T is α-included in $f(\mathcal{F})$. On the other hand, suppose S^c is α-included in $\mathcal{F}$, since $S^c = f^{-1}(T^c)$, we can obtain that T^c is α-included in $f(\mathcal{F})$ by the similar arguement to the above procedure. Hence $f(\mathcal{F})$ is an ultra $\bar{\alpha}$-filter on Y by Theorem 2.5 again.

3. Stable fuzzy compactness

A *fuzzy point* $p = x_\lambda$ in a set X is a fuzzy set in X given by $p(x) = \lambda$ $(0 < \lambda < 1)$ and $p(y) = 0$ for $y \neq x$. For a fuzzy set A in X, $x_\lambda \in A$ means $A(x) \geq \lambda$ and $x_\lambda \in_s A$ means $A(x) > \lambda$. In a fuzzy topological space (X, δ), a *neighborhood system* of x_λ is defined by $\mathcal{N}(x_\lambda) = \{V \in I^X | x_\lambda \in_s U \subseteq V$ for some U in $\delta\}$. This neighborhood system characterizes the fuzzy topology in X in exactly same axioms as in topology [7]. A prefilter $\mathcal{F}$ on X is said to be *convergent to* x_λ if $\mathcal{N}(x_\lambda) \subseteq \mathcal{F}$, and denote it by $\mathcal{F} \to x_\lambda$ [8].

Definition 3.1. Let $\mathcal{F}$ be a prefilter on a fuzzy topological space (X, δ), $p = x_\lambda$ a fuzzy point in X and $0 < \alpha < 1$. p is called an α-*cluster point* of $\mathcal{F}$, denoted by $\mathcal{F} \propto_\alpha p$, if for any $U \in \mathcal{N}(p)$ and any $F \in \mathcal{F}$, we have $(U \cap F)_\alpha = \phi$.

Note that if $\mathcal{F}$ has an α-cluster point, then it is an $\bar{\alpha}$-filter. Obviously, an α-cluster point of a prefilter is a cluster point. If an $\bar{\alpha}$-filter $\mathcal{F}$ converges to a fuzzy point x_α then $\mathcal{N}(x_\alpha) \subseteq \mathcal{F}$ and hence x_α is an α-cluster point of $\mathcal{F}$, since $\mathcal{F}$ is an $\bar{\alpha}$-filter.

Now, we can define a notion of stable fuzzy compactness.

Definition 3.2. Let (X, δ) be a fuzzy topological space and A a fuzzy set in X. A is called *stable fuzzy compact* (S-compact) if each $\bar{\alpha}$-filter containing A has an α-cluster point with value α which is contained in A. Specifically, when $A = 1$ is S-compact, we call (X, δ) an *S-compact* fuzzy topological space.

Theorem 3.3. Let (X, δ) and (Y, γ) be fuzzy topological spaces and $f : (X, \delta) \to (Y, \gamma)$ a fuzzy continuous onto map. If (X, δ) is S-compact then so is (Y, γ).

Proof. Let $\mathcal{F}$ be an $\bar{\alpha}$-filter on Y. Then $f^{-1}(\mathcal{F})$ is a prefilter on X since f is onto, and it is easily seen that $f^{-1}(\mathcal{F})$ is an $\bar{\alpha}$-filter on X. Hence $f^{-1}(\mathcal{F})$ has an α-cluster point x_α in X since (X, δ) is S-compact. It is sufficient to show that $\mathcal{F} \propto_\alpha f(x_\alpha)$. Let $U \in \mathcal{N}(f(x_\alpha))$ and $F \in \mathcal{F}$. Then $f^{-1}(U) \in \mathcal{N}(x_\alpha)$ since f is fuzzy continuous. So we have

$$\left(f^{-1}(U \cap F)\right)_\alpha = \left(f^{-1}(U) \cap f^{-1}(F)\right)_\alpha = \phi$$

and this implies that $(U \cap F)_\alpha = \phi$. Hence (Y, γ) is also S-compact.

For any $\bar{\alpha}$-filter $\mathcal{F}$ on X, it is routine to find an ultra $\bar{\alpha}$-filter on X which contains $\mathcal{F}$ using Zorn's lemma. Now we can characterize the S-compactness by a prefilter convergence.

Theorem 3.4. Let (X, δ) be a fuzzy topological space and A a fuzzy set in X. Then A is S-compact if and only if each ultra $\bar{\alpha}$-filter containing A converges to a fuzzy point with value α which is contained in A.

Proof. For sufficiency, let $\mathcal{F}$ be an ultra $\bar{\alpha}$-filter on X and $A \in \mathcal{F}$. Then there is a fuzzy point $x_\alpha \in A$ s.t. $\mathcal{F} \propto_\alpha x_\alpha$. Hence the set $\mathcal{B} = \{U \cap F | U \in \mathcal{N}(x_\alpha), F \in \mathcal{F}\}$ forms a prefilter base which generates an $\bar{\alpha}$-filter $\mathcal{G}$ and $\mathcal{F} \subseteq \mathcal{G}$, $\mathcal{N}(x_\alpha) \subseteq \mathcal{G}$. Since $\mathcal{F}$ is an ultra $\bar{\alpha}$-filter, $\mathcal{F} = \mathcal{G}$ and so $\mathcal{F} \to x_\alpha$.

For necessity, let $\mathcal{F}$ be an $\bar{\alpha}$-filter on X and $A \in \mathcal{F}$. Then there is an ultra $\bar{\alpha}$-filter $\mathcal{G}$ s.t. $\mathcal{F} \subseteq \mathcal{G}$. By condition, there is a fuzzy point $x_\alpha \in A$ s.t. $\mathcal{G} \to x_\alpha$ i.e. $\mathcal{N}(x_\alpha) \subseteq \mathcal{G}$. For any $U \in \mathcal{N}(x_\alpha)$ and any $F \in \mathcal{F}$, since $U, F \in \mathcal{G}$ and $\mathcal{G}$ is an $\bar{\alpha}$-filter, $(U \cap F)_\alpha = \phi$. Therefore $\mathcal{F} \propto_\alpha x_\alpha$, and A is S-compact.

4. Tychonoff theorem

Theorem 4.1. Let (X, δ) and (Y, γ) be fuzzy topological spaces and $f : (X, \delta) \to (Y, \gamma)$ a fuzzy continuous map. Let $\mathcal{F}$ be a prefilter on X. If $\mathcal{F}$ converges to x_α in X, then $f(\mathcal{F})$ converges to $f(x_\alpha)$ in Y.

Proof. Let $U \in \mathcal{N}(f(x_\alpha))$, then $f^{-1}(U) \in \mathcal{N}(x_\alpha) \subseteq \mathcal{F}$. Hence $f(f^{-1}(U)) \in f(\mathcal{F})$ and since $f(f^{-1}(U)) \subseteq U$, we have $U \in f(\mathcal{F})$.

Theorem 4.2. Let $(X_i, \delta_i)_{i \in I}$ be a family of fuzzy topological spaces and let $(X, \delta) = \prod_{i \in I}(X_i, \delta_i)$. Let x_α be a fuzzy point $(0 < \alpha < 1)$ in X. Then

(1) $\mathcal{F} \to x_\alpha$ if and only if $\pi_i(\mathcal{F}) \to \pi_i(x_\alpha)$ for each $i \in I$.
(2) If $\mathcal{F} \propto x_\alpha$ then $\pi_i(\mathcal{F}) \propto \pi_i(x_\alpha)$ for each $i \in I$.
(3) If $\mathcal{F} \propto_\alpha x_\alpha$ then $\pi_i(\mathcal{F}) \propto_\alpha \pi_i(x_\alpha)$ for each $i \in I$.

Proof. (1) Sufficiency: Since each projection map is fuzzy continuous, the result follows immediately by Theorem 4.1.

Necessity: Let $U \in \mathcal{N}(x_\alpha)$ then there are $U_{i_k} \in \delta_{i_k}$, $k = 1, 2, \cdots, n$ such that

$$\pi_{i_1}^{-1}(U_{i_1}) \cap \pi_{i_2}^{-1}(U_{i_2}) \cap \cdots \cap \pi_{i_n}^{-1}(U_{i_n}) \subseteq U$$

and $\pi_{i_k}(x_\alpha) \in_s U_{i_k}$ for each $k = 1, 2, \cdots, n$. Since $\pi_{i_k}(\mathcal{F}) \to \pi_{i_k}(x_\alpha)$, we have $U_{i_k} \in \pi_{i_k}(\mathcal{F})$ and hence $\pi_{i_k}(F_k) \subseteq U_{i_k}$ for some $F_k \in \mathcal{F}$. Therefore $F_k \in \pi_{i_k}^{-1}(U_{i_k})$ for each $k = 1, 2, \cdots, n$, and this implies that

$$F_1 \cap F_2 \cap \cdots \cap F_n \subseteq \pi_{i_1}^{-1}(U_{i_1}) \cap \pi_{i_2}^{-1}(U_{i_2}) \cap \cdots \cap \pi_{i_n}^{-1}(U_{i_n}) \subseteq U$$

hence we obtain $U \in \mathcal{F}$ and conclude that $\mathcal{F} \to x_\alpha$.

(2) Let $\mathcal{F} \propto x_\alpha$. Then we can find a prefilter $\mathcal{G}$ s.t. $\mathcal{F} \subseteq \mathcal{G}$ and $\mathcal{G} \to x_\alpha$ by taking $\mathcal{G} = \langle \{F \cap U | F \in \mathcal{F}, U \in \mathcal{N}(x_\alpha)\} \rangle$. Combining this and (1) gives the result.

(3) Let $\mathcal{F} \propto_\alpha x_\alpha$. Put $\mathcal{G} = \langle \{F \cap U | F \in \mathcal{F}, U \in \mathcal{N}(x_\alpha)\} \rangle$. Then $\mathcal{G}$ is an $\bar{\alpha}$-filter, $\mathcal{F} \subseteq \mathcal{G}$ and $\mathcal{G} \to x_\alpha$. Hence $\pi_i(\mathcal{F}) \subseteq \pi_i(\mathcal{G})$ and and $\pi_i(\mathcal{G}) \to \pi_i(x_\alpha)$ i.e. $\mathcal{N}(\pi_i(x_\alpha)) \in \pi_i(\mathcal{G})$. Since $\pi_i(\mathcal{G})$ is an $\bar{\alpha}$-filter by Theorem 1.6, $\pi_i(\mathcal{F}) \propto_\alpha \pi_i(x_\alpha)$.

Now we prove the Tychonoff theorem for the S-compactness as follows.

Theorem 4.3. (Tychonoff Theorem). Let $(X_i, \delta_i)_{i \in I}$ be a family of fuzzy topological spaces and let $(X, \delta) = \prod_{i \in I}(X_i, \delta_i)$. Then (X, δ) is S-compact if and only if (X_i, δ_i) is S-compact for each $i \in I$.

Proof. The sufficiency is obvious by Theorem 3.3.

For necessity, let $\mathcal{F}$ be an ultra $\bar{\alpha}$-filter on X then $\pi_i(\mathcal{F})$ is an ultra $\bar{\alpha}$-filter on X_i for each $i \in I$ by Theorem 2.6. Since (X_i, δ_i) is S-compact, there is a fuzzy point x_α^i in X_i s.t. $\pi_i(\mathcal{F}) \to x_\alpha^i$ for each $i \in I$. Let x_α be the fuzzy point in X s.t. $\pi_i(x_\alpha) = x_\alpha^i$ for each $i \in I$. Then by Theorem 4.2 (1), we have $\mathcal{F} \to x_\alpha$. Hence (X, δ) is S-compact.

5. Comparison with other notions

In this section we introduce a relationship of the notion of S-compactness with other notions.

Definition 5.1. [9,11] Let X be a set. A *fuzzy net* $S = \{S(n)|n \in D\}$ in X is a function $S : D \to \vartheta$ where D is a directed set with a relation $\leq$ and ϑ the collection of all fuzzy points in X.

Definition 5.2. Let $S = \{S(n)|n \in D\}$ be a fuzzy net in a fuzzy topological space (X, δ) and x_α a fuzzy point in X. A fzzzy net S is said to *converge to* x_α, denoted by $S \to x_\alpha$, if for any $U \in \mathcal{N}(x_\alpha)$, S is eventually in U, i.e. there is $N \in D$ such that $S(n) \in U$ whenever $n \geq N$. A fuzzy net S is said to *have a cluster point* x_α, denoted by $S \propto x_\alpha$, if for any $U \in \mathcal{N}(x_\alpha)$, S is frequently in U, i.e. for any $N \in D$, there is $n \in D$ such that $n \geq N$ and $S(n) \in U$.

Definition 5.3. [11] Let $S = \{S(n)|n \in D\}$ be a fuzzy net in a set X. The crisp net $\mathrm{val}(S) = \{\mathrm{val}(S(n))|n \in D\}$ in $(0, 1)$ is called the *value net* of S, where $\mathrm{val}(S(n))$ is the value of $S(n)$.

If the value net of S converges to α in $(0, 1]$ equipped with the usual topology, then S is called an *α-net*. Specifically, if $\mathrm{val}(S(n)) = \alpha$ for all $n \in D$, S is called a *constant α-net*.

Now we characterize the notion of S-compactness by fuzzy net in order to compare it with other notions of fuzzy compactness.

Theorem 5.4. Let (X, δ) be a fuzzy topological space and A a fuzzy set in X. Then A is S-compact if and only if each constant α-net in A has a cluster point with value α contained in A.

Proof. For sufficiency, let $S = \{S(n) | n \in D\}$ be a constant α-net $(0 < \alpha < 1)$ in A. Put $B_n = \sup\{S(m) | m \in D, m \geq n\}$ for each $n \in D$. Then $\mathcal{B} = \{B_n | n \in D\}$ is a prefilter base and $\mathcal{F} = \langle \mathcal{B} \rangle$ is an $\bar{\alpha}$-filter such that $A \in \mathcal{F}$. Since A is S-compact, there is a fuzzy point $x_\alpha \in A$ such that $\mathcal{F} \propto_\alpha x_\alpha$. For any $U \in \mathcal{N}(x_\alpha)$ and any $n \in D$, $(B_n \cap U)_\alpha \neq \phi$, therefore there is $m \in D$ such that $m \geq n$ and $S(m) \in U$ i.e. $S \propto x_\alpha$.

For necessity, let $\mathcal{F}$ be an $\bar{\alpha}$-filter on X with $A \in \mathcal{F}$. For each $F \in \mathcal{F}$, since $F \cap A \in \mathcal{F}$, we have $(F \cap A)_\alpha \neq \phi$ and there is a crisp point $x^F \in X$ such that $(F \cap A)(x^F) \geq \alpha$ i.e. the fuzzy point $x_\alpha^F \in F \cap A$. Define an order relation $\ll$ on $\mathcal{F}$ by

$$F_1 \ll F_2 \quad \text{iff} \quad F_2 \subseteq F_1$$

then $(\mathcal{F}, \ll)$ is a directed set. Put $S(F) = x_\alpha^F \in F \cap A$ for each $F \in \mathcal{F}$ then $S = \{S(F) | F \in \mathcal{F}\}$ is a constant α-net in A. By the condition, there is a fuzzy point $x_\alpha \in A$ such that $S \propto x_\alpha$. Let $U \in \mathcal{N}(x_\alpha)$ and $F \in \mathcal{F}$ be given. Since $S \propto x_\alpha$, there is $G \in \mathcal{F}$ s.t. $G \gg F$ i.e. $G \subseteq F$ and $S(G) \in U$. Notice that $S(G) \in G \subseteq F$ hence $x_\alpha^G \in F \cap U$ i.e. $(F \cap U)_\alpha \neq \phi$. So $\mathcal{F} \propto_\alpha x_\alpha$ and this means that A is S-compact.

In Theorems 5.1 and 5.6 in [11], it is easy to check that arguments for the case of $\alpha = 1$ is not necessary. Hence we restate these theorems as follows.

In a fuzzy topological space X, a closed fuzzy set P is called a *remoted-neighborhood* [11] of a fuzzy point e if $e \notin P$. Let $\eta(e)$ be the collection of all remoted neighborhoods of a fuzzy point e.

Theorem 5.5. [11] Let (X, δ) be a fuzzy topological space. Then (X, δ) is strong fuzzy compact if and only if for each constant α-net $S = \{S(n) | n \in D\}$ in X, there is a fuzzy point x_α such that for any $P \in \eta(x_\alpha)$ and any $N \in D$, there is $n \in D$ such that $n \geq N$ and $S(n) \notin P$.

Theorem 5.6 [11] Let (X, δ) be a fuzzy topological space. Then (X, δ) is fuzzy compact if and only if for each α-net $S = \{S(n) | n \in D\}$ and any $\varepsilon \in (0, \alpha)$, there is a fuzzy point $x_{\alpha-\varepsilon}$ in X such that for any $P \in \eta(x_{\alpha-\varepsilon})$ and any $N \in D$, there is $n \in D$ such that $n \geq N$ and $S(n) \notin P$.

Theorem 5.7. If a fuzzy topological space (X, δ) is strong fuzzy compact then it is S-compact.

Proof. Let $S = \{S(n) | n \in D\}$ be a constant α-net. For each $n \in D$, let $S(n)'$ be the fuzzy point defined by

$$\text{supp}(S(n)') = \text{supp}(S(n)), \quad \text{val}(S(n)') = 1 - \alpha.$$

Then $S' = \{S(n)'|n \in D\}$ is a constant $1-\alpha$-net $(0 < 1-\alpha < 1)$ in X. Since (X, δ) is strong fuzzy compact, there is a fuzzy point $x_{1-\alpha}$ such that for any $P \in \eta(x_{1-\alpha})$ and any $N \in D$, there is $n \in D$ such that $n \geq N$ and $S(n)' \notin P$. Consider the fuzzy point x_α. We will show that x_α is a cluster point of S. Let $U \in \mathcal{N}(x_\alpha)$ and $N \in D$ be given. Take $V \in \delta$ such that $x_\alpha \in_s V \subseteq U$, then $1 - V \in \eta(x_{1-\alpha})$ since $1 - \alpha > 1 - V(x)$. Hence there is $n \in D$ such that $n \geq N$ and $S(n)' \notin 1 - V$. Therefore

$$1 - \alpha > 1 - V(\mathrm{supp}(S(n)')) = 1 - V(\mathrm{supp}(S(n)))$$
$$\alpha < V(\mathrm{supp}(S(n))).$$

Hence $S(n) \in V \subseteq U$, so $S \propto x_\alpha$ which implies that (X, δ) is S-compact.

Theorem 5.8. If a fuzzy topological space (X, δ) is S-compact then it is fuzzy compact.

Proof. Let $S = \{S(n)|n \in D\}$ be an α-net in X and $\varepsilon \in (0, \alpha)$, and let $0 < \varepsilon' < \varepsilon$. For each $n \in D$, let $S(n)'$ be the fuzzy point defined by

$$\mathrm{supp}(S(n)') = \mathrm{supp}(S(n)), \quad \mathrm{val}(S(n)') = 1 - \alpha + \varepsilon'.$$

Then $S' = \{S(n)'|n \in D\}$ is a constant $1 - \alpha + \varepsilon'$-net. Since (X, δ) is S-compact, there is a fuzzy point $x_{1-\alpha+\varepsilon'}$ such that $S' \propto x_{1-\alpha+\varepsilon'}$ by Theorem 5.4. Consider the fuzzy point $x_{\alpha-\varepsilon}$. Let $P \in \eta(x_{\alpha-\varepsilon})$ and $N \in D$ be given. Since $\alpha - \varepsilon > P(x)$, $1 - \alpha + \varepsilon' < 1 - \alpha + \varepsilon < 1 - P(x)$ and $1 - P \in \delta$. Hence $1 - P \in \mathcal{N}(x_{1-\alpha+\varepsilon'})$. $S' \propto x_{1-\alpha+\varepsilon'}$ implies that there is $n \in D$ such that $n \geq N$ and $S(n)' \in 1 - P$ i.e.

$$1 - \alpha + \varepsilon' \leq 1 - P(\mathrm{supp}(S(n)')) = 1 - P(\mathrm{supp}(S(n)))$$
$$\alpha - \varepsilon' \geq P(\mathrm{supp}(S(n)))$$
$$\alpha > P(\mathrm{supp}(S(n))).$$

Hence $S(n) \notin P$ and (X, δ) is fuzzy compact by Theorem 5.6.

From the results in the above Theorems, we can obtain the following implications in which none of the arrows is reversible.

$$\text{ultra fuzzy compact}$$
$$\Downarrow$$
$$\text{N-compact}$$
$$\Downarrow$$
$$\text{strong fuzzy compact} \quad \Rightarrow \quad \alpha\text{-compact}$$
$$\Downarrow$$
$$\text{S-compact}$$
$$\Downarrow$$
$$\text{fuzzy compact} \quad \Rightarrow \quad \text{weakly fuzzy compact}$$

To prove that none of the arrows is reversible we use some examples from [4].

Counterexamples. 1. *A fuzzy compact fuzzy topological space which is not S-compact.*

Let $X = I$ be the unit interval. For each $x \in X$, let 1_x denote the charateristic function of $\{x\}$. For each $x \in X \cap Q$, let $x = \frac{p}{q}$ in smallest terms where $p, q \geq 0$ and then put

$$U_x^s = \frac{s}{q} + \frac{1}{q} 1_x \quad \text{for all } s \in N, \quad 0 \leq s \leq q - 1.$$

Let

$$\sigma_1 = \{1_x | x \in X \cap Q^c\}$$
$$\sigma_2 = \{U_x^s | x \in X \cap Q, x = \frac{p}{q}, s \in N, 0 \leq s \leq q - 1\}.$$

Let δ be the fuzzy topology on X generated by

$$\sigma = \{\alpha | \alpha \text{ is a constant}\} \cup \sigma_1 \cup \sigma_2$$

then (X, δ) is fuzzy compact [4]. Now we show that (X, δ) is not S-compact. Let $x^k = \frac{1}{k}$ for each $k \in N$ and take $\alpha \in I \cap Q^c$. Let $S(k)$ be the fuzzy point with support x^k and value α i.e. $S(k) = x_\alpha^k$ for each $k \in N$. Then $S = \{S(k) | k \in N\}$ is a constant α-net (in fact, a constant α-sequence). Let $y \in X$ then either $y \in X \cap Q^c$ or $y \in X \cap Q$. If $y \in X \cap Q^c$ or $y = 0$, $1_y \in \mathcal{N}(y_\alpha)(U_0^s = 1_0)$ hence y_α is not a cluster point of S. If $y \in X \cap Q$ and $y \neq 0$, put $y = \frac{p}{q}$ in the smallest terms where $p, q \geq 0$ and take $s \in N$ such that $\frac{s}{q} < \alpha < \frac{s+1}{q}$. Then $U_y^s \in \mathcal{N}(y_\alpha)$ since $U_y^s \in \delta$ and

$$U_y^s(y) = \frac{s}{q} + \frac{1}{q} 1_y(y) = \frac{s+1}{q} > \alpha.$$

Take $k^* \in N$ such that $\frac{1}{k^*} < y$. If $k \geq k^*$ then $\frac{1}{k} \neq y$ and hence

$$U_y^s(x^k) = U_y^s(\frac{1}{k}) = \frac{s}{q} < \alpha.$$

This means that $S(k) = x_\alpha^k \notin U_y^s$ if $k \geq k^*$ i.e. y_α is not a cluster point of S. Therefore (X, δ) is not S-compact.

2. *An S-compact fuzzy topological space which is not strong fuzzy compact.*

Let $X = I$ and δ be the fuzzy topology on X generated by

$$\{1_0\} \cup \{\alpha | \alpha \text{ is a constant}\} \cup \{U \in I^X | U(x) \leq x \text{ for all } x \in X\}.$$

Then

$$\iota_0(\delta) = \{\{0\}\} \cup \{X\} \cup \{A \subseteq X | A \subseteq (0, 1]\}$$

hence $(X, \iota_0(\delta))$ is not compact and so (X, δ) is not strong fuzzy compact. Now we show that (X, δ) is S-compact. Let $S = \{S(n) | n \in D\}$ be a constant α-net in X and take $x \in (0, \alpha)$. For any open fuzzy set V such that $x_\alpha \in_s V$, there is a constant fuzzy set β such that $x_\alpha \in_s \beta \subseteq V$ since $V(x) > \alpha > x$. Hence $S(n) \in \beta \subseteq V$ for all $n \in D$, and this implies that $S \propto x_\alpha$ (in fact, $S \to x_\alpha$). So (X, δ) is S-compact.

Since the strong fuzzy compactness and the fuzzy compactness are good extensions of compactness in topology and the S-compactness lies between them, it is obvious that the S-compactness is also a good extension of compactness in topology.

Remark: In theories of fuzzy compactness in terms of prefilter convergence, Katsaras [1] dealt with pointwise neighborhood system by Warren [12] and Wang [11] with R-neighborhood system which has a same spirit of Q-neighborhood system by Pu and Liu [9]. In this paper we have developed our theory with a neighborhood system by Srivastava and et al. [10]. In view of these facts, a general theory of fuzzy compactness in terms of prefilter convergence can be developed in a unified manner. Our result on this direction will be appeared elsewhere.

References

1. Katsaras, A. K. , Convergence of fuzzy filters in fuzzy topological spaces, *Bull. Math. de la Soc. Sci. Math. de la R. S. de Roumanie*, Yome 27 (75), nr. 2 (1983), 131-137.

2. Lowen, R. , Fuzzy topological spaces and fuzzy compactness, *J. Math. Anal. Appl.* 56 (1976), 621-633.

3. Lowen, R. , Initial and final fuzzy topologies and the fuzzy Tychonoff theorem, *J. Math. Anal. Appl.* 58 (1977), 11-21.

4. Lowen, R., A comparison of different compactness notions in fuzzy topological spaces, *J. Math. Anal. Appl.* 64 (1978), 445-454.

5. Lowen, R., Convergence in fuzzy topological spaces, *General Topology and its Applications*, 10 (1979), 147-160.

6. Macho Stadler, M. and de Prada Vicente, M. A., On N-convergence of fuzzy nets, *Fuzzy Sets and Systems* V.51, No 2 (1992), 203-217.

7. Min, K. C. and et al, Fuzzy Topology and Neighborhood Systems, Research Report (1986).

8. de Prada Vicente, M. A. and Saralegui Aranguren, M., Fuzzy filters, *J. Math. Anal. Appl.* 129 (1988), 560-568.

9. Pu, P-M. and Liu, Y-M., Fuzzy topology I; Neighborhood structure of a fuzzy point and Moore-Smith convergence, *J. Math. Anal. Appl.* 76 (1980), 571-599.

10. Srivastava, R., Lal, S. N. and Srivastava, A. K., Fuzzy Hausdorff topological spaces, *J. Math. Anal. Appl.* 81 (1981), 497-506.

11. Wang, G. J., A new fuzzy compactness defined by fuzzy nets, *J. Math. Anal. Appl.* 94 (1983), 1-23.

12. Warren, R. H., Neighborhoods, bases and continuity in fuzzy topological spaces, *Rocky Mountain J. Math.* V.8, No.3 (1978), 459-470.

FUZZY MATRICES AND RELATIONAL EQUATIONS

HAN HYUK CHO
Department of Mathematics Education
Seoul National University
Seoul, KOREA (hancho@alliant.snu.ac.kr)

1. Introduction

Let $\mathcal{F} = [0, 1]$ be the fuzzy algebra with operations $(+, \cdot)$ and the standard order $\leq$: $a + b = max\{a, b\}$ and $a \cdot b = min\{a, b\}$. The subalgebra $\mathcal{B} = \{0, 1\}$ of $\mathcal{F}$ is called the Boolean algebra, and the set F_n of all $n \times n$ matrices over $\mathcal{F}$ (fuzzy matrices) and the set B_n of all $n \times n$ matrices over $\mathcal{B}$ (Boolean matrices) form partially ordered multiplicative matrix semigroups under these fuzzy operations and order. For A and B in F_n, if the (i, j)-entry A_{ij} of A is less than or equal to B_{ij} for each i and j, then we say B dominates A and is denoted by $A \leq B$. If A dominates a permutation matrix, then A is called a Hall matrix and the set H_n of all $n \times n$ Hall matrices is a subsemigroup of F_n. $A \in F_n$ is a regular fuzzy matrix of F_n if there exists $X \in F_n$ such that $AXA = A$. Note that every invertible matrix is also a regular matrix.

Fuzzy relational equations are widely used in many science and engineering researches (refer to Di Nola [3] for an overview), and there are many researches on fuzzy equations nowadays. Finite fuzzy relation equations can be expressed in the form of fuzzy matrix equations, and Guo Si-Zhong et al. [10] studied fuzzy equations in the general framework of fuzzy matrix theory. Now consider a fuzzy relational equation $x \cdot A = b$, where $A \in F_n$ and b is a fuzzy vector of length n. Then $x \cdot A = b$ has a solution if and only if b can be expressed as a linear combination of some rows of A. Furthermore there is an important relationship between regular fuzzy matrices and fuzzy relational equations. If A is regular in F_n with a generalized inverse X, then $x \cdot A = b$ is solvable if and only if $b \cdot X$ is a solution.

Z. Bien and K. C. Min (eds.),
Fuzzy Logic and its Applications, Information Sciences, and Intelligent Systems, 287–296.
© 1995 *Kluwer Academic Publishers.*

2. Fuzzy Matrices and Ranks

For each M in the semigroup R_n of $n \times n$ real matrices, the row rank and the column rank of M have the same value. Also they are equal to the smallest integer r such that $M = L \cdot R$, where L and R are $n \times r$ and $r \times n$ real matrices respectively. In this section, we study the rank properties of fuzzy matrices.

Definition 2.1. Let A be an $n \times n$ fuzzy matrix. The row space $R(A)$ of A is the subspace generated by the rows of A and the row rank of A is the smallest possible size of a spanning set for $R(A)$. Column space $C(A)$ and column rank of A are defined in dual fashion. The fuzzy rank of A is the smallest integer r such that $A = B \cdot C$, where B and C are $n \times r$ and $r \times n$ fuzzy matrices respectively (the fuzzy rank of a zero matrix is 0).

Example 2.2. In the semigroup F_n of fuzzy matrices, familiar rank properties of field based matrices do not hold as follows:

$$A = \begin{pmatrix} 1 & 1 & 1 & 1 & 0 & 0 \\ 0 & 1 & 1 & 1 & 1 & 0 \\ 0 & 0 & 1 & 1 & 1 & 1 \\ 1 & 0 & 0 & 1 & 1 & 1 \\ 1 & 1 & 0 & 0 & 1 & 1 \\ 1 & 1 & 1 & 0 & 0 & 1 \end{pmatrix}, \quad B = \begin{pmatrix} 1 & 1 & 0 & 0 & 0 \\ 0 & 1 & 0 & 1 & 0 \\ 0 & 0 & 1 & 1 & 0 \\ 0 & 0 & 0 & 1 & 1 \\ 1 & 0 & 0 & 0 & 1 \\ 1 & 0 & 1 & 0 & 0 \end{pmatrix},$$

$$C = \begin{pmatrix} 1 & 1 & 0 & 0 & 0 & 0 \\ 0 & 1 & 1 & 1 & 0 & 0 \\ 0 & 0 & 1 & 0 & 0 & 1 \\ 0 & 0 & 0 & 1 & 1 & 0 \\ 1 & 0 & 0 & 0 & 1 & 1 \end{pmatrix}, \quad D = \begin{pmatrix} 1 & 1 & & & & \\ & 1 & 1 & & & \\ & & \ddots & \ddots & & \\ & & & & 1 & 1 \\ 1 & \cdots & \cdots & & 1 & 1 \end{pmatrix}.$$

Then $A = B \cdot C$ and the fuzzy rank of A is 5, whereas the row rank and the column rank of A are both 6 (so the row (column) rank of A is not equal to the fuzzy rank of A). Also the row rank of D is $n - 1$ and the column rank of D is n (so the row rank of D is not equal to the column rank of D).

For a given $A \in F_n$, let the row rank of A be r and the column rank of A be c. Then, there are permutation matrices P and Q so that

$$PAQ = \begin{pmatrix} M & MC \\ RM & RMC \end{pmatrix},$$

where M is an $r \times c$ matrix of full row rank and full column rank, and R and C are $(n-r) \times r$ and $c \times (n-c)$ fuzzy matrices respectively. In general,

r may not be equal to c, but for a regular matrix A in F_n, it is known that r is equal to c (Refer to Kim 12]). Furthermore, we have the following Theorem.

Theorem 2.3. *For $A \in F_n$, P, Q, and M mentioned in the above paragraph, we have*

(i) *If A is regular, then the row rank of A and the column rank of A are all equal to the fuzzy rank of A.*

(ii) *A is regular iff M is regular.*

Proof. (i) Refer to Cho [8].

(ii) Without loss of generality, we may assume P and Q are identy permutation matrix. Let A be regular and $AXA = A$ for some $X \in F_n$ and

$$X = \begin{pmatrix} N & U \\ V & W \end{pmatrix},$$

where N is an $r \times c$ matrix and W is an $(n-r) \times (n-c)$ matrix (in fact, $r = c$). Then, $M(N + CV + UR + CWR)M = M$ and M is a regular matrix. Now let M be regular and $MZM = M$ for some $Z \in F_r$. Then $\begin{pmatrix} Z & 0 \\ 0 & 0 \end{pmatrix}$ is a generalized inverse of A.

Definition 2.4. Let G be a multiplicative semigroup and A be an element of G. A is regular in G if $AXA = A$ for some generalized inverse $X \in G$ of A (such X is called a semi-inverse of A if $XAX = X$ holds also). Now let G have an identity element. Then non-zero non-invertible element A is prime in G if A cannot be expressed as a product of two non-invertible elements of G. A is called factorizable in G if A is not prime in G.

A semiring is an algebraic system satisfying all the axioms of a ring except that of requiring an additive inverse of each element. In other word, a semiring is an algebraic system $(\mathcal{R}, +, \cdot)$ such that $(\mathcal{R}, +)$ is a commutative semigroup with the identity 0, and $(\mathcal{R}, \cdot)$ is a semigroup with the identity 1, and the distributive law holds in $\mathcal{R}$ (see Gregory and Pullman [9]). Some examples of semirings are: any Boolean algebras especially the two-element Boolean algebra $\mathcal{B}$, fuzzy numbers $\mathcal{F}$ in $[0, 1]$, the nonnegative real numbers R^*, the nonnegative integers Z^*. For each semiring $\mathcal{R}$, let $M_n(\mathcal{R})$ denote the set of all $n \times n$ matrices over $\mathcal{R}$. We may consider regular matrices and prime matrices in $M_n(\mathcal{R})$.

For the semigroup of all $n \times n$ matrices with integer entries, the units are the matrices with determinant ± 1 and the primes are the matrices with determinant a prime number. Now let P_n denote the set of all $n \times n$

permutation materices. Then for the multiplicative semigroups $M_n(\mathcal{B})$, $M_n(Z^*)$, and $M_n(\mathcal{F})$, the set of all the units is the $n \times n$ permutation matrices. Primes in R_n^+ have been studied by Richman and Schneider [13], and by Borosh, Hartfiel and Maxson [1]. Primes in B_n have been studied by D. de Caen and Gregory [2] and by the author [5]. Also regular matrices over general semirings are studied by Zhao [14] and Cho [8].

Definition 2.5. For $\alpha \in [0,1]$, $\pi_\alpha : F_n \to B_n$ is a semigroup homomorphism such that $\pi_\alpha(A)$ is an α-cut A_α of A. Thus, every (i,j)-entry A_{ij} of A is greater than or equal to α iff $(A_\alpha)_{ij} = 1$. For each integer i, A_{i*} and A_{*j} denote respectively the i-th row and the j-th column of A.

Note that α-cut map π_α is a semiring homomorphism. Because regualr matrices and primes in B_n are well-known, using this α-cut map we can study regular matrices and primes in other matrix semirings easily. For example, if A is regular (factorizable) in F_n, then every α-cut A_α of A is also regular (factorizable) in B_n respectively. But the converse is not true as shown in the following example.

Example 2.6. Let $0 < \alpha < \beta < 1$. Then the following matrix A is a prime fuzzy matrix (refer to [7]),

$$A = \begin{pmatrix} 1 & \alpha & \beta & 0 \\ 0 & 1 & \alpha & \beta \\ \beta & 0 & 1 & \alpha \\ \alpha & \beta & 0 & 1 \end{pmatrix}.$$

But the α-cut of A is not a prime Boolean matrix since

$$\pi_\alpha(A) = \begin{pmatrix} 1 & 1 & 1 & 0 \\ 0 & 1 & 1 & 1 \\ 1 & 0 & 1 & 1 \\ 1 & 1 & 0 & 1 \end{pmatrix} = \begin{pmatrix} 1 & 1 & 0 & 0 \\ 0 & 1 & 1 & 0 \\ 0 & 0 & 1 & 1 \\ 1 & 0 & 0 & 1 \end{pmatrix} \begin{pmatrix} 1 & 1 & 0 & 0 \\ 0 & 1 & 1 & 0 \\ 0 & 0 & 1 & 1 \\ 1 & 0 & 0 & 1 \end{pmatrix}.$$

Thus there is a fuzzy prime matrix such that its α-cut is a Boolean factorizable matrix. From the relationship between F_n and B_n through the α-cut maps we have the following Theorem.

Theorem 2.7. *For $A \in F_n$, let the fuzzy rank of $\pi_1(A)$ be n. Then*

    (i) *A is a Hall matrix.*
    (ii) *A is regular in F_n iff A is regular in H_n.*
    (iii) *A is prime in F_n iff A is prime in H_n.*

Proof. (i) Let S be $\pi_1(A)$. Note that $S = B \cdot C$ implies $S = \sum_{i=1}^r R_i$, where $R_i = B_{*i} \cdot C_{i*}$. Therefore the fuzzy rank of S is the minimum number of rank-one dominated submatrices of S whose fuzzy sum is S. It follows from the *König's Theorem* that the term rank of S is greater than or equal to its fuzzy rank. Thus S dominates a permutation matrix and so does A.

(ii) Note that $AXA = A$ means $\pi_1(A) \, \pi_1(X) \, \pi_1(A) = \pi_1(A)$. Thus if the fuzzy rank of $\pi_1(A)$ is n, then the fuzzy rank of $\pi_1(X)$ is also n and A and X are Hall matrices by (i). Thus A is regular in H_n. The converse is obvious.

(iii) By the same reasoning as (ii).

Example 2.8. The non-trivial implication of (ii) in the above Theorem does not hold when the matrix $\pi_1(A)$ does not have full fuzzy rank. For example let

$$A = \begin{pmatrix} 1 & 0 & 0 & 1 \\ 0 & 1 & 0 & 1 \\ 0 & 0 & 1 & 1 \\ 0 & 1 & 0 & 1 \end{pmatrix}, \quad X = \begin{pmatrix} 1 & 0 & 0 & 0 \\ 0 & 1 & 0 & 0 \\ 0 & 0 & 1 & 0 \\ 0 & 0 & 0 & 0 \end{pmatrix}.$$

Then the fuzzy rank of $A = \pi_1(A)$ is 3, and A is a Hall matrix. Also, A and $X = \pi_1(X)$ are regular in B_n since $AXA = A$ and $XAX = X$. But A is not regular in H_n because the first row A_{1*} of A dominates only one row of A even though $|A_{1*}|=2$ (refer to Cho [6]).

It is well known that any nonsingular real matrix in the semigroup R_n of $n \times n$ real matrices can be written as a product of elementary matrices. Similarly every $n \times n$ Boolean matrix of fuzzy rank n can be expressed as a product of prime matrices and elementary matrices (refer to Cho [5]). Note that any prime matrix A in F_n has full row and column and fuzzy ranks.

3. Regular Matrices and Relational Equations

For a Boolean matrx $B \in B_n$, $R(B)$ has a unique basis consisted of some rows of B. But this familiar property of Boolean matrices does not always hold for fuzzy matrices. Instead there exists a unique standard basis for its row space $R(A)$ of $A \in F_n$. This is a major difference between Boolean matrix theory and fuzzy matrix theory. Recall that fuzzy vectors $r_1, ..., r_d$ form a standard basis provided that $r_i = \sum c_{ij} r_j$ implies $r_i = c_{ii} r_i$.

Definition 3.1. $A \in F_n$ is called a row-standard matrix if some rows $r_1, ..., r_d$ of A forms a standard basis of $R(A)$ (i.e. $r_i = \sum c_{ij} r_j$ implies $r_i = c_{ii} r_i$). Column-standard matrix is defined in dual fashion, and a matrix is a stadard matrix if it is both row-standard and column-standard.

Note that every Boolean matrix and every prime fuzzy matrix is a standard matrix. It is well known that for each $A \in F_n$ there is a row-standard matrix B such that $R(A) = R(B)$. Note that if $A \in F_n$ is a regular row-standard matrix of row rank n, then there is a permutation matrix Q such that $AQA = A$ (refer to Kim [11]). In this case, a fuzzy equation $x \cdot A = b$ has a solution if and only if $b \cdot Q$ is a solution.

Example 3.2. There is a regular fuzzy matrix that is row-standard but not column-standard as we see from the following idempotent matrix

$$E = \begin{pmatrix} 1 & 1 & & \\ & \alpha & & \\ & & \ddots & \\ & & & \alpha \end{pmatrix}.$$

Here, unspecified entries of E are all 0.

Definition 3.3. The permanent $per(A)$ of an $n \times n$ fuzzy matrix A is a fuzzy sum $\sum_{\sigma \in S_n} a_{1\sigma(1)} a_{2\sigma(2)} \cdots a_{n\sigma(n)}$, where S_n denotes the set of all permutations on $\{1, ..., n\}$. The adjoint $adj(A)$ of A is an $n \times n$ fuzzy matrix such that $adj(A)_{ij}$ is equal to $per(A(j|i))$ for each i and j, where $A(j|i)$ denotes an $(n-1) \times (n-1)$ fuzzy matrix obtained from A by deleting the j-th row and the i-th column of A.

Lemma 3.4. *For $A \in H_n$, the following statements are all equivalent.*

    (i) *A is regular in H_n.*
    (ii) *$adj(A)$ is the unique semi-inverse of A.*
    (iii) *$adj(A)$ is the unique maximal generalized inverse of A.*
    (iv) *$adj(adj(A)) = A$.*

Proof. Refer to Cho [4].

Consider a fuzzy equation $x \cdot A = b$ and its solution set $\Omega(A, b) = \{x \in \mathcal{F}^n | \ x \cdot A = b\}$. Then there is a unique maximal element in $\Omega(A, b)$, but there may not be a unique minimal element in $\Omega(A, b)$. Using the results of Guo Si-Zhong [10], Li Jian-Xin [12] gave some necessary and sufficient conditions for the existence of a unique minimal element in $\Omega(A, b)$.

Theorem 3.5. *Let $A \in H_n$ be a row-standard matrix of row rank n. If A is a regular matrix of F_n, then*

    (i) *There exists a unique permutation matrix P such that X is a generalized inverse of A if and only if $P^t \leq X \leq adj(A)$.*
    (ii) *If $x \cdot A = b$ has a solution, then $b \cdot P$ is the unique maximal solution.*

Proof. (i) Let P be a permutation matrix dominated by A. Then for $Y \in F_n$, $AYA = A$ if and only if $P^t A \cdot YP \cdot P^t A = P^t A$. Thus we may assume that P is an identity permutation matrix. Now let X be any generalized inverse of A. We will show that X also dominates an identity permutation matrix P. If this is false, without loss of generality, we may assume that $X_{11} < P_{11}(= 1)$. Since $A_{1*} = \sum_{i=1}^{n} (\sum_{j=1}^{n} A_{1j} X_{ji}) A_{i*}$ and $A_{11} = 1$, we have $A_{1*} = (\sum_{j=1}^{n} A_{1j} X_{j1}) A_{1*}$ because the rows of A forms a stadard basis of $R(A)$. Thus, there exists k ($\neq 1$) such that $A_{1k} = X_{k1} = 1$, and $A_{1*} < A_{k*}$ since $A_{kk} = X_{k1} = 1$ and $A_{k*} = \sum_{i=1}^{n} (\sum_{j=1}^{n} A_{kj} X_{ji}) A_{i*}$. Therefore $X_{1k} < 1$ and $X_{kk} < 1$ since either $X_{1k} = 1$ or $X_{kk} = 1$ implies $A_{k*} < A_{1*}$. Therefore $A_{ks} = X_{sk} = 1$ for some s ($\neq 1, k$) since $A_{k*} = (\sum_{j=1}^{n} A_{kj} X_{jk}) A_{k*}$ implies $\sum_{j=1}^{n} A_{kj} X_{jk} = 1$. Then we have $A_{k*} < A_{s*}$ since $A_{ss} = X_{sk} = 1$. Thus $X_{ks} < 1$ and $X_{ss} < 1$ since either $X_{ks} = 1$ or $X_{ss} = 1$ implies $A_{s*} < A_{k*}$. By the similar reasoning, we can show that $X_{1s} < 1$ and $A_{st} = X_{ts} = 1$ for some t ($\neq 1, k, s$). Moreover we can show that $A_{1*} < A_{k*} < A_{s*} < A_{t*}$. This procedure must stop within finite steps, and we have a contradiction. Therefore if $AXA = A$, then X dominates P^t and $adj(A)$ dominates X since $adj(A)$ is the unique maximal generalized inverse of A by Lemma 3.2. Now consider $X \in F_n$ with $P^t \leq X \leq adj(A)$. Then $A = AP^t A \leq AXA \leq Aadj(A)A = A$ and X is a generalized inverse of A.

(ii) Let P be a permutation matrix dominated by A. If $c \cdot A = b$ for some c, then $c \cdot AP^t A = b$ and $b \cdot P^t A = b$. Note that $P^t A$ contains an identity permutation matrix. Thus $c \cdot A \leq b$ implies $c \cdot P \leq b$, and $b \cdot P^t$ is the unique maximal solution.

References

[1] Borosh, Hartfiel and Maxson, Answer to the questions posed by Richman and Schneider, Linear and Multilinear Algebra 3 (1976) 255-258.

[2] D. de Caen and D. A. Gregory, Primes in the semigroup of Boolean matrices, Linear Algebra and Appl. 37 (1981) 119-134.

[3] A. Di Nola et al., Fuzzy relation equations as a basis of fuzzy modelling: An overview, Fuzzy Sets and Systems 40 (1991) 415-429.

[4] H. H. Cho, Fuzzy matrices of permanent one, Fuzzy Sets and Systems 56 (1993) 291-296.

[5] H. H. Cho, Prime Boolean matrices and factorizations, Linear Algebra Appl. 190 (1993) 87-98.

[6] H. H. Cho, Regular matrices in the semigroup of Hall matrices, Linear Algebra Appl. 191 (1993) 151-163.

[7] H. H. Cho and E. Y. Oh, On the regular fuzzy matrices and prime fuzzy matrices, Proc. Workshops in Pure Math. 13 (1993) 129-135.

[8] H. H. Cho, Regular matrices over semirings, Preprint.

[9] D. A. Gregory and N. J. Pullman, Semiring rank: Boolean rank and nonnegative rank factorizations, J. Comb. Inf. and Syst. Sci. 8 (1983) 223-233.

[10] Guo Si-Zhong et al., Further contribution to the study of finite fuzzy relational equations, FSS 26 (1988) 93-104.

[11] K. H. Kim and F. W. Roush, Generalized Fuzzy Matrices, FSS 4 (1980) 293-315.

[12] Li Jian-Xin, The smallest solution of max-min fuzzy equations, FSS 41 (1990) 317-327.

[13] D. J. Richman and H. Scneider, Primes in the semigroup of nonnegative matrices, Linear and Multilinear Algebra 2 (1974) 135-140.

[14] C. -K. Zhao, Inverses of L-fuzzy matrices, FSS 34 (1990) 103-116.

EMBEDDINGS OF Z-DOMAINS IN CUBES

XIAO-QUAN XU

Mathematics Department, Jiangxi Teachers University,
Nanchang, Jiangxi 330027, P.R. China

Abstract: The structure theory of domains and their general-
izations plays a crucial role in denotational semantics as
developed by Scott and his followers. This paper is mainly
devoted to the embeddings of Z-domains in cubes. We present
a direct approach to constructing Z-morphisms of strongly
precontinuous Z-domains, a generalization of continuous domains (i.e. continuous posets), into the unit interval, and
show that a strongly precontinuous Z-domain possesses enough
Z-morphisms into the unit interval to separate the points.

1. Introduction

Domain theory is an area which has evolved from two separate
impetuses. The first and most prominent has been denotational semantics of high-level programming languages. It was the
pioneering work of Scott [13,14] which led to the discovery
that algebraic lattices, and their generalization, continuous lattices could be used to assign meanings to programs
written in high-level programming languages. When Plotkin

Project supported by the National Natural Science Foundation
of China and the Natural Science Foundation of Jiangxi Prov-
ince, China.

Z. Bien and K. C. Min (eds.),
Fuzzy Logic and its Applications, Information Sciences, and Intelligent Systems, 333–342.
© 1995 *Kluwer Academic Publishers.*

[10] pointed out the need for more general objects to use as mathematical models, the notion of a domain was formulated, and the structure theory of domains has been a focal point for research in denotational semantics ever since.

On the purely mathematical side, research into the structure theory of compact semilattices led Lawson [8] and others [6,7] to consider the category of those compact semilattices which admit a basis of subsemilattice neighborhoods at each point. It was discovered by Hofmann and Stralka [7] that those objects are precisely the continuous lattices of Scott. One of the most important features of continuous lattices is that they admit sufficiently many homomorphisms into the unit interval to separate the points. The topological form of this result is due to Lawson [8].

In [15] Wright et al. developed a "uniform approach" method which enable them to generalize the concept of an algebraic poset. Following them, some authors introduced Z-continuous posets, a more general concept of continuous posets (see, for example, [2,9]).

For a poset P and a family Z of subsets of P, we call P a Z-domain if every $S \in Z$ has a join $\bigvee S$ in P. We say that a map $f : P \longrightarrow Q$ between a Z-domain P and a poset Q is a (resp. strong) Z-morphism if f preserves arbitrary (existing) meets and Z-joins (resp. f preserves Z-joins and has a lower adjoint).

This paper is mainly devoted to the following two questions:

(i) which Z-domains can be embedded isomorphically into a cube via a Z-morphism? Namely, which Z-domains admit enough Z-morphisms into the unit interval [0,1] to strongly separate the points?

(ii) which Z-domains admit enough strong Z-morphisms into [0,1] to strongly separate the points?

First, in the theory $ZFDC_\omega$ (ZF denotes the Zermelo-Fraenkel set theory and DC_ω denotes the axiom of ω dependent choices) we present a direct approach to constructing

(1) Z-morphisms of strongly precontinuous Z-domains (see Definition 2.4 below) into $[0,1]$, and

(2) strong Z-morphisms of those strongly precontinuous Z-domains into $[0,1]$ in which every countable chain C has a join $\bigvee C$ and the least element 0 exists.

Next, we solve the questions (i) and (ii).

Finally, we give some applications of the results about strongly precontinuous Z-domains to complete lattices.

The general idea of constructing morphisms of the right kind into chains by using maximal complete strict chains relative to suitable auxiliary relations dates from the 1950s; forerunners are to be found in Raney's classical papers on completely distributive lattices [11,12]. The classical technique of Raney and Bruns [5], based on the investigations of maximal complete strict chains of suitable auxiliary relations, is somewhat complicated and requires the axiom of choice. Moreover, it cannot be applied to posets. Comparatively, the technique presented in this paper is simpler, requires only the axiom of ω dependent choices, and can be directly applied to the case of posets.

Finally we point out that this article has its genesis in [16] and [17].

2. Precontinuous Z-Domains and Z-Morphisms

In this section, for the reader's convenience, first we briefly review some basic definitions, notations and results concerning posets. The reader wishing more details can consult [2,6,9].

Let P be a poset and let $x \in P$. We define $\downarrow x = \{y \in P \mid y \leqslant x\}$. As usual $\downarrow Y = \{x \in P \mid x \leqslant y \text{ for some } y \in Y\}$ denotes the lower set generated by Y. By a system of subsets of P we mean a family Z of subsets of P, called Z-sets, such that for every $x \in P$ there exists $S \in Z$ with $x = \bigvee S$. The lower set generated by Z-sets are called Z-ideals and the set $I_Z(P)=\{\downarrow Y \mid Y \in Z\}$ is called the system of all Z-ideals of P. Important examples of such Z are

$\mathscr{P}(= \mathscr{P}(P))$, the family of all subsets of P,

$\mathscr{D}(= \mathscr{D}(P))$, the family of all directed subsets of P, and

$\mathscr{F}(= \mathscr{F}(P))$, the family of all finite subsets of P.

The following conditions on Z are relevant:

($\cdot$) $\downarrow x \in I_Z(P)$ for all $x \in P$,

($\cdot\cdot$) $\{x\} \in Z$ for all $x \in P$,

(*) if $\{S_i \mid i \in I\} \subset Z$, $\bigvee S_i$ exists for each $i \in I$ and $\{\bigvee S_i \mid i \in I\} \in Z$, then there exists $S \in Z$ such that $S \subset \downarrow \bigcup_{i \in I} S_i$ and $\bigvee S = \bigvee \bigcup_{i \in I} S_i$,

(**) if $\{S_i \mid i \in I\} \subset Z$, $\bigvee S_i$ exists for each $i \in I$ and $\{\bigvee S_i \mid i \in I\} \in Z$, then $\bigcup_{i \in I} S_i \in Z$.

2.1. Definition. Let ω denote the ordinal number of the set of natural numbers. A poset P is called ω-chain complete if every countable chain $C \subset P$ has a join $\bigvee C$ in P. We say that P is a Z-domain if Z is a system of subsets of P and every $S \in Z$ has a join $\bigvee S$ in P.

For a Z-domain P, an important binary relation on P depending on Z is the Z-below relation $\ll_Z$: $x \ll_Z y$ holds if and only if for each $S \in Z$, if $y \leq \bigvee S$, then there exists $s \in S$ with $x \leq s$. In the case $Z = \mathscr{D}$ the relation $\ll_Z$ is the well-known way below relation $\ll$. For $x \in P$, let $\downarrow_Z x = \{y \in P \mid y \ll_Z x\}$. Clearly, $\downarrow_Z x = \bigcap \{A \in I_Z(P) \mid x \leq \bigvee A\}$.

A map $f : P \longrightarrow Q$ between two posets is said to preserve arbitrary (existing) meets if whenever $A \subset P$ has a meet $\bigwedge A$ in P, then $\bigwedge f(A)$ exists in Q and equals $f(\bigwedge A)$. We say that f preserves Z-joins if P is a Z-domain and $\bigvee f(S)$ exists and equals $f(\bigvee S)$ for every $S \in Z$.

A pair (f,g) of order preserving maps $f : S \longrightarrow T$ and $g : T \longrightarrow S$ between two posets is called a Galois connection or an adjunction between S and T provided that the relations $f(s) \geq$

t and $s \geqslant g(t)$ are equivalent for all pairs of elements (s,t) $\in S \times T$. In an adjunction (f,g), the map f is called the upper adjoint and g the lower adjoint.

2.2. Definition. Let P be a Z-domain and Q a poset. A map f : $P \longrightarrow Q$ is called a (resp. strong) Z-morphism if f preserves arbitrary meets and Z-joins (resp. f preserves Z-joins and has a lower adjoint). Let $IZ^{\uparrow}(P,Q) = \{f : P \longrightarrow Q \mid f$ is a Z-morphism$\}$ and $\text{Hom}_Z(P,Q) = \{f : P \longrightarrow Q \mid f$ is a strong Z-morphism$\}$.

Obviously, $\text{Hom}_Z(P,Q) \subset IZ^{\uparrow}(P,Q)$.

Suppose we are given a poset S, a family $\{T_i \mid i \in I\}$ of posets and a family of maps $\mathcal{F} = \{f_i : S \longrightarrow T_i \mid i \in I\}$. We say that $\mathcal{F}$ (resp. strongly) separates the points of S if for every pair of elements $x,y \in S$ with $x \neq y$ (resp. with $x \nleqslant y$) there exists a map $f_i \in \mathcal{F}$ such that $f_i(x) \neq f_i(y)$ (resp. $f_i(x)$ $> f_i(y)$).

2.3. Definition. A binary relation $\sqsubset$ on a set X is said to satisfy the interpolation property, if the following condition is satisfied:

(INT) For all $x,y \in X$ with $x \sqsubset y$, there is a $z \in X$ such that $x \sqsubset z \sqsubset y$.

2.4. Definition. A Z-domain P is called precontinuous if the Z-below relation $\ll_Z$ on P is approximating, i.e. $x = \bigvee \!\Downarrow_Z x$ for all $x \in P$; P is called strongly precontinuous if the relation $\ll_Z$ satisfies, in addition, (INT).

2.5. Remark. If P is (resp. strongly) precontinuous and $\Downarrow_Z x \in I_Z(P)$ for all $x \in P$, then P is called (resp. strongly) continuous (see [2,9]).

For a Z-domain P, we define $Z^+ = \{A \subset P \mid \bigvee A$ exists in P and for each $x \in P$, $x \ll_Z \bigvee A$ implies there is a $a \in A$ with $x \leqslant a\}$. One can easily prove that the following two conditions are equivalent:

(1) P is a (resp. strongly) precontinuous Z-domain;

(2) $Z \subset Z^+$ and P is a (resp. strongly) continuous Z^+-domain.

3. Embeddings of Strongly Precontinuous Z-Domains in Cubes

In what follows ZF denotes the Zermelo-Fraenkel set theory and DC_ω denotes the axiom of ω dependent choices, which was proposed by Bernays [4] in 1942 as a weaker version of AC (the axiom of choice).

One can easily prove the following

3.1. Lemma ($ZFDC_\omega$). Let $\sqsubset$ be a binary relation on a set X and let B be the set of dyadic rational numbers in $[0,1]$. If $\sqsubset$ is transitive and satisfies (INT) (that is, $\sqsubset$ is idempotent with respect to the relational product), then for $x,y \in$ X with $x \sqsubset y$, there is a family $\{x(b) \mid b \in B\} \subset X$ such that
(1) $x(0)=x$, $x(1)=y$, and
(2) $x(s) \sqsubset x(t)$ whenever $s < t$.

3.2. Remark. Let X be a topological space and $O(X)$ be the complete lattice of all open sets of X. Define a relation $\sqsubset_t$ on $O(X)$ by $U \sqsubset_t V$ if $\overline{U} \subset V$. Obviously, $\sqsubset_t$ is transitive. If X is a normal space, then $\sqsubset_t$ satisfies (INT). For a pair F,G of disjoint closed subsets of a normal space X, there is an open set U such that $F \subset U \subset \overline{U} \subset X \setminus G$. The main technique used in the proof of Urysohn's lemma is to construct a family $\{U(b) \mid b \in B\} \subset O(X)$ such that
(1) $U(0)=U$, $U(1)=X \setminus G$, and
(2) $U(s) \sqsubset_t U(t)$ whenever $s < t$.

3.3. Theorem ($ZFDC_\omega$). Let P be an ω-chain complete Z-domain in which the relation $\ll_Z$ satisfies (INT). Suppose that $x \not\leq y$ in P. Then the following two conditions are equivalent:
(1) There is a map $f \in \text{Hom}_Z(P,[0,1])$ with $f(x)=1$ and $f(y)=0$;
(2) P has the least element 0 and there is a $u \in P$ with $u \ll_Z x$ and $u \not\leq y$.
Proof. (1)$\Longrightarrow$(2). Let $g : [0,1] \longrightarrow P$ be the lower adjoint of f. Then $0=g(0)$ is the least element of P and $g(s) \ll_Z g(t)$

whenever $s<t$. Let $u=g(\frac{1}{2})$. Then $u\lll_Z x$ and $u\not\leqslant y$.

$(2)\Longrightarrow(1)$. By Lemma 3.1, there is a family $\{x(b) \mid b\in B\}\subset P$ such that (i) $x(0)=u$, $x(1)=x$, and (ii) $x(s)\lll_Z x(t)$ whenever $s<t$. Define $f : P\longrightarrow[0,1]$ by

$$f(z)=\bigvee\{b\in B \mid x(b)\leqslant z\}. \tag{1}$$

Then $f(x)=1$ and $f(y)=0$. We show that f preserves Z-joins. For each $S\in Z$, if $b\in B$ and $x(b)\leqslant\bigvee S$, then for each $c\in B$ with $c<b$, we have $x(c)\lll_Z\bigvee S$. Therefore, there is $s\in S$ such that $x(c)\leqslant s$; and hence $c\leqslant f(s)$. It follows that $f(\bigvee S)=\bigvee f(S)$. Define $d : [0,1]\longrightarrow P$ by

$$d(t)=\begin{cases}\bigvee\{x(b) \mid b\in B \text{ and } b<t\}, & \text{if } 0<t,\\ 0, & \text{if } t=0.\end{cases} \tag{2}$$

Then (f,d) is a Galois connection between P and $[0,1]$. Hence $f\in\text{Hom}_Z(P,[0,1])$.

3.4. Corollary ($ZFDC_\omega$). Let P be a $\mathfrak{D}$-domain in which the way below relation $\ll$ satisfies (INT). Suppose that $x\not\leqslant y$ in P. Then the following two conditions are equivalent:
  (1) There is a map $f\in\text{Hom}_{\mathfrak{D}}(P,[0,1])$ with $f(x)=1$ and $f(y)=0$;
  (2) P has the least element and there is a $u\in P$ with $u\ll x$ and $u\not\leqslant y$.

Corollary 3.4 is an improvement of Proposition IV-2.20 of [6].

3.5. Theorem ($ZFDC_\omega$). Let P be an ω-chain complete Z-domain in which the least element 0 exists. Consider the following conditions:
  (1) P is strongly precontinuous,
  (2) $Z\subset Z^+$ and P is a strongly continuous Z^+-domain,
  (3) $\text{Hom}_Z(P,[0,1])$ strongly separates the points of P,
  (4) There is a system Z^* of subsets of P such that $Z\subset Z^*$ and P is a strongly continuous Z^*-domain,
  (5) P is precontinuous.
Then $(1)\Longleftrightarrow(2)\Longrightarrow(3)\Longleftrightarrow(4)\Longrightarrow(5)$. If $\bigcup\{\Downarrow_Z b \mid b\in\Downarrow_Z a\}\in I_Z(P)$ for all $a\in P$, then $(5)\Longrightarrow(1)$ is true, and hence all five

conditions are equivalent.

Proof. (1)$\Longrightarrow$(2). Recall that $Z^+ = \{A \subset P \mid \bigvee A$ exists in P and for each $x \in P$, $x \ll_Z \bigvee A$ implies there is $a \in A$ with $x \leqslant a\}$. Obviously, $Z \subset Z^+$ and $\ll_Z = \ll_{Z^+}$. It is easy to prove that for $x \in P$, $x = \bigvee \downarrow_Z x$ implies $\downarrow_{Z^+} x = \downarrow_Z x \in Z^+$. Whence (1)$\Longleftrightarrow$(2).

(1)$\Longrightarrow$(3). By Theorem 3.3.

(3)$\Longrightarrow$(4). Define $Z^* = \{A \subset P \mid \bigvee A$ exists in P and $f(\bigvee A) = \bigvee f$ (A) for all $f \in \mathrm{Hom}_Z(P, [0, 1])\}$. Then $Z \subset Z^*$ and $\mathrm{Hom}_{Z^*}(P, [0, 1]) = \mathrm{Hom}_Z(P, [0, 1])$. We show that P is a strongly continuous Z^*-domain. (i) For $x, y \in P$, if $x \not\leqslant y$, then by (3), there is $f_0 \in \mathrm{Hom}_{Z^*}(P, [0, 1])$ such that $f_0(x) > f_0(y)$. Let g_0 be the lower adjoint of f_0 and select a $t \in [0, 1]$ with $f_0(y) < t < f_0(x)$. Let $u = g_0(t)$. Then $u \ll_{Z^*} x$ and $u \not\leqslant y$. Therefore, $x = \bigvee \downarrow_{Z^*} x$; (ii) For each $f \in \mathrm{Hom}_{Z^*}(P, [0, 1])$, let g be the lower adjoint of f. For $x \in P$ and $t \in [0, 1]$, if $t < f(x)$, then $g(t) \ll_{Z^*} g(f(x)) \leqslant x$ and $t \leqslant f(g(t))$. Therefore, $f(x) = f(\bigvee \downarrow_{Z^*} x) = \bigvee f(\downarrow_{Z^*} x)$; and hence $\downarrow_{Z^*} x \in Z^*$; (iii) Let $x, y \in P$ and let $A = \bigcup \{\downarrow_{Z^*} z \mid z \in \downarrow_{Z^*} y\}$. Then by (i) and (ii), $y = \bigvee A$ and $A \in Z^*$. If $x \ll_{Z^*} y$, then $x \in \downarrow_{Z^*} y \subset A$. It follows that there is a $z \in P$ such that $x \ll_{Z^*} z \ll_{Z^*} y$. By (i), (ii) and (iii), P is a strongly continuous Z^*-domain.

(4)$\Longrightarrow$(3). By the implication (1)$\Longrightarrow$(3).

(4)$\Longrightarrow$(5). Trivial.

(5)$\Longrightarrow$(1). Suppose that P is precontinuous and $\bigcup \{\downarrow_Z b \mid b \in \downarrow_Z a\} \in I_Z(P)$ for all $a \in P$. We show that the relation $\ll_Z$ on P satisfies (INT). For $x, y \in P$ with $x \ll_Z y$, since $y = \bigvee \bigcup \{\downarrow_Z z \mid z \in \downarrow_Z y\}$ and $\bigcup \{\downarrow_Z z \mid z \in \downarrow_Z y\} \in I_Z(P)$, we have that $x \in \downarrow_Z y \subset \bigcup \{\downarrow_Z z \mid z \in \downarrow_Z y\}$. It follows that there is a $z \in P$ such that $x \ll_Z z \ll_Z y$. Whence, P is strongly precontinuous.

3.6. Theorem (ZFDC$_\omega$). Let P be a Z-domain in which the relation $\ll_Z$ satisfies (INT). Suppose $u \ll_Z x$ and $u \not\leqslant y$ in P. Then there is $f \in IZ^\uparrow(P, [0, 1])$ with $f(x) = 1$ and $f(y) = 0$.

Proof. By the proof of Theorem 3.3 we only need to prove that the map f defined by (1) preserves arbitrary meets.

Suppose that $\{a_i\}_{i\in I}\subseteq P$ has a meet $\bigwedge_i a_i$ in P. For each $i\in I$, set $B_i=\{b\in B \mid x(b)\leqslant a_i\}$. Then $\bigwedge_i f(a_i)=\bigwedge_i \bigvee B_i=\bigvee\{\bigwedge\varphi(I) \mid \varphi \in \prod_i B_i\}$. For $\varphi\in\prod_i B_i$ and $b\in B$ with $b\leqslant\bigwedge\varphi(I)$, we have $x(b)\leqslant\bigwedge_i a_i$. Therefore, $b\leqslant f(\bigwedge_i a_i)$; and hence $f(\bigwedge_i a_i)=\bigwedge_i f(a_i)$.

3.7. Corollary (ZFDC$_\omega$). Let P be a strongly precontinuous Z-domain. Then $IZ^\uparrow(P,[0,1])$ strongly separates the points of P. Therefore, in ZF+AC, P can be embedded isomorphically into a cube $[0,1]^\lambda$ via an $IZ^\uparrow$-map.

3.8. Remark. For a Z-domain P, the condition that $IZ^\uparrow(P,[0,1])$ strongly separates the points of P does not imply that P is precontinuous (comparing this with the implication $(3)\Longrightarrow(5)$ in Theorem 3.5). For an infinite set I, let $P=\{(a_i)\in[0,1]^I \mid 0<a_i$ for all but finitely many $i\in I\}$ and let $Z=\mathcal{P}(P)\setminus\{\phi\}$. Then the inclusion map $P\longrightarrow[0,1]^I$ is an $IZ^\uparrow$-map; and hence $IZ^\uparrow(P,[0,1])$ strongly separates the points of P. One can easily see that P is not a precontinuous Z-domain. In fact, $x\ll_Z y$ holds for no pair of elements $x,y\in P$.

Finally, we point out that the results in this paper can be directly applied to complete lattices, especially Z-distributive lattices (please see [16,17]). The work in this paper has unified, generalized and improved the corresponding work done in [1,3,5,6,8,9,11,12,16,17].

References

1. Bandelt, H.-J.: M-distributive lattices, **Arch. Math.** 39(1982), 436-442.

2. Bandelt, H.-J. and Erné, M.: The category of Z-continuous posets, **J. Pure and Appl. Algebra** 30(1983), 219-226.

3. Bandelt, H.-J. and Erné, M.: Representations and embeddings of M-distributive lattices, **Houston J. Math.** 10(1984), 315-324.

4. Bernays, P.: A system of axiomatic set theory III, **J. Symbolic Logic** 7(1942), 65-89.

342

5. Bruns, G.: Distributivität und subdirekte Zerlegbarkeit
   vollständiger Verbände, **Arch. Math.** 12(1961), 61-66.
6. Gierz, G. et al.: A Compendium of Continuous Lattices,
   Springer-Verlag, Berlin, 1980.
7. Hofmann, K.H. and Stralka, A.R.: The algebraic theory of
   Lawson semilattices—Applications of Galois connections
   to compact semilattices, **Diss. Math.** 137(1976), 1-54.
8. Lawson, J.D.: Topological semilattices, **J. London Math.**
   Soc. 2(1969), 719-724.
9. Novak, D.: Generalization of continuous posets, **Trans.**
   Amer. Math. Soc. 272(1982), 645-667.
10. Plotkin, G.D.: A powerdomain construction, **SIAM J. Comp.**
    5(1976), 452-487.
11. Raney, G.N.: A subdirect-union representation for com-
    pletely distributive complete lattices, **Proc. Amer.**
    Math. Soc. 4(1953), 518-522.
12. Raney, G.N.: Tight Galois connections and complete dis-
    tributivity, **Trans. Amer. Math. Soc.** 97(1960), 418-426.
13. Scott, D.S.: Outline of a mathematical theory of compu-
    tation, in **Proceedings of the 4th Annual Princeton Con-**
    ference on Information Science and Systems, Princeton
    University, 1970, pp. 169-176.
14. Scott, D.S.: Continuous lattices, in F.W. Lawvere (ed.),
    Toposes, Algebraic Geometry and Logic, Lecture Notes in
    Mathematics, vol. 274, Springer-Verlag, Berlin, 1972,
    pp. 97-136.
15. Wright, J.B., Wagner, E.G. and Thatcher, J.W.: A uniform
    approach to inductive posets and inductive closure, **The-**
    oretical Computer Science, 7(1978), 57-77.
16. Xu, X.Q.: Embedding M-continuous lattices in cubes, in
    Proceedings of the 5th IFSA World Congress, Seoul, 1993,
    pp. 347-377.
17. Xu, X.Q.: Construction of homomorphisms of M-continuous
    lattices, **Trans. Amer. Math. Soc.** (in press).

ON FUZZY QUANTIFIERS

HELMUT THIELE
University of Dortmund, Department of Computer Science 1,
D-44221 Dortmund, Germany

Abstract

First, A.MOSTOWSKI'S *concept of generalized quantifiers introduced in 1957 has been translated into fuzzy logic and has been compared with the concept of fuzzy quantifiers that was introduced by* L.A.ZADEH *in 1983 and has been investigated by him, by* R.R.YAGER *and others in the following years.*

Secondly, using the notions of T-norm and S-norm a new class of fuzzy universal and existential quantifiers, respectively, has been created.

Thirdly, there are approaches for defining fuzzy quantifiers like "almost all," "most," and "many" in arbitrary universes, where the quantifier "almost-all" is used as basic quantifier.

1. Introduction

First of all, we introduce some notations. The set of real numbers r with $0 \le r \le 1$ is denoted by $\langle 0, 1 \rangle$ its power set by $I\!P \langle 0, 1 \rangle$. For an arbitrary set $S \subseteq \langle 0, 1 \rangle$ we denote the infimum and the supremum of S in $\langle 0, 1 \rangle$ by $inf S$ and $sup S$ respectively. Furthermore, $\langle 0, 1 \rangle^n$ denotes the n-fold cartesian product of $\langle 0, 1 \rangle$ (n an integer, $n \ge 1$).

If U is an arbitrary set, then $card\, U$ is said to be the cardinal number of U. A fuzzy set F on U is a mapping of the form $F : U \to \langle 0, 1 \rangle$, i. e. we do not distinguish between a fuzzy set and its membership function. The set of all fuzzy sets on U is denoted by $\langle 0, 1 \rangle^U$.

For arbitrary fuzzy sets F and G on U we define the ("crisp") inclusion relation $F \subseteq G$ to be equivalent to the fact that $F(x) \le G(x)$ holds for every $x \in U$.

For a mapping $\varphi : U \to U$ we define the transformed fuzzy set as follows.

$$\varphi(F)(x) =_{\text{def}} F(\varphi(x)) \qquad (x \in U)$$

As usual, a mapping $\varphi : U \to U$ is said to be a bijection on U if and only if φ is surjective (i. e. a mapping onto U) and φ is injective (i. e. $\varphi(x) = \varphi(y)$ implies $x = y$ for every $x, y \in U$).

Z. Bien and K. C. Min (eds.),
Fuzzy Logic and its Applications, Information Sciences, and Intelligent Systems, 343–352.
© 1995 *Kluwer Academic Publishers.*

Finally, for arbitrary $c \in \langle 0, 1 \rangle$, $x, y \in U$, and $F : U \rightarrow \langle 0, 1 \rangle$ we define

$$(F < x := c >)(y) =_{\text{def}} \begin{cases} c & \text{if} \quad y = x \\ F(y) & \text{if} \quad y \neq x \end{cases}$$

MOSTOWSKI [10] and RESCHER [13] have introduced quantifiers as second-order predicates of two-valued and multiple-valued logic, respectively. Also ZADEH formulates the idea that a fuzzy quantifier is to be considered as a second-order predicate (see [23], page 756).

Following these intentions, we define

Definition 1.1
QUANT is said to be a general fuzzy quantifier on U $=_{\text{def}}$

$$QUANT : \langle 0, 1 \rangle^U \rightarrow \langle 0, 1 \rangle.$$

Reading papers of ZADEH, YAGER, and others on the subject "fuzzy quantifiers" one can very often find the assumption that the universe U is finite or, more general, that for a fuzzy set F on U the set $\{x \mid F(x) \neq 0 \wedge x \in U\}$ is finite (" Let A be a finite fuzzy subset of the universe of discourse U", see [23], page 756).

The reason for this restriction is that fuzzy quantifiers are associated with certain concepts of cardinality of a fuzzy set.

Among these possible concepts, the "nonfuzzy cardinality" of a fuzzy set F on U, expressed as sigma-count in the form

$$\Sigma Count(F) =_{\text{def}} \sum_{x \in U} F(x)$$

is very simple and convenient, but it works only in the "finite case", in general.

So, one could think that one should restrict all considerations to the "finite case", in particular, also with respect to the applications.

To this assumption we present the following counter-arguments: If we restrict the logical investigations to finite universes, the we must expect the same difficulties as in the two-valued case, namely as KALMÁR and TRAKHTENBROT have proved (see [8, 17]) that the computational complexity of logic in finite domains is significantly higher than the computational complexity of logic in arbitrary (non-empty) domains.

Furthermore, also the applications demand an infinite universe U, in many cases.

Both in theory and in applications certain restrictions of the concept of general fuzzy quantifier play an important role.

In order to define these restrictions we introduce the following three equivalence relations $\approx_{\text{iso}}$, $\approx_{\text{card}}$, and $\approx_{\text{val}}$ for arbitrary fuzzy sets F and G on U.

Definition 1.2
1. F and G are said to be **isomorphic** *($F \approx_{\text{iso}} G$) if and only if there exists a bijection φ on U such that $\varphi(F) = G$.*

2. F aud G are said to be **cardinality equivalent** *($F \approx_{\text{card}} G$) if and only if for every real number $r \in \langle 0, 1 \rangle$, the equation*

$$card\{x \mid x \in U \wedge F(x) = r\} = card\{x \mid x \in U \wedge G(x) = r\}$$

holds.

3. F and G are said to be **value equivalent** *($F \approx_{\text{val}} G$) if and only if the equation*

$$\{F(x) \mid x \in U\} = \{G(x) \mid x \in U\}$$

holds.

Conclusion 1.1

For arbitrary fuzzy sets F and G on U we have
1. *$F \approx_{\text{iso}} G$ if and only if $F \approx_{\text{card}} G$.*
2. *If $F \approx_{\text{card}} G$, then $F \approx_{\text{val}} G$.*
3. *The inversion of 2 does not hold, in general.*

Using definition 1.2 we introduce the following important restrictions of the concept of general fuzzy quantifier, following the intentions of MOSTOWSKI, RESCHER, and ZADEH.

Definition 1.3

1. A general fuzzy quantifier QUANT on U is said to be a **cardinal quantifier** *if and only if for every fuzzy set F and G on U, the condition $F \approx_{\text{card}} G$ implies $QUANT(F) = QUANT(G)$.*

2. QUANT is said to be **extensional** *if and only if for every fuzzy set F and G on U, the condition $F \approx_{\text{val}} G$ implies $QUANT(F) = QUANT(G)$.*

The relation $F \approx_{\text{card}} G$ can be interpreted in the sense that F and G have the same "cardinality". The intuitive motivation of the definition of $F \approx_{\text{card}} G$ is the conception that the family

$$(card\{x \mid x \in U \wedge F(x) = r\})_{r \in \langle 0, 1 \rangle}$$

characterizes a "natural size" of the fuzzy set F as GOTTWALD has proposed [5].

THIELE [16] has considered the inversion of the mapping

$$r \mapsto card\{x \mid x \in U \wedge F(x) = r\} \qquad (r \in \langle 0, 1 \rangle)$$

and introduced this innversion as "size" $SIZE(F)$ of the fuzzy set F, where we underline that $SIZE(F)$ is a "multiple" fuzzy set on the domain of all (classical) cardinal numbers. In [16], an "arithmetic" for $SIZE(F)$ is developed. Furthermore, the problem under which restrictions to F the multiple fuzzy set $SIZE(F)$ is an "ordinary" (i. e. not multiple) fuzzy set and, moreover, is a fuzzy cardinal number is discussed.

Using this terminology, we can state that for a cardinal quantifier $QUANT$, the real number $QUANT(F)$ only depends on the "size" of F, i. e. that there exists a function φ from the domain of all "sizes" of fuzzy sets on U into the set $\langle 0, 1 \rangle$ such that the equation

$$QUANT(F) = \varphi(SIZE(F))$$

holds for all fuzzy sets F on U.

Conclusion 1.2

For an arbitrary fuzzy quantifier on U we have

1. $QUANT$ is a cardinal quantifier if and only if it is "invariant" with respect to isomorphic fuzzy sets, i. e. if for every fuzzy set F and G on U, the condition $F \approx_{\text{iso}} G$ implies $QUANT(F) = QUANT(G)$.

2. If $QUANT$ is an extensional quantifier, then $QUANT$ is a cardinal quantifier. But not vice versa, in general.

3. For a given extensional quantifier $QUANT$ on U there exists a mapping $\psi : I\!\!P\langle 0, 1 \rangle \to \langle 0, 1 \rangle$ such that for every fuzzy set F on U, the equation

$$QUANT(F) = \psi(\{F(x) \mid x \in U\})$$

holds.

Using this characterization, we can define the well-known quantifiers ALL and EX (on U) as extensional quantifiers for an arbitrary fuzzy set F on U by

$$
\begin{aligned}
ALL(F) \quad &=_{\text{def}} \quad inf\{F(x) \mid x \in U\} \\
EX(F) \quad &=_{\text{def}} \quad sup\{F(x) \mid x \in U\}.
\end{aligned}
$$

2. On T-Quantifiers and S-Quantifiers

We start with the (trivial) observation that the extensional fuzzy quantifiers ALL and EX can be generated by iteration of the T-norm min and the S-Norm max, respectively. Basis of this iterative construction is the fact that both of the functions min and max are commutative and associative.

If we accept that beside the functions min and max other T- and S-norms play an important role both in theory and practice we are faced with the problem to consider and to apply iterations of such norms. Infinite iterations of T- and S-norms lead to the concept of T-quantifier and S-quantifier, respectively [15].

Let τ be an arbitrary function with $\tau : \langle 0, 1 \rangle^2 \to \langle 0, 1 \rangle$, in special cases let τ be a T-norm or an S-norm. By induction on the integer $n \geq 1$ we define an n-ary function $\tau^n : \langle 0, 1 \rangle^n \to \langle 0, 1 \rangle$ and after this an "ALL_τ-quantifier" and an "EX_τ-quantifier" generated by τ as follows, where $r_1, \ldots, r_n, \ldots \in \langle 0, 1 \rangle$.

Definition 2.1

1. $\tau^1(r_1) =_{\text{def}} r_1$
2. $\tau^{n+1}(r_1, \ldots, r_n, r_{n+1}) =_{\text{def}} \tau(\tau^n(r_1, \ldots, r_n), r_{n+1})$
3. $ALL_\tau(F) =_{\text{def}} inf\{\tau^n(F(x_1), \ldots, F(x_n)) \mid n \geq 1 \wedge x_1, \ldots, x_n \in U\}$
4. $EX_\tau(F) =_{\text{def}} sup\{\tau^n(F(x_1), \ldots, F(x_n)) \mid n \geq 1 \wedge x_1, \ldots, x_n \in U\}$

Obviously, we get

$$ALL = ALL_{min} \quad \text{and} \quad EX = EX_{max}.$$

For shortness, we consider only the "T-case" in the following.

Theorem 2.1
If τ is a T-norm, then ALL_τ satisfies the following axioms.
TQ$_0$ ALL_τ *is a cardinal quantifier*
TQ$_1$ *For every fuzzy set F on U and every $x \in U$, if $F(y) = 1$ for every $y \in U$*
       with $y \neq x$, then $ALL_\tau(F) = F(x)$.
TQ$_2$ *For every fuzzy set F on U, if there is an $x \in U$ with $F(x) = 0$,*
       then $ALL_\tau(F) = 0$.
TQ$_3$ *For every fuzzy set F and G on U, if $F \subseteq G$, then $ALL_\tau(F) \leq ALL_\tau(G)$.*
TQ$_4$ *For every fuzzy set F on U and every mapping $\beta : U \to U$,*
       if β is a bijection on U, then $ALL_\tau(\beta(F)) = ALL_\tau(F)$.
TQ$_5$ *For every fuzzy set F and G on U and for every $x, y \in U$, the equation*
       $ALL_\tau(F < x := ALL_\tau(G) >) = ALL_\tau(G < y := ALL_\tau(F < x := G(y) >) >)$
       holds.

This theorem leads to the following

Definition 2.2
A general fuzzy quantifier $QUANT$ on U is said to be a T-quantifier if and only if $QUANT$ satisfies the axioms TQ$_1$, TQ$_2$, TQ$_3$, TQ$_4$, and TQ$_5$.

In order to formulate the following theorem, we define the weakly drastic ALL-quantifier ALL^{wd} and the drastic ALL-quantifier ALL^d as follows, where F is an arbitrary fuzzy set on U.

Definition 2.3
1. $ALL^{wd}(F) =_{\text{def}} \begin{cases} F(y) & \text{if } F(x) = 1 \text{ for every } x \in U \text{ with } x \neq y \\ 0 & \text{else} \end{cases}$

2. $ALL^d(F) =_{\text{def}} \begin{cases} 1 & \text{if } F(x) = 1 \text{ for every } x \in U \\ 0 & \text{else} \end{cases}$

Obviously, we get that ALL^{wd} is a cardinal T-quantifier. Furthermore, ALL^d is extensional, but not a T-quantifier, and for every fuzzy set F on U, the inequation

$$ALL^d(F) \leq ALL^{wd}(F)$$

holds.

If we define the weakly drastic fuzzy conjunction et^{wd} and drastic fuzzy conjunction et^d for $r, s \in \langle 0, 1 \rangle$ by

$$et^{wd}(r, s) =_{\text{def}} \begin{cases} r & \text{if } s = 1 \\ s & \text{if } r = 1 \\ 0 & \text{if } 0 \leq r < 1 \text{ and } 0 \leq s < 1 \end{cases}$$

$$et^d(r, s) =_{\text{def}} \begin{cases} 1 & \text{if } r = s = 1 \\ 0 & \text{if } 0 \leq r < 1 \text{ or } 0 \leq s < 1 \end{cases}$$

then we get

$$ALL^{wd} = ALL_{et\,wd} \quad \text{and} \quad ALL^d = ALL_{et\,d}.$$

Theorem 2.2
If $QUANT$ is a T-quantifier on U, then for every fuzzy set F on U,

$$ALL^{wd}(F) \leq QUANT(F) \leq ALL(F).$$

Definition 2.4
A general quantifier $QUANT$ on U is said to be idempotent if and only if for every $c \in \langle 0, 1 \rangle$, the equation $QUANT(F) = c$ holds for every fuzzy set F on U with $F(x) = c$ for every $x \in U$.

Theorem 2.3
If $QUANT$ is an idempotent T-quantifier on U, then for every fuzzy set F on U,

$$QUANT(F) = ALL(F).$$

3. On the Mutual Definability of T-Norms and T-Quantifiers

From Theorem 2.1 the following two questions arise [15].
Question 1. Can every extensional T-quantifier be generated from a suitable T-norm?
Question 2. Is the generation procedure described by definition 2.1 uniquely reversible, i. e. can a T-quantifier be generated from at most one T-norm?

In the following we give a positive answer to both questions.
For this end we construct a T-norm τ_Q from a given cardinal T-quantifier Q on U, where F is a fuzzy set on U and $r, s \in \langle 0, 1 \rangle$.

Definition 3.1
$\tau_Q(r, s) =_{\text{def}} Q(F)$ *if there exists $x, y \in U$ with $F(x) = r$ and $F(y) = s$, and for every $z \in U$, if $z \neq x$ and $z \neq y$, then $F(z) = 1$.*

Conclusion 3.1
1. *τ_Q is total if $card\,U \geq 2$.*
2. *τ_Q is uniquely defined and does not depend on U if Q is a cardinal quantifier.*

Theorem 3.2
If $card\,U \geq 2$ and Q is a cardinal T-quantifier, then τ_Q is a T-norm

From this theorem we get the following three further questions:

Question 3. Can every T-norm be generated from a suitable extensional T-quantifier?

Question 4. Is the generation procedure described by definition 3.1 uniquely reversible, i. e. can a T-norm be generated from at most one T-quantifier?

Question 5. Which result will we get if we execute the generation procedure described by the definitions 2.1 and 3.1 one after the other?

The following two theorems give a complete answer to the five questions asked above.

Theorem 3.3

If $card\,U \geq 2$ and τ is a T-norm, then

1. ALL_τ *is a cardinal T-quantifier,*
2. τ_{ALL_τ} *is a T-norm, and*
3. $\tau_{ALL_\tau} = \tau$

Theorem 3.4

If $card\,U \geq 2$ and Q is a cardinal T-quantifier, then

1. τ_Q *is a T-norm,*
2. ALL_{τ_Q} *is a cardinal T-quantifier, and*
3. $ALL_{\tau_Q} = Q$.

4. The Quantifiers ALMOST-ALL and INF-EX as Basic Quantifiers

In two-valued logic and in classical (crisp) mathematics, for instance in analysis, the *ALMOST-ALL* plays an important role.

For a two-valued predicate P on U, i. e. for $P : U \to \{0, 1\}$, the *ALMOST-ALL* quantifier is defined by the condition

$ALMOST\text{-}ALL\,(P) = 1$ if and only if there exists a finite set $V \subseteq U$

such that the equation $P(x) = 1$ holds for all $x \in U \setminus V$.

Furthermore, from two-valued logic we know that the quantifier *ALMOST-ALL* is dual to the quantifier *INF-EX* which can be characterized by the condition

$INF\text{-}EX = 1$ if and only if there exists an infinite set $V \subseteq U$

such that the equation $P(x) = 1$ holds for all $x \in V$.

Now, we transform these characterizations equivalently into the following form that can be immediately taken over into fuzzy logic [14].

For definiteness we recall the following definitions. Let F be an arbitrary fuzzy set on U. We denote the Lukasiecicz complement of F by $\bar{F}$, i. e. the fuzzy set defined by

$$(\bar{F})(x) =_{\text{def}} 1 - F(x) \qquad (x \in U).$$

Furthermore, let Q_1 and Q_2 be general fuzzy quantifiers on U. These quantifiers are said to be dual if and only if for every fuzzy set F on U, the equation

$$Q_1(F) = 1 - Q_2(\bar{F})$$

holds.

Definition 4.1

1. $ALMOST\text{-}ALL(F) =_{\mathrm{def}} sup\{inf\{F(x) \mid x \in U \setminus V\} \mid V \subseteq U$ and V is finite$\}$
2. $INF\text{-}EX(F) =_{\mathrm{def}} sup\{inf\{F(x) \mid x \in V\} \mid V \subseteq U$ and V is infinite$\}$

Conclusion 4.1

1. *The quantifiers ALMOST-ALL and INF-EX are dual.*
2. *For every finite universe U and every fuzzy set F on U, we have*
    $ALMOST\text{-}ALL(F) = 1$ *and* $INF\text{-}EX(F) = 0$.

Obviously, conclusion 4.1.2 describes a very bad modelling of the meaning of "almost all" in natural languages, because even for the empty fuzzy set E on a finite universe (i. e. $E(x) = 0$ for all $x \in U$) we get

$$ALMOST\text{-}ALL(E) = 1.$$

For this reason we restrict definition 4.1 to infinite universes and, hence, we have to ask for a better definition in the case that U is finite. We propose for non-empty **finite** U and an arbitrary fuzzy set F on U.

Definition 4.2

1. $ALMOST\text{-}ALL(F) =_{\mathrm{def}} \frac{1}{card\,U} \cdot \sum_{x \in U} F(x)$

2. $INF\text{-}EX(F) =_{\mathrm{def}} \frac{1}{card\,U} \cdot \sum_{x \in U} F(x)$

In order to motivate definition 4.2.1 one can think that this definition which is closely related to the relative sigma count, is a good fuzzy modelling of the meaning of "allmost all" in natural languages in the case that U is finite. As one can easily prove, definition 4.2.2 is a logical consequence of 4.2.1 if we additionally demand the quantifiers *ALMOST-ALL* and *INF-EX* to be dual in the case of finite U.

Conclusion 4.2

For every non-empty finite U, the quantifier ALMOST-ALL (and also INF-EX) is a cardinal quantifier (even independent of its definition according to definition 4.1 or 4.2).

Using *ALMOST-ALL* as basic quantifier, we propose to define further (fuzzy) quantifiers as follows where U is an arbitrary (non-empty) universe and F is an arbitrary fuzzy set on U.

Definition 4.3

1. $MOST(F) =_{\mathrm{def}} \sqrt{ALMOST\text{-}ALL(F)}$
2. $MANY(F) =_{\mathrm{def}} \sqrt{MOST(F)}$
3. $ALMOST\text{-}NONE(F) =_{\mathrm{def}} 1 - ALMOST\text{-}ALL(F)$
4. $FEWEST(F) =_{\mathrm{def}} 1 - MOST(F)$
5. $FEW(F) =_{\mathrm{def}} 1 - MANY(F)$

Remark: Other approaches to study quantifiers as defined in definition 4.3 can be found in [1, 2, 9].

Acknowledgement

The author wishes to thank Stephan Lehmke and Norbert Schmechel for useful discussions on the subject and their help in preparing the manuscript.

References

[1] J. BARWISE and R. COOPER: Generalized Quantifiers and Natural Language. Linguistics and Philosophy 4 (1980), 159-219.

[2] J. BARWISE and F. FEFERMAN (Eds.): Model-Theoretic Logics. Series: Perspectives in Mathematical Logics. Springer-Verlag, 1985.

[3] D. DUBOIS and H. PRADE: On fuzzy syllogisms. Computational Intelligence 4 (1988), 171–179.

[4] K. GOEDEL: Die Vollstaendigkeit der Axiome des logischen Funktionenkalkuels. Monatshefte fuer Mathematik und Physik 37 (1930), 349–360.

[5] S. GOTTWALD: A Note on Fuzzy Cardinals. Kybernetika 16 (1980), 156-158.

[6] S. GOTTWALD: Fuzzy Sets and Fuzzy Logic. Foundations of Application — from a Mathematical Point of View. Vieweg 1993.

[7] A. DeLUCA and S. TERMINI: A definition of non–probabilistic entropy in the setting of fuzzy sets. Information and Control 20 (1972), 301–312.

[8] L. KALMÁR: Constristions to the reduction theory of the decision problem. Fourth paper: Reduction to the case of a finite set of individuals. Acta Mathematica Academiae Scientiarum Hungaricae 2 (1951), 125–142.

[9] H.J. KEISLER: Logic with the Quantifier "there exists uncountably many". Ann. Math. Logic, 1 (1970), 1-93.

[10] A. MOSTOWSKI: On a generalization of quantifiers. Fundamenta mathematicae 44 (1957), 12-36.

[11] V. NOVÁK: Fuzzy Sets and their Applications. Adam Hilger 1989.

[12] P. PETERSON: On the Logic of Few, Many and Most. Notre Dame J. Formal Logic 20 (1979), 155-179.

[13] N. RESCHER: Many-valued Logic. McGraw-Hill Book Company 1969.

[14] H. THIELE: On Fuzzy Quantifiers. Fifth International Fuzzy Systems Association World Congress '93, July 4–9, 1993, Seoul, Korea. Conference Proceedings. Volume I, 395–398.

[15] H. THIELE: On T-Quantifiers and S-Quantifiers. Twenty-Fourth International Symposium on Multiple-Valued Logic, May 25–27, 1994, Boston, Massachusetts. Conference Proceedings, 264–269.

[16] H. THIELE: On the Concept of Cardinal Number for Fuzzy Sets. Invited Paper, EUFIT'94 (European Congress on Fuzzy and Intelligent Technologies), Aachen, Germany, September 20-23, 1994. Conference Proceedings, vol. 1, 504-516.

[17] B.A. TRAKHTENBROT: On the algorithmic unsolvability of the decision problem in finite domains. (in Russian). Dokl. Akad. Nauk SSSR 70 (1950), 569–572.

[18] R.R. YAGER: Reasoning with fuzzy quantified statements, part I. Kybernetes 14 (1985), 233-240.

[19] R.R. YAGER: Connectives and Quantifiers in Fuzzy Sets. Fuzzy Sets and Systems 40 (1991), 39-75.

[20] L.A. ZADEH: A computational approach to fuzzy quantifiers in natural language. Comp. Math. Appl. 9 (1983), 149–184.

[21] L.A. ZADEH: A computational theory of dispositions. In: Proc. 1984 Int. Conference Computational Linguistics (1984), 312-318. See also: [ZAD87A].

[22] L.A. ZADEH: Syllogistic reasoning as a basis for combination of evidence in expert systems. In: Proceedings of IJCAL, Los Angeles, CA (1985), 417–419.

[23] L.A. ZADEH: Syllogistic reasoning in fuzzy logic and its application to usuality and reasoning with dispositions. IEEE Transactions on Systems, Man, and Cybernetics 15 (1985), 754–763.

[24] L.A. ZADEH: On computational theory of dispositions. International Journal of Intelligent Systems 2 (1987), 39–63.

[25] L.A. ZADEH: Dispositional logic and commonsense reasoning. In: Proceedings of the Second Annual Artificial Intelligence Forum, NASA–Ames Research Center, Moffett–Field, CA (1987), 375-389.

[26] L.A. ZADEH: Dispositional logic. Appl. Math. Lett. 1 (1988), 95–99.

TOWARDS FORMALIZED INTEGRATED THEORY
OF FUZZY LOGIC

VILÉM NOVÁK
University of Ostrava
Faculty of Natural Sciences, Dept. of Mathematics
Bráfova 7, 70100 Ostrava 1, Czech Republic

and

Institute of Information and Automation Theory
Academy of Sciences of the Czech Republic
Pod vodárenskou věží 4, 186 02 Praha 8, Czech Republic

Abstract. The paper presents the idea to establish the theory of fuzzy logic in broader sense (approximate reasoning) to integrate the results of fuzzy logic in narrow sense, parts of the theory of linguistic semantics and possibly also other branches. The goal is to make it a precise formal theory. Some formal issues towards this goal presented elsewhere are recalled and a way to formulation of theorems is outlined.

1. Introduction

Fuzzy logic can be generally be characterized as a theory of finding imprecise conclusions from imprecise knowledge. Historically, there are two meanings of it: fuzzy logic in *narrow sense* and in *broader sense*[1]. Fuzzy logic in narrow sense is a special case of many-valued logic presented in [7, 8, 13, 15] which is rich enough to include other systems of fuzzy logic and to offer a formal background for graded approach to vagueness.

Fuzzy logic in broader sense is its extension to the approximate reasoning. Two main directions in the development of this theory may be distinguished.

---

[1] There is also fuzzy logic in broad sense, which means any kind of application of fuzzy concept. However, this interpretation will not be considered in this paper.

Z. Bien and K. C. Min (eds.),
Fuzzy Logic and its Applications, Information Sciences, and Intelligent Systems, 353–363.
© 1995 *Kluwer Academic Publishers.*

The first direction assumes that an imprecise and incomplete information about a some (precise) function $Y = F(X)$ is given. Our aim is to find an unknown functional value from imprecisely given argument. As has been demonstrated (see e.g. [6, 12]), such a function can be described by means of a disjunction of linguistic conjunctions

$$\mathcal{R}_i := X \text{ is } \mathcal{A}_i \text{ AND } Y \text{ is } \mathcal{B}_i \qquad i = 1, \ldots, m \tag{1}$$

where $\mathcal{A}_i, \mathcal{B}_i$, are some linguistic (i.e., imprecise) expressions. Equivalently, it can also be described by conjunction of implications

$$\mathcal{R}_i := \text{IF } X \text{ is } \mathcal{A}_i \text{ THEN } Y \text{ is } \mathcal{B}_i \qquad i = 1, \ldots, m \tag{2}$$

which, however, are *classical* implications consisting of imprecise statements. The formula for computation of an unknown "functional" value of F is the well known Max–Min (Sup–Inf) projection of a fuzzy set $A' \subseteq U$ in the fuzzy relation $R \subseteq U \times V$ given by the formula

$$B'y = \bigvee_{x \in U} (A'x \wedge R\langle x, y \rangle) = \bigvee_{x \in U} \left(A'x \wedge \bigvee_{i=1}^{m} (A_i x \wedge B_i y) \right) \tag{3}$$

for all $y \in V$ where $A', A_i \subseteq U$, $B_i \subseteq V$ are fuzzy sets being interpretations of the linguistic expressions 'X is $\mathcal{A}'$', 'X is $\mathcal{A}_i$', 'Y is $\mathcal{B}_i$', $i = 1, \ldots, m$, respectively. The infimum ($\wedge$) operation in (3) may be replaced by a general T-norm. We may see that, strictly speaking, approximate reasoning is the imprecise (fuzzy) interpolation of imprecisely known precise function.

The second direction aims at constitution of approximate reasoning as a logical theory. In the frame of it we deal with a set of linguistically specified special axioms

$$\mathcal{R} = \{\mathcal{R}_1, \mathcal{R}_2, \ldots, \mathcal{R}_m\} \tag{4}$$

where the rules $\mathcal{R}_i$, $i = 1, \ldots, m$ have the form (2) and they are interpreted as fuzzy logical implications. On the basis of this assumption we obtain a formula analogous to (3)

$$B'y = \bigvee_{x \in U} \left(A'x \otimes \bigwedge_{i=1}^{m} (A_i x \to B_i y) \right) \tag{5}$$

where $\otimes$ is the operation of Łukasiewicz product and $\to$ that of residuation.

This direction requires elaboration of the theory of approximate reasoning as a formal logical theory. Its basic formal frame could be fuzzy logic in narrow sense. Recall that this kind of logic has been proved to preserve many formal properties of classical logic (probably the most possible ones). Among most important ones is the syntactico–semantical completeness property which assures us that the formal deduction is in a good balance with its semantics.

In [10, 11], the formal theory of approximate reasoning has been proposed which is based on first–order fuzzy logic in narrow sense. In this paper, we propose a program for fuzzy logic to be established as a general formal theory which would integrate both meanings mentioned above and selected parts of linguistics into the *theory of human reasoning* with good mathematical foundations. Of course, influence from other branches such as probability theory, category theory, theory of algorithms, evolutionary programming, but also psychology and others is necessary for its successful development.

Since fuzzy logic in narrow sense has been described already in several papers and books, we demonstrate our view on the foundations of the second part of fuzzy logic beginning with the integration of linguistics and the logic in narrow sense.

2. Formal Theory of Fuzzy Logic In Broader Sense

Classical logic became metalanguage of precise formal reasoning. Analogously, fuzzy logic in narrow sense should be metalanguage of the imprecise formal reasoning. This statement is supported by the long discussion arguing that graded approach is a sound, well justified view on the vagueness phenomenon.

Consider a formal language ^{AR}J of fuzzy logic in narrow sense extended by a set $\{c_j; j \in J\}$ of additional connectives and containing a sufficiently large set K of constants. By ^{AR}M we denote a set of all terms without variables of a language ^{AR}J. By L we denote the set of truth values. It is assumed to have a structure of the residuated lattice. Usually, $L = [0, 1]$.

Crucial role in human reasoning is played by natural language and so we have to include also some linguistic concepts into fuzzy logic in broader sense. As most distinguished feature of natural language semantics is its vagueness, we are led to the use of fuzzy logic in narrow sense. Of course, we will not attempt to model the linguistic semantics as a whole. For our purposes, we may confine only to a suitable part of natural language, namely that playing crucial role in drawing conclusions about behaviour of phenomena. We will deal with selected linguistic expressions which constitute the basis of human reasoning. These expressions will then be translated into formulas of fuzzy logic in narrow sense and finally lead to construction of a special first-order fuzzy theory ^{AR}T.

More formally, we deal with a set S of syntagms[2] of natural language

---

[2]By syntagm we will call any syntactically correct part of the sentence which is se-

whose main part consists of the so called *evaluating expressions*, i.e., the linguistic expressions characterizing the position on an ordered scale (usually on a set of real numbers). Natural language expression are, in general, names of properties φ of objects[3]. The decision whether a concrete object x_0 has a property φ, i.e., whether $\varphi(x_0)$ holds, is tantamount to the question whether $\varphi(x_0)$ is true, or not. Since the property φ is *vague*, such a question cannot be unambiguously answered. Sharp, unambiguous answer is practically impossible and thus, we should use some kind of a scale whose elements would express various degrees of truth of $\varphi(x_0)$. We have naturally come to fuzzy logic in narrow sense. It has been demonstrated many times that it is quite natural for the human mind to use a scale when an unsharp grouping is to be characterized. We will use the term *fuzzy approach* or *graded approach* for this. Of course, using a scale is somewhat superficial solution since the study of the inner structure of the property φ is replaced by some grading. On the other hand, fuzzy approach is satisfactory from many points of view and proved to be successful in applications.

Let $\mathcal{A}$ be a natural language expression. It can be assigned a formula $A(x)$ of the language ^{AR}J. However, the formula itself does not characterize the vagueness of $\mathcal{A}$. To characterize it, we will consider the set of evaluated formulas

$$\underline{A} = \{[A_x[t];\ a_t] \mid t \in\ ^{AR}M\}. \tag{6}$$

The $A_x[t]$ denotes the formula $A(x)$ after replacing the variable x by the term $t \in\ ^{AR}M$. These terms are *names* of the objects which have the property $A_x[t]$ named by the linguistic expression $\mathcal{A}$.

Note that the set (6) is equivalent with the fuzzy set

$$\underline{A} = \{\ ^{a_t}/A_x[t] \mid t \in\ ^{AR}M\}. \tag{7}$$

In general, each syntagm $\mathcal{A} \in \mathcal{S}$ is assigned a formula $A(x)$ and a set (6), i.e., it is assigned the couple

$$\langle A(x), \underline{A}\rangle. \tag{8}$$

More exact characterization of the set of syntagms $\mathcal{S}$ is difficult. The following set of syntagms should at least be a subset of $\mathcal{S}$:

1. Let $\mathcal{A} := [\langle\text{linguistic modifier}\rangle]\langle\text{adjective}\rangle$ where $\langle$linguistic modifier$\rangle$ is an intensifying adverb with narrowing or extending effect and $\langle$adjective$\rangle$ characterizes a position on some ordered scale, e.g., "small", "medium", "big", etc. Then the clause

    '$\langle$noun$\rangle$ is $\mathcal{A}' \in \mathcal{S}$.

mantically meaningful.

[3]Of course, there are connectives, exclamations, etc. for which this assumption does not hold. However, these are not in our considerations now.

is a *simple syntagm*. Simple syntagms are syntagms.

2. Let $\mathcal{B}_1, \ldots, \mathcal{B}_n$ be syntagms. Then

$$\mathcal{B}_1 \text{ AND } \mathcal{B}_2 \text{ AND } \cdots \text{ AND } \mathcal{B}_n$$
$$\mathcal{B}_1 \text{ OR } \mathcal{B}_2 \text{ OR } \cdots \text{ OR } \mathcal{B}_n$$

are syntagms.

3. Let $\mathcal{B}$ and $\mathcal{C}$ be syntagms. Then

$$\text{IF } \mathcal{B} \text{ THEN } \mathcal{C}$$

is a syntagm.

The expression $\langle \ldots \rangle$ denotes a metavariable for the kind of the word given inside. Note that the clause

$$\text{'}\langle\text{noun}\rangle \text{ is } \langle\text{linguistic modifier}\rangle)]\langle\text{adjective}\rangle\text{'}$$

is tantamount to the syntagm

$$[\langle\text{linguistic modifier}\rangle]\langle\text{adjective}\rangle\langle\text{noun}\rangle.$$

The $\langle\text{noun}\rangle$ is usually replaced by the variable x.

Syntagms defined by the items 1. to 3. above belong to $\mathcal{S}$. We do not exclude the possibility to extend $\mathcal{S}$ by more kinds of syntagms. So far, however, the theory is well developed only for the above defined ones.

Let us now introduce the *translation rules*.

(a) The $\langle\text{adjective}\rangle$ is assigned a predicate symbol $p \in {}^{AR}J$.

(b) The $\langle\text{linguistic modifier}\rangle$ is assigned a unary connective $c \in {}^{AR}J$.

(c) The $\langle\text{noun}\rangle$ is assigned a variable $x \in {}^{AR}J$.

(d) A simple syntagm is assigned the couple (8).

(e) The connective AND is assigned a logical conjunction $\hat{\&}$ being interpreted by a logically fitting t-norm (see [8, 10, 13]). We very often put $\hat{\&} = \wedge$ but due to the intrinsic properties of the corresponding linguistic phenomenon called *close coordination*, this does not always work. Therefore, in various situations we should use other connectives.

(f) The connective OR leads to similar problems as AND. It is often interpreted by the logical disjunction $\vee$ but this may not always work. Therefore, OR should generally be interpreted by some logically fitting t-conorm.

(g) The connective 'IF ... THEN ...' should be understood as a linguistically expressed logical implication $\Rightarrow$ assigned the basic Łukasiewicz implication operation. There are also implication-like connectives constructed on the basis of various t-norms. The use of them, however, is not clear. Furthermore, due to the necessity to be logically fitting, they are isomorphs of the Lukasiewicz one (cf. [7, 15]).

Assume now that the set (4) consisting of the statements of the form (2) is given. This set will be called the *linguistic description*. It represents the expert knowledge and from our point of view it can be understood as a set of *linguistically expressed special axioms*.

Using the translation rules, each $\mathcal{R}_i$, $i = 1, \ldots, m$ is assigned the couple

$$\langle A_i(x) \Rightarrow B_i(y), \underline{A_i(x) \Rightarrow B_i(y)} \rangle$$

where

$$\underline{A_i(x) \Rightarrow B_i(y)} = \{[A_{i,x}[t] \Rightarrow B_{i,y}[s]; c_{ts}] \mid t, s \in {}^{AR}M\}. \tag{9}$$

Of course, we need not always suppose the $\mathcal{R}_i$ to have the form (2). They can be any syntagms from $\mathcal{S}$. As a special case, they can have the form (1), i.e., they are linguistic conjunctions. In this case, each $\mathcal{R}_i$ is assigned the couple

$$\langle A_i(x) \wedge B_i(y), \underline{A_i(x) \wedge B_i(y)} \rangle$$

where

$$\underline{A_i(x) \wedge B_i(y)} = \{[A_{i,x}[t] \wedge B_{i,y}[s]; c_{ts}] \mid t, s \in {}^{AR}M\}. \tag{10}$$

Disjunction of (1) can be interpreted as linguistic description of fuzzy function (fuzzy graph) (cf. [7, 16]).

Unlike (1), (2) is linguistically expressed logical implication. Note that implication (see [9], Section 6.1) is the most general form how a relation between phenomena can be characterized.

To be able to manipulate with expressions, we define the *inference rules*. In fuzzy logic in broader sense, they may be written in the following form:

$$\mathcal{R} := \frac{A_1, \ldots, A_n}{B} \left[\frac{\underline{A_1}, \ldots, \underline{A_n}}{\underline{B}} \right] \tag{11}$$

where B is a syntagm assigned to $A_1, \ldots, A_n$ using the rule $\mathcal{R}$. The square brackets contain analogous assignment of sets of evaluated formulas of the form (6) (or, equivalently, fuzzy sets of formulas of the form (7)). For example, the inference rule of *modus ponens* in fuzzy logic can be written as

$$\mathcal{R}_{MP} := \frac{A, \text{ IF } A \text{ THEN } B}{B} \left[\frac{\underline{A}, \underline{A \Rightarrow B}}{\underline{B}} \right], \tag{12}$$

or in detail,

$$\mathcal{R}_{MP} := \frac{A, \text{ IF } A \text{ THEN } B}{B}$$

$$\left[\frac{\{[A_x[t];\ a_t] \mid t \in {}^{AR}M\}, \{[A_x[t] \Rightarrow B_y[s];\ c_{ts}] \mid t, s \in {}^{AR}M\}}{\{[B_y[s];\ \bigvee_{t \in {}^{AR}M}(a_t \otimes c_{ts})]\}} \right]. \quad (13)$$

By this rule, we derive the linguistic statement B from the linguistic statements A and 'IF A THEN B'. The statement A is interpreted by a fuzzy set of formulas $\underline{A}$, 'IF A THEN B' by $\underline{A \Rightarrow B}$ and B by $\underline{B}$. Looking inside the rule (13), given a truth degree a_t of a formula $A_x[t]$ (from the (fuzzy) set $\underline{A}$) and a truth degree c_{ts} of a formula $A_x[t] \Rightarrow B_y[s]$ (from the fuzzy set $\underline{A \Rightarrow B}$), we compute the truth value b_s of the resulting formula $B_y[s]$ (from the (fuzzy) set $\underline{B}$) using

$$b_s = \bigvee_{t \in {}^{AR}M} (a_t \otimes c_{ts}). \quad (14)$$

The use of the supremum operation in (14) follows from the following theorem ([8, 15]).

Theorem 1

$$T \vdash_a A \text{ iff } a = \bigvee \{\text{Val}_T(w) \mid w \text{ is a proof of } A\}$$

holds for any formula A where $\text{Val}_T(w)$ is a value of the proof w.

Note that the formula (14) corresponds to the formula for the weak composition of fuzzy relations (cf. [7]) $B = A \overset{\circ}{\times} R$. However, the formula (14), has been derived purely on the basis of the fuzzy logic in narrow sense.

The necessary condition for inference rules is their soundness (see, e.g., [8, 15]). The rule (12) may easily be proved to be sound. Its soundness is not harmed even when considering a conjunction of implications.

However, when trying to explain Max-Min rule as a deduction rule from the point of view of fuzzy logic, we come to the following:

$$\mathcal{R}_{MM} := \frac{A_k, \text{ OR}_{i=1}^m(A_i \text{ AND } B_i)}{B_k} \left[\frac{A_k, \bigvee_{i=1}^m(A_i \wedge B_i)}{\underline{B_k}} \right] \quad (15)$$

where $1 \leq k \leq m$. Analogously as in the previous case, we derive the truth b_s of the formula $B_{k,y}[s]$ using

$$b_s = \bigvee_{t \in {}^{AR}M} (a_{kt} \wedge c_{ts}). \quad (16)$$

Note that (16) is exactly the formula for composition of fuzzy relations $B = A \circ R$ but, again, here it is derived on the basis of fuzzy logic. Unfortunately, considered as a formal rule, it is sound only if $\vdash A_i \wedge A_j \Leftrightarrow \mathbf{0}$ for $i \neq j$ (cf. [12]).

As has already been pointed out (cf. [6]), Max-Min rule is not the logical inference rule but the interpolation rule which, from the logical point of view represents equivalence of formulas (see [14]).

We may also define the concept of a formal proof in fuzzy logic in broader sense which is a sequence

$$\mathcal{B}_1, \ldots, \mathcal{B}_n$$

of linguistic statements each of which is a linguistically formulated axiom (logical or special), or it is derived using some inference rule defined above. This definition is analogous to the formal definition of the proof in classical logic. In approximate reasoning we, moreover, should seek inference rules modelling human reasoning as closely as possible. Due to the capability to grasp vagueness, this might be successful. We also do not exclude some probabilistic or possibilistic considerations at this problem.

Consider a structure (model) $\mathcal{D}$. Then $\mathcal{D}(t)$ is a concrete object and $\mathcal{D}(A_x[t])$ is a truth degree in which the object $\mathcal{D}(t)$ has the property A. This yields the fuzzy set

$$\{\, \mathcal{D}(A_x[t])/\mathcal{D}(t) \mid t \in {}^{AR}M \,\}.$$

Of course, when $A(x, y)$ is a formula with two or more variables then we obtain a (binary) fuzzy relation. We see that the common formulas used for computation of truth degrees which are based on various operations with fuzzy sets and fuzzy relations may be explained on the basis of syntactic and semantic manipulation with formulas of fuzzy logic, namely those from the support of the fuzzy sets of the form (7). When considering a model $\mathcal{D}$, we obtain a common fuzzy set/fuzzy relation representation. In approximate reasoning we work on syntactic level and interpret results in some model afterwards.

To summarize, some chosen syntagms $\mathcal{T} = \{\mathcal{A}_0, \ldots, \mathcal{A}_m\}$ take the role of linguistically expressed special axioms being the basis of approximate reasoning at the given moment. Then we obtain a certain first–order fuzzy theory ${}^{AR}T$ given by a fuzzy set of special axioms

$$^{AR}A_S = \underline{A_0} \cup \cdots \cup \underline{A_m}$$

(cf. the formal definition of the fuzzy theory in [8]). Hence, fuzzy logic in broader sense can be understood as a *multiple deduction in a fuzzy theory* ${}^{AR}T$ whose result are sets of evaluated formulas of the form (6).

However, Theorem 1 has also the following consequence. Finding a proof of a formula does not ensure us that we know its provability degree. To obtain it, we have to seek a sequence of still better proofs. In approximate reasoning, moreover, we have to find all the proofs with targets $A_x[t]$ for all $t \in {}^{AR}M$. Therefore, most formulas known for manipulation with fuzzy sets and fuzzy relations in approximate reasoning are only *lower* estimations of the required provability (truth) degree. An important but quite difficult task is to study the conditions under which they give also the highest possible provability degrees. Theorems below give us information about the quality of the formulas used in the theory of approximate reasoning.

The underlying formal representation of simple syntagm is a *simple formula* defined as follows:

(i) An atomic formula which is no $\boldsymbol{a}$, $a \in L$ is simple.

(ii) Let $\boldsymbol{c}$ be a surjective connective (i.e., the function $c : L \longrightarrow L$ assigned to it is surjective) and A a simple formula. Then $\boldsymbol{c}(A)$ and $\neg A$ are simple formulas.

Simple formulas A and B are *independent* if they have no common subformulas.

Using the completeness theorem of fuzzy logic (in narrow sense) we can prove the following theorem.

Theorem 2 *Let*

$$
{}^{AR}T = \{ \ a_{k,t}/A_{k,x}[t] \mid c_{ts}/ \bigwedge_{j=1}^{m} (A_{j,x}[t] \Rightarrow B_{j,y}[s]) \mid
$$

$$
t \in M_1, s \in M_2\}
$$

for some k, $1 \leq k \leq m$, where $A_j(x)$, $j = 1, \ldots, m$ are simple independent formulas and $M_1, M_2 \subseteq {}^{AR}M$, $M_1 \cap M_2 = \emptyset$. Then

$$
T \vdash_{b_s} B_{k,y}[s], \qquad s \in M_2
$$

where $b_s = \bigvee_{t \in M_1} (a_{k,t} \otimes c_{ts})$.

On the basis of Theorem 2 we may formulate the following one:

Theorem 3 *Let a theory of fuzzy logic be given by a linguistic description*

$$
\mathcal{T} = \{\mathcal{A}_k, \underset{j=1}{\overset{m}{\text{AND}}}(\text{IF } \mathcal{A}_j \text{ THEN } \mathcal{B}_j)\}
$$

for some k, $1 \leq k \leq m$ where $\mathcal{A}_j, \mathcal{B}_j$, $j = 1, \ldots, m$ are simple syntagms. Furthermore, $\mathcal{A}_j$ and $\mathcal{B}_j$ have no common noun and either $\mathcal{A}_j$ and $\mathcal{A}_i$ for $i \neq j$ have no common adjective or one of them has intensifying modifier with stronger effect than the other. Similarly, $\mathcal{B}_j$ and $\mathcal{B}_i$ for $j \neq i$ have no common adjective or one of them has weakening modifier with weaker effect than the other. Let the syntagms be interpreted by fuzzy sets of formulas $\underline{A_k}$, $\wedge_{j=1}^m (A_j \Rightarrow B_j)$ respectively, i.e., the theory $\mathcal{T}$ is assigned a fuzzy theory ^{AR}T given by the fuzzy set of special axioms

$$^{AR}A_S = \{\, ^{a_{kt}}/A_{k,x}[t]\} \cup \{\, ^{c_{ts}}/(\wedge_{j=1}^m (A_j \Rightarrow B_j))_{x,y}[t,s] \mid t \in M_1, s \in M_2\}$$

where M_1, M_2 are disjoint sets of terms. Then there is a conclusion $\mathcal{B}_k$ whose interpretation is

$$\underline{B_k} = \{\, ^{b_s}/B_k[s] \mid s \in M_2\}$$

where

$$b_s = \bigvee_{t \in M_1} (a_{k,t} \otimes c_{ts})$$

are the provability degrees $^{AR}T \vdash_{b_s} B_k[s]$, $s \in M_2$.

It follows from this theorem that when confining ourselves to the linguistic syntagms commonly used in approximate reasoning and taking care of different linguistic statements, the commonly used formula of approximate reasoning for IF–THEN statements considered as logical implications gives us the highest possible truth degrees. Analogous statement can also be proved for fuzzy approximation of the function.

3. Further development

We may see that the proposed formal apparatus is capable to express various tasks of fuzzy logic. The presentation of fuzzy logic in broader sense was oriented towards the theory of approximate reasoning, whose part is also elaboration of some natural language expressions. However, the reader has certainly noticed that the crucial moment is translation of linguistic expressions into sets of evaluated formulas (fuzzy sets of formulas). Therefore, we may restrict ourselves to the latter case only and elaborate calculus of sets of evaluated formulas. This is, of course, motivated by modeling of linguistic semantics but can be developed more or less independently. Hence, fuzzy logic in broader sense may be considered as the theory of approximate reasoning and a certain kind of second order fuzzy logic.

The future development of the formal theory of approximate reasoning and fuzzy logic in broader sense can be seen, besides other, in the following directions:

– Develop the theory of of sets of evaluated formulas, formal deduction with them, characterization of provability.

- Extend the theory of semantics of selected linguistic expressions to more complex kinds of them and study their connection with the laws of human reasoning.
- Study the connection of the theory due to previous item with the theory of sets of evaluated formulas, especially with the stress to formal reasoning, looking for other, more sophisticated inference rules including probabilistic and possibilistic reasoning.
- Develop the theory of generalized quantifiers.

A non-negligible fact is the demonstration of the close connection between fuzzy logics in narrow and broader sense. This makes us possible to exploit the results of the former and obtain general results which might have direct impact to applications.

References

1. Butnariu, D. and E. P. Klement: *Triangular Norm-based Measures and Games with Fuzzy Coalitions.* Kluwer, Dordrecht, 1993.
2. Dubois, D. and H. Prade: *Possibility Theory. An Approach to Computerized Processing of Uncertainty.* Plenum, New York, 1988.
3. Goguen, J. A.: The logic of inexact concepts, *Synthese* **19**(1968-69), 325–373.
4. Gottwald, S.: *Mehrwertige Logik.* Akademie-Verlag, Berlin, 1989.
5. Hájek, P.: Fuzzy Logic and Arithmetical Hierarchy. Preprint. Institute of Informatics, Academy of Sciences of Czech Republic, Prague, 1994.
6. Klawonn, F. and V. Novák: The Relation between Inference and Interpolation in the Framework of Fuzzy Systems. *Fuzzy Sets and Systems* (submitted).
7. Novák, V.: *Fuzzy Sets and Their Applications,* Adam–Hilger, Bristol, 1989.
8. Novák, V.: On the Syntactico-Semantical Completeness of First-order Fuzzy Logic. Part I — Syntactical Aspects; Part II — Main Results. *Kybernetika* **26**(1990), 47–66; 134–154.
9. Novák, V.: *The Alternative Mathematical Model of Linguistic Semantics and Pragmatics,* Plenum, New York, 1992.
10. Novák, V.: On the logical basis of approximate reasoning, in V. Novák, J. Ramík, M. Mareš, M. Černý and J. Nekola, Eds.: *Fuzzy Approach to Reasoning and Decision Making.* Academia, Prague and Kluwer, Dordrecht, 1992.
11. Novák, V.: Fuzzy Logic As a Basis of Approximate Reasoning. In: Zadeh, L. A., Kacprzyk, J. *Fuzzy Logic for the Management of Uncertainty.* J. Wiley & Sons, New York, 1992.
12. Novák, V.: Logical analysis of Max–Min rule of inference. *Proc. of Int. Conference EUFIT'93,* RWTH Aachen, 1993.
13. Novák, V.: Paradigm, Formal Properties and Limits of Fuzzy Logic. *Int. J. of General Systems* (to appear).
14. Novák, V.: *Fuzzy Control from the Point of View of Fuzzy Logic.* Fuzzy Sets and Systems **66**(1994), 159–173.
15. Pavelka, J.: On fuzzy logic I, II, III, *Zeit. Math. Logic. Grundl. Math.* **25**(1979), 45–52; 119–134; 447–464.
16. Zadeh, L.A.: The concept of a linguistic variable and its application to approximate reasoning I, II, III, *Inf. Sci.,* **8**(1975), 199–257, 301–357; **9**(1975), 43–80.

Chapter 4.

INFORMATION SCIENCES

RECENT DEVELOPMENT IN FUZZY DATABASE SYSTEMS AND APPLICATIONS

S. MIYAMOTO
Institute of Information Sciences and Electronics
University of Tsukuba, Ibaraki 305, Japan

AND

M. UMANO
Department of Systems Engineering
Faculty of Engineering Science
Osaka University, Osaka 560, Japan

Abstract. Queries for database retrieval and data in the real world have fuzziness which cannot or need not be precisely defined. In order to deal with them we have two research fields of fuzzy database systems and fuzzy information retrieval. They have different histories and methodologies. In this paper we survey works on fuzzy database systems in the view of fuzzy query processing and fuzzy data model as well as those in fuzzy information retrieval. Moreover we discuss models of fuzzy database systems and those for fuzzy information retrieval, comparing characteristics of the models.

1. Introduction

Queries for retrieving database and data in the real world have fuzziness which cannot or need not be precisely defined. In order to deal with them we have two approaches of fuzzy database and fuzzy information retrieval. Our first question in this paper is: "Is it necessary to distinguish between fuzzy database and fuzzy information retrieval?" Our current answer is "yes," because they have different histories and methodologies.

Let us consider non-fuzzy fields for these concepts. Researchers on ordinary database systems do not consider information retrieval as an independent subject but as application programs which are a part of database sys-

367

Z. Bien and K. C. Min (eds.),
Fuzzy Logic and its Applications, Information Sciences, and Intelligent Systems, 367–376.
© 1995 *Kluwer Academic Publishers.*

tems. On the other hand, there is a group of researchers who are interested in information retrieval alone. While computer scientists consider database systems, chemists and librarians discuss information retrieval. Two groups use the same word *database* but computer scientists mean database architecture and data models for representing and retrieving data, while chemists and librarians mean information contents for practical use.

In Salton and McGill [1] five types of information search and management are shown: information retrieval, database management systems, decision support systems, management information systems, and question answering systems, whereas ordinary computer scientists consider that a database system can cover the above five categories. Thus, we must be careful in our terminology of fuzzy databases. In this paper we divide our topics into fuzzy database systems and fuzzy information retrieval, because historically there is little overlap between these two research areas.

It is not difficult to speak of main researches of fuzzy database systems up to now, since they are concentrated on fuzzy relational database systems and related problems, which include (1) fuzzy data model, (2) query processing, and (3) implementation of query languages.

On the other hand, research tendency in fuzzy information retrieval includes (1) appropriate or sophisticated mathematical model, (2) evaluation of information contents using fuzzy measures, (3) comparison between fuzzy and crisp weighted retrieval, (4) efficient algorithms for fuzzy retrieval, and (5) the use of fuzzy dictionaries (the simplest is a fuzzy thesaurus), and (6) development of a practical document retrieval system using fuzziness.

In the next two sections we overview typical studies in these research areas.

2. Fuzzy Database Systems

In order to store and retrieve fuzzy data, fuzzy databases have been proposed, most of which are extensions of the crisp relational model.

2.1. FUZZY QUERIES

Let us first consider a fuzzy set as a condition in a query for a conventional relational model. Tahani [2] has proposed a processing method for it.

For a database shown in Table 1, consider a query [2]: "Retrieve the names of persons who are either young or employed recently but highly paid."

This query requests computation of the degree of consistency with each condition for all tuples of the Employee database. The degree of consistency with the fuzzy set is computed as the membership value and the connectives *and* and *or* are interpreted as the min and max, respectively. Table 2 shows

TABLE 1. Employee database

Name	Age	Salary	Emp. Year
Anderson	30	20000	1974
Brown	30	15000	1974
Long	25	40000	1972
Nelson	55	20000	1950
Smith	25	25000	1975

TABLE 2. Consistency for the fuzzy query

Condition	Age=young	Emp.Year=recent	Salary=high	Whole query
(Anderson,30,20000,1974)	0.5	0.6	0.5	0.5
(Brown,30,15000,1974)	0.5	0.6	0	0
(Long ,25,40000,1972)	1	0	1	1
(Nelson,55,20000,1950)	0	0	0.6	0
(Smith,25,25000,1975)	1	0.8	0.8	0.8

the degrees for each tuple, where fuzzy sets *young*, *recent*, and *high* are defined for the respective attributes. The answer to the query is a fuzzy set for the selected domain:

$$\{0.5/\text{Anderson}, 1/\text{Long}, 0.8/\text{Smith}\}.$$

2.2. EXTENSION OF DATA MODELS

Since conventional data models are insufficient for representing fuzzy data, we extend them as follows.

(a) Fuzzy relational models. The simplest is to extend an ordinary relation to a fuzzy relation by adding grades of membership. Thus a fuzzy database DB_{F1} is defined by a set of fuzzy relations:

$$DB_{F1} = \{R_1, R_2, ..., R_m\}$$

where R_i is a fuzzy relation characterized by a membership function

$$\mu_{R_i} : U_{i1} \times U_{i2} \times \cdots \times U_{im} \to [0, 1]$$

in which U_{ij} is the domain of j-th attribute of relation R_i.

TABLE 3. A relation with possibility distributions

Name	Age	Child's name
Tom	23	Ted
Susan	35	John
Susan	35	Mike
Richard	40	$\{Judy, Anna\}_p$
Raymond	*young*	*unknown*
Victor	*unknown*	*undefined*
Smith	$\{50,51\}_p$	*null*

This type of fuzzy databases is used in several systems, e.g., in Zadeh [3]. Raju and Majumdar [4] study a theory of schema design for this type of fuzzy relational model.

(b) Unification by similarity relation. In the query processing, a set of tuples matching the query is retrieved and equal tuples are unified into one. Buckles and Petry [5] have proposed a fuzzy database where tuples are unified on a similarity relation [6]. The relational model is extended so that an entry in the relation can be a set.

(c) Possibility distribution-relational models. Fuzziness of data values is represented by a possibility distribution [7]. Several models have been proposed of which attribute values are possibility disributions.

A fuzzy database DB_{F2} is defined by a set of relations R_i such that

$$R_i \subseteq \mathcal{P}(U_{i1}) \times \mathcal{P}(U_{i2}) \times \cdots \times \mathcal{P}(U_{im})$$

where $\mathcal{P}(U_{ij})$ represents the family of all possibility distributions in the domain U_{ij}.

Umano et al. [8, 9] first proposed the possibility distribution-relational model of fuzzy database and a retrieval method for it.

Consider a relation containing the possibility distributions shown in Table 3 [9]. The name of Richard's child $\{Judy, Anna\}_p$ ia a possibility distribution, indicating that the value is either Judy or Anna. Note that if he has two children, his name appears twice like Susan. Raymond's age is a possibility distribution of *young*. The *unknown*, *undefined* and *null* (we do not know even whether a value is unknown or undefined) are defined as possibility distributions $\{1/u : u \in U\}_p$, $\{0/u : u \in U\}_p$, and $\{1/unknown, 1/undefined\}_p$, respectively.

A query retrieves elements certainly satisfying the condition as well as possibly satisfying it. Thus, the query "find the persons over 25 years old" retrieves

$$\text{clearly satisfying} = \{1/\text{Susan}, 1/\text{Richard}, 1/\text{Smith}\}$$
$$\text{possibly satisfying} = \{1/\text{Raymond}, 1/\text{Victor}\}.$$

Since the query does not contain fuzzy conditions, the resulting degrees are either 0 or 1. If the query is "find young persons," the degrees may be between 0 and 1.

Prade and Testemale [10] have proposed a retrieval method using a possibility measure and a necessity measure. Let F be a fuzzy condition in a query and A_i be a possibility distribution of i-th data. We have a possibility measure:

$$\Pi(F|A_i) = \sup_{u \in D}\{\mu_F(u) \wedge \pi_{A_i}(u)\}$$

and a necessity measure:

$$N(F|A_i) = \inf_{u \in D}\{\mu_F(u) \vee (1 - \pi_{A_i}(u))\}$$

where D is the domain of F and A_i. The measures Π and N are obtained as the retrieval results.

Zemankova-Leech and Kandel [11] have developed a fuzzy database system in which query processing is similar to that in [10], but they consider another possbility measure, using the multiplication instead of the min:

$$p_{A_i}(F) = \sup_{u \in D}\{\mu_F(u) \cdot \pi_{A_i}(u)\}$$

Moreover they indicate that if a possibility distribution is not normalized, the necessity measure may be larger than the possibility measure, and hence they introduce a certainty measure:

$$c_{A_i}(F) = \inf_{u \in D'}\{\mu_F(u) \cdot \pi_{A_i}(u)\},$$

where $D' = \{u \in D : \mu_F(u) \cdot \pi_{A_i}(u) > 0\}$.

(d) Possibility-distribution-fuzzy-relational model. In addition to fuzziness of data values, there is another kind of fuzziness in a relation. A possibility-distribution-fuzzy-relational model has been proposed by Umano [12]. The fuzzy database DB_{F3} is defined by a set of fuzzy relations in which a fuzzy relation R_i is characterized by the membership function

$$\mu_{R_i} : \mathcal{P}(U_{i1}) \times \mathcal{P}(U_{i2}) \times \cdots \times \mathcal{P}(U_{im}) \to \mathcal{P}([0,1]),$$

where $\mathcal{P}([0,1])$ is a family of all possibility distributions on the unit interval.

3. Fuzzy Information Retrieval

An early study on fuzzy information retrieval by Negoita [13] discusses the construction of fuzzy document clusters and retrieval of them using the separation theorem of fuzzy convex sets.

Fuzzy thesaurus is a basic and practical tool for realizing fuzzy retrieval. A method for obtaining a documentary set using a fuzzy thesaurus has been proposed by Radecki [14]. His method includes reduction of the number of documents for increasing retrieval efficiency using level fuzzy sets.

Studies of fuzzy information retrieval started in Japan in 1980s. Nakamura and Iwai [15] have proposed a structure of a fuzzy thesaurus. They called their method for constructing the structure "analogical inference." Miyamoto et al. [16] have proposed a fuzzy set model for a thesaurus and provided an algorithm for automatic generation of a thesaurus. Miyamoto and Nakayama [17] have presented an algorithm for fuzzy retrieval with a fuzzy thesaurus. Retrieval with a documentary citation network has been studied by Nomoto et al. [18].

Theoretical and methodological aspects have been studied, e.g., by Kohout and Bandler [19] who discuss properties of retrieval operators. Applications of a knowledge base to information retrieval and utilization of the Dempster-Shafer theory of evidence have been proposed by Biswas et al. [20].

3.1. MODELS OF FUZZY INFORMATION RETRIEVAL

The word of a *model* is used in a variety of meanings in general. When we consider database systems (including fuzzy databases), this word is used to refer to data models. On the other hand, in information retrieval, a *model* is a mathematical model, which often has a simpler structure than a data model and is suitable for mathematical analysis. We show here two types of models for fuzzy information retrieval.

(e) A block diagram model. Miyamoto [21] proposes a block diagram model of fuzzy information retrieval such as the one shown in Fig. 1, in which F is a fuzzy thesaurus and U is the index relating a set of terms and a set of documents. F and U are fuzzy relations. Given a query q: a fuzzy set in the set of terms, Fq is an extended query using the fuzzy thesaurus. $r = UFq$ is the fuzzy set of documents retrieved from the query q.

The block diagram model itself is not new. Heaps [22] considered the diagram of the same form, although Heaps used the ordinary algebra of calculating the diagram, whereas the fuzzy algebra of max for addition and min for multiplication has been used in Miyamoto [21] who shows advantages of the fuzzy algebra. Namely,

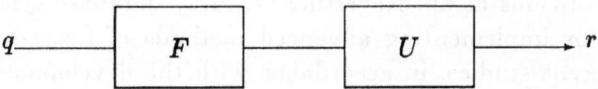

Figure 1. A block diagram model for information retrieval

(i) this model combines the traditional vector model and the logical model for information retrieval by modifying the ordinary algebra into the fuzzy algebra.

(ii) Feedback in a diagram leads to hierarchical clustering of documents in this model. That is, automatic classification of documents is shown to have a close relationship with the feedback by using the fuzzy algebra.

(f) Matching by fuzzy set models. Simple set-theoretical models are often useful in fuzzy information retrieval. Let us consider a set X, possibly a set of documents or terms. A and B are assumed to be subsets of X. Simple measures of matching between A and B are $|A \cap B|/|A \cup B|$ and $|A \cap B|/|A|$, where $|\cdot|$ is the cardinality. Suppose that A is the set of terms included in an information request and B is the terms indexed to a document. If these measures are large, the document is well matched to the request. This simple set-theoretical model has been used to generate a fuzzy thesaurus structure by Miyamoto et al. [16, 21] and an efficient algorithm for a large amount of data has been given. The same type of the model is used in Li and Liu [23] for retrieval in fuzzy relational databases.

A generalization of the set-theoretical model has been discussed by Miyamoto et al. [24] for application to picture retrieval, in which picture contents are represented by *fuzzy propositions*: $\{(p(x_1, x_2, ..., x_m), \mu)\}$. For example, when a picture displays a person in front of a house and a flag on the top of the house (see, Sakauchi [25]), the fuzzy propositions are

{ (in-front-of(person,house),1.0),   (on-the-top-of(flag,house),0.8) }

where the degrees represent relative importance, ambiguity, or other factors of relevance in the picture. When the index of a picture is given in such an abstract form, matching degrees of two pictures are calculated by the above set-theoretical model by regarding a set of fuzzy propositions as an ordinary fuzzy set. Retrieval of similar pictures to a given picture has been studied, e.g., by Katoh et al. [26]. It is straightforward to apply the above model to retrieval of similar pictures. Miyamoto et al. [24] discuss algorithms for calculating the matching degrees.

A method of fuzzy information retrieval does not necessarily assume a fuzzy data model. An ordinary crisp database system can be used for realizing a method of fuzzy retrieval. Thus, the relationship between the both is not obvious at all. Nevertheless, fuzzy database systems are more often used for implementing advanced methods of fuzzy information retrieval in recent studies, in accordance with the development of practical fuzzy database systems, since fuzzy database systems are more convenient for realizing fuzzy retrieval, especially when a variety of fuzziness exists in the data.

4. Overview of recent researches

Information retrieval models frequently include features of expert-knowledge systems[27]. Murai et al. [28] have a motivation of using modality, which they call doxastic retrieval. Nakata [29] considers integrity constraints in updating fuzzy relational databases. Shirai et al. [30] propose a graph-oriented fuzzy data model motivated from the entity-relationship approach. A link connecting entities is with a membership value in their method.

As in most fields of fuzzy systems, theoretical studies have been actively pursued overseas, while applications have been actively studied and implemented in Japan. A system developed by Ogawa et al. [31] has been put into commercial use and a system proposed by Miyamoto et al. (see [21]) can be used for large-scale documentary databases. A consultation system by Maeda and Yoshida [32] provides information of companies, which is used by university students for finding employment. Ito et al. [33] have developed a fuzzy retrieval system for historical earthquakes. Recent earthquakes have been recorded by numerics while historical earthquakes were recorded by linguistic expressions in old documents, and therefore a fuzzy database system provides a useful means for the retrieval. Their system include graphic display of geographical distributions of fuzzily retrieved earthquakes.

Object-oriented database is a new paradigm of database systems. Tanaka [34] defines fuzzy objects and develops a system for retrieving movies. He has also developed a hypertext system using fuzzy concepts.

Nakajima et al. [35] have developed a library for a fuzzy SQL language for a fuzzy relational database of practical use. The fuzzy database is built on a crisp database system using the library that interpretes a fuzzy SQL program into an ordinary SQL and vice versa. This method provides a fuzzy view of the database system even when the underlying system is crisp.

5. Conclusion

We have given an overview of researches and developments of fuzzy database systems and fuzzy information retrieval, especially in Japan. Researches of fuzzy database systems include representation, integrity, and basic retrieval methods of fuzzy data, while those of fuzzy information retrieval discuss advanced techniques, frequently depending on specific fields of applications.

Object-oriented and/or multimedia fuzzy databases are main subjects that we must study in the near future. Another important topic is the develepment of a large-scale fuzzy database and retrieval system. A big project should be scheduled and carried out for realization of such systems.

References

1. Salton, G. and McGill, M. J.: *Introduction to Modern Information Retrieval*, McGraw-Hill, New York, 1983.
2. Tahani, V.: A conceptual framework for fuzzy query processing – a step toward very intelligent database systems, *Information Processing and Management* **13** (1977), 289-303.
3. Zadeh, L. A.: PRUF – a meaning representation language for natural languages, *International Journal of Man-Machine Studies* **10** (1978), 395-460.
4. Raju, K.V.S.V.N. and Majumdar, A.K.: Fuzzy functional dependencies and lossless join decomposition of fuzzy relational database systems, *ACM Transactions on Database Systems* **13** (1988), 129-166.
5. Buckles, B. and Petry F.E.: A fuzzy model for relational databases, *Fuzzy Sets and Systems* **7** (1982), 213-226.
6. Zadeh, L.A.: Similarity relations and fuzzy orderings, *Information Sciences* **3** (1971), 177-200.
7. Zadeh, L.A.: Fuzzy sets as a basis for a theory of possibility, *Fuzzy Sets and Systems* **1** (1978), 3-28.
8. Umano, M., Fukami, S., Mizumoto, M., and Tanaka, K.: Retrieval processing from fuzzy database, *IEICE Technical Reports*, **80**, 204 (1980), 45-54, AL-80 (in Japanese).
9. Umano, M.: FREEDOM-0: a fuzzy database system, in M.M. Gupta and E. Sanchez (eds.): *Fuzzy Information and Decision Processes*, North-Holland, Amsterdam, 1982, 339-347.
10. Prade, H. and Testemale, C.: Generalizing database relational algebra for the treatment of incomplete or uncertain information and vague queries, *Information Sciences*, **34** (1984), 115-143.
11. Zemankova-Leech M. and Kandel A.: *Fuzzy Relational Data Bases - A Key to Expert Systems*, Verlag TÜV Rheinland, Köln, 1984.
12. Umano, M.: Retrieval from fuzzy database by fuzzy relational algebra, *Proc. of IFAC Symposium on Fuzzy Information, Knowledge Representation, and Decision Analysis*, Marseille, July 19-21, 1983, 1-6.
13. Negoita, C.V.: On the application of the fuzzy sets separation theorem for automatic classification in information retrieval systems, *Information Sciences*, **5** (1973), 279-286.
14. Radecki T.: Mathematical model of information retrieval system based on the concept of fuzzy thesaurus, *Information Processing and Management* **12** (1976), 313-318.
15. Nakamura, K. and Iwai, S.: A representation of analogical inference by fuzzy sets and its application to information retrieval system, in M.M. Gupta and E. Sanchez (eds.): *Fuzzy Information and Decision Processes*, North-Holland, Amsterdam, 1982, 373-386.
16. Miyamoto, S., Miyake, T., and Nakayama, K.: Generation of a pseudothesaurus for information retrieval based on co-occurrences and fuzzy set operations, *IEEE Trans. on Syst., Man, and Cybern.* **SMC-13** (1983), 62-70.
17. Miyamoto, S. and Nakayama, K.: Fuzzy information retrieval based on a fuzzy pseudothesaurus, *IEEE Trans. on Syst., Man, and Cybern.* **SMC-16** (1986), 278-282.

18. Nomoto, K., Wakayama, S., Kirimoto, T., and Kondo, M.: A fuzzy retrieval system based on citations, *Second IFSA Congress*, Tokyo, July 20-25, 1987, 723-726.

19. Kohout, L.J. and Bandler W.: The use of fuzzy information retrieval techniques in construction of multi-center knowledge-based systems, in B. Bouchon and R.R. Yager (eds.): *Uncertainty in Knowledge-Based Systems*, Lecture Notes in Computer Science, 286, Springer, Berlin, 1987, 257-264.

20. Biswas, G., Bezdek, J.C., Marques, M., and Subramanian, V.: Knowledge-assisted document retrieval: I. the natural language interface, *Journal of the American Society for Information Science*, **38** (1987), 83-96.

21. Miyamoto, S.: *Fuzzy Sets in Information Retrieval and Cluster Analysis*, Kluwer Academic Publishers, Dordrecht, 1990.

22. Heaps, H. S.: *Information Retrieval: Computational Aspects*, Academic Press, New York, 1976.

23. Li, D. and Liu D.: *A Fuzzy PROLOG Database System*, Wiley, New York, 1990.

24. Miyamoto, S., Konishi, N., and Miyake, T.: Document retrieval and image retrieval based on fuzzy propositional index, *Fourth IFSA Congress*, Brussels, July 7-12, Vol. Computer, Management and Systems Science, 1991, 157-160.

25. Sakauchi, M.: Image retrieval technology, *The Journal of the Institute of Electronics, Information, and Communication Engineers of Japan*, **71, 9** (1988), 911-914 (in Japanese).

26. Katoh, T. and Kurita, T.: Visual interaction with image database systems, *Johoshori* **33** (1992), 466-477 (in Japanese).

27. Kohout, L.J., Kalantar, H., and Anderson, J.: Use of fuzzy relational information retrieval techniques in management of multiple streams of diagnostic knowledge in knowledge-based system CLINAID, *Fourth IFSA Congress*, Brussels, July 7-12, Vol. Computer, Management and Systems Science, 1991, 117-120.

28. Murai, T., Miyakoshi, M., and Shimbo, M.: Doxastic document retrieval, *Fifth IFSA World Congress*, Seoul, Korea, July 4-9, 1993, 541-544.

29. Nakata, M.: Updating under integrity constraints in fuzzy databases, *9th Fuzzy System Symposium*, Sapporo, Japan, May 19-21, 1993, 205-208 (in Japanese).

30. Shirai, Y., Fujishiro, I., and Kunii, T.L.: A link-oriented language for fuzzy databases, *Fourth IFSA Congress*, Brussels, July 7-12, Vol. Computer, Management and Systems Science, 1991, 252-255.

31. Ogawa, Y., Morita, T., and Kobayashi, K.: A fuzzy document retrieval system and learning method based on the dynamic keyword connection method, *International Workshop on Fuzzy System Applications*, Iizuka, Japan, Aug. 22-24, 1988, 143-144.

32. Maeda, H. and Yoshida, K.: Ability of a natural language interface for fuzzy database retrieval, *Journal of Japan Society for Fuzzy Theory and Systems*, 4 (1992), 361-368 (in Japanese).

33. Ito, H., Wakayama, A., and Hoshiba, M.: A fuzzy information processing system of seismic data, *International Fuzzy Engineering Symposium*, Yokohama, Japan, Nov. 13-15, 1991, 919-928.

34. Tanaka, K.: Fuzzy object-oriented database using Versant/Ontos, *Second Obase Symposium*, Senri, Japan, Oct. 1-2, 1992 (in Japanese).

35. Nakajima, H., Sogoh, T., and Arao, M.: Development of an efficient fuzzy SQL for large scale fuzzy relational database, *Fifth IFSA World Congress*, Seoul, Korea, July 4-9, 1993, 517-520.

FACTOR SPACES AND FUZZY TABLES

Wang Pei-zhuang
Institute of Systems Science,
Natioanal University of Singapore,
Singapore 0511

Abstract

This is a brief interpretation of factor space theory. A intuitive background of why do we need to study factor spaces theory can be found in paragraph 1. The basic mathematical definitions about factor spaces are introduced in paragraph 2. In paragraph 3, as an example, we can see how to use factor space theory to represent concepts towards to concept automatic generation. Finally, the concept of fuzzy table is introduced in paragraph 4.

1.   Introduction

Concept is the base of thinking, to describe concepts, there three main approaches:

Identification: Searching the relationship between a concept and its properties. So called *intension* of a concept α consists of those properties who are essentially occupied by α. Which is the definition of α.

Connotation: Searching the relationship between a concept and other concepts.

Denotation: Searching the relationship between a concept and objects from real world.

Z. Bien and K. C. Min (eds.),
Fuzzy Logic and its Applications, Information Sciences, and Intelligent Systems, 377–386.
© *1995 Kluwer Academic Publishers.*

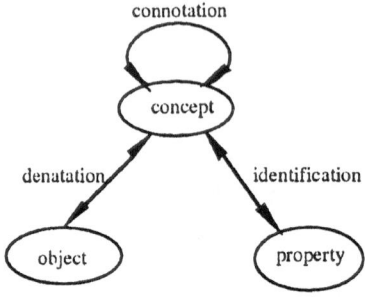

Fig. 1

So called *extension* of a concept α consists of those objects who are indicated by α.

Existent methods on conceptual representation in AI belong to the first and second approaches. Along them, computer cannot automatically generate new concept out side the closure of known concepts. If we want to make a computer automatically generate new concepts like brain creates a new concept whenever it catch a new class of objects, we have to consider the third approach.

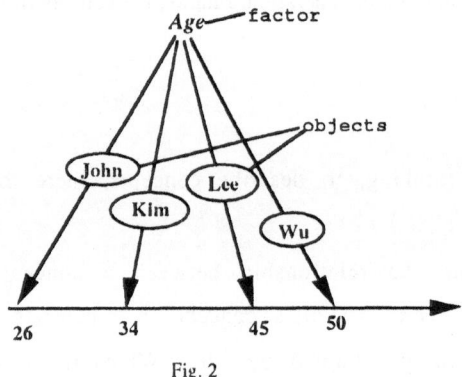

Fig. 2

In mathematics, a concept α is described by means of a subset **A** in a related universe of discourse. The subset **A** is exactly the extension of concept α . Unfortunately, this approach cannot be received by computer scientists because, the objects cannot be directly reflected and stored into computers from the real world. Factor spaces theory aims to provide a framework to construct generalized coordinate systems to describe real objects, then to represent concepts along both three approaches such that we can enable computer automatically generate new concepts.

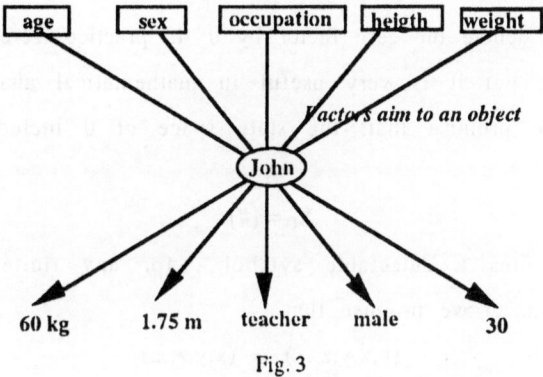

Fig. 3

As shown in Fig. 4, a person can be viewed as a point in a generalized Cartesian coordinate system provided we can put down appropriate axes in there. The names of those axes are called *factors*.

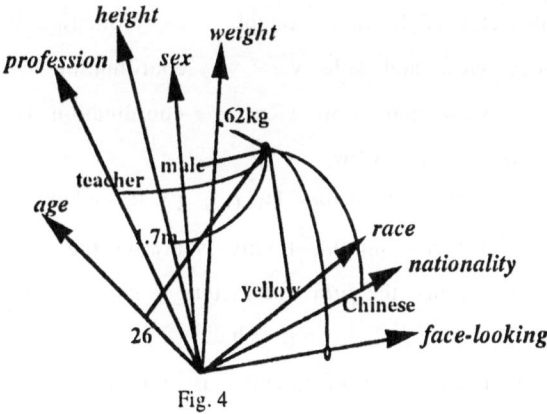

Fig. 4

A factor is an aspect by which we can analytically recognize an object. Like genes are the analytical elements of living's organs, factors are the analytical elements of human being's thinking

2. Factor space.

There are some basic relations and operations within a class of factors can be founded

1) Sub-factor. The factor Height is a sub-factor of the factor Shape when we observe a person.

2). A factor is called a *zero factor* if it is a sub-factor of any factor involved. We denote the zero factor by 0. In practice, zero factor is an empty factor, but it is very useful in mathematical analysis. As an exception, we promise that: the state space of 0 includes only one element:

$$X_0 = \{\#\}$$

where # denotes a delectable symbol , for any finite or numeral sequence (x,y,z,...) we promise that

$$(\#,x,y,z,...) = (x,y,z,...)$$

this means that the Cartesian product space of X_0 and Y is equal to Y:

$$X_0 \times Y = Y$$

3) Meet. Factor f is called the *meet* of factors g and h, denoted as

$$f = g \wedge h,$$

if f is sub-factor of both g and h and any other common sub-factor of g and h is a sub-factor of f. For example, x-coordination is the meet of factors bird's-eye view and side view, y-coordination is the meet of factors bird's-eye view and front view, z-coordination is the meet of factors front view and side view.

4) Irrelevant A family of factors $\{f_t\}$ $(t \in T)$ is called irrelevant if the meet of any pair of factors in this family is always the zero factor.

5) Join Factor f is called the *join* of factors g and h, denoted by

$$f = g \vee h,$$

if g and h are both sub-factor of f, and f is sub-factor of other factor for which g and h are both sub-factor. For example, bird's-eye view is the union of factors x-coordination and y-coordination, side view is the union of factors x-coordination and z-coordination, front view is the union of factors y-coordination and z-coordination.

6) Subtraction and complementary. Suppose that factor h is a sub-factor of g, we call f the *subtraction* of g, denoted as

$$f = g - h$$

and h   if h and f are irrelevant, and the join of f and h is g.   We call 1 the complete factor with respect to F, a set of factors,   if every factor in F is a sub-factor of it.   Giving a factor f in F , denoting

$$f^c = 1 - f$$

which is called the *complementary* of f with respect to 1

We can get an axiomatic definition of factor space as follows:

DEFINITION A *factor space* is a family of sets $\{X(f)\}_{(f \in F)}$ where F is a complete Boolean algebra $F=(F, \vee, \wedge, {}^c, 1, 0)$ satisfying

F.1) $X_0 = \{\#\}$;

F.2)   For any $f, g \in F$, if $f \wedge g = o$, then

$$X(f \vee g) = X(f) \times X(g)$$

where ×   is the Cartesian product operator.

The concept of factor space generalize the concepts of the states space in control theory,   the feature space, character space or parameter space in pattern recognition, and the phase space in physics, and so on, but a factor space is not constrained in a fixed states space, it is a family of state spaces. Factor space theory emphasizes the variety of dimension of state space; a factor space can be viewed as a state space with variable dimension, This is an essential progress in constructing a description systems for knowledge description.

3.   Conceptual   description   and   conceptual   automatically generation

Let O be a class of objects, it concerns with a factor space $\{X(f)\}_{(f \in F)}$
. For each f in F, there corresponds a mapping

$$r_f: O \to X_f$$

$$r_f(o) = \text{the state of } o.$$

r_f is called the *representation mapping* on O.

382

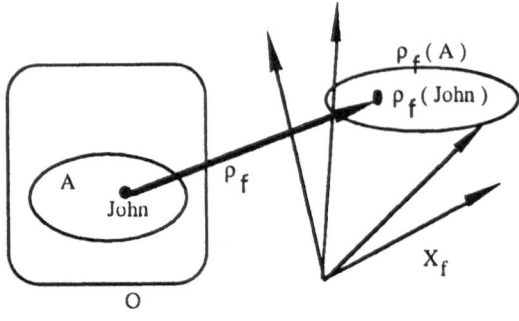

Fig. 7

Suppose that $f,g \in F$ and $f>g$. Since $f=g+(f-g)$, any point z in X_f can be written as $z=(x,y)$ where $x \in X_g$ and $y \in X_{f-g}$. Define the mapping

$$\downarrow_g^f : X_f \rightarrow X_g$$
$$\downarrow_g^f (z)=x$$

it is called the *projecting mapping* from f to g.

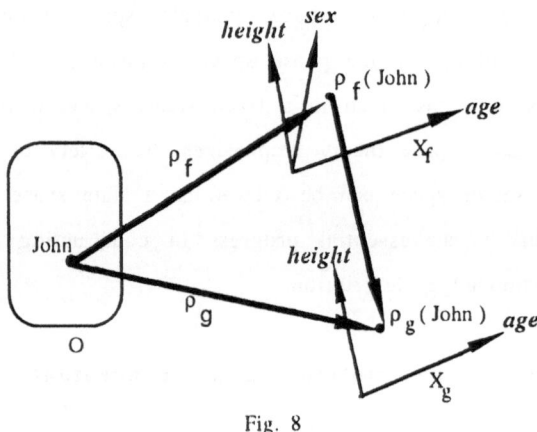

Fig. 8

Any concept a can be represented as a subset of an appropriate universe U which consists of objects concerned. By means of representation mappings, a concept can be viewed as a (fuzzy or non-fuzzy) subset in a factor space.

Recognition is indeed taken on an appropriate factor space. The higher the dimension, the easier to distinguish. But, the higher the dimension, the more expensive in information. How to select an appropriate factor to recognize an object is the most important problem.

We call that factor f is *surplus* for concept A under factor g if

$$r_{f \vee g}(A) = \bigwedge_g^{f \vee g} r_g(A) = \{(x,y) | x \in X_f, y \in r_g(A)\}$$

Factor g is called *sufficient* for A if `g is surplus for A under g.

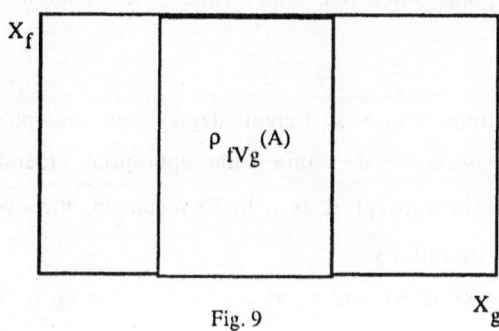

$$X_f$$

$$\rho_{f \vee g}(A)$$

$$X_g$$

Fig. 9

Definition   Factor

$$r(A) = \wedge \{g | g \text{ is sufficient for A}\}$$

is called the *rank* of concept A. It is the minimum of dimension conserving complete information about A.

In Actual applications, r(A) requires too high for dimension,   there is an axiomatic definition of sufficiency measure:

Definition A mapping $s:F \times F(U) \rightarrow [0,1]$ is called a *sufficient measure* if it satisfies:

(σ.1) $(\forall f \in F) (\rho_f(A) \in P_0(X_f) \Rightarrow \sigma(f,A) = 1)$;

(σ.2) $(\forall f \in F) (\rho_f(A) \equiv 0.5 \Rightarrow \sigma(f,A) = 0)$.

(σ.3) $(\forall f \in F) (\sigma(f,A) = \sigma(f,A^c))$;

(σ.4) $(\forall f,g \in F) (f \geq g \Rightarrow \sigma(\phi,A) \geq \sigma(g,A))$

here $P_0(U) = P(U) \backslash \{U, \varnothing\}$

Two concrete examples:

Amplitude measure

$$Sa(f,A) = 1/2[\vee \{[r_f(A)](x) | x \in X_f\} - \wedge \{[r_f(A)](x) | x \in X_f\}]$$

$$+ 1/2[\vee \{[r_f(A^c)](x) | x \in X_f\} - \wedge \{[r_f(A^c)](x) | x \in X_f\}]$$

Entropy measure

$$Se(f,A) =$$
$$1+1/M \int \left\{ \mu_{r_f(A)}(x)\log_2[\mu_{r_f(A)}(x)] + (1-\mu_{r_f(A)}(x))\log_2[1-\mu_{r_f(A)}(x)] \right\} dx$$

An extension of a concept gives us a conceptual image. Conceptual image generates conceptual intension statements. It concerns concept auto-generating.

Select those factors whose sufficient degree are enouph high. Put the extension of a concept α into the appropriate factor space. If the conceptual image of concept α is a hyper-rectangle, then we can state the definition of a as follows:

$$\text{If} \quad x_1 \text{ is } A_1 \text{ and } x_2 \text{ is } A_2 \text{ and } \quad \text{ and } x_k \text{ is } A_k$$
$$\text{then} \quad (x_1, x_2,..., x_k) \text{ obeys concept } \alpha$$

If the conceptual image is an union of hyper-rectangles, then we can state the definition of α as follows:

If x_1 is A_{11} and x_2 is A_{12} and and x_k is A_{1k}

 x_1 is A_{21} and x_2 is A_{22} and and x_k is A_{2k}

 x_1 is A_{n1} and x_2 is A_{n2} and and x_k is A_{nk}

$$\text{then} \quad (x_1, x_2,..., x_k) \text{ obeys concept } \alpha$$

Factor space theory can unify intension and extension of a concept

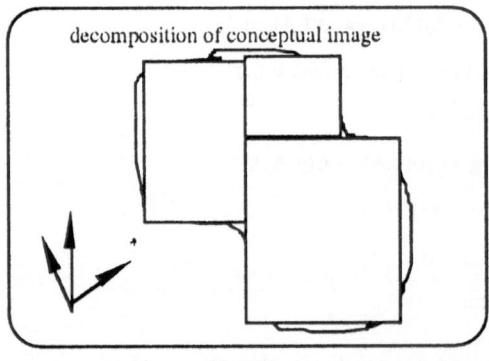

decomposition of conceptual image

Fig. 10

4. Fuzzy tables

Whenever we try to apply factor space theory in computer's applications, we have to mathematically constrain ourselves in finite universes. In this case, the concept of fuzzy table play a very important role.

Giving two groups of linguistic variables x_i ($i =1,...,n$) and y_j ($j=1,...,m$), denote the concerned phase spaces as $X=\Xi_1\times\Xi_2\times ...\times\Xi_v$ and $Y=\Psi_1\times\Psi_2\times ...\times\Psi_\mu$ respectively. For each i, let U_i be the set of linguistic values with respect to x_i i.e., $U_i = \{A_{ik}\}$($k =1,...,K_i$) with $A_{ik}\in F(X_i)$, and for each j, let V_j be the set of linguistic values with respect to y_j , i.e., $V_j= \{B_{js}\}$($s =1,...,S_j$) with $B_{js}\in F(Y_j)$. A mapping T: $D \to \varsigma_1\times \varsigma_2\times ...\times\varsigma_\mu$ with $D\subseteq Y_1\times Y_2\times ...\times Y_v$ is called a *fuzzy table* from X to Y.

A fuzzy table is a table with linguistic values as its elements. The table of fuzzy rules in a fuzzy controller is a fuzzy table; The tables in use of fuzzy decision making can be viewed as fuzzy tables; The tables in use of fuzzy relation data base are fuzzy tables; A group of fuzzy sample points in fuzzy aggregation can be viewed as a fuzzy table. A fuzzy relation matrix can be also viewed as a fuzzy table.

Several interested topic including fuzzy interpolation, fuzzy information compress, operations of fuzzy tables, ...atc. can be found in fuzzy table theory.

Reference

[1]   Wang, P.Z. and M. Sugeno (1982) The factors fields and the background structures of fuzzy subsets, Fuzzy Mathematics (in Chinese), No. 2, 45-54

[2]   Wang, P.Z. and E. Sanchez (1982) Treating a fuzzy subset as a projectable random set, in Fuzzy Information and decision, M.M. Gupta, E. Sanchez (eds.), Pergamon Press, 212-219 ; Memor No.UCB/ERL, M82/35(1982) Univ. of California, Berkeley

[3]   Chen, Y.Y. and Y.F. Liu and P.Z. Wang (1983) The model of synthesis evaluation, Fuzzy Mathematics (in CHinese), No.1, 43-54

[4]   Zhang, D.Z. and P.Z.Wang (1985) On the mathematical description of concepts, judgments, and reasonings (in Chinese) SI WEI KE XUE TONG XUN, Beijing Institute of Technology, 39-49

[5]   Wang, P.Z. and K. Chuan and D.Z. Zhang (1986) Degree analysis and its application in plan choosing problem in the construction of

hyperpower stations, in Reports of Seminar on Soft Science Models ,
Beijing

[6]   Wang, P.Z.(1990) Factor space and knowledge representation, in
Approximate Reasoning Tools for Artificial Intelligence, **Verdegay and**
Delgado(eds.) Verlag T U Rheinland , 62-79

[7]   Wang, P.Z. (1990), A factor spaces approach to **knowledge**
representation, Fuzzy Sets and Systems, Vol.36,113-124

[8]   Kandel, A. and X.T.Peng and Z.Q.Cao and P.Z.Wang (1990)
Representation of concepts by factor spaces, Cybernetics and **Systems:**
An International Journal, 21,43-57

[9]   Peng, X.T. and A.Kandel, P.Z.Wang, (1991) Concepts, rules, and
fuzzy reasoning: a factor space approach, IEEE-SMC Vol. 21, No. 1,194-
205

[10]   Yuan,X.H.,and P.Z.Wang and E.S.Lee (1992) Factor space and its
algebraic representation theory, J. of Mathematical Analysis and
Applications, Vol. 171, No.1 ,   256-276

[11]   Liu, Z.L. (1992) Factors Neural Networks, Beijing **Normal**
University Press, Beijing.

[12].   Zadeh L.A.(1978) Fuzzy sets as a basis for a theory of
possibility, Fuzzy Sets and   Systems, 1, 3-28.

FUZZY CLUSTERING WITH MULTIPLE OBJECTIVE FUNCTIONS

M. SATO

Hokkaido Musashi Women's Junior College
Kita 22, Nishi 13, Kita-ku, Sapporo 001, Japan,
E-mail : mika@huie.hokudai.ac.jp

AND

Y. SATO

Hokkaido University
Kita 13, Nishi 8, Kita-ku, Sapporo 060, Japan,
E-mail : ysato@huie.hokudai.ac.jp

Abstract. In recent years, researches for 3-way data have become interested in data analysis by improving the technique of observing data and highly progress of the computer capacity. 3-way data is composed of objects, attributes and situations, that is, it is expressed by 3-dimensional array, objects × attributes × situations. The clustering problem for such 3-way data can be regarded as a multicriteria optimization problem. In this optimization, the practical problem is to get a Pareto efficient solution. In a hard (non-fuzzy) clustering, there are some difficulties to find Pareto efficient clusters for the combinational optimization. But we show that fuzzy clustering has a merit to obtain Pareto efficient clusters. Using this property, the method of fuzzy clustering for 3-way data is offered.

1. Introduction

In a clustering problem, if several clustering criteria are given, the optimum solution for each criterion is not always the same. Then, we have to find a solution that is consistent with all criteria and satisfies each criterion as much as possible. This problem has been discussed as a multicriteria optimization problem, whose practical solution is called Pareto efficient solution, which

Z. Bien and K. C. Min (eds.),
Fuzzy Logic and its Applications, Information Sciences, and Intelligent Systems, 387–396.
© 1995 *Kluwer Academic Publishers.*

defined that if it can not be improved on any criteria without sacrificing some other criteria. In this sense, we will define that the Pareto efficient clustering is a Pareto efficient solution of the multicriteria clustering problem. On the hard (non-fuzzy) multicriteria clustering, the algorithm to find the Pareto efficient clustering has not been completed yet [1], because of the combinational optimization problem, but on multicriteria fuzzy clustering in this paper, we show that the Pareto efficient clustering can be obtained by solving a nonlinear optimization problem.

Usually, the observations consist of the following three phases, objects (O), attributes (A), and situations (S). Concerning these phases, 3-way data is classified into the following three types:

(i) O × A × S
(ii) A × A × O,   O × O × A,   O × O × S, etc
(iii) O × O × O,   A × A × A,   S × S × S,

Namely, the type (i) consist of all three phases O, A, S, and the type (ii) is the similarity or dissimilarity (distance) between the same two phases for each fixed other phase. And the last type (iii) shows the mutual relation of triplet for one phase.

In the analysis of the data (i), the Tucker's threemode factor analysis [2] is well known, and INDSCAL [3] or INDCLUS [4] are typical methods for the data (ii). On the data (iii), the method to solve PARAFAC-model has been suggested in [5].

The clustering problem could be discussed for all types of 3-way data (i),(ii) and (iii). In this paper, we deal with type (i) data. In this case, each phase, either O, A or S, is considered to be the classified subject. Since, in any case, the clustering process are the same except for the definition of the similarity between the pair of classified subjects, hereafter the objects (O) are considered to be classified. The purpose of this paper is to perform the clustering for 3-way data using the concept of multicriteria optimization.

2. The Multicriteria Optimization Problem

Optimization problem (Φ, F) can be characterized formally as to determine the solution x^* satisfying the following,

$$F(x^*) = \min_{x \in \Phi} F(x),$$

where $F : \Phi \rightarrow R$ is the (single) criterion and Φ is a set of feasible solutions. In the multicriteria optimization problem $(\Phi, F_1, \cdots, F_T)$, our purpose is to determine the solution $x^* \in \Phi$ in such a way that

$$F_t(x^*) = \min_{x \in \Phi} F_t(x), \quad t = 1, 2, \cdots, T,$$

where $F = (F_1, F_2, \cdots, F_T) : \Phi \to R^T$ are the criterion functions over the set of feasible solutions Φ. Hereafter we assume $F_t(x) > 0$ $(t = 1, \cdots, T)$ without loss of generality.

In the problem $(\Phi, F_1, \cdots, F_T)$, the following useful concepts have been introduced.

Definition 1. (dominant solution): The solution x_0 is a dominant solution if for any solution $x \in \Phi$ and for each criterion F_t it holds,

$$F_t(x_0) \leq F_t(x), \quad t = 1, 2, \cdots, T.$$

But, in most cases, the dominant solution does not exist, then the following concept of dominate and Pareto efficient solution have been used.

Definition 2. (dominate): For two solutions x_1, x_2, if $F_t(x_1) \leq F_t(x_2)$ for all t, and the strict inequality $F_s(x_1) < F_s(x_2)$ holds for at least one $s \in \{1, 2, \cdots, T\}$, then we state that solution x_1 dominates solution x_2.

Definition 3. (Pareto efficient solution): The solution $x^* \in \Phi$ is a Pareto efficient solution if there exists no other solution $x \in \Phi$ dominates x^*, that is,

$$\{x \mid \exists s \ F_s(x) < F_s(x^*) \text{ and } F_t(x) \leq F_t(x^*), \text{ for } t \neq s\} = \phi$$

where ϕ is empty set.

Definition 4. (local Pareto efficient solution): The solution $x^* \in \Phi$ is a local Pareto efficient solution if there exists no other solution $x \in \Phi \cap N(x^*)$ dominates x^*, that is,

$$\{x \mid \exists s \ F_s(x) < F_s(x^*) \text{ and } F_t(x) \leq F_t(x^*), \text{ for } t \neq s\} = \phi,$$

where $N(x^*)$ denotes the neighborhood of x^*.

A Pareto efficient solution of multicriteria problem $(\Phi, F_1, \cdots, F_T)$ can be obtained by solving the following weighting method $P(w)$, which transform a multicriteria problem into single criteria problem, minimizes the weighted summation of the criteria $(F_1, \cdots, F_T)$, i.e.

$$\sum_{t=1}^{T} w_t F_t(x) \to \min,$$

under $w_t \geq 0$, $(t = 1, 2, \cdots, T)$.

On the relationship between the Pareto efficient solution and the optimum solution of $P(w)$, the following two effective theorems have been well known:

Theorem 1. [6]: Let $x^* \in \Phi$ be a optimum solution of $P(w)$ with the weight $w_t > 0$, $(t = 1, 2, \cdots, T)$. Then x^* is a Pareto efficient solution of the multicriteria problem $(\Phi, F_1, \cdots, F_T)$.

Theorem 2. [6]: Let $F_t(x)$, $(t = 1, 2, \cdots, T)$ be convex functions. If $x^* \in \Phi$ is a Pareto efficient solution of the multicriteria problem $(\Phi, F_1, \cdots, F_T)$, then x^* is a optimum solution of $P(w)$ with the weight $w_t \geq 0$.

3. Multicriteria Clustering Problem for 3-way Data

The 3-way data, which is observed by the values of p variables with respect to n objects for T times, is denoted by the following

$$X^{(t)} = (x_{ia}^{(t)}). \qquad \left(\begin{array}{ll} i & = 1, 2, \cdots, n; \ a = 1, 2, \cdots, p \\ t & = 1, 2, \cdots, T. \end{array} \right) \qquad (3.1)$$

The purpose of this clustering is to classify n objects into K fuzzy clusters $C_1, \cdots, C_K$ which are denoted by the fuzzy subset of the set of objects $O = \{o_1, o_2, \cdots, o_n\}$. The value of the membership functions of K fuzzy subsets, namely, the degree of belonging of each object to the cluster are denoted by

$$U = (u_{ik}), \quad u_{ik} \geq 0, \quad \sum_{k=1}^{K} u_{ik} = 1. \qquad (3.2)$$

$$(i = 1, 2, \cdots, n, \quad k = 1, 2, \cdots, K.)$$

And we denote the centroid of the cluster at t-th time

$$v_k^{(t)} = (v_{ka}^{(t)}). \qquad \left(\begin{array}{ll} t & = 1, 2, \cdots, T; \ k = 1, 2, \cdots, K \\ a & = 1, 2, \cdots, p. \end{array} \right)$$

The goodness of clustering in the t-th time is given by the sum of extended within-class dispersion [7],

$$J^{(t)}(U, v^{(t)}) = \sum_{k=1}^{K} \sum_{i=1}^{n} (u_{ik})^m \sum_{a=1}^{p} (x_{ia}^{(t)} - v_{ka}^{(t)})^2. \qquad (3.3)$$

The parameter m plays a role of determining the degree of fuzziness of the cluster. We should note that the clusters, fuzzy subset of O, is determined uniquely through the times $t = 1, \cdots, T$. But the centroid of clusters $v_k^{(t)}$ may differ with times.

If there exist the solution (U, v) which minimize all $J^{(t)}$ $(t = 1, \cdots, T)$, then it is the best solution or dominant solution. But usually such a solution does not exist. Then this problem, a clustering for 3-way data, becomes a multicriteria optimization problem. Again we assume that Φ is a set of feasible solutions (U, v). So we define a single clustering criterion by the weighted sum of $J^{(t)}$, that is, for $w^{(t)} > 0$,

$$J(U, v) = \sum_{t=1}^{T} w^{(t)} J^{(t)}(U, v^{(t)}). \tag{3.4}$$

According to the theorem 1 and theorem 2, (3.4) shows that the problem to get a Pareto efficient solution of $(\Phi, J_1, \cdots, J_T)$ is reduced to usual nonlinear optimization problem. Since $J(U, v)$ in (3.4) is regarded as multivariable continuous function of (U, v), the necessary condition to obtain the local Pareto efficient solution on the multicriteria clustering problem $(\Phi, J_1, \cdots, J_T)$ is described as the following theorem:

Theorem 3. Suppose that the data (3.1) are given and K is the number of clusters. Let $U = (u_{ik})$ be a grade of membership of fuzzy clusters, and let $v_k^{(t)} = (v_{ka}^{(t)})$ be a centroid of cluster at t-th time. If $(U, v) \in \Phi$ is at least a local Pareto efficient solution of multicriteria optimization problem $(\Phi, J_1, \cdots, J_T)$, i.e. minimization problem of (3.4), then (U, v) is satisfied the following (3.5) and (3.6). Let us suppose

$$I_i = \{k \mid 1 \le k \le K;\ \sum_{t=1}^{T} d(x_i^{(t)}, v_k^{(t)}) = \sum_{t=1}^{T} \sum_{a=1}^{p} (x_{ia}^{(t)} - v_{ka}^{(t)})^2 = 0\},$$

$$\tilde{I}_i = \{1, 2, \cdots, K\} - I_i.$$

If $I_i = \phi$ and $v_k^{(t)}$ is given, then

$$u_{ik} = \left[\sum_{l=1}^{K} \left\{ \frac{g_{ik}}{g_{il}} \right\}^{\frac{1}{m-1}} \right]^{-1}, \quad \left(\begin{array}{l} i = 1, 2, \cdots, n \\ k = 1, 2, \cdots, K, \end{array} \right) \tag{3.5}$$

where we put $g_{ik} \equiv \sum_{t=1}^{T} w^{(t)} d(x_i^{(t)}, v_k^{(t)})$.

If $I_i \ne \phi$, then we define

$$u_{ik} = 0 \quad \forall k \in \tilde{I}_i, \quad \sum_{k \in I_i} u_{ik} = 1.$$

For fixed u_{ik}, the centroid of each cluster is given by

$$v_{ka}^{(t)} = \frac{\sum_{i=1}^{n} (u_{ik})^m x_{ia}^{(t)}}{\sum_{i=1}^{n} (u_{ik})^m}. \quad \left(\begin{array}{l} t = 1, 2, \cdots, T;\ k = 1, 2, \cdots, K \\ a = 1, 2, \cdots, p. \end{array} \right) \tag{3.6}$$

4. The Feature of Pareto Efficient Fuzzy Clustering

To illustrate the feature of Pareto efficient fuzzy clustering, we suppose now two-criteria clustering problem (Fig. 1). In Fig. 1, we assume $\{P_1, P_2, P_3, P_4\}$ to be a subset of feasible clusterings, whose coordinates represent the values for criterion (J_1, J_2). P_1 is the optimum clustering with respect to the first criterion J_1, and P_2 is the optimum clustering under the criterion J_2. The clustering P_0 is a dominant solution because P_0 is the optimum clustering for the both criteria J_1 and J_2 (Definition 1). But if such a solution does not exist, $S = \{P_1, P_2, P_4\}$ is the set of Pareto clustering. P_3 is not Pareto clustering because P_4 dominates P_3 (Definition 2). Next, using the artificial 3-way data in Table 1, we shall investigate the property of the proposed multicriteria fuzzy clustering. The data described by the values of two variables (X, Y) are observed for twenty objects $\{o_1, \cdots, o_{20}\}$ at two time points. In this case, to simplify, we denote the clustering criterion (3.4) as follows:

$$J(U, \boldsymbol{v}) = wJ_1 + (1 - w)J_2, \quad (1 \geq w \geq 0). \tag{4.1}$$

The local Pareto clustering under the criterion (4.1) is obtained by using (3.5) and (3.6) in Theorem 3. Fig. 2 shows the values (J_1, J_2) of the local Pareto clusterings. In this figure, w denotes the weight of the criterion (4.1), and the upper row of points show the results of $m = 1.5$. The lower points show that $m = 2.0$. The symbols, "minJ1" and "minJ2", denote the minimum value of each criterion, J_1 and J_2, respectively. From this figure, we can find the variation of each value of criterion with the change of weight w. Fig. 3 shows the state of local Pareto clustering for both T_1 and T_2 and also show the change of the state of clustering with the value of w, when m is 2.0. In this figure, the largeness of symbols, circle and square, represents the fuzzy grade of the membership of cluster, the larger is the greater grade. From this figure, in the case of $w = 1.0$, we can find that clusters of T_1 are clearly classified. At T_2, however, the detection of clusters is difficult and these clusters seem to have no meanings, because the value of J_2 is the maximum. On the other hand, in the case of $w = 0.0$, we can find the opposite situation, namely the clusters located in the upper and lower sides at T_2 are clear, and the state of T_1 is not clear. In the case of $w = 0.5$, the clusters are obtained as an intermediate state. From this we know that the clustering of both cases $w = 1.0$ and $w = 0.0$ are not the same. Therefore, we have to select the suitable solution among the Pareto efficient clustering. In practical situations of data analysis, such a selection seems to be efficient, because there is the case that the weights of (4.1) are given by the exterior condition or by the subjective condition. On the other hand, it is an important factor whether the result of the clustering could be interpreted or not. So the feasible selection should be effective.

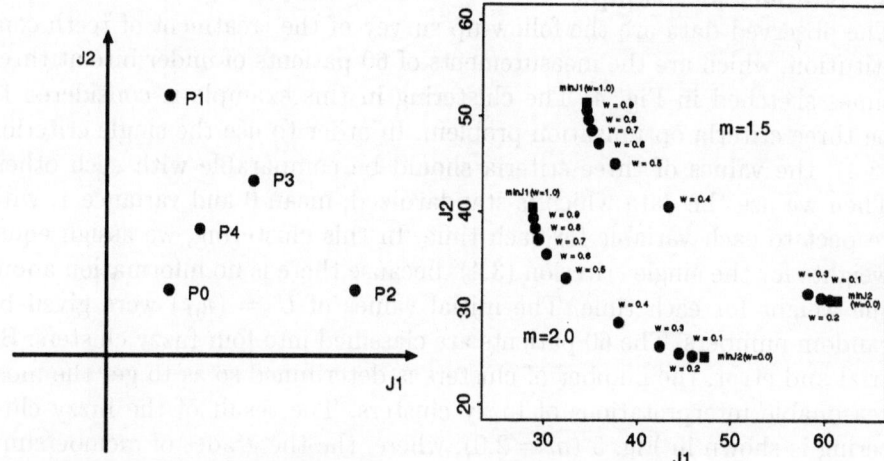

Fig. 1. Two-criteria clustering.

Fig. 2. The value of each criterion, J_1 and J_2, of Pareto efficient clustering with the weight w.

Table 1. 20 objects at two time points.

Object	T_1 (1st time)		T_2 (2nd time)	
	X	Y	X	Y
o_1	0.42	0.51	0.42	1.51
o_2	1.08	-0.71	1.08	-0.71
o_3	0.14	2.70	0.14	3.70
o_4	-0.78	0.38	-0.78	1.38
o_5	0.23	-1.22	0.23	-1.22
o_6	0.79	-0.07	0.79	0.92
o_7	-0.29	-0.80	-0.29	-0.80
o_8	-0.41	-0.82	-0.41	-0.82
o_9	0.86	0.31	0.86	1.31
o_{10}	1.10	-1.33	1.10	-1.33
o_{11}	3.68	2.09	2.18	3.09
o_{12}	2.29	0.82	0.79	1.82
o_{13}	2.82	-0.79	1.32	-0.79
o_{14}	3.72	-1.29	2.22	-1.29
o_{15}	5.20	1.55	3.70	1.55
o_{16}	3.18	1.14	1.68	2.14
o_{17}	2.12	-0.56	0.62	-0.56
o_{18}	4.50	-0.57	3.00	-0.57
o_{19}	3.40	0.17	1.90	0.17
o_{20}	2.64	0.52	1.14	1.52

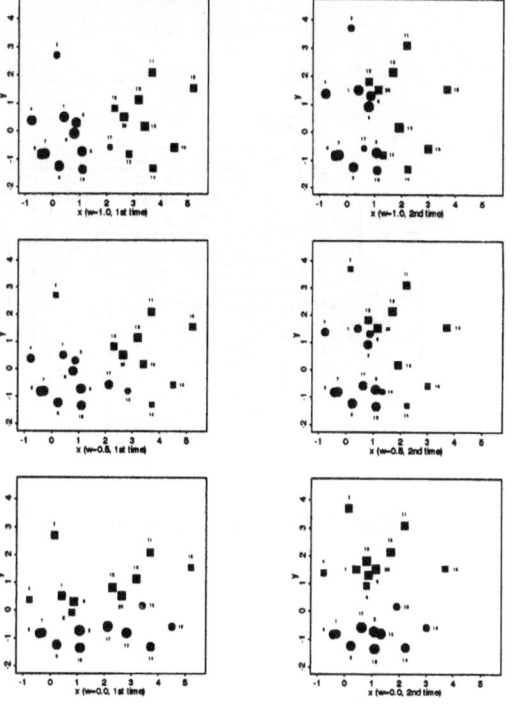

Fig. 3 The state of Pareto efficient clustering.

5. Numerical Example

The observed data are the follow-up survey of the treatment of teeth constitution, which are the measurements of 60 patients of under bite at three times sketched in Fig. 4. The clustering in this example is considered to be three criteria optimization problem. In order to use the single criterion (3.4), the values of three criteria should be comparable with each other. Then we use the data which is standardized, mean 0 and variance 1, with respect to each variable for each time. In this clustering, we assign equal weights for the single criterion (3.4), because there is no information about the weight for each time. The initial values of $U = (u_{ik})$ were given by random numbers. The 60 patients are classified into four fuzzy clusters. By trial and error, the number of clusters is determined so as to get the most reasonable interpretations of fuzzy clusters. The result of the fuzzy clustering is shown in Fig. 5 ($m = 2.0$), where, the the grades of memberships are represented for each cluster. We can find the characteristic feature of each cluster, inquiring into Fig. 5 and the standardized values of centroids represented in Table 2.

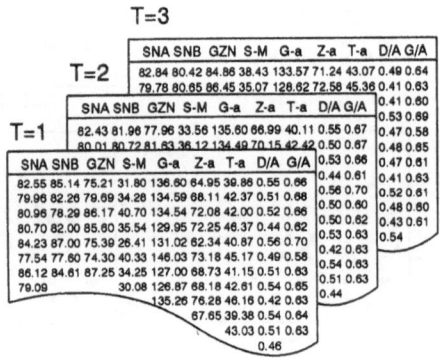

Fig. 4. Constitution of 60 patients

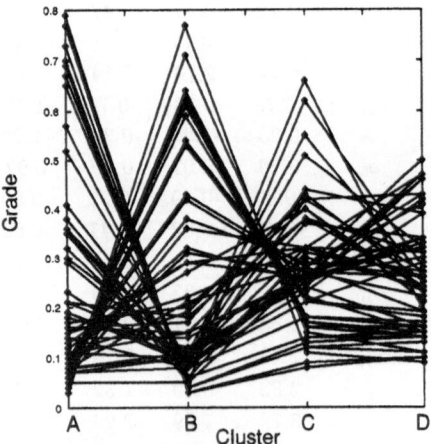

Fig. 5. The grade of clustering
(m=2.0).

Table 2. The centroids of clusters.

C	T	SNA	SNB	GZN	S-M	G-a	Z-a	T-a	D/A	G/A
						Attribute				
A	1	-.997	-.965	.423	.960	.494	.938	.868	-.739	-.328
	2	-1.032	-1.213	.758	1.109	.369	1.195	.936	-.813	-.406
	3	-.672	-.905	1.084	.896	-.058	1.273	1.413	-1.395	-.551
B	1	.730	1.427	-.903	-1.031	-.193	-1.469	-.927	1.080	1.076
	2	.756	.960	-.477	-.820	-.336	-1.105	-.903	1.001	.994
	3	1.059	.870	.087	-.786	-.731	-.723	-.445	.307	.589
C	1	-.248	.021	-.258	-.215	.013	-.303	-.273	.389	.300
	2	-.217	-.303	.015	-.056	-.058	-.027	-.234	.314	.176
	3	.146	-.017	.281	-.279	-.449	.136	.327	-.430	-.257
D	1	-.023	.109	-.509	.151	.512	-.135	-.325	.328	-.225
	2	.000	-.222	-.216	.324	.438	.169	-.258	.220	-.412
	3	.364	.032	.097	.185	.083	.338	.252	-.462	-.714

C=Cluster, T=Time

From the centroid of cluster A, we find that the patients in A have an
under bite with a negative tooth angle through all the times. Conversely,
the centroid of cluster B shows an under bite with a positive tooth angle.
Both centroids of cluster C and cluster D mean the similar constitution. But
there is a difference in the few attributes, namely for attributes SNA, GZN,
S-M, G-a, G/A, the difference between cluster C and D are so conspicuously,
comparing other attributes. This means that the difference of an under bite,
namely, the patients who belongs to cluster C have a tendency of an over
bite. Consequently, we found the reasonable clusters of the constitution by
taking into account the change pattern.

6. Concluding Remarks

The hard clustering for 3-way data has been discussed from the following
viewpoints. The first point is to decompose 3-way similarity array by the
method of INDCLUS [8]. The other point is to find clusters as the Pareto
optimum solution in a multicriteria optimization problem [1]. In this paper,
we discussed the fuzzy clustering from the latter standpoint. The data used
in numerical example can be applied to hard (non-fuzzy) clustering, but it
seems that the results of fuzzy clustering are more suitable, since the feature
of teeth constitution consists of many composite factors. In the formulation of multicriteria fuzzy clustering, the single criterion, using weighting
method, can be considered as the usual nonlinear continuous function of

multivariables. Then once the weight vector $w = (w_1, \cdots, w_T)$ is given, the Pareto efficient solution is obtained by solving usual nonlinear optimization problems. But there is no algorithm to determine the best solution among the Pareto efficient solutions. Therefore, the suitable selection of the weight is important. But this problem has difficulties because we must evaluate previously the ranges of the criterion functions and the proportion of each criterion. Then if we have no exterior information for the weights, we should search for the best weight heuristically.

References

1. A. Ferligoj and V. Batagelj, "Direct Multicriteria Clustering Algorithms", *J. Class.*, **9**(1992) 43-61.
2. L.R. Tucker, "Relation between multidimensional scaling and three-mode factor analysis", *Psychometrika*, **37**(1972) 3-27.
3. J.D. Carroll and J.J. Chang, "Analysis of individual differences in multidimensional scaling via an N-way generalization of "Eckart-Young" decomposition", *Psychometrika*, **35**(1970) 283-319.
4. J.D. Carroll and P. Arabie, "INDCLUS: An individual differences generalization of the ADCLUS model and MAPCLUS algorithm", *Psychometrika*, **48**(1983) 157-169.
5. R.A. Harshman, "PARAFAC: Foundation of the PARAFAC procedure: Models and conditions for an 'explanatory' multi-modal factor analysis", *Working Papers in Phonetics* **16** (University of California at Los Angeles, 1970).
6. N.C. Da Cunha and E. Polak, "Constrained Minimization under Vector valued Criteria in Finite-Dimensional Space", *J. Math. Anal. and Appl.***19**, No.1(1967) 103-124.
7. J.C. Bezdek, *Pattern Recognition with Fuzzy Objective Function Algorithms*, (Plenum Press, 1987).
8. P. Arabie, J.D. Carroll and W.S. DeSarbo, *Three-way scaling and clustering*, (Newbury Park, CA. Sage 1987).

THE EFFECTIVENESS OF FUZZY OPERATORS
IN INFORMATION RETRIEVAL

JOON HO LEE
Korea Research and Development Information Center
Korea Institute of Science and Technology
P.O. Box 1, Yusung, Taejon, Korea

AND

WON YONG KIM, MYOUNG HO KIM AND YOON JOON LEE
Department of Computer Science
Korea Advanced Institute of Science and Technology
373-1, Kusung-dong, Yusung-gu, Taejon 305-701, Korea

1. Introduction

An important component of Information Retrieval (IR) systems is the ranking facility that can rank documents in decreasing order of query-document similarities. Users can reduce their time spent to find relevant documents by reading the top-ranked documents first. Boolean retrieval systems have been most widely used among commercially available IR systems. Conventional Boolean retrieval systems, however, do not support the document ranking facility because they cannot calculate the similarities between Boolean queries and documents.

The fuzzy set model was developed to calculate the similarities between Boolean queries and documents [1, 2, 3]. Though the fuzzy set model is an elegant approach, it generates in some circumstances incorrect document rankings not to agree with humans' intuition [4, 5]. This is because the fuzzy set model uses MIN and MAX operators to evaluate AND and OR operations in Boolean queries. The MIN and MAX operators generate the resulting values depending on only one operand without considering the other, which results in incorrect document rankings.

Since the first introduction of fuzzy set theory a variety of fuzzy operators have been proposed for AND and OR operations, and different fuzzy

Z. Bien and K. C. Min (eds.),
Fuzzy Logic and its Applications, Information Sciences, and Intelligent Systems, 397–405.
© 1995 *Kluwer Academic Publishers.*

operators have different operational characteristics. The fuzzy operators can be classified into two groups such as T-operators and averaging operators [6]. In this paper we first describe that the fuzzy set model gives incorrect ranked output in some circumstances even though MIN and MAX operators are replaced with any types of T-operators. We then analyse the behavioral properties of averaging operators, and define a class of averaging operators called positively compensatory operators that are suitable for achieving high retrieval effectiveness. We also show through performance evaluation that the fuzzy set model using positively compensatory operators provides the higher retrieval effectiveness.

The remainder of this paper is organized as follows. Section 2 describes the conventional fuzzy set model and its problems. In section 3 we present the effect of various fuzzy operators on retrieval effectiveness. Performance evaluation and concluding remarks are given in sections 4 and 5, respectively.

2. The Fuzzy Set Model

The conventional fuzzy set model based on MIN and MAX operators has been proposed in the past to support the document ranking facility for Boolean retrieval systems [1, 2, 3]. It uses document term weights reflecting the importance of individual terms to a document. Formally, an IR system based on the conventional fuzzy set model can be defined by the quadruple $< T, Q, D, F >$, where

- T is a set of index terms used to represent queries and documents.
- Q is a set of Boolean queries that can be recognized by the system. A query $q \in Q$ is constructed from the terms in T and logical operators AND, OR and NOT.
- D is a set of documents. Each document $d \in D$ is represented by $\{(t_1, w_1), \ldots, (t_n, w_n)\}$ where w_i designates the weight of term t_i in document d and w_i takes a value between zero and one, i.e. $0 \le w_i \le 1$.
- F is a retrieval function taking a pair (d, q) in $D \times Q$ to a *document value* in the closed interval [0,1]. The document value means the similarity between document d and query q. The retrieval function $F(d, q)$ is defined as follows :

  1. When a query q is composed of an index term t_i, the function $F(d, q)$ is defined as the weight of term t_i in document d, i.e. w_i.

  2. Given a complex query with logical operators, it is evaluated based on the formulas given in Table 1. The evaluation proceeds recursively from the innermost clause.

TABLE 1. Similarity evaluation formulas in the conventional fuzzy set model

Boolean formulation	Evaluation formula
$F(d, q_1 \text{ AND } q_2)$	$\text{MIN}(F(d, q_1), F(d, q_2))$
$F(d, q_1 \text{ OR } q_2)$	$\text{MAX}(F(d, q_1), F(d, q_2))$
$F(d, \text{NOT } q_1)$	$1 - F(d, q_1)$

The conventional fuzzy set model has been criticized to generate incorrect document values in some circumstances [4, 5]. This is because MIN and MAX operators have the property of single operand dependency that makes the resulting value depend on only one operand without considering the other. The problem resulting from the single operand dependent property is illustrated in the following example. Although we explain only problems incurred by AND operations, it should be noted that OR operations cause similar problems.

Single Operand Dependency Problem: Suppose that we have two documents d_1 and d_2 shown below. The documents are represented by two pairs of an index term and its weight.

$d_1 = \{(\text{Thesaurus}, 0.40), (\text{Clustering}, 0.40)\}$

$d_2 = \{(\text{Thesaurus}, 0.99), (\text{Clustering}, 0.39)\}$

$q_1 = \text{Thesaurus AND Clustering}$

When the MIN operator is used for the AND operation, the document values of d_1 and d_2 for query q_1 are evaluated as 0.40 and 0.39, respectively. Hence, d_1 is retrieved with a higher rank than d_2. Most people, however, will obviously decide that d_2 rather than d_1 is more similar to q_1.

3. Effect of Fuzzy Operators on Retrieval Effectiveness

3.1. CLASSIFICATION OF FUZZY OPERATORS

The development of fuzzy set theory has allowed researchers to apply set theoretic concepts to sets of objects whose membership values vary in the interval [0, 1]. This applicability has been achieved by defining new operators for classical set theory operators. Often there is more than one fuzzy operator corresponding to a given classical operator, and different operators have different operational characteristics. A variety of fuzzy operators have been developed for AND and OR operations in the literature. They can be classified into two groups such as T-operators and averaging operators [6], which will be denoted by T and A, respectively, for simplicity.

TABLE 2. T-Operators

	AND	OR	
T_1	$\text{MIN}(x,y)$	$\text{MAX}(x,y)$	
T_2	$x \cdot y$	$x + y - xy$	
T_3	$\text{MAX}(x + y - 1, 0)$	$\text{MIN}(x + y, 1)$	
T_4	$\frac{xy}{x+y-xy}$	$\frac{x+y-2xy}{1-xy}$	
T_5	$\begin{cases} x & \text{if } y = 1 \\ y & \text{if } x = 1 \\ 0 & \text{otherwise} \end{cases}$	$\begin{cases} x & \text{if } y = 0 \\ y & \text{if } x = 0 \\ 1 & \text{otherwise} \end{cases}$	
T_6	$\frac{\lambda xy}{1-(1-\lambda)(x+y-xy)}$	$\frac{\lambda(x+y)+xy(1-2\lambda)}{\lambda+xy(1-\lambda)}$	$0 \le \lambda \le \infty$
T_7	$\text{MAX}(1 - ((1-x)^p + (1-y)^p)^{\frac{1}{p}}, 0)$	$\text{MIN}((x^p + y^p)^{\frac{1}{p}}, 1)$	$1 \le p \le \infty$
T_8	$\dfrac{1}{1+\left(\left(\frac{1}{x}-1\right)^{\lambda}+\left(\frac{1}{y}-1\right)^{\lambda}\right)^{\frac{1}{\lambda}}}$	$\dfrac{1}{1+\left(\left(\frac{1}{x}-1\right)^{-\lambda}+\left(\frac{1}{y}-1\right)^{-\lambda}\right)^{-\frac{1}{\lambda}}}$	$0 \le \lambda \le \infty$
T_9	$\frac{xy}{\text{MAX}(x,y,\lambda)}$	$1 - \frac{(1-x)(1-y)}{\text{MAX}(1-x,1-y,\lambda)}$	$0 \le \lambda \le 1$
T_{10}	$\text{MAX}((1+\lambda)(x+y-1) - \lambda xy, 0)$	$\text{MIN}(x + y + \lambda xy, 1)$	$-1 \le \lambda \le \infty$

TABLE 3. Averaging Operators

A_1	$(x \cdot y)^{1-\gamma} \cdot (x + y - x \cdot y)^{\gamma}$	$0 \le \gamma \le 1$
A_2	$(1 - \gamma) \cdot \text{MIN}(x,y) + \gamma \cdot \text{MAX}(x,y)$	$0 \le \gamma \le 1$
A_3	$(1 - \gamma) \cdot (x \cdot y) + \gamma \cdot (x + y - x \cdot y)$	$0 \le \gamma \le 1$
$A_{4.\text{AND}}$	$\gamma \cdot \text{MIN}(x,y) + \frac{(1-\gamma)(x+y)}{2}$	$0 \le \gamma \le 1$
$A_{4.\text{OR}}$	$\gamma \cdot \text{MAX}(x,y) + \frac{(1-\gamma)(x+y)}{2}$	$0 \le \gamma \le 1$

The T-operators consisting of T-norms and T-conorms originated from the studies of probabilistic metric spaces. Later, it was proposed that T-norms and T-conorms could be used for AND and OR operations, respectively in fuzzy set theory. The MIN operator belongs to T-norms and the MAX operator belongs to T-conorms. Table 2 shows some well-known T-operators [7].

In fuzzy decision theory the "decision" has been viewed as the intersection or the union of fuzzy sets, and T-operators have been used to model human decisions in many cases. However, it has been noted that neither T-norms nor T-conorms are appropriate to model "managerial decisions". Averaging operators have been developed to overcome this problem. Table 3 shows some well-known averaging operators [8, 9].

3.2. PROBLEMS OF T-OPERATORS

The T-operators except MIN and MAX have the following two common properties [10]: First, when one operand value of the two is zero or one, the resulting value is equal to one of two operand values. Second, they allow some compensation between two operand values in other cases, and the resulting value is less than the lower value of the two, or greater than the higher value. Even though they are used in the fuzzy set model, the first common property still causes the single operand dependency problem. On the other hand, the second common property, i.e. the effect of compensation alleviates the problem. For example, when the product operator is used instead of the MIN operator in the example of the single operand dependency problem, the document values of d_1 and d_2 are evaluated as 0.16 and 0.39 respectively, and hence d_2 is retrieved with a higher rank than d_1. However, the compensated value, that is less than the lower value or greater than the higher value, may result in the additional problem shown in the following example.

Negative Compensation Problem: Suppose a document d_3 and two queries q_1 and q_2 are given as follows:

$$d_3 = \{(\text{Thesaurus}, 0.70), (\text{Clustering}, 0.70), (\text{System}, 0.70)\}$$

$$q_1 = \text{Thesaurus AND Clustering}$$

$$q_2 = \text{System}$$

Using other types of T-norms except the MIN operator will decide that d_3 is more similar to q_2 than q_1. For instance, if the product operator is applied, the document value of d_3 is 0.49 for q_1 and 0.70 for q_2. Note that the similarity between q_1 and d_3 is less than that between q_2 and d_3, which clearly does not agree with most people's decision.

3.3. POSITIVELY COMPENSATORY OPERATORS

When one operand value of the two is zero, the averaging operator A_1 results in the single operand dependency problem because it constantly gives zero as the resulting value. The averaging operators A_1 and A_3 also result in the negative compensation problem in some circumstances because they give the resulting value lower than the two operand values or higher than the two operand values for some operand values.

The problems of single operand dependency and negative compensation can be overcome if the fuzzy operators incorporated in the fuzzy set model have the property generating a resulting value between the lower operand and the higher operand. The positively compensatory operator having the aforementioned property is defined as follows:

Definition of Positively Compensatory Operators: An operator θ is *positively compensatory* if $\text{MIN}(x, y) < \theta(x, y) < \text{MAX}(x, y)$ for all $x, y \in [0, 1]$ with the exception that $\theta(x, x) = x$.

TABLE 4. N-ary forms of averaging operators

$A_{1.N}$	$(1 - (1 - w_1) \cdot \ldots \cdot (1 - w_n))^\gamma \cdot (w_1 \cdot w_2 \cdot \ldots \cdot w_n)^{1-\gamma}$	$0 \leq \gamma \leq 1$
$A_{2.N}$	$\gamma \cdot \text{MAX}(w_1, w_2, \ldots, w_n) + (1 - \gamma) \cdot \text{MIN}(w_1, w_2, \ldots, w_n)$	$0 \leq \gamma \leq 1$
$A_{3.N}$	$\gamma \cdot (1 - (1 - w_1) \cdot \ldots \cdot (1 - w_n)) + (1 - \gamma) \cdot (w_1 \cdot w_2 \cdot \ldots \cdot w_n)$	$0 \leq \gamma \leq 1$
$A_{4.N.AND}$	$\gamma \cdot \text{MIN}(w_1, \ldots, w_n) + \frac{(1-\gamma)(w_1 + \ldots + w_n)}{n}$	$0 \leq \gamma \leq 1$
$A_{4.N.OR}$	$\gamma \cdot \text{MAX}(w_1, \ldots, w_n) + \frac{(1-\gamma)(w_1 + \ldots + w_n)}{n}$	$0 \leq \gamma \leq 1$

Since the averaging operators A_2 and A_4 belong to positively compensatory operators, they do not cause either the single operand dependency problem or the negative compensation problem. A_2 and A_4 are mathematically equivalent even though they are independently developed by different researchers at different time [11].

AND operations combine terms into term phrases and OR operations treat two terms as effectively synonymous. Thus, we often use consecutive AND operations such as "fuzzy AND set AND theory" and consecutive OR operations such as "human OR people OR person OR man". Unfortunately, The four averaging operators A_1 through A_4 do not satisfy associativity. This implies that different document values can be obtainable for logically equivalent queries such as $q_1 = t_1$ AND $(t_2$ AND $t_3)$ and $q_2 = (t_1$ AND $t_2)$ AND t_3. For example, suppose a document d is represented by $\{(t_1, 1), (t_2, 0.7), (t_3, 0.5)\}$. The fuzzy set model using A_4 gives 0.721 and 0.607 as document values of document d with respect to queries q_1 and q_2, respectively. Hence, we should use the n-ary forms of the averaging operators shown in Table 4 [12]. Note that $A_{2.N}$ and $A_{4.N}$ are not equivalent whereas their binary forms, i.e. A_2 and A_4 are.

It has been known in the IR literature that the P-norm model provides high quality of retrieval effectiveness [13]. The difference between the P-norm model and the fuzzy set model is the formulas used for the evaluation of AND and OR operations. The following are the evaluation formulas of the P-norm model, and their binary forms belong to positively compensatory operators.

$$(P_{AND}) \quad 1 - \left(\frac{(1-w_1)^p + \ldots + (1-w_n)^p}{n} \right)^{1/p} \qquad 1 \leq p \leq \infty$$

$$(P_{OR}) \quad \left(\frac{w_1^p + \ldots + w_n^p}{n} \right)^{1/p} \qquad 1 \leq p \leq \infty$$

4. Performance Evaluation

We use two different experimental collections covering documents in library science and computer science [13]. The library science collection, designated as CISI, covers highly cited items extracted from the Social Science Citation Index. The other collection called CACM covers articles published between 1959 and 1979 in the Communications of the ACM. Queries were formulated first in natural language for each of these collections, and later in Boolean form by graduate students and staff of Cornell University. The CISI collection consists of 1460 documents and 35 Boolean queries. The CACM collection consists of 3204 documents and two sets of 52 Boolean queries, and we used only the 50 queries that do not request documents by particular authors. The two sets of Boolean queries are designated as BL1 and BL2. Both collections also contain relevance assessment of each document with respect to each query.

The fuzzy set model exploits document term weights to calculate document values. The CISI and CACM collections do not have specific information about the actual importance of document terms. In this case, the weights of document terms can be derived from their occurrence frequency [14]. We calculated the weight of term t_k in document d_i, i.e. w_{ik} as follows:

$$w_{ik} = \begin{cases} (0.5 + 0.5 \cdot \frac{tf_{ik}}{maxtf_i}) \cdot \frac{\log(N/n_k)}{\log(N)} & \text{if } tf > 0 \\ 0 & \text{if } tf = 0 \end{cases}$$

where tf_{ik} is the occurrence frequency of term t_k in document d_i, $maxtf_i$ is the maximum frequency of any term (excluding function words) in document d_i, N is the number of documents in the collection, and n_k is the number of documents to which term t_k is assigned.

To evaluate retrieval effectiveness of an IR system, it is customary to compute values of the *recall* and *precision* [15]. The recall represents the proportion of relevant documents retrieved out of the total number of relevant documents in the whole collection, and the precision represents the proportion of relevant documents retrieved out of the total number of documents retrieved. In general, users want to retrieve most relevant documents and to reject most extraneous documents - to get high recall and high precision.

Table 5 shows the experimental results, in which % change is given with respect to the effectiveness of the P-norm model. For easy comparison we calculated a single precision value representing the average precision at three typical recall levels, including a low recall level of 0.20, a medium recall of 0.50, and a high recall of 0.80. When an operator has a parameter, we computed the precision value as changing the corresponding parameter in its legal range and then chose the maximum value. Table 5 shows that the operators P, $A_{2.N}$ and $A_{4.N}$, whose binary forms belong to

TABLE 5. Retrieval effectiveness of fuzzy operators

	CACM (BL1)	CACM (BL2)	CISI
P	0.3857	0.3517	0.1983
T_1	0.2262 (-41.4%)	0.1568 (-55.4%)	0.1352 (-31.8%)
T_2	0.2327 (-39.7%)	0.1591 (-54.8%)	0.1509 (-23.9%)
T_3	0.0284 (-92.6%)	0.0306 (-91.3%)	0.0415 (-70.1%)
T_4	0.2327 (-39.7%)	0.1627 (-53.7%)	0.1443 (-27.2%)
T_5	0.0313 (-91.9%)	0.0242 (-93.1%)	0.0968 (-51.2%)
T_6	0.2324 (-39.7%)	0.1606 (-54.3%)	0.1491 (-24.8%)
T_7	0.2271 (-41.1%)	0.1599 (-54.5%)	0.1415 (-28.6%)
T_8	0.2330 (-39.6%)	0.1626 (-53.8%)	0.1492 (-24.7%)
T_9	0.2327 (-39.7%)	0.1591 (-54.8%)	0.1509 (-23.9%)
T_{10}	0.2327 (-39.7%)	0.1591 (-54.8%)	0.1509 (-23.9%)
$A_{1.N}$	0.1153 (-70.1%)	0.0799 (-77.3%)	0.0399 (-79.9%)
$A_{2.N}$	0.3772 (-2.2%)	0.3020 (-14.1%)	0.1856 (-6.4%)
$A_{3.N}$	0.3856 (0.0%)	0.3485 (-0.9%)	0.1958 (-1.2%)
$A_{4.N}$	0.3902 (+1.2%)	0.3597 (+2.3%)	0.2034 (+2.6%)

positively compensatory operators, provide relatively higher retrieval effectiveness than the other operators except $A_{3.N}$. The binary form of $A_{3.N}$ causes the negative compensation problem, but does not result in the single operand dependency problem. Hence, this exceptional case suggests that the single operand dependency problem might be more adverse to retrieval effectiveness than the negative compensation problem.

5. Concluding Remarks

It has been argued that the conventional fuzzy set model based on MIN and MAX operators is not appropriate as a model of IR systems. This is because the MIN and MAX operators have properties adverse to effective calculation of document values. In recent years a variety of fuzzy operators have been developed, which can replace MIN and MAX operators. We have analysed the behavioral properties of various fuzzy operators, and have described the two properties, namely single operand dependency and negative compensation that may decrease retrieval effectiveness. We have then defined an operator class called positively compensatory operators, which can avoid the problems of single operand dependency and negative compensation. We have also shown through experiments that positively compensatory operators provide better retrieval effectiveness than the other operators.

References

1. D.A. Buell, "A General Model of Query Processing in Information Retrieval System," *Information Processing & Management*, Vol. 17, No. 5, pp. 127-136, 1981.
2. T. Radecki, "Fuzzy Set Theoretical Approach to Document Retrieval," *Information Processing & Management*, Vol. 15, No. 5, pp 247-259, 1979.
3. W.M. Sachs, "An Approach to Associative Retrieval through the Theory of Fuzzy Sets," *Journal of the American Society for Information Science*, Vol. 27, pp. 85-87, 1976.
4. A. Bookstein, "Fuzzy Requests: An Approach to Weighted Boolean Searches," *Journal of the American Society for Information Science*, Vol. 31, No. 4, pp. 240-247, 1980.
5. S.E. Robertson, "On the Nature of Fuzz: A Diatribe," *Journal of the American Society for Information Science*, Vol. 29, No. 6, pp. 304-307, 1978.
6. H.J. Zimmermann, *Fuzzy Set Theory and Its Applications*, 2nd ed., Kluwer Academic Publishers, 1991.
7. M.M. Gupta and J. Oi, "Theory of T-Norms and Fuzzy Inference Methods," *Fuzzy Sets and Systems*, Vol. 40, No. 3, pp. 431-450, 1991.
8. M.H. Kim, J.H. Lee and Y.J. Lee, "Analysis of Fuzzy Operators for High Quality Information Retrieval," *Information Processing Letters*, Vol. 46, No. 5, pp. 251-256, 1993.
9. H.J. Zimmermann and P. Zysno, "Latent Connectives in Human Decision Making," *Fuzzy Sets and Systems*, Vol. 4, No. 1, pp. 37-51, 1980.
10. G.J. Klir and T.A. Folger, *Fuzzy Sets, Uncertainty, and Information*, Prentice-Hall, 1992.
11. J.H. Lee, W.Y. Kim, M.H. Kim and Y.J. Lee, "On the evaluation of Boolean Operators in the Extended Boolean Retrieval Framework," *Proceedings of the 16th Annual International ACM SIGIR Conference on Research and Development in Information Retrieval*, pp. 291-297, 1993.
12. J.H. Lee, "Properties of Extended Boolean Models in Information Retrieval," *Proceedings of the 17th Annual International ACM SIGIR Conference on Research and Development in Information Retrieval*, pp. 182-190, 1994.
13. G. Salton, E.A. Fox, and H. Wu, "Extended Boolean Information Retrieval," *Communications of the ACM*, Vol. 26, No. 11, pp. 1022-1036, 1983.
14. G. Salton and C. Buckley, "Term-Weighting Approaches in Automatic Text Retrieval," *Information Processing & Management*, Vol. 24, No. 5, pp. 513-523, 1988.
15. G. Salton, *Automatic Text Processing: The Transformation, Analysis, and Retrieval of Information by Computer*, Addison Wesley, 1989.

A FUZZY QUERY EVALUATION IN RELATIONAL DATABASES

Soon C. Park[1], Chang Suk Kim[2] and Dae Su Kim[3]

[1]*Dept. of Infomation and Telecommunication Engineering*
Chonbook National University, Chonju, KOREA

[2]*Dept. of Computer Engineering*
Pusan University of Foreign Studies, Pusan, KOREA

[3]*Dept. of Computer Science*
Hanshin University, Osan, KOREA

Abstract

A fuzzy query evaluation in relational databases is presented in this paper. Average method, N-Root method and Weight method are introduced to determine the tuple membership value for fuzzy query processing. The tuple membership value of these new methods is reflected by all domain membership values in a tuple, while that of existing methods, so called MIN function method and B-P method, by only one value which is the lowest one among the domain membership values. Also, we introduce a filtering mechanism which makes fuzzy query processing efficient.

1. Introduction

Relational data model has been extensively studied and widely used because of its simplicity in designing databases and its clarity based on mathematical background. This model usually takes care of only well-defined data and precise queries. However, in real world applications queries are often imprecise or incomplete. For example, The query such as "retrieve a set of single women who are about 25 years old and possess attainments of high school graduate", is hardly processed in relational database systems. In order to deal with such imprecise query, fuzzy database systems have been proposed by Buckles and Petry, Shenoi, Melton and Motro and etc.[1, 6, 7, 12].

In this paper, three types of calculation methods of the tuple membership value

407

Z. Bien and K. C. Min (eds.),
Fuzzy Logic and its Applications, Information Sciences, and Intelligent Systems, 407–415.
© 1995 *Kluwer Academic Publishers.*

for fuzzy query processing are introduced, so called Average method, N-Root method and Weight method. These new methods provide more precise information to users. Moreover, the fuzzy query processing with these methods uses SQL as a host language that has three levels as filters to process fuzzy queries effectively.

This paper is organized as follows. Section 2 deals with some basic theoretical backgrounds of proximity relations and new calculation methods of the tuple membership value. In Section 3, fuzzy query language and an fuzzy query processing algorithm are presented. Section 4 shows an example of the three newly proposed calculation methods of tuple membership value. Finally, conclusions are discussed.

2. Theoretical backgrounds

2.1. PROXIMITY RELATION

In fuzzy theory, ambiguity such as "young" or "old" can be represented by a membership value[1, 2, 7, 13]. The membership value is represented with the similarity relation or the proximity relation[2, 7]. However, the similarity relation emphasizes transitivity too much to represent precise membership values in some cases. Therefore, we use the proximity relation that is more generous than the similarity relation.

In the ordinary relational database, a tuple t_i has the form $(d_{i1}, d_{i2}, \cdots, d_{im})$. Each component d_{ij} of the tuple is an element of a domain set D_j, i.e., $d_{ij} \in D_j$. A proximity relation s_j is defined over the set elements:

$$s_j : D_j \times D_j \to [0,1] \tag{1}$$

In Equation (1), [0,1] means a real number between 0 and 1. If $x, y \in D_j$ then s_j is as follows.

$$
\begin{aligned}
s_j(x,x) &= 1, \\
s_j(x,y) &= s_j(y,x)
\end{aligned}
\tag{2}
$$

The proximity relation includes and generalizes the similarity relation.

2.2. MEMBERSHIP VALUE

Three types of methods to represent a membership value in this paper are as follows.

• *Tabular form*

User defines membership values for an attribute in a fuzzy query using a tabular form[1, 4].

• *Computational method*

Computational method use a relation function to determine the membership value. The function can be defined of user's own accord. The normal distribution function that is used to get a contiguous set of membership values is shown in Equation (3).

$$\mu(x, a) = e^{-\frac{1}{2}\left[\frac{(x-a)}{\alpha}\right]^2} \tag{3}$$

In Equation (3), x is a variable, a is the mean value and α is a standard deflection.

• *Binary method*

Equation (4) can be used when the membership values for a domain set are manifest [7]. For example, the membership value of male is 0 for female and that of female is 1 in a set of {male, female}.

$$s_j : D_j \times D_j \rightarrow \{0,1\} \tag{4}$$
$$s_j(x, y) = \begin{cases} 1 & if \quad x = y, \\ 0 & otherwise, \end{cases}$$

where {0, 1} means the value is an integer which is 0 or 1.

2.3. CALCULATION METHOD OF THE TUPLE MEMBERSHIP VALUE

In this section, we show how to determine a membership value using the methods in Section 2.2. Suppose the tuple $t_j = (a'_{j1}, a'_{j2}, \cdots, a'_{jm})$, where $k \leq m$, for the query $Q(a_i, a_h, \cdots, a_k)$. Then the algorithm for the tuple membership value μ_t is as follows.

FindTupleMembershipValue()
step 1. determine membership values of domain elements, $\mu_{a_{il}}(a'_{il})$,
         for the query $Q(a_i, a_h, \cdots, a_k)$ using one of the methods in Section 2.2.

step 2. Calculate a tuple membership value μ_t using $\mu_{a_{il}}(a'_{il})$ which is the result of step 1.

The following methods are to calculate a tuple membership value with several domain membership values. MIN function method and B-P method are widely recognized, but Average method, N-Root method and Weight method are newly proposed in this paper.

• *MIN function method*

MIN function method [1, 2, 7, 8] are very widely used in fuzzy database systems. It selects the minimum membership value among domain membership values. MIN function method is simple but difficult to calculate precise result for the given queries.

$$\mu_t = MIN(\mu_1, \mu_2, \cdots, \mu_n) \tag{5}$$

In Equation (5), μ_t is the tuple membership value and $\mu_1, \mu_2, \cdots, \mu_n$ are the domain membership values.

• *B-P method*

B-P method [2, 11] includes two functions, $DIL()$ and $CON()$. $DIL()$ emphasizes and $CON()$ deemphasizes a domain tuple membership value. Sometimes those functions overpower to emphasize or deemphasize the value.

$$CON(\mu_i) = \mu_i^2$$
$$DIL(\mu_i) = \mu_i^{1/2} \tag{6}$$

• *Average Method*

Average method calculates the average of a set of domain membership values which can be obtained by a tabular form or the computational method and then the average should be multiplied by each domain membership value obtained by the binary method to get the tuple membership value, μ_t.

$$\mu_t = \frac{1}{m} \cdot \left(\sum_{i=1}^{m} \mu_i \right) \cdot (\mu_{m+1} \times \mu_{m+2} \times \cdots \times \mu_n), \quad (1 \le m \le n) \tag{7}$$

• *N-Root Method*

N-Root method calculates n-square root of a set of domain membership values which can be obtained by a tabular form or the computational method and then the result should be multiplied by each domain membership value obtained by the binary method.

$$\mu_t = \sqrt[n]{\mu_1 \times \mu_2 \times \cdots \times \mu_m}\,(\mu_{m+1} \times \mu_{m+2} \times \cdots \times \mu_n),\ (1 \le m \le n) \quad (8)$$

• *Weight Method*

Each domain membership value obtained by a tabular form or the computational method is multiplied by a weight, each result is added, and then the final result should be multiplied by each domain membership value obtained by the binary method.

$$\mu_t = (\alpha_1 \cdot \mu_1 + \alpha_2 \cdot \mu_2 + \cdots + \alpha_m \cdot \mu_m) \cdot (\mu_{m+1} \times \mu_{m+2} \times \cdots \times \mu_n),$$

$$(1 \le m \le n)$$

(9)

where $\alpha_1 + \alpha_2 + \cdots + \alpha_m = 1$.

In Equation's (7), (8) and (9), $\mu_1, \mu_2, \cdots, \mu_m$ are the domain membership values obtained by a tabular form or the computational method described in Section 2.2. $\mu_{m+1}, \mu_{m+2}, \cdots, \mu_n$ are the domain membership values obtained by the binary method in Section 2.2. If any value of $\mu_{m+1}, \mu_{m+2}, \cdots, \mu_n$ is 0, μ_t is 0. In real world, if one of the domain membership values, $\mu_{m+1}, \mu_{m+2}, \cdots, \mu_n$, obtained by the binary method is found to be 0 during processing, processing can stop immediately and the result, μ_t, is 0. In the above equations, n is the number of attributes in a query.

3. Fuzzy query processing

Suppose that the query "Retrieve a set of single women who are about 25 years old and have high-school grade" is given. The corresponding query is then

> **Q**(age, education, sex, marriage)
> [who is] **NEARLY** 25[years old]
> **AND** [who has a] **NEARLY** high school [diploma]
> **AND** [who is] a female
> **AND** [who is] not_married.

The query can be rewritten by SQL-type language as follows.

Select NAME, TLEVEL **with** TLEVEL ≥ 0.6
        from   personal_characteristics
        where   age = "25" **and**
                    education = "high school" **and**
                    sex = "female" **and**
                    marriage ="not_married"
        with   FLEVEL(age) ≥ 0.6 **and**
                    FLEVEL(education) ≥ 0.6 **and**
                    LEVEL(sex) = 1 **and**
                    LEVEL(marriage) = 1;

This SQL has three types of levels, LEVEL, FLEVEL and TLEVEL. These levels serve as filters when the query is processed. LEVEL selects tuples whose every domain member ship value obtained by Binary method is 1. FLEVEL cuts tuples whose indicated domain membership values are low. TLEVEL serves as filter for tuple membership values.

The following algorithm of the fuzzy query processing is reflected by three levels, LEVEL, FLEVEL and TLEVEL.

      step 1.   read a query as an input
      step 2.   **IF** (it is a fuzzy query)
                WHILE (not_end_of_table) **DO BEGIN**
                    retrieve a tuple of the table in a relational database;
                    get the domain membership values "FLEVEL's" and
                    "LEVEL's"
                    using the methods in Section 2.1;
                    IF ((all LEVEL =1) **and** (all FLEVEL $\geq$ required level))
                        TLEVEL = *FindTupleMembershipValue()*;
                    IF (TLEVEL $\geq$ required level)
                        print(information of the tuple, TLEVEL);
                END WHILE
      step 3.   **ELSE**
                do normal query processing;
      step 4.   **RETURN**

4. Example and Analysis

In this section, we show an example of the three newly proposed calculation methods of tuple membership value, Average method, N-Root method and Weight method. And we analyze that the proposed query processing algorithm with new methods provides the better results to users than that with existing methods, MIN method and B-P method.

Example 1   Consider the Personal_characteristic relation (TABLE 1) and the Membership values of the Personal_characteristic relation (TABLE 2). Calculate the tuple membership values for the query, "retrieve a set of unmarried women who are about 25 years old and possess attainments of high school graduate", using the newly proposed methods and existing methods.

TABLE 1. Personal_characteristic relation

Name	Age	Education	Sex	Marriage
Mary	30	Univ.	f	married
Jane	26	Univ. halfway	f	single
Smith	19	High school	m	single
Anne	18	High school	f	single
Shilvia	28	High school	f	single
Tom	45	Univ.	m	married
Anna	35	Univ halfway	f	single
John	27	High school	m	married
Lyne	24	High school	f	married
Ammy	28	Middle	f	single

TABLE 2. Membership values of Personal_characteristic

Name	Age	Education	Sex	Marriage
Mary	0.2	0.2	1	0.0
Jane	0.9	0.6	1	1.0
Smith	0.1	1.0	0	1.0
Anne	0.0	1.0	1	1.0
Shilvia	0.6	1.0	1	1.0
Tom	0.0	0.2	0	0.0
Anna	0.0	0.6	1	1.0
John	0.8	1.0	0	0.0
Lyne	0.9	1.0	1	0.0
Ammy	0.6	0.6	1	1.0

Solution and analysis   The tuple membership values for the Personal_characteristic relation is as follow (TABLE 3). The MIN operator pick up the smallest value of the attribute membership values. It is dependent on smallest value. The Average method makes up for the weak point in the MIN method. It is reflected in all attribute membership values.

TABLE 3. Result membership values

Name	MIN	Average	N-Root	B-P		Weight	
				a	b	a	b
Mary	0	0	0	0	0	0	0
Jane	0.6	0.75	0.73	0.77	0.36	0.72	0.78
Smith	0	0	0	0	0	0	0
Anne	0	0.5	0	0	0	0.6	0.6
Shilvia	0.6	0.8	0.77	0.36	0.77	0.84	0.76
Tom	0	0	0	0	0	0	0
Anna	0	0.3	0	0	0	0.36	0.24
John	0	0	0	0	0	0	0
Lyne	0	0	0	0	0	0	0
Ammy	0.6	0.6	0.6	0.36	0.36	0.6	0.6

N-Root method is similar to Average method. But, this method produces zero if any one value is zero among the attribute values.

Weight method gives a weight factor in each attribute value. In TABLE 3, The case a of Weight or B-P method means that 'Education' attribute has weight factor 0.6 and 'Age' attribute has weight factor 0.4. The case b is vice versa.

Consider the query processing algorithm in Section 3. The query processing algorithm produces following results.

MIN	: {(Jane, 0.6), (Shilvia, 0.6), (Ammy, 0.6)}
Average	: {(Jane, 0.75), (Shilvia, 0.8), (Ammy, 0.6)}
N-Root	: {(Jane, 0.73), (Shilvia, 0.77), (Ammy, 0.6)}
B-P.a	: {(Jane, 0.77)}
B-P.b	: {(Shilvia, 0.77)}
Weight.a	: {(Jane, 0.72), (Shilvia, 0.84), (Ammy, 0.6)}
Weight.b	: {(Jane, 0.78), (Shilvia, 0.76), (Ammy, 0.6)}

We can see that the newly proposed methods generate more discriminate results than the existing methods, MIN and B-P method.

5. Conclusions

This paper concerns with fuzzy query processing in relational databases. The main characteristics of our study is to deal with three newly proposed calculation methods of the tuple membership value, so called Average method, N-Root method and Weight method. Proposed fuzzy query processing algorithm with new methods provides the better results to users than that with existing methods, MIN method and B-P method. And also, the algorithm including three levels as filters, LEVEL, FLEVEL and TLEVEL, do fuzzy query processing effectively. The relational database involving fuzzy query processing is to support an extended SQL which not only subsumes the standard SQL but also deals with imprecise queries.

References

1. Buckles, B. P. and Petry, F. E. (1982) A Fuzzy Representation of Data for Relational Database, *Fuzzy Sets and Systems* **7**, 213-226.
2. Buckles, B. P. and Petry, F. E. (1983) Information-Theoretical Characterization of Fuzzy Relational Databases, *IEEE Trans. on Systems, Man, and Cybernetics* **SMC-13**, 74-77.
3. Date, C. J. (1985) *An Introduction to Database Systems* 1, Addison Wesley.

4. Klir G. J. and Folger, T. A. (1988) *Fuzzy Sets, Uncertainty, and Information,* Prentice-Hall International (UK) Limited, London.

5. Korth H. F. and Silberschatz, A. (1991) *Database System Concepts,* McGRAW-HILL, N. Y.

6. Motro, A. (1988) VAGUE: A User Interface to Relational Databases that Permit Vague Queries, *ACM Trans. on Office Information System* **6**, 187-214.

7. Shenoi, S. and Melton, A. (1989) Proximity Relations in the Fuzzy Relational Database Model, *Fuzzy Sets and Systems* **31**, 285-296.

8. Shenoi, S., Melton, A. and Fan, L. T. (1990) An Equivalence Classes Model of Fuzzy Relational Databases, *Fuzzy Sets and Systems* **38**, 153-170.

9. Keller, G., Warrack, B. and. Bartel, H (1988) *Statistics for Management and Economics,* Wadsworth, California.

10. Zadeh, L. A. (1965) Fuzzy Sets, *Information Control* **8**, 338-353.

11. Zadeh, L. A. (1973) Outline of a New Approach to the Analysis of Complex Systems and Decision Process, *IEEE Trans. on Systems, Man, and Cybernetics* **SMC-3.**

12. Zemankova, M. and Kandel, A. (1985) Implementing Imprecision in Information Systems, *Information Science* **37**, 107-141.

13. Lee K. H. and Oh, G. R. (1991) *Fuzzy Sets and Applications* **2**, Hongreung Science Publishing Company.

A SOLUTION CONCEPT IN MULTIOBJECTIVE MATRIX GAMES WITH FUZZY PAYOFFS AND FUZZY GOALS

MASATOSHI SAKAWA

Department of Industrial and Systems Engineering,
Faculty of Engineering, Hiroshima University
Kagamiyama 1-4-1, Higashi-Hiroshima 724 Japan

AND

ICHIRO NISHIZAKI

Faculty of Business Administration and Informatics,
The Setsunan University
Ikeda-Nakamachi 17-8, Neyagawa, Osaka 572 Japan

1. Introduction

In this paper we consider multiobjective matrix games with fuzzy payoffs and fuzzy goals. Games considered here and conventional matrix games differ by the following three points. First, we employ representing entries of a payoff matrix as fuzzy numbers for expressing ambiguity and imprecision in information which is utilized in modeling of games because such information is not always accurate. Second, multiple payoffs are considered in games because most of decision making problems under conflict possess multiple objectives such as cost, time and productivity. Third, we assume that each player has a fuzzy goal for each objective in order to incorporate ambiguity of human judgment.

In section 2, a fuzzy expected payoff is defined, and a degree of attainment of a fuzzy goal is considered in games with fuzzy payoff matrices. The max-min solution with respect to a degree of attainment of a fuzzy goal is also defined. In section 3, the methods for computing the solution of a multiobjective game are proposed when membership functions of fuzzy goals and shape functions of L-R fuzzy numbers for fuzzy payoffs are linear. An original problem for computing the max-min solution is formulated as a nonlinear programming problem, but it can be transformed to a linear programming problem by making use of the bisection method and phase one of the simplex method (Sakawa, 1983), the variable

417

Z. Bien and K. C. Min (eds.),
Fuzzy Logic and its Applications, Information Sciences, and Intelligent Systems, 417–426.
© 1995 *Kluwer Academic Publishers.*

transformation (Charnes and Cooper, 1962), and the relaxation procedure (Shimizu and Aiyoshi, 1980).

2. Problem Formulation and Solution Concepts

Let $i \in \{1, 2, \ldots, m\}$ be a pure strategy of Player I and $j \in \{1, 2, \ldots, n\}$ be a pure strategy of Player II.

Definition 1 (Zero-sum fuzzy matrix game)   When Player I chooses a pure strategy i and Player II chooses a pure strategy j, let $\tilde{a}_{ij}$ be a fuzzy payoff for Player I and $-\tilde{a}_{ij}$ be a fuzzy payoff for Player II. The fuzzy payoff $\tilde{a}_{ij}$ is represented by the L-R fuzzy number:

$$\tilde{a}_{ij} = (a_{ij}, \alpha_{ij}, \beta_{ij})_{LR}, \tag{1}$$

where a_{ij} is a mean value, α_{ij} is a right spread and β_{ij} is a left spread. The two-person zero-sum fuzzy matrix game can be represented as a fuzzy payoff matrix:

$$\tilde{A} = \begin{bmatrix} \tilde{a}_{11} & \tilde{a}_{12} & \cdots & \tilde{a}_{1n} \\ \tilde{a}_{21} & \tilde{a}_{22} & \cdots & \tilde{a}_{2n} \\ \vdots & \vdots & \ddots & \vdots \\ \tilde{a}_{m1} & \tilde{a}_{m2} & \cdots & \tilde{a}_{mn} \end{bmatrix}. \tag{2}$$

Games defined by (2) are called two-person zero-sum fuzzy matrix games.

When each of the players chooses a strategy, a payoff for each of them is represented as a fuzzy number, but the outcome has a zero-sum structure such that, when one player receives a gain, the other player suffers an equal loss.

Two-person zero-sum multiobjective fuzzy matrix games can also be represented by multiple fuzzy payoff matrices

$$\tilde{A}^1 = \begin{bmatrix} \tilde{a}_{11}^1 \cdots \tilde{a}_{1n}^1 \\ \vdots \ddots \vdots \\ \tilde{a}_{m1}^1 \cdots \tilde{a}_{mn}^1 \end{bmatrix}, \tilde{A}^2 = \begin{bmatrix} \tilde{a}_{11}^2 \cdots \tilde{a}_{1n}^2 \\ \vdots \ddots \vdots \\ \tilde{a}_{m1}^2 \cdots \tilde{a}_{mn}^2 \end{bmatrix}, \ldots, \tilde{A}^r = \begin{bmatrix} \tilde{a}_{11}^r \cdots \tilde{a}_{1n}^r \\ \vdots \ddots \vdots \\ \tilde{a}_{m1}^r \cdots \tilde{a}_{mn}^r \end{bmatrix}, \tag{3}$$

where we assume that each of the two players has r objectives. Then a fuzzy expected payoff can be represented by an L-R fuzzy number. A fuzzy payoff can be extended to a fuzzy expected payoff by using mixed strategies in a procedure similar to the extension from a payoff to an expected payoff.

Let $x \in X \triangleq \{x = (x_1, x_2, \ldots, x_m) \mid \sum_{i=1}^{m} x_i = 1, x_i \geq 0\}$ be a mixed strategy for Player I and let $y \in Y \triangleq \{y = (y_1, y_2, \ldots, y_n) \mid \sum_{j=1}^{n} y_j = 1, y_j \geq 0\}$ be a mixed strategy for Player II.

Definition 2 (Fuzzy expected payoff)   For any pair of mixed strategies $x \in X$ and $y \in Y$, the kth fuzzy expected payoff of Player I is defined as the fuzzy number

$$\tilde{E}^k(x, y) = \left(\sum_{i=1}^{m} \sum_{j=1}^{n} a_{ij}^k x_i y_j, \sum_{i=1}^{m} \sum_{j=1}^{n} \alpha_{ij}^k x_i y_j, \sum_{i=1}^{m} \sum_{j=1}^{n} \beta_{ij}^k x_i y_j \right)_{LR} \tag{4}$$

characterized by the membership function

$$\mu_{\tilde{E}^k(x,y)} : D^k \rightarrow [0, 1], \tag{5}$$

where $D^k \in R$ is the domain of the kth payoff for Player I.

Addition and scalar multiplication on L-R fuzzy numbers are used in the definition of a fuzzy expected payoff (4).

Definition 3 (Fuzzy goal)   Let the domain of the kth payoff for Player I be denoted $D^k \in R$. Then the fuzzy goal $\tilde{G}^k$ with respect to the kth payoff for Player I is defined as the fuzzy set on the set D^k characterized by the membership function

$$\mu_{\tilde{G}^k} : D^k \rightarrow [0, 1]. \tag{6}$$

A membership function value of a fuzzy goal can be interpreted as a degree of attainment of the fuzzy goal. Then we assume that, for any pair of payoffs, a player prefers the payoff having the greater degree of attainment of the fuzzy goal to the other payoff.

Definition 4 (A degree of attainment of a fuzzy goal)   For any pair of mixed strategies (x, y), let the kth fuzzy expected payoff for Player I be denoted $\tilde{E}^k(x, y)$ and let the kth fuzzy goal for Player I be denoted $\tilde{G}^k$. Then a fuzzy set expressing an attainment state of the fuzzy goal is represented by the intersection of the fuzzy expected payoff $\tilde{E}^k(x, y)$ and the fuzzy goal $\tilde{G}^k$. The membership function of the fuzzy set is represented as

$$\mu^k_{a(x,y)}(p) = \min \left(\mu_{\tilde{E}^k(x,y)}(p), \ \mu_{\tilde{G}^k}(p) \right), \tag{7}$$

where $p \in D^k$ is a payoff for Player I. A degree of attainment of the kth fuzzy goal is defined as the maximum of the membership function (7), i.e.,

$$\begin{aligned} \hat{\mu}^k_{a(x,y)}(p^*) &= \max_p \mu^k_{a(x,y)}(p) \\ &= \max_p \{\min(\mu_{\tilde{E}^k(x,y)}(p), \ \mu_{\tilde{G}^k}(p))\}. \end{aligned} \tag{8}$$

A degree of attainment of a fuzzy goal can be considered to be a concept similar to a degree of satisfaction of the fuzzy decision by Bellman and Zadeh (1970) when the fuzzy constraint can be replaced by the fuzzy expected payoff. When Players I and II choose strategies $\hat{x}$ and $\hat{y}$, respectively, the degree of attainment of the kth fuzzy goal $\hat{\mu}^k_{a(\hat{x},\hat{y})}(p^*)$ is determined by (8).

We assume that Player I supposes that Player II chooses a strategy $\hat{y}$ so as to minimize Player I's degree of attainment of the fuzzy goal $\hat{\mu}^k_{a(\hat{x},\hat{y})}(p^*)$, i.e., Player I's degree of attainment of the fuzzy goal, assuming he uses $\hat{x}$, will be $e^k(x) = \min_{y \in Y} \hat{\mu}^k_{a(\hat{x},\hat{y})}(p^*)$. Hence, Player I chooses a strategy so as to maximize his degree of attainment of the fuzzy goal $e^k(x)$. In short, we assume that Player I behaves according to the max-min principle in terms of a degree of attainment of his fuzzy goal.

We usually consider the vector optimization for multiple objectives, but each of the measures for the objectives can be transformed to the measure of the degree of attainment of the fuzzy goal as a commensurable measure. Thus, we can consider max-min problems in terms of maximization and minimization of the degree of attainment of the aggregated fuzzy goal.

Definition 5 (A max-min solution with respect to a degree of attainment of an aggregated fuzzy goal) For any pair of mixed strategies (x, y), let the degree of attainment of the aggregated fuzzy goal for Player I be denoted $\hat{\mu}_{a(x,y)}(p^*)$. Then Player I's max-min value with respect to a degree of attainment of the aggregated fuzzy goal is

$$\max_{x \in X} \min_{y \in Y} \hat{\mu}_{a(x,y)}(p^*), \tag{9}$$

and such a strategy x is called the max-min solution with respect to the degree of attainment of the aggregated fuzzy goal.

The max-min solution can be considered to be the solution maximizing the function, which is the minimal value of the function with respect to the opponent's decision variables. We assume that a player has no information about his opponent or the information is not useful for the decision making if he has. Player I supposes that Player II chooses the strategy which minimizes Player I's degree of attainment of the fuzzy goal, and then Player I maximizes his degree of attainment of the fuzzy goal with respect to his decision variables.

We can also consider Player II's min-max solution with respect to a degree of attainment of the aggregated fuzzy goal in a similar way.

3. Computational Methods

We propose a method for computing the max-min solution of a multiobjective game. Consider two-person zero-sum multiobjective fuzzy matrix games, i.e., two-person zero-sum games with multiple fuzzy payoff matrices $\tilde{A}^k$, $k = 1, 2, \ldots, r$. We assume that a player has a fuzzy goal for each of the objectives, which expresses the player's degree of satisfaction for a payoff. Let Player I's membership function of the fuzzy goal for the kth objective be denoted $\mu_{\tilde{G}^k}(p^k)$ for any payoff p^k.

When the membership function $\mu_{\tilde{G}^k}(p^k)$ of the fuzzy goal is linear, it can be represented as

$$\mu_{\tilde{G}^k}(p^k) = \begin{cases} 0 & \text{if } p^k < \underline{a}^k \\ 1 - \dfrac{\overline{a}^k - p^k}{\overline{a}^k - \underline{a}^k} & \text{if } \underline{a}^k \le p^k \le \overline{a}^k \\ 1 & \text{if } \overline{a}^k < p^k, \end{cases} \tag{10}$$

where, for the kth objective, $\underline{a}^k$ is the payoff giving the worst degree of satisfaction for Player I and $\overline{a}^k$ is the payoff giving the best degree of satisfaction for Player I.

Moreover, when the membership function $\mu_{\tilde{a}^k_{ij}}(p^k)$ of the entry $\tilde{a}^k_{ij}$, which is a fuzzy number, of the fuzzy payoff matrix $\tilde{A}^k$ for the kth objective is linear, it can be represented as

$$\mu_{\tilde{a}^k_{ij}}(p^k) = \begin{cases} 0 & \text{if } p^k < a^k_{ij} - \alpha^k_{ij} \\ (p^k - a^k_{ij} + \alpha^k_{ij})/\alpha^k_{ij} & \text{if } a^k_{ij} - \alpha^k_{ij} \leq p^k < a^k_{ij} \\ (a^k_{ij} + \beta^k_{ij} - p^k)/\beta^k_{ij} & \text{if } a^k_{ij} \leq p^k \leq a^k_{ij} + \beta^k_{ij} \\ 0 & \text{if } a^k_{ij} + \beta^k_{ij} < p^k. \end{cases} \tag{11}$$

In general, the degree of attainment of the fuzzy goal can be represented as a vector. For such a problem, we employ the fuzzy decision rule by Bellman and Zadeh (1970), which is often used in decision making problems in fuzzy environments, as an aggregation rule of multiple fuzzy goals. Then the membership function of the aggregated fuzzy goal is expressed as

$$\hat{\mu}_{a(x,y)}(p^*) = \min_k \max_{p^k} \min(\mu_{\tilde{E}^k(x,y)}(p^k), \mu_{\tilde{G}^k}(p^k)), \tag{12}$$

where $p^* = (p^{*1}, p^{*2}, \ldots, p^{*r})$.

When membership functions are linear, the max-min strategy with respect to a degree of attainment of the aggregated fuzzy goal can be obtained by solving the mathematical programming problem in the following theorem.

Theorem 1　Let membership functions of fuzzy goals and a shape function of L-R fuzzy numbers for fuzzy payoffs be linear such as (10) and (11). A solution for the max-min problem with respect to the degree of attainment of the fuzzy goal

$$\max_{x \in X} \min_{y \in Y} \min_k \max_{p^k} \min(\mu_{\tilde{E}^k(x,y)}(p^k), \mu_{\tilde{G}^k}(p^k)) \tag{13}$$

is equal to an optimal solution of the following nonlinear programming problem:

$$\begin{aligned} & \underset{(x,\sigma)}{\text{maximize}} && \sigma \\ & \text{subject to} && \frac{\displaystyle\sum_{i=1}^{m}\sum_{j=1}^{n}(a^k_{ij} + \beta^k_{ij})x_i y_j - \underline{a}^k}{\displaystyle\sum_{i=1}^{m}\sum_{j=1}^{n}\beta^k_{ij}x_i y_j + \overline{a}^k - \underline{a}^k} \geq \sigma, \ \forall y \in Y, \ k = 1, 2, \ldots, r \\ & && \sum_{i=1}^{m} x_i = 1, \end{aligned} \tag{14}$$

when the optimal solution σ^* satisfies $0 \leq \sigma^* \leq 1$. The problem (14) is a nonlinear programming problem which has decision variables x_i, $i = 1, 2, \ldots, m$ and σ, and has an infinite number of inequality constraints and one equality constraint.

Proof When the optimal solution σ^* satisfies $0 \le \sigma^* \le 1$, for any pair of mixed strategies x and y, Player I's degree of attainment of the fuzzy goal is represented as

$$
\begin{aligned}
\mu_{a(x,y)}(p^*) &= \min_k \max_{p^k} \min(\mu_{\tilde{E}^k(x,y)}(p^k), \mu_{\tilde{G}^k}(p^k)) \\
&= \min_k \frac{\displaystyle\sum_{i=1}^{m}\sum_{j=1}^{n} a_{ij}^k x_i y_j + \sum_{i=1}^{m}\sum_{j=1}^{n} \beta_{ij}^k x_i y_j - \underline{a}^k}{\displaystyle \overline{a}^k - \underline{a}^k + \sum_{i=1}^{m}\sum_{j=1}^{n} \beta_{ij}^k x_i y_j}.
\end{aligned}
\tag{15}
$$

Therefore, the max-min problem with respect to the degree of attainment of the fuzzy goal is represented as

$$
\begin{aligned}
&\max_{x \in X} \min_{y \in Y} \min_k \max_{p^k} \min(\mu_{\tilde{E}^k(x,y)}(p^k), \mu_{\tilde{G}^k}(p^k)) \\
&= \max_{x \in X} \min_{y \in Y} \min_k \frac{\displaystyle\sum_{i=1}^{m}\sum_{j=1}^{n} a_{ij}^k x_i y_j + \sum_{i=1}^{m}\sum_{j=1}^{n} \beta_{ij}^k x_i y_j - \underline{a}^k}{\displaystyle \overline{a}^k - \underline{a}^k + \sum_{i=1}^{m}\sum_{j=1}^{n} \beta_{ij}^k x_i y_j}.
\end{aligned}
\tag{16}
$$

Since the constraints of maximizing decision variable x and the minimizing decision variable y in the problem (16) are separated each other, the max-min solution can be determined by solving the following mathematical programming problem by introducing an auxiliary variable σ:

$$
\begin{aligned}
&\underset{(x,\sigma)}{\text{maximize}} \quad \sigma \\
&\text{subject to} \quad \min_{y \in Y} \min_k \frac{\displaystyle\sum_{i=1}^{m}\sum_{j=1}^{n}(a_{ij}^k + \beta_{ij}^k) x_i y_j - \underline{a}^k}{\displaystyle\sum_{i=1}^{m}\sum_{j=1}^{n} \beta_{ij}^k x_i y_j + \overline{a}^k - \underline{a}^k} \ge \sigma \\
&\qquad\qquad\quad \sum_{i=1}^{m} x_i = 1.
\end{aligned}
\tag{17}
$$

Since the first condition in (17) is equivalent to the following conditions:

$$
\frac{\displaystyle\sum_{i=1}^{m}\sum_{j=1}^{n}(a_{ij}^k + \beta_{ij}^k) x_i y_j - \underline{a}^k}{\displaystyle\sum_{i=1}^{m}\sum_{j=1}^{n} \beta_{ij}^k x_i y_j + \overline{a}^k - \underline{a}^k} \ge \sigma, \qquad \forall y \in Y, \ k = 1, 2, \ldots, r,
\tag{18}
$$

the problem (17) is equivalent to the problem (14). ∎

Sufficient conditions such that the optimal solution σ^* of the problem (14) satisfies $0 \leq \sigma^* \leq 1$ are $\min_{(i,j)}(a_{ij}^k + \beta_{ij}^k) \geq \underline{a}^k, k = 1, 2, \ldots, r$ and $\max_{(i,j)} a_{ij}^k \leq \overline{a}^k, k = 1, 2, \ldots, r$.

The constraints of maximizing decision variable x and the minimizing decision variable y in the problem (16) are separated each other, so we can calculate the max-min solution with respect to a degree of attainment of a fuzzy goal by applying the method based on the relaxation procedure by Shimizu and Aiyoshi (1980). However, although the problem (16) is still a max-min problem, it has an extra min operator. Thus we have to revise the algorithm by the relaxation procedure.

Consider the following relaxed problem for the original problem (14) by taking L points $y_j^l, l = 1, 2, \ldots, L$, satisfying $\sum_{j=1}^n y_j^l = 1$:

$$
\begin{aligned}
&\underset{(x,\sigma)}{\text{maximize}} \quad \sigma \\
&\text{subject to} \quad \frac{\sum_{i=1}^m \sum_{j=1}^n (a_{ij}^k + \beta_{ij}^k) x_i y_j^l - \underline{a}^k}{\sum_{i=1}^m \sum_{j=1}^n \beta_{ij}^k x_i y_j^l + \overline{a}^k - \underline{a}^k} \geq \sigma, \; l = 1, 2, \ldots, L, \; k = 1, 2, \ldots, r \\
&\qquad\qquad \sum_{i=1}^m x_i = 1.
\end{aligned}
$$

$$(19)$$

Let $\sigma = \hat{\sigma}$, where $\hat{\sigma}$ is a constant value in $[0, 1]$. Then the constraints of the relaxed problem (19) become as follows:

$$
\sum_{i=1}^m \sum_{j=1}^n (a_{ij}^k + \beta_{ij}^k) x_i y_j^l - \underline{a}^k \geq \hat{\sigma} \left(\sum_{i=1}^m \sum_{j=1}^n \beta_{ij}^k x_i y_j^l + \overline{a}^k - \underline{a}^k \right),
$$
$$l = 1, 2, \ldots, L, \; k = 1, 2, \ldots, r \qquad (20)$$
$$\sum_{i=1}^m x_i = 1.$$

The test for feasibility (i.e., whether the problem with the constraints (20) is feasible or not) can be accomplished by using phase one of the simplex method. If it is feasible, renew the constant value $\hat{\sigma}$ as $\hat{\sigma} \leftarrow \hat{\sigma} + \frac{1}{2}\hat{\sigma}$. If it is not feasible, renew the constant value $\hat{\sigma}$ as $\hat{\sigma} \leftarrow \hat{\sigma} - \frac{1}{2}\hat{\sigma}$. Then the test for feasibility is executed again after renewing the constant value $\hat{\sigma}$. We can find the maximal constant value $\hat{\sigma}$ satisfying the constraints (20) by repeating this procedure in a finite number of iterations, so it follows that the pair of the feasible solution x^* and the maximal constant value $\hat{\sigma}$ must be an optimal solution $(x^*, \sigma^* = \hat{\sigma})$ of the relaxed problem (19).

The r minimization problems for the test of feasibility and the generation of the most violated constraint are represented as follows:

$$
\underset{y}{\text{minimize}} \quad
\left.
\begin{aligned}
&\dfrac{\displaystyle\sum_{i=1}^{m}\sum_{j=1}^{n}(a_{ij}^{k}+\beta_{ij}^{k})x_{i}^{L}y_{j}-\underline{a}^{k}}{\displaystyle\sum_{i=1}^{m}\sum_{j=1}^{n}\beta_{ij}^{k}x_{i}^{L}y_{j}+\overline{a}^{k}-\underline{a}^{k}} \\[4mm]
&\text{subject to}\quad \sum_{j=1}^{n}y_{j}=1
\end{aligned}
\right\}, \qquad k=1,2,\ldots,r. \tag{21}
$$

The above minimization problems (21) can be reduced to linear programming problems by using the following variable transformations. Set

$$
1\Big/\!\left(\sum_{i=1}^{m}\sum_{j=1}^{n}\beta_{ij}^{k}x_{i}^{L}y_{j}+\overline{a}^{k}-\underline{a}^{k}\right)=t^{k},\ k=1,2,\ldots,r, \tag{22}
$$

and

$$
y_{j}t^{k}=z_{j}^{k},\ k=1,2,\ldots,r. \tag{23}
$$

The minimization problem can be represented as the following r linear programming problems:

$$
\underset{(z^{k},\,t^{k})}{\text{minimize}} \quad
\left.
\begin{aligned}
&\sum_{i=1}^{m}\sum_{j=1}^{n}(a_{ij}^{k}+\beta_{ij}^{k})x_{i}^{L}z_{j}^{k}-\underline{a}^{k}\,t^{k} \\[2mm]
\text{subject to}\quad &\sum_{j=1}^{n}z_{j}^{k}=t^{k} \\[2mm]
&\sum_{i=1}^{m}\sum_{j=1}^{n}\beta_{ij}^{k}x_{i}^{L}z_{j}^{k}+(\overline{a}-\underline{a})\,t^{k}=1.
\end{aligned}
\right\}, \qquad k=1,2,\ldots,r. \tag{24}
$$

The kth problem in (24) is a linear programming problem which has decision variables z_{j}^{k}, $j=1,2,\ldots,n$ and t^{k}, and has two equality constraints. Since there are r problems, the test for feasibility for the original problem and the generation of the most violated constraint can be accomplished by solving the r linear programming problems and finding the problem having the smallest optimal value.

The algorithm for computing the max-min solution of fuzzy multiobjective matrix games can be summarized in the following steps.

Algorithm 1
[Step 1] Identify r fuzzy goals for payoffs. Choose any initial point $y^{1}\in Y$ and set $l=1$. Then formulate a relaxed problem (19), which is a linear fractional programming problem.
[Step 2] Formulate the constraints (20) by setting $\sigma=\hat{\sigma}$ in the constraints of the relaxed problem (19). Compute an optimal solution (x^{*},σ^{*}) by making use of the bisection method and phase one of the simplex method. Then set $x^{L}=x^{*}$.

[Step 3] Formulate r minimization linear programming problems (24) with x^L.

[Step 4] Solve r problems (24) and obtain r optimal solutions (z^{k*}, t^{k*}), $k = 1, 2, \ldots, r$. Let each of the minimal objective function values be denoted $\phi^k(z^{k*}, t^{k*})$, $k = 1, 2, \ldots, r$ and then let $\phi^{\hat{k}}(z^{\hat{k}*}, t^{\hat{k}*}) = \min_k \phi^k(z^{k*}, t^{k*})$.

[Step 5] If $\phi^{\hat{k}}(z^{\hat{k}*}, t^{\hat{k}*}) \geq \sigma^* + \varepsilon$, terminate, where ε is a predetermined constant. Then x^L is a max-min solution with respect to a degree of attainment of a fuzzy goal. Otherwise, i.e., if $\phi^{\hat{k}}(z^{\hat{k}*}, t^{\hat{k}*}) < \sigma^* + \varepsilon$, set $l = l + 1$ and go back to [Step 2].

Theorem 2 For any given $\varepsilon > 0$, the above algorithm for the max-min problem (13) terminates in a finite number of iterations.

Proof The theorem can be proved by a procedure similar to the proof of the theorem by Shimizu and Aiyoshi (1980). ∎

We can also obtain Player II's min-max solution with respect to a degree of attainment of a fuzzy goal in a similar way.

We can also present the other method for the computing solutions. From the property of the constraints of the linear programming problem (24), the problem (14) is equivalent to the following problem:

$$
\begin{array}{ll}
\underset{(x,\sigma)}{\text{maximize}} & \sigma \\
\text{subject to} & \dfrac{\displaystyle\sum_{i=1}^{m}(a_{i1}^k + \beta_{i1}^k)x_i - \underline{a}^k}{\displaystyle\sum_{i=1}^{m}\beta_{i1}^k x_i + \overline{a}^k - \underline{a}^k} \geq \sigma, \ k = 1, 2, \ldots, r \\
& \quad\quad\vdots \\
& \dfrac{\displaystyle\sum_{i=1}^{m}(a_{in}^k + \beta_{in}^k)x_i - \underline{a}^k}{\displaystyle\sum_{i=1}^{m}\beta_{in}^k x_i + \overline{a}^k - \underline{a}^k} \geq \sigma, \ k = 1, 2, \ldots, r \\
& \displaystyle\sum_{i=1}^{m} x_i = 1.
\end{array}
\tag{25}
$$

The number of the constraints of the problem (25) is $nr + 1$, which becomes larger as the numbers of Player II's strategies and objectives increase. Therefore, the method that includes the relaxation procedure, Algorithm 1, is considered to be efficient when the numbers of Player II's strategies and objectives are large.

4. Concluding Remarks

In two-person zero-sum multiobjective matrix games, we have represented entries of payoff matrices as fuzzy numbers in order to express ambiguity and imprecision of information about decision making problems under conflict, and we have employed fuzzy goals to handle the imprecise nature of human judgment.

When membership functions of fuzzy goals and a shape function of fuzzy number entries in a fuzzy payoff matrix can be constructed as linear functions, two methods for computing the solutions have been developed.

Finally, we briefly discuss related properties for Pareto equilibrium solutions. Equilibrium solutions multiobjective matrix games, which are studied by Shapley (1959), are such mixed strategies x^* and y^* that there never exist any x satisfying $xAy^* > x^*Ay^*$ and any y satisfying $x^*Ay < x^*Ay^*$. Since membership functions of fuzzy goals are assumed to be monotone nondecreasing functions, if players assess $\underline{a}^k$ sufficiently small and assess $\overline{a}^k$ sufficiently large, the proposed solution belongs to the set of Pareto equilibrium solutions.

References

Aubin, J.P. (1979) *Mathematical Methods of Game and Economic Theory*, North-Holland, Amsterdam.

Aubin, J.P. (1981) Cooperative fuzzy game, *Math. of Operations Research*, **Vol. 6**, pp. 1–13.

Bellman, R.E. and L.A. Zadeh (1970) Decision making in a fuzzy environment, *Management Science*, **Vol. 17**, pp. 209–215.

Blackwell, D. (1956) An analog of the minimax theorem for vector payoffs, *Pacific Journal of Mathematics*, **Vol. 98**, pp. 1–8.

Butnariu, D. (1978) Fuzzy games; a description of the concept, *Fuzzy Sets and Systems*, **Vol. 1**, pp. 181–192.

Butnariu, D. Stability and shapley value for an *n*-persons fuzzy game, *Fuzzy Sets and Systems*, **Vol. 4**, pp. 63–72.

Campos, L. (1989) Fuzzy linear programming models to solve fuzzy matrix games, *Fuzzy Sets and Systems*, **Vol. 32**, pp. 275–289.

Charnes, A. and W. Cooper (1962) Programming with linear fractional function, *Naval Research Logistics Quarterly*, **Vol. 9**, pp. 181–186.

Charnes, A., Z. Huang, J. Rousseau and Q. Wei (1990) Cone extremal solutions of multi-payoff games with cross-constrained strategy set, *Optimization*, **Vol. 21**, pp. 51–69.

Contini, M., I. Olivtti and C. Milano (1966) A decision model under certainty with multiple payoffs, in A. Mensch, Ed., *Theory of games; Techniques and Applications*, American Elsevier, New York, pp. 50–63.

Cook, W.D. (1976) Zero-sum games with multiple goals, *Naval Research Logistics Quarterly*, **Vol. 23**, pp. 615–622.

Nishizaki, I. and M. Sakawa (1992) Two-person zero-sum games with multiple fuzzy goals, *Journal of Japan Society for Fuzzy Theory and Systems*, **Vol. 4**, pp. 504–511. (in Japanese)

Sakawa, M. (1983) Interactive computer program for fuzzy linear programming with multiple objectives, *International Journal of Man-Machine Studies*, **Vol. 18**, pp. 489–503.

Sakawa, M. and I. Nishizaki (1994) A Lexicographical solution concept in an *n*-person cooperative fuzzy game, *Fuzzy Sets and Systems*, **Vol. 61**, pp. 265–275.

Shapley, L.S. (1959) Equilibrium Points in Games with Vector Payoff, *Naval Research Logistics Quarterly*, **Vol. 6**, pp. 57–61.

Shimizu, K. and E. Aiyoshi (1980) Necessary conditions for min-max problems and algorithm by a relaxation procedure, *IEEE Transaction on Automatic Control*, **Vol. AC-25**, pp. 62–66.

Zeleny, M. (1975) Games with multiple payoffs, *International Journal of Game Theory*, **Vol. 4**, pp. 179–191.

Zhao, J. (1991) The equilibria of a multiple objective game, *International Journal of Game Theory*, **Vol. 20**, pp. 171–182.

ON CLASSICALIZING FUZZY DATABASES

SUJEET SHENOI

Department of Mathematical and Computer Sciences
University of Tulsa, Tulsa, Oklahoma 74104-3189, U.S.A.

Abstract. This paper describes a formal mechanism for classicalizing fuzzy relational databases. The mechanism employs partitions or "contexts" to capture and articulate the subjective equivalences underlying fuzziness. The semantics-based context formalism is used to specify database integrity constraints which support the uniform representation and processing of precise and fuzzy information. The resulting context-based model generalizes the classical relational database model and a family of fuzzy relational database models. A key advantage of the context formalism is that it considerably simplifies the implementation of fuzzy relational databases and eases their integration with existing classical relational database systems.

1. Introduction

Fuzzy relational database models extend their classical counterparts by supporting fuzzy information storage and facilitating information retrieval based on fuzzy queries [1,3,4,11,14]. They are appealing because they offer significant increases in expressive power and in the ability to abstractly manipulate database information. Unfortunately, most fuzzy relational database models are too complex and unwieldy for efficient implementation and integration with popular classical database systems. This has retarded the commercial development of fuzzy relational database systems and their application in business and industry.

This paper describes a formal mechanism for classicalizing fuzzy relational databases, thereby rendering them feasible for real-world implementation. The methodology employing partitions or "contexts" [10] to capture and articulate subjective equivalences underlying fuzziness is motivated by Zadeh's classic work on similarity relations and fuzzy orderings [12]. The

427

Z. Bien and K. C. Min (eds.),
Fuzzy Logic and its Applications, Information Sciences, and Intelligent Systems, 427–435.

semantics-based context formalism is used to specify database integrity constraints which support the uniform representation and processing of precise and fuzzy information. The resulting database model generalizes the classical relational model and a family of fuzzy relational database models based on similarity and proximity relations [1,2,7-9]. In fact, the new model can be viewed as a high-level specification of a relational database for uniformly dealing with precise and fuzzy (or abstract) information. The classical relational model and the well-known similarity-based and proximity-based fuzzy models are merely implementations of this high-level database specification. A key bonus is that the context-based formalism simplifies the implementation of other prominent fuzzy relational models [3-5,11,14] and eases their integration with existing classical relational database systems. A version of the context-based relational model is currently implemented as an upward extension of the popular ORACLE Relational Database Management System.

2. Contexts and Domain Integrity

This section describes the notion of a context and uses it to define a generalized domain integrity constraint restricting tuple components legally stored in database relations. Legal tuple components can be precise values or "meaningful" fuzzy sets. The term "meaningful" is defined with respect to prevailing contexts which are constructed from data semantics. The contexts also control the maximal imprecision permitted in stored components.

A context C is a partition on a set $\hat{D}$ generated by an equivalence relation ρ on $\hat{D}$. $\hat{D}$ is a subset of an underlying scalar domain D. It represents a "frame of reference" or "focus of attention." The equivalence relation captures semantic equivalences between domain elements. The equivalence classes in the corresponding context are collections of "closely related" (indistinguishable) elements. The set of all equivalence relations on subsets of D is denoted by $\mathcal{R}_D$; the corresponding context set is $\mathcal{C}_D$.

Definition: Let ρ and ρ' be equivalence relations in $\mathcal{R}_D$. Then, ρ' is *coarser than* ρ, i.e., $\rho' \sqsubseteq_\rho \rho$, whenever $\rho \subseteq \rho'$. Further, if C and C' are the contexts induced by ρ and ρ', respectively, then C' is *coarser than* C, i.e., $C' \sqsubseteq_C C$.

A "coarser" equivalence relation contains more equivalences than a "finer" equivalence relation. It employs a weaker notion of indistinguishability and yields a correspondingly "coarser" context with larger equivalence classes. Each equivalence class in a finer context is a subset of an equivalence class in a coarser context. The coarsest context in $\mathcal{C}_D$ is $\{D\}$ containing a single (largest) equivalence class. The finest is the "vacuous context" induced by the empty equivalence relation.

In enforcing a domain constraint a context acts as a "sieve" controlling the quality of the information consistently stored as tuple components. Note that precise information is expressed as a singleton classical set, or more generally, a fuzzy set with a spiked characteristic function. More imprecise information is expressed using broader fuzzy sets with wider possibility distributions [13]. An information chunk which passes through a sieve opening, i.e., an equivalence class, at a certain α-level is "consistent" with respect to the context at level α. Since equivalence classes comprise closely related elements, consistent chunks are guaranteed to be meaningful. The null set is not a consistent value and cannot be used to specify unknown (albeit defined) values. This is because the maximally imprecise chunk from an underlying domain D is defined as D itself.

Definition: A fuzzy set t_i is *consistent* at level α with respect to a context C_i whenever its α-cut set, $(t_i)_\alpha$, is a non-empty subset of an equivalence class in C_i.

Definition: A fuzzy tuple $t = (t_1, t_2, ..., t_n)$ is *α-consistent* with respect to $(C_1, C_2, ..., C_n)$ when each t_i is consistent with context C_i at level α.

Context-based consistency clearly generalizes the domain integrity constraint governing classical tuple components. Precise information is consistent with respect to "precise" contexts containing singleton equivalence classes for all $\alpha > 0$. In fact, precise contexts only permit the storage of precise information as in the classical relational model. Tuples embodying fuzzier information can only be rendered consistent by coarser contexts.

A fuzzy tuple (and database) can be effectively classicalized by employing normal fuzzy sets and by considering consistency at the $\alpha_{1.0}$-level. Constraining only the "high water mark" ($\alpha = 1.0$) for information storage in a fuzzy database considerably simplifies the resulting model and its implementation and better bridges the gap between precise databases and fuzzy databases. More imprecise fuzzy sets have wider possibility distributions at level $\alpha_{1.0}$. Employing a single high water mark means that fuzzy sets with singleton $\alpha_{1.0}$-cut sets – including triangular fuzzy sets – are consistent with respect to precise contexts. It may therefore be necessary to also incorporate a "low" or "mid water mark", e.g., $\alpha > 0$ or $\alpha = 0.5$, or possibly multiple water marks for constraining fuzzy tuples in practical database applications.

For reasons of simplicity in the following sections we employ a single high water mark, and define all properties at the associated $\alpha_{1.0}$-level. The extension to multiple α-levels is straightforward.

Example: The exact salary $\{30K\}$ and the fuzzy concept "moderate salary" denoted by the trapezoidal fuzzy set $(30K, 35K, 45K, 50K)$ are consistent at $\alpha_{1.0}$-level with respect to the salary context $\{ [0K, 24K],$

$(24K, 48K]$, $(48K, 84K]$, $(84K+)$ }. On the other hand, the trapezoidal fuzzy set "approx. 50K": $(40K, 45K, 55K, 60K)$ is not consistent at $\alpha_{1.0}$-level because it straddles two equivalence classes. However, "approx. 50K" can be made consistent by using a coarser salary context, e.g., { $[0K, 24K]$, $(24K, 84K]$, $(84K+)$ }.

Note that (L_i, H_i) and $[L_i, H_i]$ denote open and closed intervals, respectively. Also, for convenience we express normal trapezoidal fuzzy sets on linearly-ordered domains as four-tuples: (l_0, l_1, r_1, r_0). The support of such a trapezoidal fuzzy set is the open interval (l_0, r_0) and its $\alpha_{1.0}$-cut set is the closed interval $[l_1, r_1]$.

3. Fuzzy Equivalence and Entity Integrity

This section clarifies the notion of context-based fuzzy equivalence and defines entity integrity as it applies to fuzzy database relations. Fuzzy equivalence, entity integrity, and the notion of a fuzzy database relation reduce to the corresponding classical definitions in precise contexts. They can also be shown to generalize their counterparts in several fuzzy database models [1,5,11,14]. Some of these fuzzy database models (e.g., 5,11,14]) allow tuples to be tagged with membership values (e.g., to specify the association between tuple components or the degree to which a tuple belongs to a relation). In this work we assume that tuple membership is a "crisp" property. The extension of tuple membership to values in [0,1] or even fuzzy sets defined on [0,1] is accomplished by treating tuple membership as a separate attribute.

A context acts as a sieve for determining fuzzy equivalence. Since an equivalence class in a context comprises indistinguishable elements, all consistent fuzzy information chunks passing through the same sieve opening are considered equivalent in the context. As with consistency, the notion of equivalence is qualified by an α-level. For simplicity we adopt a single high water mark ($\alpha_{1.0}$-level) in the definitions below.

Definition: Fuzzy sets t and t' are *equivalent* at level α with respect to a context C, denoted by $t \sim_{(C,\alpha)} t'$, whenever $(t)_\alpha$ and $(t')_\alpha$ are non-empty subsets of the same equivalence class in C.

In precise contexts containing singleton equivalence classes fuzzy equivalence ($\sim_{(C,\alpha)}$) reduces to classical equality ($=_{(C,\alpha)}$) for $\alpha > 0$. Thus, precise information chunks are equivalent only when they are identical. Note also that the notion of fuzzy equivalence captures the fuzzy resemblance measure EQUAL described in [5].

Definition: Two tuples t and t' are *redundant* with respect to contexts $C = (C_1, C_2, ..., C_n)$, i.e., $t \sim_C t'$, whenever $t_i \sim_{(C_i, \alpha=1.0)} t'_i$ for each component i.

Tuple redundancy reduces to the classical notion for precise contexts C_i. In precise contexts (classical) tuples are redundant only when they are identical.

A relation scheme is a collection of attributes and associated contexts. A classical relation scheme has precise contexts on all its attributes. Its extension, i.e., classical database relation, can only hold precise information. A fuzzy relation scheme would have coarser contexts on some or all of its attributes. This enables the corresponding fuzzy relation to consistently hold fuzzy information.

Definition: A *relation scheme* $R_{(A,C)}$ is a collection of attributes $A = (A_1, A_2, ..., A_n)$ with contexts $C = (C_1, C_2, ..., C_n)$.

Definition: A *database relation* r on scheme $R_{(A,C)}$ is a set of non-redundant tuples with respect to the contexts in C.

The entity integrity property in the classical relational model mandates that a relation be a set of tuples. Thus, no two tuples in a relation can be identical. We retain this notion of entity integrity for fuzzy databases. As in the classical model a fuzzy database relation is a "set" of tuples and no two tuples in a relation can be "identical". In this case, however, "identical" is defined as equivalence with respect to contexts, and a "set" cannot contain multiple equivalent objects.

To maintain entity integrity it is necessary to define an operation for subsuming redundant tuples. In a classical relation each block of identical tuples is simply replaced by one of the tuples. However, as redundancy in a fuzzy relation is defined in terms of equivalence, it is possible to have redundant fuzzy tuples which are different from each other. Entity integrity is maintained by "merging" a block of redundant fuzzy tuples into a single non-redundant tuple [1,8,9].

Definition: The *merge* of two tuples t and t' is given by $u = (u_1, u_2, ..., u_n)$ where $u_i = t_i \cup t'_i$.

Merging fuzzy tuples which are redundant with respect to contexts C always gives rise to a tuple which is consistent with respect to C. Therefore, merging redundant classical tuples gives rise to a classical tuple. This reduces to the classical technique for maintaining entity integrity.

Example: The relation below is consistent with respect to a precise context for *Profession* and the coarser context $\{ [0K, 24K], (24K, 48K], (48K, 84K], (84K+) \}$ for *Salary*. The fuzzy concepts *approx. 100K* and *approx. 150K* are expressed using the trapezoidal fuzzy sets, $(80K, 90K, 120K, 130K)$ and $(120K, 130K, 170K, 180K)$, respectively.

Profession	Salary
{Accountant}	{36K}
{Engineer}	{40K}
{Manager}	{50K}
{Physician}	approx. 100K
{Scientist}	{36K}
{Surgeon}	approx. 150K

Associating a coarser context { {Accountant, Manager}, {Engineer, Scientist}, {Physician, Surgeon} } with the *Profession* attribute causes tuples 2 and 5, and tuples 4 and 6 to become redundant. The new relation with merged redundant tuples is given below. Note that *approx.* 100K *to* 150K: (80K, 90K, 170K, 180K) is the union of the original trapezoidal fuzzy sets.

Profession	Salary
{Accountant}	{36K}
{Engineer, Scientist}	{36K, 40K}
{Manager}	{50K}
{Physician, Surgeon}	approx. 100K to 150K

4. Fuzzy Queries and Fuzzy Dependencies

This section shows how fuzzy equivalence can be employed in uniformly querying precise and fuzzy relations. It also clarifies the context-based notion of a fuzzy functional dependency. Since context-based fuzzy equivalence generalizes classical equality, the corresponding context-based query language and data dependency generalize their classical counterparts.

Precise or fuzzy relations can be "fuzzily" queried by associating contexts C_Q with values in a query specification and engaging the underlying equivalences in information retrieval. Each context in C_Q behaves as a "sieve" controlling information retrieval. For example, a *Select* operation specifies certain sieve openings (i.e., equivalence classes containing the specified values) and only information passing through these openings is retrieved. The coarser the "query contexts" C_Q, the weaker the equivalence and the fuzzier the corresponding query. Classical operations, e.g., the relational algebra, are associated with precise contexts containing singleton equivalence classes. They formulate precise queries based on equality of atomic values (equivalence in precise contexts). On the other hand, a fuzzy query employs coarser contexts. The weaker equivalence extracts information which "fuzzily" matches the specifications.

The classical relational algebra operations are readily extended to their context-based counterparts. Set operations such as *Union* and *Difference* are based on equivalence in the associated contexts, and union compatibility is defined in terms of context-based relation schemes (i.e., attributes and their contexts). The *Select* operation uses equivalence instead of equality and is the natural extension of its classical counterpart. The *Cartesian Product* and *Project* operations are obvious; however, projection must be followed by the merging of redundant tuples.

Note that database users are free to adjust their query contexts C_Q as desired; thus they can specify arbitrarily fuzzy queries. However, it is clear that an "effective" query cannot be finer than the actual information stored in relations and seen by users. The requirement that an effective query be no finer than stored/viewed information is enforced by computing "effective query contexts" as the *greatest lower bound* (*glb*) of corresponding C_S and C_Q contexts. The *glb* context is the finest context which is coarser than both its arguments.

Evaluating all queries with respect to effective contexts C_E yields a single flexible mechanism for fuzzily querying precise databases or for fuzzily querying fuzzy databases. It also fields precise queries submitted to fuzzy databases with the expected fuzzy results. Context-based query processing is readily incorporated within existing precise query facilities. The upgraded facilities better support abstract information retrieval in AI/expert systems and security-control applications [6].

Example: The relations in Section 3 use the coarse context $C_S = \{ [0K, 24K], (24K, 48K], (48K, 84K], (84K+) \}$ for the *Salary* attribute. C_S specifies the finest precision for queries — users can only distinguish between "low", "moderate", "high" and "very high" salaries. For example, the precise query,

     Select where $Salary = 40K$

with precise $C_Q = \{ \{40K\} \}$ is evaluated as:

     Select where $Salary \sim_{C_E} 40K$,

where $C_E = glb(C_S, C_Q) = C_S$. This corresponds to the query:

     Select where $Salary =$ "moderate."

Likewise, the fuzzy query:

     Select where $Salary = approx. 50K$

with $C_Q = \{ [45K, 55K] \}$, corresponding to *approx. 50K*: $(40K, 45K, 55K, 60K)$, is evaluated as:

     Select where $Salary \sim_{C_E} approx. 50K$,

where $C_E = \{ [0K, 24K], (24K, 84K], (84K+) \}$. Precision is lost because the query specification straddles two equivalence classes in C_S. The resulting query is:

Select where $Salary = $ "moderate" or "high."

The context-based notion of equivalence defines data dependencies which convey "value oblivious" constraints. The idea is that if some tuple components satisfy certain equivalences, other tuple components must exist and their values must be equivalent. Below we define the notion of a functional dependency (FD). It extends the classical FD from a notion based on equality to one based on equivalence with respect to contexts. The resulting generalized fuzzy FD can specify constraints on precise and/or fuzzy information. It also generalizes some other versions of fuzzy functional dependencies defined in the literature [5,9].

Definition: An *FD:* $X(C_X) \to Y(C_Y)$ holds in scheme $R_{(A,C)}$ if for all tuples t, t' in every extension r of $R_{(A,C)}$, $t \sim_{C_X} t'$ implies $t \sim_{C_Y} t'$.

Definition: $X(C_X) \to Y(C_Y)$ is a classical FD when C_X and C_Y are precise contexts.

Proposition: FDs satisfy Armstrong's Axioms:

$Y(C_Y) \subseteq X(C_X) \subseteq U(C_U)$ implies $X(C_X) \to Y(C_Y)$ $\qquad$ (reflexivity)

$X(C_X) \to Y(C_Y)$, $Z(C_Z) \subseteq U(C_U)$ implies $XZ(C_XC_Z) \to YZ(C_YC_Z)$
$\qquad\qquad\qquad\qquad\qquad\qquad\qquad\qquad\qquad\qquad$ (augmentation)

$X(C_X) \to Y(C_Y)$, $Y(C_Y) \to Z(C_Z)$ implies $X(C_X) \to Z(C_Z)$
$\qquad\qquad\qquad\qquad\qquad\qquad\qquad\qquad\qquad\qquad$ (transitivity).

The property that context-based FDs satisfy Armstrong's Axioms follows directly from the FD definition. Note that U is the set of attributes in the universal relation scheme and C_U is the set of associated contexts.

Since fuzzy FDs satisfy Armstrong's Axioms it is possible to define the notion of a relation key, and also to redefine entity integrity in terms of keys. These definitions reduce to their classical counterparts for precise contexts.

5. Conclusions

Contexts provide an elegant semantics-based mechanism for classicalizing fuzzy databases and rendering them feasible for implementation. They help define key relational database integrity constraints which support the uniform representation and processing of precise/fuzzy information. The resulting context-based model generalizes the classical relational model and a family of well-known fuzzy relational database models. It simplifies the implementation of other prominent fuzzy database models and facilitates their integration with commercially-available classical relational database systems.

A version of the context model is currently implemented using the popular ORACLE RDMS. The prototype employs underlying classical relations as the main thrust of the project was to effectively field fuzzy queries submitted to precise databases. It includes the *Miquel* language and its translator to ORACLE SQL. The *Miquel* system demonstrates that the context formalism can be employed in significantly enhancing the power and flexibility of existing classical databases. Moreover, it is possible to develop a single readily-integratable facility for fuzzily querying precise databases or for fuzzily querying fuzzy databases. Such as facility would be crucial to upgrading existing classical databases to better support artificial intelligence and expert systems and certain security-control applications.

Acknowledgement

This research was supported by NSF Grant IRI-9110709 and OCAST Grants AR-2-002/9-010.

References

1. Buckles, B.P. and Petry, F.E. (1982) A fuzzy representation of data for relational databases, *Fuzzy Sets and Systems*, **7**, 213-226.
2. Buckles, B.P. and Petry, F.E. (1984) Extending the fuzzy database with fuzzy numbers, *Information Sciences*, **34**, 145-155.
3. Prade, H. and Testemale, C. (1984) Generalizing database relational algebra for the treatment of imprecise or uncertain information and vague queries, *Information Sciences*, **34**, 115-143.
4. Raju, K.V. and Majumdar, A.K. (1987) The study of joins in fuzzy relational databases, *Fuzzy Sets and Systems*, **21**, 19-34.
5. Raju, K.V. and Majumdar, A.K. (1988) Fuzzy functional dependencies and lossless join decomposition of fuzzy relational database systems, *ACM Transactions on Database Systems*, **13**, 129-166.
6. Shenoi, S. (1993) Multilevel database security using information clouding, *Proceedings of the Second IEEE International Conference on Fuzzy Systems*, pp. 483-488.
7. Shenoi, S. and Melton, A. (1990) An extended version of the fuzzy relational database model, *Information Sciences*, **52**, 35-52.
8. Shenoi, S., Melton, A., and Fan, L.T. (1990) An equivalence classes model of fuzzy relational databases, *Fuzzy Sets and Systems*, **38**, 153-170.
9. Shenoi, S., Melton, A., and Fan, L.T. (1992) Functional dependencies and normal forms in the fuzzy relational database model, *Information Sciences*, **60**, 1-28.
10. Shenoi, S., Shenoi, K., and Melton, A. (1991) Contexts and abstract information processing, *Proceedings of the Fourth International Conference on Industrial and Engineering Applications of AI and Expert Systems*, pp. 44-50.
11. Umano, M. (1984) Retrieval from fuzzy database by fuzzy relational algebra, in E. Sanchez (ed.), *Fuzzy Information, Knowledge Representation and Decision Analysis*, Pergamon Press, Oxford, U.K., pp. 1-6.
12. Zadeh, L.A. (1970) Similarity relations and fuzzy orderings, *Information Sciences*, **3**, 177-200.
13. Zadeh, L.A. (1978) Fuzzy sets as a basis for a theory of possibility, *Fuzzy Sets and Systems*, **1**, 3-28.
14. Zemankova, M. and Kandel, A. (1984) *Fuzzy Relational Databases: A Key to Expert Systems*, Verlag TUV Rheinland, Cologne, Germany.

A COOPERARIVE FUZZY GAMES
FOR INTERNATIONAL CONFLICT SOLVING

FUMIKO SEO,* MASATOSHI SAKAWA,** ICHIRO NISHIZAKI*
* Faculty of Business Administration and Informatics,
  The Setsunan University, Ikeda-Nakamachi, Neyagawa, Osaka, 572
  Japan
** Faculty of Engineering, Hiroshima University, Kagamiyama,
  Higashi-Hiroshima, 724, Japan

Abstract. This paper concerns an effective formation of an international concord for international conflict solving under the fuzzy decision environments. For treating this problem, an *n*-person cooperative fuzzy game in characteristic function form is constructed, where the characteristic function is assessed as the fuzzy number embodying diversified evaluation. The nucleolus as the solution concept of the game is derived also in fuzzy terms by solving a fuzzy linear programming problem which comes to formulate a parametric programming problem.

1. Introduction

In this paper, an effective formation of an international concord for international conflicting solving under the fuzzy environments is discussed, using an *n*-person cooperative fuzzy game in characteristic function form.

Game theoretic approaches, such as two-person zero sum games (Rogers 1969) and differential games (Gillespie, Zinnes & Tahim 1978), have been used for international conflict analysis by many researchers. An n-person cooperative game has been also constructed for a cost allocation problem in a regional water resource development program (Suzuki and Nakamura 1976, Young, Okada & Hashimoto 1982, Young 1985). The utility function has been also used in constructing the characteristic function of cooperative games (Nagao, Kuroda & Wakai 1983). The authors have also discussed about the construction of an international cooperative game for effective formation of an international concord (Seo and Sakawa 1990). The international decision environments, however, include various ambiguous factors for decision making due to the existence of diversified value standards among countries. This situation will bring some fuzziness in the evaluation of the characteristic functions of cooperative games. The authors have proposed to treat such ambiguousness as the fuzziness and to construct an international cooperative game in the form of a fuzzy game (Seo and Nishizaki 1992). In that paper, the authors have used the concepts of the external and internal coalitions and the variation of both evaluations have been evaluated with diversified possibility distribution functions.

In the present paper, we treat the fuzziness in the evaluation of the characteristic

437

Z. Bien and K. C. Min (eds.),
Fuzzy Logic and its Applications, Information Sciences, and Intelligent Systems, 437–445.
© 1995 *Kluwer Academic Publishers.*

function with an interval and propose to construct a fuzzy game in a resemblant form to the parametric program, which will lead to find appropriate solutions for an intended policy program from among a wider range of optional solutions of the game.

We define the n-person cooperative fuzzy game in which the characteristic function is assessed with the membership function in fuzzy set theory. The fuzzy game is constructed on the evaluation of the fuzzy excess of coalitions based on the fuzzy characteristic function, which is assessed with an admissible interval. The fuzzy game is formulated as a fuzzy linear programming problem with the fuzzy constraints and the solution of the game is evaluated as the fuzzy nucleolus. The fuzzy game can be solved in the form of parametric programming which is known as equivalent to a fuzzy linear programming formulation.

2. An N-Person Cooperative Fuzzy Game

Let each country l, $l = 1, ..., n$, be a player of a game. Denote a set of all players as $N \underline{\Delta} \{1, 2, ..., n\}$. An international coalition $S \subseteq N$ is a coalition in a cooperative game and $v(S)$ denotes the characteristic function of the game. The characteristic function is assessed in terms of the pay-off values. An n-person cooperative game in characteristic function form is represented by $\Gamma \underline{\Delta} (S, v(S))$.

An n-person cooperative fuzzy game in characteristic function form is defined as $F\Gamma \underline{\underline{\Delta}} (S, \tilde{v}(S))$, $S \subseteq N$, where $\tilde{v}(S)$ is the fuzzy characteristic function of a nonfuzzy coalition S. The fuzzy excess of a coalition S is defined as:

$$\tilde{e}(S, z) \underline{\Delta} \tilde{v}(S) - \sum_{l \in S} z_l, \quad S \subset N,$$

$$(1)$$

where z denotes a pay-off vector to be allocated among countries as a result of participating in the game.

The core of the fuzzy game is called as the fuzzy core and is defined with (i) fuzzy coalitional rationality:

$$\tilde{e}(S, z) \leq 0 \quad \text{for all } S \neq \phi, N.$$

$$(2)$$

and (ii) fuzzy collective rationality:

$$\sum_{l \in N} z_l = \tilde{v}(N).$$

$$(3)$$

The fuzzy quasi-core, or the fuzzy ε-core, of the fuzzy game is defined with (ii) and (iii):

$$\tilde{e}(S, z) \leq \varepsilon \quad \text{for all } S \neq \phi, N.$$

$$(4)$$

The fuzzy nucleolus is defined with the fuzzy least core concept (Eq. (5)) on the (fuzzy) lexicographic ordering of the fuzzy excess in Eq. (4).

$$\tilde{\xi}_0(z) = \min_{z=\chi} \max_{S \neq \phi, N} \tilde{e}(S, z),$$

$$(5)$$

where χ is a set of the preimputation that satisfies the fuzzy collective rationality (3). The fuzzy linear programming problem for solving the fuzzy games is formulated as

$(\tilde{P})$

$$\underset{z_l, \varepsilon}{\text{minimize}} \quad \varepsilon$$

$$\text{subject to} \quad \tilde{v}(S) - (z(S) + \varepsilon) \leq 0, \qquad S \subset N,$$

$$\tilde{v}(N) - z(N) = 0, \qquad z_l \geq 0, \ l \in N. \tag{6}$$

3. Construction of the Fuzzy Game in Parametric Programming Form

The fuzzy characteristic functions $v(S)$ and $v(N)$ can be represented with the interval and the Eq.(6) can be reformulated as

$(\tilde{P}^*)$

$$\underset{z_l, \varepsilon}{\text{minimize}} \quad \varepsilon$$

$$\text{subject to} \quad z(S) + \varepsilon \geq [\underline{v}(S); \overline{v}(S)] \quad S \subset N,$$

$$z(N) = [v(N); \underline{v}(N); \overline{v}(N)] \quad z_l \geq 0, \ l \in N, \tag{7}$$

Using the L-R type fuzzy number, the membership functions that define the fuzzy constraints sets, $\mu_S(g_S(z, \varepsilon))$, $\mu_N(g_N(z))$, can be assessed as

$$\mu_S(g_S(z, \varepsilon)) = \begin{cases} 1, & \text{if } g_S(z, \varepsilon) > \overline{v}(S), \\[2mm] 1 - \dfrac{\overline{v}(S) - g_S(z, \varepsilon)}{d}, & \\[2mm] & \text{if } \underline{v}(S) < g_S(z, \varepsilon) \leq \overline{v}(S), \\[2mm] 0, & \text{if } g_S(z, \varepsilon) \leq \underline{v}(S), \end{cases}$$

$$(8)$$

$$
\mu_N(g_N(z,\ \varepsilon)) = \begin{cases} 0, & \text{if} \quad g_N(z) \leq \underline{v}(S), \\[2mm] 1 - \dfrac{v(N) - g_N(z)}{\underline{d}}, & \text{if} \quad \underline{v}(N) < g_N(z) < v(N), \\[2mm] 1, & \text{if} \quad g_N(z) = v(N), \\[2mm] 1 - \dfrac{g_N(z) - v(N)}{\overline{d}}, & \text{if} \quad v(N) < g_N(z) \leq \overline{v}(N), \\[2mm] 0, & \text{if} \quad g_N(z) > \overline{v}(N), \end{cases}
\tag{9}
$$

where $d = \overline{v}(S) - \underline{v}(S)$ shows an interval of acceptable levels for S. $\underline{v}(S)$ is an minimum admissible level for S. $\overline{v}(S)$ is an internal evaluation level, or a desirable level, of $v(S)$ for S. $d = \underline{d} + \overline{d} = (v(N) - \underline{v}(N)) + (\overline{v}(N) - v(N))$ is an interval of acceptable values for N, where $\underline{v}(N)$ is a minimum admissible level for N and $\overline{v}(N)$ is a maximum possible level for N. These membership functions are depicted in Figure 1.

Let $\alpha \in [0, 1]$ be a level of the membership functions. Using the concept of a nonnegative α-level set of the membership functions, the fuzzy linear programming problem $(\widetilde{P}^{\,*})$ can be converted to a nonfuzzy parametric programming problem (Chanas 1983, Verdegay 1982, Leung 1988).

Define a nonnegative solution set based on the α-level set of μ_S and μ_N in Eqs. (8)(9).

$$
C_\alpha^S \triangleq \bigcap_S \{z \mid z \in R^S,\ z \geq 0,\ \mu_S(g_S(z,\ \varepsilon)) \geq \alpha\}
$$

$$
= \bigcap_S \{z \mid z \in R^S,\ z \geq 0,\ g_S(z,\ \varepsilon) \geq r_S(\alpha)\}
\tag{10}
$$

$$
C_\alpha^N \triangleq \{z \mid z \in R^N,\ \mu_N(z) \geq \alpha\}
$$

$$
= \{z \mid z \in R^N,\ r_N'(\alpha) \geq g_N(z) \geq r_N(\alpha)\},
\tag{11}
$$

where $\alpha \in [0, 1]$, $r_S(\alpha) \triangleq \mu_S^{-1}(\alpha)$ and $\{r_N(\alpha),\ r_N'(\alpha)\} \triangleq \mu_N^{-1}(\alpha)$. The function

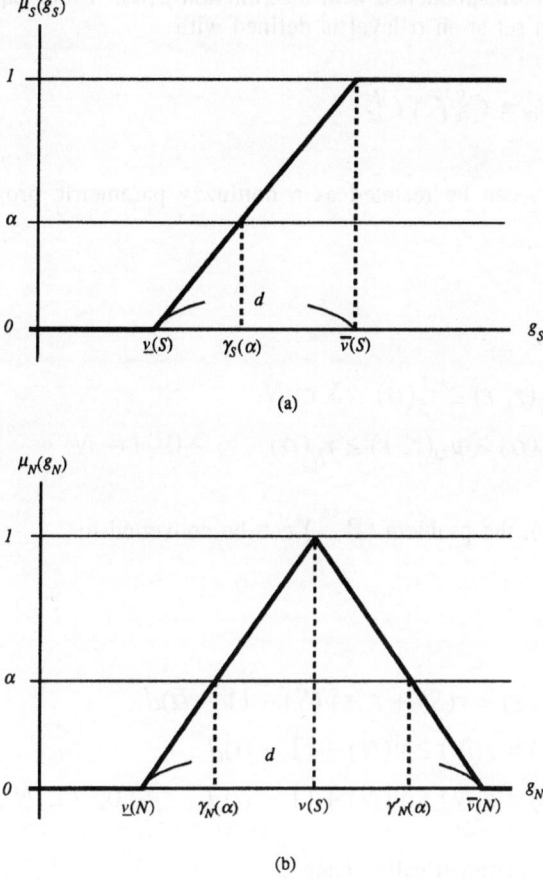

Figure 1. The membership functions defining the fuzzy constraints

$r(\alpha)$ has the one to one correspondence with the function $\mu(\alpha)$. From Eqs.(10) and (11), the feasible solution set at an α-level is defined with

$$C_\alpha = C_\alpha^S \cap C_\alpha^N \tag{12}$$

The fuzzy program $(\widetilde{P}\,^*)$ can be restated as a nonfuzzy parametric program with $\alpha \in [0, 1]$.

$(\widetilde{P}\,^{**})$

$$\begin{array}{ll}
\underset{z_l,\, \varepsilon}{\text{minimize}} & \varepsilon \\[1ex]
\text{subject to} & g_S(z,\, \varepsilon) \geq r_S(\alpha), \quad S \subset N \\[1ex]
& r_N'(\alpha) \geq g_N(z,\, \varepsilon) \geq r_N(\alpha), \quad z_l \geq 0, \ l \in N.
\end{array} \tag{13}$$

Using Eqs. (8)(9)(10)(11), the problem $(\widetilde{P}\,^{**})$ can be converted to

$(\widetilde{P}\,^{***})$

$$\begin{array}{ll}
\underset{z_l,\, \varepsilon}{\text{minimize}} & \varepsilon \\[1ex]
\text{subject to} & g_S(z,\, \varepsilon) = z(S) + \varepsilon \geq \overline{v}(S) - (1 - \alpha)d \\[1ex]
& g_N(z) = z(N) \geq v(N) - (1 - \alpha)\underline{d} \\[1ex]
& g_N(z) = z(N) \leq v(N) + (1 - \alpha)\overline{d}, \quad z_l \geq 0, \ l \in N,
\end{array} \tag{14}$$

where $(1 - \alpha) = \theta$ can be parametrically changed.

The cooperative fuzzy game is known to be equivalent to the parametric programming formulation of a nonfuzzy game. In both formulations, the variant assessment for the characteristic function $v(S)$-values is included, which may show a diversification of the evaluation as a result of the existence of group decision making in a coalition S. This interpretation will be particularly meaningful in examining the robustness of solutions in diversified value environments.

4. A Numerical Example

Let $ER_l(\theta)$, $ER_S(\theta)$ be expected risk values from a risky event θ having an international spill-over effect for a country l and a coalition S. Then the characteristic function $v(S)$ as the coalition value for S is defined with a decreased risk value due to the formation of a coalition S:

$$v(S) = \sum_{l \in S} \frac{1}{S} ER_l(\theta) - ER_S(\theta), \quad S \subseteq N, \tag{15}$$

where s is the number of countries participating in the coalition S. A method for assessing a risk function is discussed in Seo (1992). As an alternative to Eq.(15), a method to assess the utility functions of the coalitions S in constructing the characteristic functions v(S) is discussed in Seo (1991). Table 1 shows a numerical example of the coalition values with the interval using the risk function as in Eq.(15). Table 2 shows a result of the parametric changes in the coalition values. Table 3 shows the pay-off values as the nucleolus obtained from the parametric changes. As seen, wider ranges for selecting effective solutions of the game have been presented.

5. Concluding Remarks

Similarity of the parametric programming formulation to the fuzzy linear programming leads to construct a cooperative fuzzy game in the solvable terms. The characteristic of the parametric programming method to solve a fuzzy linear programming is its capability to find all feasible ranges of solutions for a coalition S and to check their behavioral properties, which is different from the optimal fuzzy decision rule based on the maximin criterion. As a result, large variants of the solution behaviors can be discerned. This device can be particularly useful in group decision environments where diversified value evaluation is included.

References

Chanas, S. (1983), The Use of parametric programming in fuzzy linear programming, *Fuzzy Sets and Systems,* **11,** 243-251.

Gillespie,J.V., Zinnes, D.A. and Tahim, G.S. (1978) Embedded games analysis of a problem in international relations, *IEEE Transactions on Systems, Man and Cybernetics,* SMC-8(8), 612-621..

Leung, Y. (1988), *Spatial Analysis and Planning Under Imprecision,* North-Holland.

Nagao, Y., Kuroda, K. and Wakai, I. (1983) Decision making under conflict in project evaluation, in W. Isard and Y. Nagao (ed.) *International and Regional Conflict,* Ballinger Publishing Co.. 73-90.

Rogers, P. (1969) A game theory approach to the problems of international river basin, *Water Resource research,* **5**(4), 740-760.

Schmeidler, D. (1969), The nucleolus of a characteristic function game, *SIAM Journal of Applied Mathematics,* **17**(6), 1163-1170.

Sakawa, M. (1992), *The Software Package* : a supplement to *Foundation of Mathematical Systems* (1991). (in Japanese) The Morikita publishing Co. Tokyo.

Seo, F. and Sakawa, M. (1990), A game theoretic approach with risk assessment for international conflict solving, *IEEE Transactions on Systems. Man. and Cybernetics* **20**(1), 141 -148.

Seo, F. (1991) Utilization of mathematical programming for public systems: an application to effective formation of integrated regional information networks, *Mathematical Programming,* **52**, 71-98.

Seo, F. (1992), On Construction of the Fuzzy Multiattribute Risk Function for Group Decision Making, *Decision Making Using Fuzzy Set and Possibility Theory,* edited by J. Kacprzyk and M. Fedrizzi, Kluwer Academic Publishers, Dordrecht pp. 198-218.

Seo, F. and Nishizaki, I.: (1992), On construction of a cooperative fuzzy game embodying fuzzy decision environments: a possibilistic approach, *Control and Cybernetics,* **20**(1) 277-294.

Shapley, L. S. and Shubik, M. (1966), Quasi-Cores in a monetary economy with nonconvex preferences, *Econometrica,* **34**(4), 805-827.

Simonnard, M. (1966), *Linear Programming* (translated by W. S. Jewell), Prentice-Hall.

Suzuki, M. and Nakamura, K. (1976) *Social Systems: A Game Theoretic Approach,* (in Japanese) Kyoritsu Publishing Co. Tokyo.

Verdegay, J. L.: (1982), "Fuzzy mathematical programming", , in M. M. Gupta and E. Sanchez (eds.), *Fuzzy Information and Decision Processes ,* North-Holland. pp. 231-237.

Young, H. P. (1985) *Cost Allocation: Methods, Principles, Applications,* North-Holland.

Young, H. P., Okada, N. and Hashimoto, T. (1982), Cost allocation in water resources development, *Water Resources Research,* **18**(3), 463-475.

TABLE 1. Coalitional values with the interval

S	$\underline{v}(S)$	$\bar{v}(S)$	d	$\bar{v}(S) - (1 - \alpha)d$
1	0	0	0	0
2	0	0	0	0
3	0	0	0	0
4	0	0	0	0
5	0	0	0	0
1, 2	9	11	2	$9 + \alpha$
1, 3	10	14	4	$10 + 4\alpha$
1, 4	12	17	5	$12 + 5\alpha$
1, 5	11	14	3	$11 + 3\alpha$
2, 3	12	16	4	$12 + 4\alpha$
2, 4	15	20	5	$15 + 5\alpha$
2, 5	7	10	3	$7 + 3\alpha$
3, 4	13	17	4	$13 + 4\alpha$
3, 5	12	15	3	$12 + 3\alpha$
4, 5	17	22	5	$17 + 5\alpha$
1, 2, 3	16	22	6	$16 + 6\alpha$
1, 2, 4	23	30	7	$23 + 7\alpha$
1, 2, 5	15	20	5	$15 + 5\alpha$
1, 3, 4	31	40	9	$31 + 9\alpha$
1, 3, 5	16	22	6	$16 + 6\alpha$
1, 4, 5	21	28	7	$21 + 7\alpha$
2, 3, 4	24	32	8	$24 + 8\alpha$
2, 3, 5	15	20	5	$15 + 5\alpha$
2, 4, 5	18	25	7	$18 + 7\alpha$
3, 4, 5	25	35	10	$25 + 10\alpha$
1, 2, 3, 4	38	48	10	$38 + 10\alpha$
1, 2, 3, 5	26	35	9	$26 + 9\alpha$
1, 2, 4, 5	30	40	10	$30 + 10\alpha$
1, 3, 4, 5	39	50	11	$39 + 11\alpha$
2, 3, 4, 5	38	49	11	$38 + 11\alpha$
1, 2, 3, 4, 5	47	60	13	$47 + 7\alpha$
	$v(N) = 54$			$60 - 6\alpha$

TABLE 2. Coalitional values with paramertric changes

S	v(S)					
	α = 1.0	α = 0.8	α = 0.6	α = 0.4	α = 0.2	α = 0.0
1	0	0	0	0	0	0
2	0	0	0	0	0	0
3	0	0	0	0	0	0
4	0	0	0	0	0	0
5	0	0	0	0	0	0
1, 2	11.0	10.6	10.2	9.8	9.4	9.0
1, 3	14.0	13.2	12.4	11.6	10.8	10.0
1, 4	17.0	16.0	15.0	14.0	13.0	12.0
1, 5	14.0	13.4	12.8	12.2	11.6	11.0
2, 3	16.0	15.2	14.4	13.6	12.8	12.0
2, 4	20.0	19.0	18.0	17.0	16.0	15.0
2, 5	10.0	9.4	8.8	8.2	7.6	7.0
3, 4	17.0	16.2	15.4	14.6	13.8	13.0
3, 5	15.0	14.4	13.8	13.2	12.6	12.0
4, 5	22.0	21.0	20.0	19.0	18.0	17.0
1, 2, 3	22.0	20.8	19.6	18.4	17.2	16.0
1, 2, 4	30.0	28.6	27.2	25.8	24.4	23.0
1, 2, 5	20.0	19.0	18.0	17.0	16.0	15.0
1, 3, 4	40.0	38.2	36.4	34.6	32.8	31.0
1, 3, 5	22.0	20.8	19.6	18.4	17.2	16.0
1, 4, 5	28.0	26.6	25.2	23.8	22.4	21.0
2, 3, 4	32.0	30.4	28.8	27.2	25.6	24.0
2, 3, 5	20.0	19.0	18.0	17.0	16.0	15.0
2, 4, 5	25.0	23.6	22.2	20.8	19.4	18.0
3, 4, 5	35.0	33.0	31.0	29.0	27.0	25.0
1, 2, 3, 4	48.0	46.0	44.0	42.0	40.0	38.0
1, 2, 3, 5	35.0	33.2	31.4	29.6	27.8	26.0
1, 2, 4, 5	40.0	38.0	36.0	34.0	32.0	30.0
1, 3, 4, 5	50.0	47.8	45.6	43.4	41.2	39.0
2, 3, 4, 5	49.0	46.8	44.6	42.4	40.2	38.0
1, 2, 3, 4, 5	54.0	52.6	51.2	49.8	48.4	47.0
1, 2, 3, 4, 5	54.0	55.2	56.4	57.6	58.8	60.0

TABLE 3. Pay-off vectors (nucleolus)

α	0.0	0.2	0.4	0.6	0.8	1.0
Player 1	11.67	10.35	9.00	8.00	7.00	6.25
Player 2	9.67	8.25	7.40	7.00	6.00	5.25
Player 3	12.00	13.95	16.00	16.60	15.80	15.25
Player 4	17.00	16.75	15.80	16.20	18.60	20.00
Player 5	9.67	9.50	9.40	8.60	7.80	7.25
ε	-9.67	-8.25	-6.20	-3.80	-1.40	1.25

CONSENSUS DEGREES UNDER FUZZINESS VIA ORDERED WEIGHTED AVERAGE (OWA) OPERATORS

JANUSZ KACPRZYK

Systems Research Institute
Polish Academy of Sciences
ul. Newelska 6, 01 - 447 Warsaw, Poland
Email: `kacprzyk@ibspan.waw.pl`

AND

MARIO FEDRIZZI

Institute of Computer and Management Sciences
University of Trento
Via Inama 5, 38100 Trento, Italy
Email: `fedrizzi@cs.unitn.it`

Abstract. The use of Yager's (1988) OWA (ordered weighted average) operators is proposed for handling fuzzy linguistic quantifiers (many, most, almost all, ...) which are used for the formalization of a fuzzy majority in the derivation of a "soft" degree of consensus under fuzzy preferences proposed in the authors' works (Fedrizzi, 1988; Kacprzyk, 1987; Kacprzyk and Fedrizzi, 1986, 1988, 1989) in which the classical Zadeh's (1983) and Yager's (1983) fuzzy-logic-based calculi of linguistically quantified proposition were employed. It seems that the use of the OWA operators can contribute to a high computational efficiency and intuitive appeal.

Key words: fuzzy logic, linguistic quantifier, fuzzy preference relation, fuzzy majority, group decision making, social choice, consensus, OWA (ordered weighted average) operator

Z. Bien and K. C. Min (eds.),
Fuzzy Logic and its Applications, Information Sciences, and Intelligent Systems, 447–453.
© *1995 Kluwer Academic Publishers.*

1. Introduction

This paper is a continuation of the authors' previous related works (Fedrizzi, 1988; Kacprzyk, 1987; Kacprzyk and Fedrizzi, 1986, 1988, 1989 - see also Kacprzyk, Fedrizzi and Nurmi, 1992) in which new definitions of "soft" degrees of consensus in a group of experts are proposed. The basic setting is as follows. There is a set of options and a set of individuals who provide their testimonies as (individual) fuzzy preference relations over the set of options. Moreover, as opposed to most works on the fuzzification of (group) decision making models, we go further by assuming a fuzzy majority examplified by *most, almost all, much more than a half, ...*).

Though in virtually all formal models of group decision making and consensus formation a strict majority is usually assumed (e.g., *at least 50%, more than 2/3, ...*), a fuzzy majority is commonly used by the humans, and not only in everyday discourse. A good example in a biological context may be found in Loewer and Laddaga (1985): "...It can correctly be said that there is a *consensus* among biologists that Darwinian natural selection is an important cause of evolution though there is currently *no consensus* concerning Gould's hypothesis of speciation. This means that there is a *widespread agreement* among biologists concerning the first matter but *disagreement* concerning the second ...". A rigid majority as, e.g., at least 75% would evidently not reflect the essence of the above statement. It should be noted that there are naturally situations when a strict majority is necessary, for obvious reasons, as in political elections. Anyway, the ability to accomodate a fuzzy majority in consensus formation models should help make them more human-consistent hence easier implementable.

A natural manifestations of a fuzzy majority are the so-called *linguistic quantifiers* exemplified by *most, almost all, much more than a half, ...* Though they cannot be handled by conventional logical calculi, fuzzy logic provides here simple and efficient tools (Yager, 1983; Zadeh, 1983). These calculi have been applied by the authors to introduce a fuzzy majority for measuring (a degree of) consensus and deriving new solution concepts in group decision making (Fedrizzi and Kacprzyk, 1988; Kacprzyk, 1985, 1986, 1987; Kacprzyk and Fedrizzi, 1986, 1988, 1989; Kacprzyk, Fedrizzi and Nurmi, 1992a, b; Nurmi and Kacprzyk, 1991) which have been implemented in a decision support system for consensus reaching (Fedrizzi, Kacprzyk and Zadrożny, 1988).

This degree is meant to overcome some "rigidness" of the conventional concept of consensus in which (full) consensus occurs only when "*all* the individuals agree as to *all* the issues". This may often be counterintuitive, and not consistent with a real human perception of the very essence of consensus (see, e.g., the citation from a biological context given above).

The new degree of consensus can be therefore equal to 1, which stands for full consensus, when, say, "*most* of the (important) individuals agree as to *almost all* (of the relevant) options".

In the derivation of these degrees of consensus using the two calculi of linguistically quantified proposition: the one due to Zadeh (1983) and the one due to Yager (1983), there is some difficulty. Namely, Zadeh's calculus is much simpler but may lead to unacceptable results mainly in case of "not fuzzy enough" fuzzy majorities (e.g, a little bit more than a half). On the other hand, Yager's calculus seems to be more general and to give "better" results but, in its original version, is not really operational for larger real-life problems.

A solution to overcome this problem may be the use of Yager's (1988) ordered weighted average (OWA) operators for the representation of fuzzy linguistic quantifiers. This seems to work well though some deeper works on the semantics of these OWA operators in the context of group decision making and consensus formation (cf. Kacprzyk and Yager, 1990 for a similar analysis within multicriteria decision making).

For clarity we will first review basic elements of calculi of linguistically quantified propositions, and their relation with the OWA operators, and then proceed to the reformulation of degrees of consensus proposed by the authors in terms of the OWA operators.

2. A fuzzy-logic-based calculus of linguistically quantified propositions, and the ordered weighted average (OWA) operators

A *linguistically quantified proposition* may be exemplified by "most experts are convinced", and may be generally written as

$$Qy\text{'s are } F \tag{1}$$

where Q is a *linguistic quantifier* (e.g., most), $V = \{y\}$ is a *set of objects* (e.g., experts), and F is a *property* (e.g., convinced). Importance can be added leading to "QBY's are F".

For our purposes, the main problem is now to find the truth of such a linguistically quantified statement, i.e. truth(Qy's are F) knowing truth(y is F), $\forall y \in \mathcal{Y}$, which can be done by using two basic calculi: due to Zadeh (1983) and Yager (1983). They have their strong and weak points as already mentioned.

The OWA operators (Yager, 1988) seems to provide good means for handling fuzzy linguistic quantifiers, and will be sketched below. An OWA operator of dimension p is a mapping $F : [0,1]^p \to [0,1]$ if associated with F is a weighting vector $W = [w_n]^T$ such that: $w_1 \in [0,1]$, $w_1 + \cdots + w_n = 1$, and

$$F(x_1, \ldots, x_n) = w_1 b_1 + \cdots + w_n b_n \tag{2}$$

where b_i is the i-th largest element in the set $\{x_1, \ldots, x_n\}$. B is called an ordered argument vector if each $b_i \in [0,1]$, and $j > i$ implies $b_i \geq b_j$, $i = 1, \ldots, p$.

Then

$$F(x_1, \ldots, x_n) = W^T B \tag{3}$$

Example 1. Let $W^T = [0.2\ 0.3\ 0.1\ 0.4]$, and calculate $F(0.6, 1.0, 0.3, 0.5)$. Thus, $B^T = [1.0\ 0.6\ 0.5\ 0.3]$, and $F(0.6, 1.0, 0.3, 0.5) = W^T B = 0.55$; and $F(0.0, 0.7, 0.1, 0.2) = 0.43$.

Some hints as to how to determine the w_i's are given in Yager (1988).

For our purposes relations between the OWA operators and fuzzy linguistic quantifiers are relevant. Due to lack of space we can only sketch the idea. Namely, under some mild assumptions (cf. Yager, 1988, Kacprzyk and Yager, 1990), a linguistic quantifier Q has the same properties as the F aggregation function, so that it is our conjecture that the weighting vector W is a manifestation of a quantifier underlying the process of aggregation of pieces of evidence. Then

$$w_k = \mu_Q(k) - \mu_Q(k-1) \tag{4}$$

with $\mu_Q(0) = 0$.

Just to give some examples of the w_i's associated with the particular quantifiers, notice that:

1. if $w_p = 1$, and $w_i = 0$, $\forall i \neq p$, then this corresponds to $Q = $ all;
2. if $w_i = 1$ for $i = 1$, and $w_i = 0$, $\forall i \neq 1$, then this corresponds to $Q = $ at least one.

The intermediate cases, which correspond to, e.g., *a half, most, much more than 75%, a few, almost all,* ... may be therefore obtained by a suitable choice of the w_i's between the above two extremes.

The OWA operators are therefore an interesting and promising class of aggregation operators which will be used here for deriving degrees of consensus under fuzzy majorities.

3. Degrees of consensus under fuzzy preferences and a fuzzy majority

The basic setting is as follows. We have a set of n options, $\mathcal{S} = \{s_1, \ldots, s_n\}$, and a set of m individuals, $\mathcal{I} = \{1, \ldots, m\}$. Each individual k provides his or her (individual) *fuzzy preference relation*, R_k, given by its membership function $\mu_{R_k} : \mathcal{S} \times \mathcal{S} \longrightarrow [0,1]$ which, if card S is small enough, may be represented by a matrix $[r_{ij}^k]$, $r_{ij}^k = \mu_{R_k}(s_i, s_j)$; $i, j = 1, \ldots, n$; $k = 1, \ldots, m$; $r_{ij}^k + r_{ij}^k = 1$.

The degree of consensus is now derived in three steps. First, for each pair of individuals we derive a degree of agreement as to their preferences between all the pair of options, next we aggregate these degrees to obtain a degree of agreement of each pair of individuals as to their preferences between $Q1$ (a linguistic quantifier as, e.g., *most, almost all, much more than 50%,* ...) pairs of options. Finally, we aggregate these degrees to obtain a degree of agreement of $Q2$ (a linguistic quantifier similar to $Q1$) pairs of individuals as to their preferences between $Q1$ pairs of options. This is meant to be the *degree of consensus* sought.

We start with the degree of a (strict) agreement between individuals $k1$ and $k2$ as to their preferences between options s_i and s_j

$$v_{ij}(k1, k2) = \begin{cases} 1 & \text{if } r_{ij}^{k1} = r_{ij}^{k2} \\ 0 & \text{otherwise} \end{cases} \tag{5}$$

where: $k1 = 1, \ldots, m-1$; $k2 = k1+1, \ldots, m$; $i = 1, \ldots, n-1$; and $j = i+1, \ldots, n$.

The degree of agreement between individuals $k1$ and $k2$ as to their preferences between all the pairs of options is

$$v(k1, k2) = \frac{2}{n(n-1)} \sum_{i=1}^{n-1} \sum_{j=i+1}^{n} v_{ij}(k1, k2) \tag{6}$$

The degree of agreement between individuals $k1$ and $k2$ as to their preferences between $Q1$ pairs of options is

$$v_{Q1}(k1, k2) = \text{OWA}_{Q1}(v(k1, k2)) \tag{7}$$

where $\text{OWA}_{Q1}(.)$ is the aggregation of $v(k1, k2)$'s with respect to $Q1$ via the OWA operator as shown in Section 2.

In turn, the degree of agreement of all the pairs of individuals as to their preferences between $Q1$ pairs of options is

$$v_{Q1} = \frac{2}{m(m-1)} \sum_{k1=1}^{m-1} \sum_{k2=k1+1}^{m} (v_{Q1}(k1, k2)) \tag{8}$$

and, finally, the degree of agreement of $Q2$ pairs of individuals as to their preferences between $Q1$ pairs of options, which is called the *degree of Q1/Q2-consensus* is

$$\text{con}(Q1, Q2) = \text{OWA}_{Q2}(v_{Q1}) \tag{9}$$

Since the strict agreement (??) may be viewed too rigid, we can use the degree of a *sufficient agreement* (at least to degree $\alpha \in [0, 1]$) of individuals $k1$ and $k2$ as to their preferences between options s_i and s_j, as well as the the degree of a *strong agreement* of individuals $k1$ and $k2$ as to their preferences between options s_i and s_j, obtaining the *degree of $\alpha/Q1/Q2$–consensus* and *$s/Q1/Q2$–consensus*, respectively.

This concludes our discussion of the reformulation of the authors' "soft" degrees of consensus (Fedrizzi and Kacprzyk, 1988; Kacprzyk, 1987a; Kacprzyk and Fedrizzi, 1986, 1988, 1989) by using the OWA operators to handle fuzzy linguistic quantifiers representing a fuzzy majority. An important issue, not yet dealt with here, is the addition of the importance of individuals and the relevance of options. This is a nontrivial problem which requires a deeper analysis, and will be discussed in a next paper.

4. Concluding remarks

We have shown how to use the ordered weighted average (OWA) operators to formally handle fuzzy linguistic quantifiers which are in turn a natural representation of a fuzzy majority. We think this may be helpful in an efficient handling of fuzzy majority in the context of degrees of consensus. Basically, the method described seems to help attain a better operationality maintaining many favorable properties of more complicated methods.

5. References

Fedrizzi M. (1986) Group decisions and consensus: a model using fuzzy sets theory (in Italian). *Rivista per le scienze econ. e soc.* A. 9 F. 1 12–20.

Fedrizzi M. and J. Kacprzyk (1988) On measuring consensus in the setting of fuzzy preference relations. In J. Kacprzyk and M. Roubens (Eds.): *Non-Conventional Preference Relations in Decision Making*, Springer-Verlag, Heidelberg, pp. 129–141.

Fedrizzi M., J. Kacprzyk and S. Zadrożny (1988) An interactive multi - user decision support system for consensus reaching processes using fuzzy logic with linguistic quantifiers. *Decision Support Systems* 4 313–327.

Kacprzyk J. (1985) Group decision-making with a fuzzy majority via linguistic quantifiers. Part I: A consensory-like pooling; Part II: A competitive-like pooling. *Cybernetics and Systems: An International Journal* 16 119–129 (Part I), 131–144 (Part II).

Kacprzyk J. (1986) Group decision making with a fuzzy linguistic majority. *Fuzzy Sets and Systems* 18 105–118.

Kacprzyk J. (1987a) On some fuzzy cores and 'soft' consensus measures in group decision making. In J.C. Bezdek (Ed.): *The Analysis of Fuzzy Information*, Vol. 2. CRC Press, Boca Raton, pp. 119–130.

Kacprzyk J. (1987b) Towards 'human consistent' decision support systems through commonsense-knowledge-based decision making and control models: a fuzzy logic approach. *Computers and Artificial Intelligence* **6** 97–122.

Kacprzyk J. and Fedrizzi M. (1986) 'Soft' consensus measures for monitoring real consensus reaching processes under fuzzy preferences. *Control and Cybernetics* **15** 309–323.

Kacprzyk J. and Fedrizzi M. (1988) A 'soft' measure of consensus in the setting of partial (fuzzy) preferences. *European Journal of Operational Research* **34** 315–325.

Kacprzyk J. and M. Fedrizzi (1989) A 'human-consistent' degree of consensus based on fuzzy logic with linguistic quantifiers. *Mathematical Social Sciences* **18** 275–290.

Kacprzyk J. and M. Fedrizzi, Eds. (1990) *Multiperson Decision Making Models Using Fuzzy Sets and Possibility Theory.* Kluwer, Dordrecht.

Kacprzyk J., M. Fedrizzi and H. Nurmi (1992a) Group decision making and consensus under fuzzy preferences and fuzzy majority. *Fuzzy Sets and Systems* **49** 21–32.

Kacprzyk J., M. Fedrizzi and H. Nurmi (1992b) Fuzzy logic with linguistic quantifiers in group decision making and consensus formation. In R.R. Yager and L.A. Zadeh (Eds.): *An Introduction to Fuzzy Logic Applications in Intelligent Systems*, Kluwer, Dordrecht pp. 263–280.

Kacprzyk J. and M. Roubens, Eds. (1988) *Non-Conventional Preference Relations in Decision Making.* Springer–Verlag, Heidelberg.

Kacprzyk J. and R.R. Yager (1990) Using fuzzy logic with linguistic quantifiers in multiobjective decision making and optimization: a step towards more human-consistent models. In R. Słowiński and J. Teghem (Eds.): *Stochastic versus Fuzzy Approaches in Multiobjective Mathematical Programming under Uncertainty.* Kluwer, Dordrecht, pp. 331–350.

Kacprzyk J., S. Zadrożny and M. Fedrizzi (1988) An interactive user-friendly decision support system for consensus reaching based on fuzzy logic with linguistic quantifiers. In M.M. Gupta and T. Yamakawa (Eds.): *Fuzzy Computing.* Elsevier, Amsterdam, pp. 307-322.

Loewer B. and R. Laddaga (1985) Destroying the consensus. In Loewer B. (Guest Ed.): Special Issue on Consensus. *Synthese* **62** (1) pp. 79–96.

Nurmi H. and J. Kacprzyk (1991) On fuzzy tournaments and their solution concepts in group decision making. *European Journal of Operational Research* **51** 223–232.

Yager R.R. (1983) Quantifiers in the formulation of multiple objective decision functions. *Information Sciences* **31** 107–139.

Zadeh L.A. (1983) A computational approach to fuzzy quantifiers in natural languages. *Computers and Mathematics with Applications* **9** 149–184.

A FUZZY PROGRAMMING APPROACH TO STRUCTURED LINEAR PROGRAMS

M. SAKAWA, M. INUIGUCHI AND K. KATO
Department of Industrial and Systems Engineering,
Faculty of Engineering, Hiroshima University,
Kagamiyama 1-4-1, Higashi-Hiroshima 724 Japan

AND

K. SAWADA
Information System Center, Matsushita Electric Works, Ltd,
Kadoma, Osaka 571 Japan

Abstract. In this paper, we focus on large-scale linear programming problems with block angular structure by assuming that the decision maker (DM) may have a fuzzy goal for the objective function and fuzzy constrains for the coupling constraints. Having elicited the corresponding linear membership functions through the interaction with the DM, we adopt the convex fuzzy decision. Then if some simple conditions are satisfied, it is shown that the formulated problem can be reduced to a number of independent linear subproblems and the satisficing solution for the DM is directly obtained just only solving the subproblems.

1. Introduction

In this paper, we focus on large-scale linear programming problems with block angular structure [1,3] which have been solved by the so-called Dantzig-Wolfe decomposition method [1]. By considering the vague nature of human judgements, it is quite natural to assume that the decision maker (DM) may have a fuzzy goal for the objective function and fuzzy constrains for the coupling constraints. These fuzzy goal and constraints can be quantified by eliciting the corresponding linear membership functions through the interaction with the DM [2,4,5]. After determining the membership functions,

455

Z. Bien and K. C. Min (eds.),
Fuzzy Logic and its Applications, Information Sciences, and Intelligent Systems, 455–464.
© 1995 *Kluwer Academic Publishers.*

if we adopt the convex fuzzy decision for combining them, under suitable conditions, it is shown that the formulated problem can be reduced to a number of independent linear subproblems and the satisficing solution for the DM is directly obtained just only solving the subproblems. Even if the conditions are violated, it is clarified that the satisficing solution for the DM is obtained by applying the Dantzig-Wolfe decomposition method.

2. Structured linear programming problems

Consider a large-scale linear programming problem of the following block angular form:

$$\left.\begin{array}{rl} \text{minimize} \quad cx = & c_1x_1 \ +\cdots+ \ c_px_p \\ \text{subject to} \quad Ax = & A_1x_1 \ +\cdots+ \ A_px_p \ \le b \\ & B_1x_1 \qquad\qquad\quad \le b_1 \\ & \qquad\qquad \ddots \\ & \qquad\qquad\quad B_px_p \ \le b_p \\ & x_i \ge 0, \ i = 1,\ldots,p \end{array}\right\} \tag{1}$$

where c_i, $i = 1,...,p$, are r_i dimensional cost factor row vectors, x_i, $i = 1,\ldots,p$, are r_i dimensional vector of decision variables, $Ax = A_1x_1 + \cdots + A_px_p \le b$ are s dimensional coupling constraints, A_i, $i = 1,...,p$, are $s \times r_i$ coefficient matrices. $B_ix_i \le b_i$, $i = 1,\ldots,p$, are t_i dimensional constraints with respect to x_i and B_i, $i = 1,\ldots,p$, are $t_i \times r_i$ coefficient matrices. Note that if the coupling constraints do not exist, the overall problem can be decomposed into the following p independent linear programming subproblems with coefficient matrix B_i, $i = 1,\ldots,p$:

$$\left.\begin{array}{l} \text{minimize} \quad c_ix_i \\ \text{subject to} \quad B_ix_i \le b_i, \ x_i \ge 0. \end{array}\right\} \tag{2}$$

Theoretically, it is possible to solve the problem by directly applying the revised simplex method for the overall problem. However the larger the overall problem becomes, the more difficult it becomes to solve due to the high dimensionality of the problem. To circumvent this difficulty, the Dantzig-Wolfe decomposition method [1] has been proposed by utilizing its special structure.

By considering the vague nature of human judgements for large-scale linear programming problems with block angular structure, it is quite natural to assume that the decision maker (DM) may have a fuzzy goal for the objective function and fuzzy constrains for the coupling constraints. These fuzzy goal and fuzzy constraints can be quantified by eliciting the corresponding membership functions through the interaction with the DM.

To elicit a linear membership function $\mu_0(cx)$ from the DM for a fuzzy goal, the DM is asked to assess a minimum value of unacceptable levels for cx, denoted by z_0^0 and a maximum value of totally desirable levels for cx, denoted by z_0^1. Then the linear membership function $\mu_0(cx)$ for the fuzzy goal of the DM is defined by:

$$\mu_0(cx) = \begin{cases} 1, & \text{if } cx \leq z_0^1 \\ \dfrac{cx - z_0^0}{z_0^1 - z_0^0}, & \text{if } z_0^1 \leq cx \leq z_0^0 \\ 0, & \text{if } cx \geq z_0^0. \end{cases} \tag{3}$$

Similarly, for each of the coupling constraints, by assessing a minimum value of unacceptable levels, denoted by z_j^0 and a maximum value of totally desirable levels, denoted by z_j^1, the following linear membership functions $\mu_j((Ax)_j)$, $j = 1, ..., s$, can be determined:

$$\mu_j((Ax)_j) = \begin{cases} 1, & \text{if } (Ax)_j \leq z_j^1 \\ \dfrac{(Ax)_j - z_j^0}{z_j^1 - z_j^0}, & \text{if } z_j^1 \leq (Ax)_j \leq z_j^0 \\ 0, & \text{if } (Ax)_j \geq z_j^0 \end{cases} \tag{4}$$

where $(Ax)_j$ denotes the jth component of the column vector Ax. These membership functions are depicted in Figures 1 and 2.

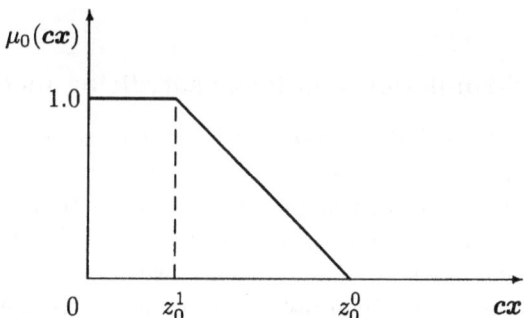

Figure 1. Linear membership function for fuzzy goal

In the followings, for notational convenience, if we set

$$\alpha_j = \frac{1}{z_j^1 - z_j^0}; \quad \beta_j = \frac{z_j^0}{z_j^1 - z_j^0}, \quad j = 0, 1, \ldots, s, \tag{5}$$

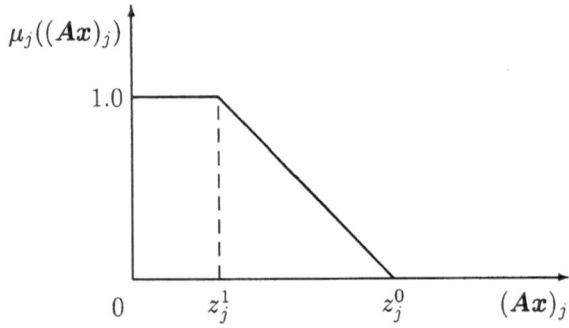

Figure 2.   Linear membership function for fuzzy constraints

these membership functions are simply expressed as:

$$\mu_0(\boldsymbol{cx}) = \begin{cases} 1, & \text{if } \boldsymbol{cx} \leq z_0^1 \\ \alpha_0 \boldsymbol{cx} + \beta_0, & \text{if } z_0^1 \leq \boldsymbol{cx} \leq z_0^0 \\ 0, & \text{if } \boldsymbol{cx} \geq z_0^0 \end{cases} \tag{6}$$

and

$$\mu_j((\boldsymbol{Ax})_j) = \begin{cases} 1, & \text{if } (\boldsymbol{Ax})_j \leq z_j^1 \\ \alpha_j(\boldsymbol{Ax})_j + \beta_j, & \text{if } z_j^1 \leq (\boldsymbol{Ax})_j \leq z_j^0 \\ 0, & \text{if } (\boldsymbol{Ax})_j \geq z_j^0 \end{cases} \tag{7}$$

for $j = 1, \ldots, s$.

3. Problem formulation and fuzzy satisficing method

Now we are ready to incorporate the fuzzy goal and fuzzy constraints of the decision maker (DM) into large-scale linear programming problems with block angular structure. In this paper, as a first attempt for incorporating the fuzzy goal and fuzzy constraints into large-scale linear programming problems with the block angular structure, we adopt the convex fuzzy decision for combining them. By adopting the convex fuzzy decision, large-scale linear programming problems incorporating the fuzzy goal and fuzzy constraints can be formulated as follows:

$$\left. \begin{aligned} \text{maximize} \quad & w_0 \mu_0(\boldsymbol{cx}) + \sum_j^s w_j \mu_j((\boldsymbol{Ax})_j) \\ \text{subject to} \quad & \boldsymbol{B}_i \boldsymbol{x}_i \leq \boldsymbol{b}_i, \ \boldsymbol{x}_i \geq 0, \ i = 1, \ldots, p, \\ & \boldsymbol{cx} \leq z_0^0, \ (\boldsymbol{Ax})_j \leq z_j^0, \ j = 1, \ldots, s \end{aligned} \right\} \tag{8}$$

where the weighting coefficients w_j, $j = 0, \ldots, s$, are normalized as $w_j \geq 0$, $j = 0, \ldots, s$; $\sum_{j=0}^{s} w_j = 1$.

In this formulation observe that the last two constraints are added since each z_j^0 denotes the minimum value of the unacceptable level. As a result, although the coupling constraints $\boldsymbol{A}\boldsymbol{x} \leq \boldsymbol{b}$ are incorporated into the objective function through fuzzy constraints, newly added constraints still remain as coupling constraints.

Here, let us consider the following linear programming problem:

$$\left. \begin{aligned} \text{maximize} \quad & w_0 \left(\alpha_0 \left(\sum_{i=1}^{p} c_i x_i \right) + \beta_0 \right) \\ & + \sum_{j=1}^{s} w_j \left(\alpha_j \left(\sum_{i=1}^{p} (A_i x_i)_j \right) + \beta_j \right) \\ \text{subject to} \quad & \boldsymbol{B}_i \boldsymbol{x}_i \leq \boldsymbol{b}_i, \boldsymbol{x}_i \geq \boldsymbol{0}, \ i = 1, \ldots, p. \end{aligned} \right\} \tag{9}$$

In this problem, the last two constraints of the problem (8) are eliminated and the membership functions $\mu_0(\boldsymbol{c}\boldsymbol{x})$ and $\mu_j((\boldsymbol{A}\boldsymbol{x})_j)$, $j = 1, \ldots, s$, are replaced by $-\alpha_0 \boldsymbol{c}\boldsymbol{x} - \beta_0$ and $-\alpha_j(\boldsymbol{A}\boldsymbol{x})_j - \beta_j$, $j = 1, \ldots, s$, which partially define $\mu_0(\boldsymbol{c}\boldsymbol{x})$ and $\mu_j((\boldsymbol{A}\boldsymbol{x})_j)$, $j = 1, \ldots, s$, within $[z_j^1, z_j^0]$, $j = 0, \ldots, s$.

Obviously, in the problem (9), there are no coupling constraints and it can be solved by decomposing into the following p independent linear programming subproblems.

$$\left. \begin{aligned} \text{maximize} \quad & w_0 \alpha_0 c_i x_i + \sum_{j=1}^{s} w_j \alpha_j (A_i x_i)_j \\ \text{subject to} \quad & \boldsymbol{B}_i \boldsymbol{x}_i \leq \boldsymbol{b}_i, \ \boldsymbol{x}_i \geq \boldsymbol{0}, i = 1, \ldots, p. \end{aligned} \right\} \tag{10}$$

where, for simplicity, constants $w_i \beta_i$, $i = 0, \ldots, p$, are omitted.

The relationships between the optimal solutions of the problems (9) and (8) can be given in the following theorem.

Theorem 1 Let $\hat{\boldsymbol{x}}_i$, $i = 1, \ldots, p$ be optimal solutions to the p linear subproblems (10). If $\hat{\boldsymbol{x}}_i$, $i = 1, \ldots, p$, satisfy

$$z_0^1 \leq \sum_{i=1}^{p} c_i \hat{x}_i \leq z_0^0; \ z_j^1 \leq (\sum_{i=1}^{p} A_i \hat{x}_i)_j \leq z_j^0, \ j = 1, \ldots, s \tag{11}$$

then $\hat{\boldsymbol{x}} = (\hat{\boldsymbol{x}}_1^T, \hat{\boldsymbol{x}}_2^T, \ldots, \hat{\boldsymbol{x}}_p^T)^T$ is an optimal solution to the problem (8).

Theorem 1 means that if optimal solutions to the p subproblems (11) satisfy the conditions (11), then the satisficing solution for the DM for large-scale linear programming problems incorporating the fuzzy goal and fuzzy constraints can be derived by just solving the p independent linear programming subproblems.

The remaining problem is to consider the case when optimal solutions to the p subproblems (10) do not satisfy (11).

For that purpose, observe that if any feasible solution x of the problem (8) satisfies

$$cx \geq z_0^1; \ (Ax)_j \geq z_j^1, \ j = 1, \ldots, s, \tag{12}$$

then the problem (8) becomes equivalently to the large-scale linear programming problem with the block angular structure

$$
\left.
\begin{array}{l}
\text{maximize} \quad w_0 \left(\alpha_0 \left(\displaystyle\sum_{i=1}^{p} c_i x_i \right) + \beta_0 \right) \\[2ex]
\qquad\qquad + \displaystyle\sum_{j=1}^{s} w_j \left(\alpha_j \left(\displaystyle\sum_{i=1}^{p} (A_i x_i)_j \right) + \beta_j \right) \\[2ex]
\text{subject to} \quad B_i x_i \leq b_i, \ x_i \geq 0, \ i = 1, \ldots, p \\[2ex]
\qquad\qquad \displaystyle\sum_{i=1}^{p} c_i x_i \leq z_0^0 \\[2ex]
\qquad\qquad \displaystyle\sum_{i=1}^{p} (A_i x_i)_j \leq z_j^0, \ j = 1, \ldots, s
\end{array}
\right\} \tag{13}
$$

and hence the Dantzig-Wolfe decomposition method is applicable. One way to guarantee these conditions (12) is to determine z_0^1 and z_j^1, $j = 1, \ldots, s$, as is given in the following theorem.

Theorem 2 Let z_0^1 be an optimal value to the linear programming problem

$$
\left.
\begin{array}{l}
\text{minimize} \quad \displaystyle\sum_{i=1}^{p} c_i x_i \\[2ex]
\text{subject to} \quad B_i x_i \leq b_i, \ x_i \geq 0, \ i = 1, \ldots, p
\end{array}
\right\} \tag{14}
$$

and $z_j^1, j = 1, \ldots, s$, be an optimal value to the linear programming problem

$$
\left.
\begin{array}{l}
\text{minimize} \quad \displaystyle\sum_{i=1}^{p} (A_i x_i)_j \\[2ex]
\text{subject to} \quad B_i x_i \leq b_i, \ x_i \geq 0, \ i = 1, \ldots, p.
\end{array}
\right\} \tag{15}
$$

Then any feasible solution x of the problem (8) satisfies the conditions (12).

From Theorem 2, even if the conditions (11) are not satisfied, by setting z_0^1 and z_j^1, $j = 1, \ldots, s$, as the optimal values of the problem (14) and (15), the Dantzig-Wolfe decomposition method is applicable for solving the problem (8). It should be noted here that the problems (14) and (15) can be decomposed respectively into the p independent linear subproblems. Hence the determination of z_0^1 and z_j^1, $j = 1, \ldots, s$, as is shown in Theorem 2 can be easily done.

One possible way to determine the weighting coefficients w_j, $j = 0, \ldots, s$, is to select them so that the relations $w_i \mu_i(z_i^{\mathrm{f}}) = w_j \mu_j(z_j^{\mathrm{f}})$, $i, j \in \{0, \ldots, s\}$ hold, i.e.,

$$
w_j = \frac{\dfrac{1}{\mu_j(z_j^{\mathrm{f}})}}{\displaystyle\sum_{i=0}^{s} \dfrac{1}{\mu_i(z_i^{\mathrm{f}})}}, \quad j = 0, \ldots, s. \tag{16}
$$

Following the above discussions, we can now construct the algorithm in order to derive the satisficing solution for the DM to the large-scale linear programming problem with the block angular structure incorporating the fuzzy goal and fuzzy constraints.

Step 1 : Calculate z_0^1 and z_j^1, $j = 1, \ldots, s$, by solving the independent linear subproblems arising from the problems (14) and (15).

Step 2 : With information of z_0^1 and z_j^1, $j = 1, \ldots, s$, elicit linear membership functions from the DM for the objective function and coupling constraints by assessing the minimum values of the unacceptable levels z_0^0 and z_j^0, $j = 1, \ldots, s$.

Step 3 : Determine the weighting coefficients w_j, $j = 0, \ldots, s$, according to the relations (16) by assessing the satisfactory levels z_j^{f}, $j = 0, \ldots, s$.

Step 4 : Solve the p linear subproblems (10) for obtaining the optimal solutions $\hat{\boldsymbol{x}} = (\hat{\boldsymbol{x}}_1^T, \hat{\boldsymbol{x}}_2^T, \ldots, \hat{\boldsymbol{x}}_p^T)^T$ to the problem (9).

Step 5 : Check whether $\hat{\boldsymbol{x}}$ satisfies (11) or not. If (11) is satisfied, stop. Then the current optimal solution $\hat{\boldsymbol{x}}$ is the satisficing solution for the DM. Otherwise, proceed to the next step.

Step 6 : Solve the large-scale linear programming problem (13) by applying the Dantzig-Wolfe decomposition method for obtaining the optimal solution $\tilde{\boldsymbol{x}}$. Then the current optimal solution $\tilde{\boldsymbol{x}}$ is the satisficing solution for the DM.

It is shown that the satisficing solution for the DM to large-scale linear programming problems with the block angular structure incorporating the fuzzy goal and fuzzy constraints can be obtained by solving a number of linear subproblems. Especially, if the condition (11) holds, the overall satisficing solution is directly derived just only solving p independent linear subproblems without using the Dantzig-Wolfe decomposition method. Even if (11) does not hold, the problem can be solved by applying the Dantzig-Wolfe decomposition method. The larger the spreads $(z_j^0 - z_j^1)$ of the fuzzy goal and fuzzy constraints are, the more often (11) holds. Thus, the proposed method will be still more powerful when the DM has very vague goal and constraints.

4. Numerical example

To illustrate the proposed method, consider the following simple numerical example:

$$
\left.
\begin{array}{llr}
\text{minimize} & -x_1 - x_2 \quad -2y_1 - y_2 & \\
\text{subject to} & x_1 + 2x_2 \quad +2y_1 + y_2 & \leq 40 \\
. & x_1 + 3x_2 & \leq 30 \\
& 2x_1 + x_2 & \leq 20 \\
& y_1 & \leq 10 \\
& \quad y_2 & \leq 10 \\
& y_1 + y_2 & \leq 15 \\
& x_1, \ x_2 \geq 0, \quad y_1, \ y_2 \geq 0 &
\end{array}
\right\}
$$

First solving

$$
\left.
\begin{array}{llr}
\text{minimize} & -x_1 - x_2 \quad -2y_1 - y_2 & \\
\text{subject to} & x_1 + 3x_2 & \leq 30, \\
& 2x_1 + x_2 & \leq 20, \\
& y_1 & \leq 10, \\
& \quad y_2 & \leq 10, \\
& y_1 + y_2 & \leq 15, \\
& x_1 \geq 0, \ x_2 \geq 0, \ y_1 \geq 0, \ y_2 \geq 0, &
\end{array}
\right\}
$$

by decomposing into 2 independent subproblems, yields $z_0^1 = -44$. Similarly, solving

$$
\left.
\begin{array}{llr}
\text{minimize} & x_1 + 2x_2 \quad +2y_1 + y_2 & \\
\text{subject to} & x_1 + 3x_2 & \leq 30, \\
& 2x_1 + x_2 & \leq 20, \\
& y_1 & \leq 10, \\
& \quad y_2 & \leq 10, \\
& y_1 + y_2 & \leq 15, \\
& x_1 \geq 0, \ x_2 \geq 0, \ y_1 \geq 0, \ y_2 \geq 0, &
\end{array}
\right\}
$$

z_1^1 becomes 0.

Then by considering these values, assume that a hypothetical decision maker (DM) assessed the values of z_0^0 and z_1^0 to be -34 and 50, respectively. Then the corresponding membership functions $\mu_0(cz)$ and $\mu_1(az)$ become as follows:

$$
\mu_0(cz) = \begin{cases}
1, & \text{if } cz \leq -44, \\
-\dfrac{cz - 34}{10}, & \text{if } -34 \leq cz \leq -44, \\
0, & \text{if } cz \geq -34,
\end{cases}
$$

$$\mu_1(az) = \begin{cases} 1, & \text{if } az \le 0, \\ -\dfrac{az - 50}{50}, & \text{if } 0 \le az \le 50, \\ 0, & \text{if } az \ge 50, \end{cases}$$

where $c = (c_1, c_2)$, $a = (a_1, a_2)$ and $z = (x^T, y^T)^T$.

Also assume that a hypothetical DM assessed the values of z_0^f and z_1^f to be -37 and 40, respectively. Then calculating the weights w_0, w_1 according to (16), yields $w_0 = \dfrac{2}{5}$, $w_1 = \dfrac{3}{5}$.

Consequently, solving linear programming problems

$$
\begin{aligned}
\text{maximize} \quad & -\frac{1}{25}c_1 x \; - \; \frac{3}{250}a_1 x \\
\text{subject to} \quad & \left.\begin{array}{rcl} x_1 + 3x_2 & \le & 30 \\ 2x_1 + x_2 & \le & 20 \\ x_1 \ge 0, \quad x_2 \ge 0 \end{array}\right\}
\end{aligned}
$$

and

$$
\begin{aligned}
\text{maximize} \quad & -\frac{1}{25}c_2 y \; - \; \frac{3}{250}a_2 y \\
\text{subject to} \quad & \left.\begin{array}{rcl} y_1 & \le & 10 \\ y_2 & \le & 10 \\ y_1 + y_2 & \le & 15 \\ y_1 \ge 0, \quad y_2 \ge 0 \end{array}\right\}
\end{aligned}
$$

yields the corresponding optimal solutions $\hat{x} = (6, 8)^T$ and $\hat{y} = (10, 5)^T$, for which both the conditions $-44 \le cz \le -34$ and $0 \le az \le 50$ are satisfied. Therefore, from Theorem 1, the satisficing solution for the DM becomes $\hat{x} = (6, 8)^T$ and $\hat{y} = (10, 5)^T$.

5. Conclusion

In this paper, we focused on large-scale linear programming problems with the block angular structure, and proposed a fuzzy satisficing method by incorporating the fuzzy goal and fuzzy constraints. By adopting the convex fuzzy decision for combining the linear membership functions, it was shown that, under some appropriate conditions, the formulated problem can be reduced to a number of independent linear subproblems and the satisficing solution for the DM is directly obtained just only solving the subproblems. Moreover, even if the appropriate conditions are not satisfied, the Dantzig-Wolfe decomposition method is applicable. An illustrative numerical example demonstrated the feasibility and efficiency of the proposed method. Extensions to large-scale multiobjective linear and nonlinear programming problems will be reported elsewhere.

References

1. Dantzig, G. B. and Wolfe, P. (1961) The decomposition algorithm for linear programming, *Econometrica* **29**, 767-778.
2. R. Słowinski and J. Teghem (eds.) (1990) *Stochastic versus Fuzzy Approaches to Multiobjective Programming under Uncertainty*, Kluwer Academic Publishers, Dordrecht.
3. Lasdon, L. S. (1970) *Optimization Theory for Large Systems*, Macmillan, New York.
4. Sakawa, M. (1993) *Fuzzy Sets and Interactive Multiobjective Optimization*, Plenum Press, New York.
5. Zimmermann, H.-J. (1991) *Fuzzy Set Theory and Its Applications*, 2nd Revised Edition, Kluwer Academic Publishers, Boston.

A NECESSARY SOFT OPTIMALITY TEST IN INTERVAL LINEAR PROGRAMMING PROBLEMS

M. INUIGUCHI AND M. SAKAWA
Department of Industrial and Systems Engineering
Faculty of Engineering, Hiroshima University
4-1 Kagamiyama 1 chome, Higashi-Hiroshima, 739
Japan

1. Introduction

So far, satisficing approaches have been mainly proposed in possibilistic programming problems (see, for example, Inuiguchi et al. [2] and Sakawa [6]). The solution obtained from such an approach is optimal to a conventional mathematical programming problem formulated by an interpretation of the possibilistic programming problem. Thus, this solution is the best solution in the sense of the interpretation but not always a good solution from the other viewpoints. Especially, in the sense of optimality extended to the possibilistic programming case, the solution is not always a good solution (see Inuiguchi and Kume [4]) since it is obtained in a satisficing view. It is significant to check the goodness of the solution from a viewpoint other than the introduced interpretation.

In this perspective, the authors [5] have proposed possible and necessary optimality tests in possibilistic linear programming problems. By those tests, the goodness of the solution can be checked from the two kinds of optimality viewpoints, the possible optimality and the necessary optimality. However, the necessary optimality test is not always powerful since no necessarily optimal solution exists in many cases. On the other hand, the possible optimality test is too weak so that a lot of feasible solutions are possibly optimal (but it may be useful as the least validity test). Therefore, a moderate optimality test is a topic of concern.

This paper is devoted to such a topic. Restricting ourself to interval objective function case, a necessary soft optimality test is proposed as a moderate between possibility and necessary optimality tests. A necessary

465

Z. Bien and K. C. Min (eds.),
Fuzzy Logic and its Applications, Information Sciences, and Intelligent Systems, 465–473.
© 1995 *Kluwer Academic Publishers. Printed in the Netherlands.*

soft optimality is a weakened concept of the necessary optimality, where the optimality is softened. The optimal solution to a certain mathematical programming problem with an objective, maximizing a function, is a feasible solution where the objective function attains the maximum value. The soft optimal solution is a feasible solution whose objective function value is roughly equal to the maximal value. In order to represent the softness, a fuzzy set is introduced. Using a necessity measure, the soft optimality in the conventional mathematical programming problem is extended to a necessary optimality in an interval programming problem. A necessary soft optimality test for a given feasible solution is discussed. It is shown that a necessary soft optimality test can be done through the linear programming techniques.

2. Possibilistic Linear Programming Problem

In this paper, the following linear programming problem with an interval linear objective function is treated:

$$\begin{aligned} \text{maximize} \quad & \boldsymbol{\gamma x}, \\ \text{subject to} \quad & \boldsymbol{Ax} \leq \boldsymbol{b}, \end{aligned} \tag{1}$$

where $\boldsymbol{x} = (x_1, x_2, \ldots, x_n)^t$, $\boldsymbol{b} = (b_1, b_2, \ldots, b_m)^t$ and $\boldsymbol{A}$ is an $m \times n$ matrix. $\boldsymbol{\gamma} = (\gamma_1, \gamma_2, \ldots, \gamma_n)$, where γ_i, $i = 1, 2, \ldots, n$ are mutually independent possibilistic variables restricted by intervals $C_i = [c_i^{\mathrm{L}}, c_i^{\mathrm{R}}]$, $i = 1, 2, \ldots, n$, respectively. For the sake of simplicity, we use a set of vectors $\boldsymbol{c}$, say Θ, defined by

$$\Theta = \left\{ \boldsymbol{c} = (c_1, c_2, \ldots, c_n) \mid c_i^{\mathrm{L}} \leq c_i \leq c_i^{\mathrm{R}}, \ i = 1, 2, \ldots, n \right\}. \tag{2}$$

3. Necessary Optimality and Necessary Soft Optimality

Let $S(\boldsymbol{c})$ be a set of optimal solutions to a linear programming problem with an objective function $\boldsymbol{cx}$, i.e.,

$$\begin{aligned} \text{maximize} \quad & \boldsymbol{cx}, \\ \text{subject to} \quad & \boldsymbol{Ax} \leq \boldsymbol{b}. \end{aligned} \tag{3}$$

Namely, $S(\boldsymbol{c})$ can be represented as

$$S(\boldsymbol{c}) = \left\{ \boldsymbol{x} \ \middle| \ \boldsymbol{cx} = \max_{\boldsymbol{Ay} \leq \boldsymbol{b}} \boldsymbol{cy}, \ \boldsymbol{Ax} \leq \boldsymbol{b} \right\}. \tag{4}$$

Using $S(c)$, a necessary optimal solution set to linear programming problem with an interval objective function can be defined as (see Inuiguchi and Sakawa [5])

$$NS = \bigcap_{c \in \Theta} S(c). \tag{5}$$

An element of NS is a feasible solution optimal for all $c \in \Theta$. Each element of NS is necessarily optimal as far as the true value of γ lies within Θ. From this point of view, an element of NS is called a "necessarily optimal solution".

Let $P(x)$ be a set of vectors c for which a feasible solution x is optimal, i.e.,

$$P(x) = \left\{ c \ \middle| \ cx = \max_{Ay \leq b} cy \right\}. \tag{6}$$

Obviously, $P(x)$ is a closed set. For any feasible solution x, we have

$$x \in NS \Leftrightarrow N_\Theta(P(x)) = 1 \Leftrightarrow \Theta \subseteq P(x), \tag{7}$$

where N is a necessity measure defined by

$$N_\Theta(P(x)) = \inf_c \max\{1 - \mu_\Theta(c), \mu_{P(x)}(c)\} \tag{8}$$

where μ_Θ and $\mu_{P(x)}$ are membership functions of Θ and $P(x)$, respectively. In the current case, $N_\Theta(P(x))$ takes 0 or 1 since Θ and $P(x)$ are crisp sets. From (7), it can be seen that necessarily optimal solutions are characterized by a necessity measure as

$$\mu_{NS}(x) = \min\left(N_\Theta(P(x)), \mu_F(x)\right). \tag{9}$$

where μ_F is a membership function of the feasible set $F = \{x \mid Ax \leq b\}$.

A necessarily optimal solution is the most reasonable optimal solution. However, as will be shown in Example 1, there is no guarantee of the existence of a necessarily optimal solution. In no necessarily optimal solution case, a necessary optimality test [5] is useless. To cope with this defect, the necessary optimality is weakened and a necessary soft optimality is proposed.

To soften the optimality, we regard a solution whose objective function value is not much smaller than the optimal value as an approximately optimal solution. From this point of view, $P(x)$ can be softened as follows;

$$\mu_{\tilde{P}(x)}(c) = \mu_D\left(\max_{Ay \leq b} cy - cx\right) \tag{10}$$

where $\mu_{\check{P}(x)}$ is a membership function of the softened set $\check{P}(x)$. μ_D is a membership function of a set having the linguistic expression "not much smaller than 0". $\mu_D : \mathbf{R} \mapsto [0,1]$ is assumed to be a non-increasing function such that $\mu_D(0) = 1$.

Using $\check{P}(x)$, the necessarily soft optimal solution set $\widetilde{NS}$ can be defined by (9) with replacing NS and $P(x)$ with $\widetilde{NS}$ and $\check{P}(x)$, respectively. In the current case, since Θ is a crisp set, we have

$$\mu_{\widetilde{NS}}(x) = \inf_{c \in \Theta} \ \mu_{\check{P}(x)}(c). \tag{11}$$

When $\delta = 0$, a necessarily soft optimal solution set $\widetilde{NS}$ is degenerated to a necessarily optimal solution set NS. Thus, $\widetilde{NS}$ is an extension of NS.

Example 1. Let us consider the following interval linear programming problem:

$$
\begin{aligned}
\text{maximize} \quad & [1.5, 2.5]x_1 + [2, 3]x_2, \\
\text{subject to} \quad & 3x_1 + 4x_2 \leq 42, \\
& 3x_1 + x_2 \leq 24, \\
& x_1 \geq 0, \ 0 \leq x_2 \leq 9.
\end{aligned}
$$

In this problem, Θ is represented by

$$\Theta = \{c = (c_1, c_2) \mid 1.5 \leq c_1 \leq 2.5, \ 2 \leq c_2 \leq 3\}.$$

Consider a solution $(x_1, x_2)^t = (6,6)^t$. The set of coefficient vectors (c_1, c_2) make $(6,6)^t$ optimal, i.e., $P((6,6)^t)$, is represented by

$$P((6,6)^t) = \{c = (c_1, c_2) \mid c_1 - 3c_2 \leq 0, 4c_1 - 3c_2 \geq 0\}.$$

As shown in FIGURE 1, $\Theta \subseteq P((6,6)^t)$ does not hold. Thus, $(6,6)^t$ is not a necessarily optimal solution.

Let us define μ_D as

$$
\mu_D(r) = \begin{cases} 1, & \text{if } r \leq 0, \\ 1 - \dfrac{r}{5}, & \text{if } 0 < r \leq 5, \\ 0, & \text{if } r > 5. \end{cases}
$$

$\check{P}((6,6)^t)$ of (10) can be represented by

$$\mu_{\check{P}((6,6)^t)}((c_1, c_2)) = \max(0, \min(1, 1.2c_1 + 1.2c_2 + 1, 1.2c_1 - 0.6c_2 + 1,$$
$$- 0.4c_1 + 1.2c_2 + 1, 0.8c_1 - 0.6c_2 + 1))$$

As shown in FIGURE 1,

$$\mu_{\widetilde{NS}}((6,6)^t) = \inf_{c \in \Theta} \ \mu_{\check{P}(x)}(c) = 0.4.$$

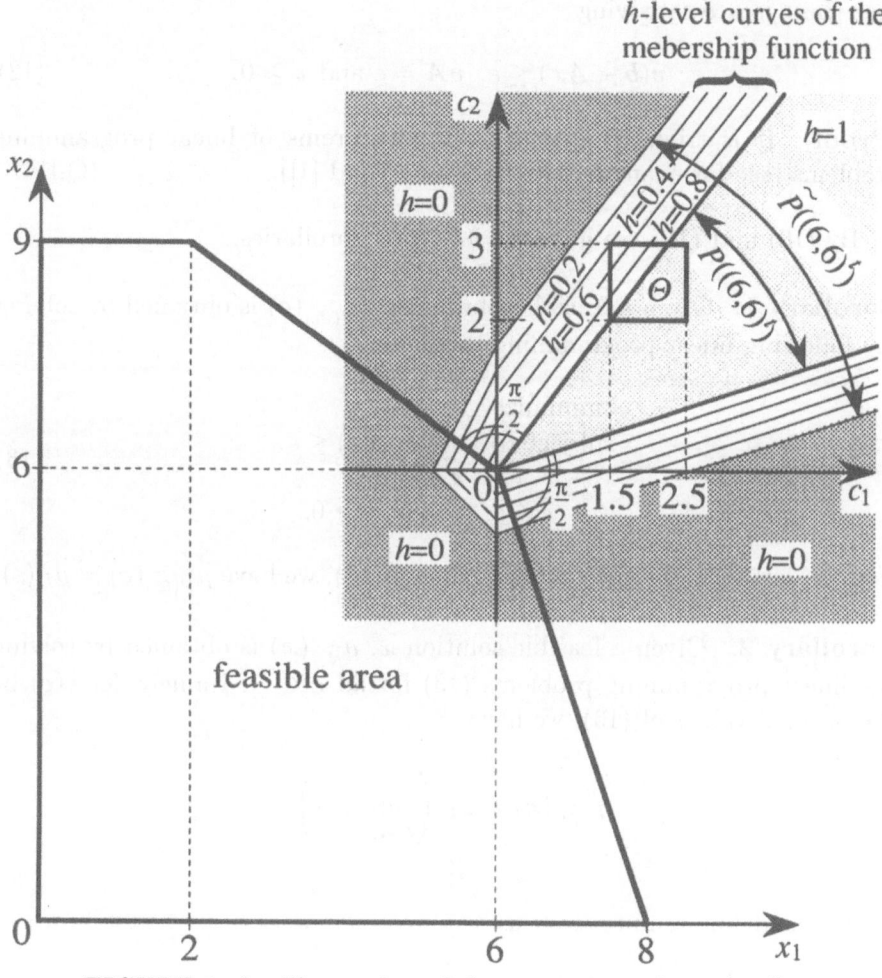

FIGURE 1. An illustration of the necessary soft optimality

Thus, $(6,6)^t$ is a necessarily soft optimal solution to the degree 0.4.

4. Necessary Soft Optimality Test

In the previous section, the necessary soft optimality has been introduced. In this section, we discuss the necessary soft optimality test for a given feasible solution. We have the following theorem.

Theorem 1. Let $\varepsilon > 0$ be a constant. Given a feasible solution $\boldsymbol{x}$, a necessary and sufficient condition for $\max_{\boldsymbol{Ay} \leq \boldsymbol{b}} \boldsymbol{cy} - \boldsymbol{cx} \leq \varepsilon$ is the fact that there is a $\boldsymbol{v}$ satisfying

$$\boldsymbol{v}(\boldsymbol{b} - \boldsymbol{Ax}) \le \varepsilon, \quad \boldsymbol{vA} = \boldsymbol{c} \text{ and } \boldsymbol{v} \ge \boldsymbol{0}. \tag{12}$$

(Proof) It is trivial from the duality theorems of linear programming problems (see for example, Goldfarb and Todd [1]). (Q.E.D.)

By (10) and (11), we have the following corollaries.

Corollary 1. Given a feasible solution $\boldsymbol{x}$, $\mu_{\tilde{P}(\boldsymbol{x})}(\boldsymbol{c})$ is obtained by solving the following linear programming problem:

$$\begin{aligned}
\text{minimize} \quad & \varepsilon, \\
\text{subject to} \quad & \boldsymbol{v}(\boldsymbol{b} - \boldsymbol{Ax}) \le \varepsilon, \\
& \boldsymbol{vA} = \boldsymbol{c}, \\
& \boldsymbol{v} \ge \boldsymbol{0}, \ \varepsilon \ge 0.
\end{aligned} \tag{13}$$

Namely, letting ε^* be the optimal value of (13), we have $\mu_{\tilde{P}(\boldsymbol{x})}(\boldsymbol{c}) = \mu_D(\varepsilon)$.

Corollary 2. Given a feasible solution $\boldsymbol{x}$, $\mu_{\widetilde{NS}}(\boldsymbol{x})$ is obtained by solving the linear programming problems (13) for all $\boldsymbol{c} \in \Theta$. Namely, let $\varepsilon(\boldsymbol{c})$ be the optimal values of (13), we have

$$\mu_{\widetilde{NS}}(\boldsymbol{x}) = \mu_D \left(\sup_{\boldsymbol{c} \in \Theta} \varepsilon(\boldsymbol{c}) \right).$$

Let us define a finite set σ as

$$\begin{aligned}
\sigma &= \{\boldsymbol{c}^1, \boldsymbol{c}^2, \dots, \boldsymbol{c}^q\} \\
&= \{\boldsymbol{c} = (c_1, c_2, \dots, c_n) \mid c_i = c_i^L \text{ or } c_i^R, \ i = 1, 2, \dots, n\}. \tag{14}
\end{aligned}$$

The following theorem is derived.

Theorem 2. Let $\delta > 0$. Given a feasible solution $\boldsymbol{x}$, $\mu_{\widetilde{NS}}(\boldsymbol{x})$ is obtained by solving the linear programming problems (13) for all $\boldsymbol{c} \in \sigma$. Namely, let $\varepsilon(\boldsymbol{c})$ be the optimal values of (13), we have

$$\mu_{\widetilde{NS}}(\boldsymbol{x}) = \mu_D \left(\max_{\boldsymbol{c} \in \sigma} \varepsilon(\boldsymbol{c}) \right).$$

(Proof) Let $h^1 = \mu_{\widetilde{NS}}(\boldsymbol{x}) = \mu_D(\sup_{\boldsymbol{c} \in \Theta} \varepsilon(\boldsymbol{c}))$ and $h^2 = \mu_D(\max_{\boldsymbol{c} \in \sigma} \varepsilon(\boldsymbol{c}))$. $h^1 \le h^2$ is obvious since $\sigma \subseteq \Theta$ and μ_D is non-increasing. Thus, we prove the converse, i.e., $h^1 \ge h^2$.

Let $(\boldsymbol{v}^i, \varepsilon^i)$ be an optimal solutions to the problem (13) with $\boldsymbol{c} = \boldsymbol{c}^j$ for $i = 1, 2, \ldots, q$. Let $\hat{\varepsilon} = \max_{i=1,2,\ldots,q} \varepsilon^i$. For $i = 1, 2, \ldots, q$, the solution $(\boldsymbol{v}^i, \hat{\varepsilon})$ is a feasible solution to the problem (13) with $\boldsymbol{c} = \boldsymbol{c}^i$, i.e.,

$$\boldsymbol{v}^i(\boldsymbol{b} - \boldsymbol{Ax}) \leq \hat{\varepsilon}, \quad \boldsymbol{v}^i \boldsymbol{A} = \boldsymbol{c}^i, \text{ and } \boldsymbol{v}^i \geq \boldsymbol{0}. \tag{$*$}$$

By the definition of σ, for an arbitrary $\boldsymbol{c} \in \Theta$, there exist λ_i, $i = 1, 2, \ldots, q$ such that

$$\boldsymbol{c} = \sum_{i=1}^{q} \lambda_i \boldsymbol{c}^i, \quad \sum_{i=1}^{q} \lambda_i = 1, \quad \lambda_i \geq 0, \quad i = 1, 2, \ldots, q.$$

From $(*)$, we have

$$\left(\sum_{i=1}^{q} \lambda_i \boldsymbol{v}^i \right) (\boldsymbol{b} - \boldsymbol{Ax}) \leq \hat{\varepsilon}, \quad \left(\sum_{i=1}^{q} \lambda_i \boldsymbol{v}^i \right) \boldsymbol{A} = \boldsymbol{c}, \text{ and } \sum_{i=1}^{q} \lambda_i \boldsymbol{v}^i \geq \boldsymbol{0}.$$

This means that the solution $(\sum_{i=1}^{q} \lambda_i \boldsymbol{v}^i, \hat{\varepsilon})$ is a feasible solution to the problem (13) with an arbitrary $\boldsymbol{c} \in \Theta$. By the optimality of $\varepsilon(\boldsymbol{c})$, we have $\varepsilon(\boldsymbol{c}) \leq \hat{\varepsilon}$ for all $\boldsymbol{c} \in \Theta$. Since Θ is closed and μ_D is non-decreasing, we obtain

$$h^1 = \mu_D \left(\max_{\boldsymbol{c} \in \Theta} \varepsilon(\boldsymbol{c}) \right) \geq \mu_D(\hat{\varepsilon}) = h^2.$$

$$\text{(Q.E.D.)}$$

By Theorem 2, given a feasible solution $\boldsymbol{x}$, we can calculate $\mu_{\widetilde{NS}}(\boldsymbol{x})$ by solving q linear programming problems,

$$
\begin{aligned}
\text{minimize} \quad & \varepsilon, \\
\text{subject to} \quad & \boldsymbol{v}(\boldsymbol{b} - \boldsymbol{Ax}) \leq \varepsilon, \\
& \boldsymbol{v}\boldsymbol{A} = \boldsymbol{c}^i, \\
& \boldsymbol{v} \geq \boldsymbol{0}, \ \varepsilon \geq 0,
\end{aligned}
\tag{15}
$$

for $i = 1, 2, \ldots, q$. Those q problems differ only the right-hand side values $\boldsymbol{c}^i$ of the second constraints. Thus, we can apply a post-optimization techniques of linear programming [1].

Example 2. Let us consider the interval linear programming problem already discussed in Example 1. Using the same μ_D of Example 1, let us check the necessary soft optimality of a feasible solution $(x_1, x_2)^t = (6, 6)^t$ via linear programming techniques. In the problem, we have

$$A = \begin{pmatrix} 3 & 4 \\ 3 & 1 \\ 0 & 1 \\ -1 & 0 \\ 0 & -1 \end{pmatrix}, \ b = \begin{pmatrix} 42 \\ 24 \\ 9 \\ 0 \\ 0 \end{pmatrix}.$$

TABLE 1. Four linear programming problems for $\mu_{\widetilde{NS}}((6,6)^t)$

maximize	ε	maximize	ε
subject to	$3v_3 + 6v_4 + 6v_5 \leq \varepsilon$	subject to	$3v_3 + 6v_4 + 6v_5 \leq \varepsilon$
	$3v_1 + 3v_2 - v_4 = 1.5$		$3v_1 + 3v_2 - v_4 = 2.5$
	$4v_1 + v_2 + v_3 - v_5 = 2$		$4v_1 + v_2 + v_3 - v_5 = 2$
	$v_1, v_2, v_3, v_4, \varepsilon \geq 0$		$v_1, v_2, v_3, v_4, \varepsilon \geq 0$

The optimal solution:
$(v_1, v_2, v_3, v_4, v_5, \varepsilon) = (\frac{1}{2}, 0, 0, 0, 0, 0)$
The optimal value: $\varepsilon^1 = 0$

The optimal solution:
$(v_1, v_2, v_3, v_4, v_5, \varepsilon) = (\frac{7}{18}, \frac{4}{9}, 0, 0, 0, 0)$
The optimal value: $\varepsilon^2 = 0$

maximize	ε	maximize	ε
subject to	$3v_3 + 6v_4 + 6v_5 \leq \varepsilon$	subject to	$3v_3 + 6v_4 + 6v_5 \leq \varepsilon$
	$3v_1 + 3v_2 - v_4 = 1.5$		$3v_1 + 3v_2 - v_4 = 2.5$
	$4v_1 + v_2 + v_3 - v_5 = 3$		$4v_1 + v_2 + v_3 - v_5 = 3$
	$v_1, v_2, v_3, v_4, \varepsilon \geq 0$		$v_1, v_2, v_3, v_4, \varepsilon \geq 0$

The optimal solution:
$(v_1, v_2, v_3, v_4, v_5, \varepsilon) = (\frac{1}{2}, 0, 1, 0, 0, 3)$
The optimal value: $\varepsilon^3 = 3$

The optimal solution:
$(v_1, v_2, v_3, v_4, v_5, \varepsilon) = (\frac{13}{18}, \frac{1}{9}, 0, 0, 0, 0)$
The optimal value: $\varepsilon^4 = 0$

The set σ is composed of the following four elements:

$$\begin{aligned} c^1 &= (1.5, 2), \\ c^2 &= (2.5, 2), \\ c^3 &= (1.5, 3), \\ c^4 &= (2.5, 3). \end{aligned}$$

Since $(x_1, x_2)^t = (6, 6)^t$, we obtain

$$b - Ax = (0, 0, 3, 6, 6)^t.$$

The four linear programming problems for $\mu_{\widetilde{NS}}((6,6)^t)$ are listed in TABLE 1. As shown in TABLE 1, the optimal values are obtained as $\varepsilon^1 = \varepsilon^2 = \varepsilon^4 = 0$ and $\varepsilon^3 = 3$. Hence, $\mu_{\widetilde{NS}}((6,6)^t)$ is calculated as

$$\mu_{\widetilde{NS}}((6,6)^t) = \mu_D \left(\max_{i=1,2,3,4} \varepsilon^i \right) = 1 - \frac{3}{5} = 0.4.$$

This coincides with the result of Example 1.

5. Concluding Remarks

In this paper, the necessary optimality is extended to a necessary soft optimality for linear programming problems with an interval objective function. A necessary soft optimality test for a given feasible solution is discussed. It is shown that the degree of the necessary soft optimality can be obtained via solving linear programming problems. A numerical example is given to illustrate the proposed method.

The concept of the necessary soft optimality can be easily introduced into linear programming problems with a possibilistic objective function. The necessary soft optimality test can also be done by linear programming techniques. This will be one topic of our future research.

References

1. Goldfarb, D. and Todd, M. J. (1989) Linear programming, in: Nemhauser, G. L. and Rinnooy Kan, A. H. G. (eds.): *Handbooks in Operations Research and Management Science*, North-Holland, Amsterdam, pp. 73–170.
2. Inuiguchi, M., Ichihashi, H. and Tanaka, H. (1990) Fuzzy programming: a survey of recent developments, in: Słowinski, R. and Teghem, J. (eds.): *Stochastic versus Fuzzy Approaches to Multiobjective Programming under Uncertainty*, Kluwer Academic Publishers, Dordrecht, pp. 45–68.
3. Inuiguchi, M. and Kume, Y. (1990) Solution concepts for fuzzy multiobjective programming problems, *Japanese Journal of Fuzzy Theory and Systems*, **Vol. 2**, pp. 1–21.
4. Inuiguchi, M. and Kume, Y. (1994) Minimax regret in linear programming problems with an interval objective function, in: Tzeng, G. H., Wang, H. F., Wen, U. P. and Yu, P. L. (eds.): *Multiple Criteria Decision Making: Proceedings of the Tenth International Conference: Expand and Enrich the Domains of Thinking and Application*, Springer-Verlag, New York, pp. 65–74.
5. Inuiguchi, M. and Sakawa, M. (1994) Possible and necessary optimality tests in possibilistic linear programming problems, *Fuzzy Sets and Systems*, **Vol. 67**, pp. 29–46.
6. Sakawa, M. (1993) *Fuzzy Sets and Interactive Multiobjective Optimization*. Plenum Press, New York and London.

Subject Index

THEORY AND DECISION LIBRARY

SERIES D: SYSTEM THEORY, KNOWLEDGE ENGINEERING AND PROBLEM SOLVING

1. E.R. Caianiello and M.A. Aizerman (eds.): *Topics in the General Theory of Structures.* 1987   ISBN 90-277-2451-2

2. M.E. Carvallo (ed.): *Nature, Cognition and System I.* Current Systems-Scientific Research on Natural and Cognitive Systems. With a Foreword by G.J. Klir. 1988   ISBN 90-277-2740-6

3. A. Di Nola, S. Sessa, W. Pedrycz and E. Sanchez: *Fuzzy Relation Equations and Their Applications to Knowledge Engineering.* With a Foreword by L.A. Zadeh. 1989   ISBN 0-7923-0307-5

4. S. Miyamoto: *Fuzzy Sets in Information Retrieval and Cluster Analysis.* 1990
                                      ISBN 0-7923-0721-6

5. W.H. Janko, M. Roubens and H.-J. Zimmermann (eds.): *Progress in Fuzzy Sets and Systems.* 1990   ISBN 0-7923-0730-5

6. R. Slowinski and J. Teghem (eds.): *Stochastic versus Fuzzy Approaches to Multiobjective Mathematical Programming under Uncertainty.* 1990
                                      ISBN 0-7923-0887-5

7. P.L. Dann, S.H. Irvine and J.M. Collis (eds.): *Advances in Computer-Based Human Assessment.* 1991   ISBN 0-7923-1071-3

8. V. Novák, J. Ramík, M. Mareš, M. Černý and J. Nekola (eds.): *Fuzzy Approach to Reasoning and Decision-Making.* 1992   ISBN 0-7923-1358-5

9. Z. Pawlak: *Rough Sets.* Theoretical Aspects of Reasoning about Data. 1991
                                      ISBN 0-7923-1472-7

10. M.E. Carvallo (ed.): *Nature, Cognition and System II.* Current Systems-Scientific Research on Natural and Cognitive Systems. Vol. 2: On Complementarity and Beyond. 1992   ISBN 0-7923-1788-2

11. R. Slowiński (ed.): *Intelligent Decision Support.* Handbook of Applications and Advances of the Rough Sets Theory. 1992   ISBN 0-7923-1923-0

12. R. Lowen and M. Roubens (eds.): *Fuzzy Logic.* State of the Art. 1993
                                      ISBN 0-7923-2324-6

13. L. Kitainik: *Fuzzy Decision Procedures with Binary Relations.* Toward a Unified Theory. 1993   ISBN 0-7923-2367-X

14. J. Fodor and M. Roubens: *Fuzzy Preference Modelling and Multicriteria Decision Support.* 1994   ISBN 0-7923-3116-8

15. E.A. Yfantis (ed.): *Intelligent Systems.* Third Golden West International Conference. Edited and Selected Papers. 1995
                                      ISBN 0-7923-3422-1 (Set of 2 volumes)